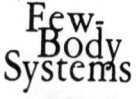

Suppl. 7

# FEW BODY XIV

Few-
Body
Systems

Editor: H. Mitter, Graz　　　　　　　Managing Editor: W. Plessas, Graz

Supplementum 7

# Few-Body Problems in Physics '93

*Proceedings of the XIVth European Conference*
*on Few-Body Problems in Physics,*
*Amsterdam, The Netherlands, August 23—27, 1993*

Edited by
*B. L. G. Bakker* and *R. van Dantzig*

Springer-Verlag Wien NewYork

Dr. B. L. G. Bakker
Department of Physics
Free University
Amsterdam, The Netherlands

Dr. R. van Dantzig
National Institute for Nuclear and High-Energy Physics (NIKHEF)
Amsterdam, The Netherlands

Printed on acid-free and chlorine-free bleached paper

With 197 Figures and 1 Frontispiece

ISSN 0177-8811
ISBN-13: 978-3-7091-9354-9      e-ISBN-13: 978-3-7091-9352-5
DOI: 10.1007/978-3-7091-9352-5

Amsterdam 23-28 Aug. 1993

## Organizing Committee

B.L.G. Bakker, Free University, Amsterdam   Chairman
R. van Dantzig, NIKHEF, Amsterdam   Scientific secretariat
E. Jans, NIKHEF, Amsterdam   Scientific secretariat
L.P. Kok, University of Groningen
J. Konijn, NIKHEF, Amsterdam   Treasurer
A. Niehaus, Utrecht University
J.J. de Swart, University of Nijmegen
J.A. Tjon, Utrecht University

## International Advisory Committee

I.R. Afnan (Flinders)
J. Arvieux (Saturne)
V.B. Belyaev (Dubna)
J. Cugnon (Liège)
H.W. Fearing (TRIUMF)
J.L. Friar (Los Alamos)
B. Loiseau (Orsay)
S.P. Merkuriev[†]
  (St Petersburg)
A.S. Rinat (Weizmann)
E.W. Schmid (Tübingen)
Yu.A. Simonov (Moscow)
P.K.A. de Witt Huberts
  (NIKHEF)

Y. Akaishi (Tokyo)
J.L. Ballot (Orsay)
J.M. Cameron (IUCF)
R.H. Dalitz (Oxford)
H.J. Fiedeldey (Pretoria)
F. Gross (W&M/CEBAF)
I. Lovas (Budapest)
J. Namyslowski (Warsaw)
W. Plessas (Graz)
D.O. Riska (Helsinki)
B. Schoch (Bonn)
I. Šlaus (Zagreb)
Yang Li-Ming (Beijing)

H. Arenhövel (Mainz)
P.D. Barnes (LAMPF)
C. Ciofi degli Atti (Perugia)
M. Fabre de la Ripelle (Orsay)
A.C. Fonseca (Lisboa)
S. Kowalski (MIT)
R.A. Malfliet (Groningen)
J. Nassalski (Warsaw)
S.G. Popov (Novosibirsk)
W. Sandhas (Bonn)
M. Simonius (Zürich)
H.Chr. Walter (PSI)

[†]Deceased

Conference Secretariat

Yvonne van der Wensch
Department of Physics and Astronomy, Free University, Amsterdam

Elise de Wit
NIKHEF, Amsterdam

**Sponsored by**

Royal Academy of Arts and Sciences of The Netherlands Amsterdam,
NIKHEF Amsterdam, Free University Amsterdam,
KVI Groningen, University of Groningen, University of Nijmegen,
FOM Utrecht, Utrecht University, Stichting Physica Utrecht
Sun Microsystems Nederland B.V. Amersfoort

# Foreword

It is apparent from the history of science, that few-body problems have an interdisciplinary character. Newton, after solving the two-body problem so brilliantly, tried his hand at the Sun-Earth-Moon system. Here he failed in two respects: neither was he able to compute the motion of the moon accurately, nor did he understand the reason for that.

It took a long time to understand the fundamental importance of Newton's failure, and only Poincaré realised what was the fundamental difficulty in Newtons programme. Nowadays, the term deterministic chaos is associated with this problem.

The deep insights of Poincaré were neglected by the founding fathers of Quantum Physics. Thus history was repeated by Bohr and his students. After quantising the hydrogen atom, they soon found that the textbook case of a three-body problem in atomic physics, the $^3$He-atom, did not yield to the Bohr-Sommerfeld quantisation methods. Only these days do people realise what precisely were the difficulties connected to this semi-classical way of treating quantum systems.

Our field, as we know it today, began in principle in the early 1950's, when Watson sketched the outlines of three-body scattering theory. Mathematical rigour was achieved by Faddeev and thereafter, at the beginning of the 1960's, the quantum three-body problem, at least as far as short-range forces were concerned, was tamed.

In the years that followed, through the work of others, who first applied Faddeev's methods, but later added new techniques, the three- and four-body problems became fully housebroken.

In this process of domestication of few-body problems, mathematical techniques were developed to handle systems in which long-range forces play an essential role, thus including atomic and molecular physics in our field.

These developments were strongly stimulated by the boom in computer hardware, that took place simultaneously. The importance of the availability of powerful computers cannot be overrated in our field, because the physics we are after manifests itself most often in tiny effects. The circumstance that the basic dynamical equations governing the behaviour of few-body systems are exactly soluble, enables us to concentrate fully on the physics. Any discrepancy that may arise between the results of calculations and the experimental data, barring simple mistakes on either side of course, must be due to an incorrect or incomplete treatment of the dynamics of the systems we are concerned with.

We see a technology push, both in theoretical and experimental few-body physics. Because the theoreticians claimed to be able to calculate physical observables to great precision, experimentalists deemed it worthwhile to check their predictions accurately. The interplay of theoretical and experimental approaches to few-body problems has led to the present situation where accurate calculations are compared with ever more ingenious experiments, involving for instance spin degrees of freedom and very small cross sections. These experiments were made feasible by the great progress we have witnessed in accelerator, beam and detector techniques.

The Fourteenth European Conference on Few-Body Problems in Physics was held in Amsterdam, at the Trippenhuis, the seat of the Royal Netherlands Academy of Arts and Sciences during August 23–27, 1993.

The previous European conferences of the series took place in Budapest (1972), Graz (1973), Tübingen (1975), Vlieland (1976), Uppsala (1977), Dubna (1979), Sesimbra (1980),

Ferrara (1981), Tbilisi (1984), Balatonfüred (1985), Fontevraud (1987), Uzhgorod (1990) and Elba (1991).

The conference was divided into sixteen plenary sessions, four parallel sessions and two presentations of posters.

This volume contains the invited talks but two, all of the contributed papers presented at either the plenary sessions or the parallel sessions.

Many people and institutions have contributed to the organisation of the conference and the preparation of the proceedings. We thank the Royal Academy of Arts and Sciences of The Netherlands, Amsterdam, NIKHEF, Amsterdam, the Free University, Amsterdam, KVI Groningen, the University of Groningen, the University of Nijmegen, FOM, Utrecht, Utrecht University, Stichting Physica, Utrecht and Sun Microsystems Nederland B.V. Amersfoort, for their generous financial support.

The main social event, a lecture by prof. dr. E. van de Wetering on the "Rembrandt Project", followed by a tour of the Gallery of Honour of the Rijksmuseum, was sponsored by the Mayor and City Council of Amsterdam, for which we are very greatful.

Our thanks are also extended to the members of the International Advisory Committee for their advise, to the chairmen of the sessions who kept the discussions in time and to the speakers who send us their manuscripts before the deadline.

We are indebted to those people who helped us before, during and after the conference to smooth out the difficulties we encountered. In particularly we want to mention Ms H. Hogerhuis, who made our stay at the Trippenhuis so enjoyable, and Yvonne van der Wensch and Elise de Wit, the conference secretaries.

Finally, the editors of the proceedings wish to thank W.J.W. Geurts and L.P Kok for their help in preparing several parts of the manuscripts for the proceedings.

Amsterdam, 23 December 1993                                 B.L.G. Bakker

for the Organising Committee

# Contents

---

[1] Paper not received

[2] Paper not received

SESSION 12
Chair: *L. Lapikas*

SESSION 13
Chair: *E.O. Alt*

## SESSION 14
### Chair: *E.L. Lomon*

## SESSION 15
### Chair: *A. Mukhamedzhanov*

## SESSION 16
### Chair: *M. Fabre de la Ripelle*

## SESSION 17
### Chair: *Th.W. Ruijgrok*

## SESSION 18
### Chair: *F. Gross*

## SESSION 19
### Chair: *T.I. Kopaleishvili*

## SESSION 20
### Chair: *W. Sandhas*

Few-Body Systems, Suppl. 7, 1—12 (1994)

Few-
Body
Systems
© Springer-Verlag 1994
Printed in Austria

# Baryon-Baryon Interactions

Th. A. Rijken

Institute for Theoretical Physics,

University of Nijmegen, The Netherlands

**Abstract**

After a short survey of some topics of interest in the study of baryon-baryon scattering, the recent Nijmegen energy dependent partial wave analysis (PWA) of the nucleon-nucleon data is reviewed. In this PWA the energy range for both $pp$ and $np$ is now $0 < T_{lab} < 350$ MeV and a $\chi^2_{d.o.f.} = 1.08$ was reached. The implications for the pion-nucleon coupling constants are discussed. Comments are made with respect to recent discussions around this coupling constant in the literature. In the second part, we briefly sketch the picture of the baryon in several, more or less QCD-based, quark-models that have been rather prominent in the literature. Inspired by these pictures we constructed a new soft-core model for the nucleon-nucleon interaction and present the first results of this model in a $\chi^2$-fit to the new multi-energy Nijmegen PWA. With this new model we succeeded in narrowing the gap between theory and experiment at low energies. For the energies $T_{lab} = 25 - 320$ MeV we reached a record low $\chi^2_{p.d.p.} = 1.16$. We finish the paper with some conclusions and an outlook describing the extension of the new model to baryon-baryon scattering.

## 1 Introduction

A review of baryon-baryon scattering and the early work by the Nijmegen group has been given in [1]. Reviews of the recent work can be found in e.g. [2] and [3]. For the nucleon-nucleon work of other groups, like Bonn and Paris, we refer the reader to [4].

Although the items we discuss here are relevant, directly or indirectly, for all baryon-baryon channels, we focus in this paper mainly on the nucleon-nucleon channels. A shopping list of the items about which we want to learn more through the analysis of the experimental data and the study of theoretical models contains for example the following subjects:

1. Long-, intermediate-, and short-range mechanisms: e.g. single meson-echange $(\pi, \rho, ...)$, double meson-exchange $(\pi \otimes \pi, \pi \otimes \rho, ...)$, quark effects.

2. Relativistic effects: e.g. off-energy-shell effects, off-mass-shell effects.

3. Chiral-symmetry and soft-pion effects.

4. $SU(2, I)$- and $SU(3, F)$-symmetry of the coupling constants.

In this paper we concentrate on the first subject *i.e.* the mechanisms behind the nuclear force. Now, it is well known that the theoretical models do not explain the $NN$-data better than with a $\chi^2_{p.d.p.} \geq 1.8$. In this paper we describe a first attempt to investigate whether the new Nijmegen multi-energy partial wave analysis (PWA) [5] allows a better theoretical description of the data. This is done by an extension of the Nijmegen soft-core model [6].

The contents of this paper is as follows. In section 2 we report on the Nijmegen $pp + np$ multi-energy PWA. In section 3 we review briefly the situation around the pion-nucleon coupling constant. In section 4 we list the popular quark models and emphasize the synthesis of these quark models and the non-relativistic quark model in the general physical picture of a baryon as advocated by the chiral-quark model. In section 5 we introduce together with its first results, a new soft-core model, which we henceforth call the extended-soft-core (ESC) model. Finally, in section 6 we offer some conclusions and an outlook. Here we indicate how the ESC-model can be extended to all baryon-baryon channels.

## 2 Multi-Energy Partial-Wave-Analysis

After the multi-energy phase shift analysis of the $pp$ data below 350 MeV [7], the Nijmegen group has recently finished a similar analysis for the $pp$ and $np$ data [5]. The $pp + np$ data base consists of 1787 $pp$-data and 2514 $np$-data. The principal method employed in this multi-energy PWA consists in a division of the internucleon distances $r_{NN}$ into three regions:

(i) $r_{NN} \geq 2.0$ fm: the long-range region. Here the potential $V = V_L$ is dominated by the well known electromagnetic and one-pion-exchange potentials, $V_L \approx V_{EM} + V_{OPE}$. The residual potential comes from the spurs of the HBE, see next item.

(ii) $b \leq r_{NN} \leq 2.0$ fm ($b = 1.4$ fm): the intermediate-range region. Here the potential is taken to be a sum of the one-pion-exchange (OPE) and the heavy-boson exchanges (HBE) from the Nijmegen [6] soft-core potential, so $V = V_{EM} + V_{OPE} + V^N_{HBE}$. For the singlet waves the following modification proved to be advantageous: $V^N_{HBE} \rightarrow f^s_{med} V^N_{HBE}$, with $f^s_{med} = 1.8$.

(iii) $r_{NN} \leq b$ fm: the short-range region. Here an energy dependent boundary condition is used in principle. In practice it appeared useful to use energy dependent square well potentials in the inner region. This is equivalent to

$$P\left(b; k^2\right) = P_{free}\left(b; k^2 - 2M_r V_S\right)$$

The parametrization of the energy dependence is as follows

$$V_{S,\beta}(k^2) = \frac{1}{2M_r} \sum_{n=0}^{N} a_{n,\beta} \, k^{2n}$$

independently for each wave $\beta = (L, S, J)$. Here, $k$ denotes the relativistic cm momentum. For each wave only a couple of 'phase parameters' $a_n$'s were needed to cover the energy interval $0 \leq T_{lab} \leq 350$ MeV unbiased. In total 21 phase parameters were used for $pp$ and 18 for $np$.

With the parametrization of the potentials completed, the radial Schrödinger equation

$$\left( \frac{d^2}{dr^2} + k^2 - \frac{L^2}{r^2} - M_r V_\beta(r) \right) \chi_\beta(r) = 0$$

is solved and the phase shifts as a function of the parameters and the energy are obtained.

Very important ingredients of this PWA are:

a. The accurate treatment of the electromagnetic interactions:

$$V_{EM} = \tilde{V}_C + V_{MM} + V_{VP}$$

where $\tilde{V}_C$ is the improved Coulomb interaction, $V_{MM}$ is the magnetic moment interaction, and $V_{VP}$ is the vacuum polarization.

b. The OPE-amplitude is treated in Coulomb-distorted-wave Born-approximation. It appeared that a simple Coulomb barrier penetration factor was not sufficiently realistic. This CDWBA-treatment is very important in the determination of the pion-nucleon coupling constant.

c. The correction of the $I = 1$ $np$-waves for the $\pi^\pm - \pi^0$-mass difference.

As a result of this PWA a $\chi^2_{d.o.f} = 1.08$ was reached. For $pp$: $\chi^2 = 1787.0$, $N_{d.o.f} = 1613$ and for $np$: $\chi^2 = 2484.2$, $N_{d.o.f} = 2332$. In a combined $pp + np$ analysis one obtained $\chi^2 = 4263.8$, $N_{d.o.f} = 4301$. As an indication of the realistic energy dependence, it was found that extrapolation to the deuteron pole results in a predicted binding energy $B = 2.2247(35)$, whereas experimentally $B = 2.224575(9)$. As a result of this PWA very accurate $I = 0$ $np$ phases are now available, the estimated errors are only slightly larger than those for the $pp$ phases. The mixing parameter $\epsilon_1$ is not small and reaches $4.57 \pm 0.25$ degrees at $T_{lab} = 350$ MeV.

The Nijmegen group has also constructed Reidlike phenomenological potential models [8], which fit the data equally well as the PWA, i.e. $\chi^2_{p.d.p} \approx 1.0$. These potentials make the results of this new PWA available for many applications in few body systems. As an example, we mention the very recent calculation of the triton using these Reidlike potentials [9]. It was found that these two-nucleon interactions predict the binding energy as $7.62 - 7.72$ MeV.

# 3  Pion-Nucleon Coupling Constant

The first accurate determination of the neutral pion-nucleon coupling constant was done by the Nijmegen group [10, 7]. When this author presented the Nijmegen determination of this pion-nucleon coupling constant at the Vancouver conference in 1989 [11] it was suggested that also the charged pion-nucleon coupling constant should be determined with the same method. With the Nijmegen 1993 PWA [5] this has been done and also the charged pion-nucleon coupling turns out to be significantly lower than that found in the Karlsruhe 1980-analysis [12]. Meanwhile, also in a recent pion-nucleon partial wave analysis by the VPI&SU group a value consistent with the Nijmegen determination was found [13]. Moreover, a PWA of the combined $pp$ and $np$ data [14] and of the antinucleon-nucleon data revealed the same result [15]. In [16] the recent determinations of the $\pi NN$ couplings are tabulated. Below in Table 1 we show Table I of ref. [16]. Here, DR refers to the use of dispersion relations, PWA to the usual phase shift or partial wave analysis. Soon after the publication of the

| Group | Year | Method | $10^3 f^2_{pp\pi^0}$ | $10^3 f^2_c$ |
|---|---|---|---|---|
| Karlsruhe-Helsinki [12] | pre-1983 | $\pi^{\pm}$ DR | | 79(1) |
| Nijmegen [7] | 1987-1990 | $pp$ PWA | 74.9(0.7) | |
| VPI&SU [13] | 1990 | $\pi^{\pm}p$ DR | | 73.5(1.5) |
| Nijmegen [14] | 1991 | combined $NN$ PWA | 75.1(0.6) | 74.1(0.5) |
| Nijmegen [15] | 1991 | $\bar{p}p$ PWA | | 75.1(1.7) |
| Nijmegen [16] | 1992 | $pp$ and $np$ PWA | 74.5(0.6) | 74.8(0.3) |

Table 1: Recent $\pi NN$-coupling constant determinations.

Nijmegen $\pi^0$-coupling constant determination [10], it was heavily criticized in the literature. Notably, the claim was made, see for example [17], that the Nijmegen group had overlooked form factor effects. Still recently, it was suggested in the panel discussion of the Adelaide conference [18] that the value of the pion coupling constant found in the Nijmegen method depends on the shape of the form factor. This was dismissed in a Nijmegen paper on the several issues raised in the literature [16]. The main points made here are:

(i) The Nijmegen PWA is statistically impeccable. The criteria used in selecting the data base are unbiased and common practice under specialists on the $NN$ phase shift analysis.

(ii) Tests show that indeed the Nijmegen method determines the pole value of the pion-nucleon coupling constant.

(iii) Neither the shape nor reasonable values of the cut-off mass have any influence.

(iv) The presently available potential models are too bad to determine $f_{\pi NN}$ with an accuracy comparable to the Nijmegen $NN$ phase shift analysis.

# 4   Baryon Structure, Chiral Quark-Models

The quark-model picture of the baryons should be of some directional value in the deduction of a realistic model for baryon-baryon scattering. Interesting bag-models are the MIT [19], the Stony Brook [20], and the TRIUMF [21] models. A particularly interesting quark-model is the chiral-quark-model [22]. This model explains the successes of the non-relativistic quark-model (NRQM) and at the same time is closely connected with the description of hadron dynamics through interactions involving mesons and baryons using effective chiral lagrangians. The general idea is that the QCD-vacuum becomes unstable at $Q^2 \leq \Lambda^2_{\chi SB} \approx (1 GeV)^2$. The vacuum goes through a phase transition, making for the quarks $\langle 0|\bar{\psi}\psi|0\rangle \neq 0$ and the gluon coupling $\alpha_S$ small. This generates the constituent quark masses and implies that the quarks move around in the core of a baryon essentially as being free, just as in the NRQM. Viewing (part of) the pion as the Goldstone boson, correlated with spontaneously broken chiral invariance, makes it natural that there is a soft-pion cloud around a constituent quark. High energy experiments indicate that the Pomeron couples to the quarks [23]. Then, a soft pion cloud around a constituent quark offers a natural explanation for the multi-peripheral component of the Pomeron. Also, the coupling of mesons to quarks dressed by a pion cloud is in accordance with the ideas that the non-linear sigma-model is relevant for the description of hadronic interactions [24]. We have drawn for a nucleon in Fig. 1 the picture that emerges from the bag-models and the chiral-quark-model, i.e. a quark-core surrounded by a meson cloud of pions and other mesons, Baryon-baryon scattering is the quantum mechanical scattering of two of such systems. The chiral-quark-model in particular, provides a natural basis for an approach to baryon-baryon scattering using only mesonic degrees of freedom in the derivation of the baryon-baryon interactions. In the next paragraph we will describe such an attempt. We construct a new $NN$-model and make a fit to the 1993 multi-energy Nijmegen PWA.

# 5   Extended Soft-Core model

The potential of this new $NN$-model, henceforth referred to as the ESC-model, consists of the contributions of

(i) The OBE-potentials of [6], which apart from the low lying pseudo-scalar-, vector-, and scalar-mesons includes also contributions of the Pomeron. The latter represents the multi-peripheral (soft)pion exchanges and multi-gluon exchanges.

(ii) The $2\pi$-potentials as given in [26]. These are two-pion-exchange potentials based on the pseudo-vector pion-nucleon coupling. We include only the so-called BW-graphs, i.e. we discard the TMO-graphs (see [26] for this nomenclature). This, because we think that the non-adiabatic expansions are not

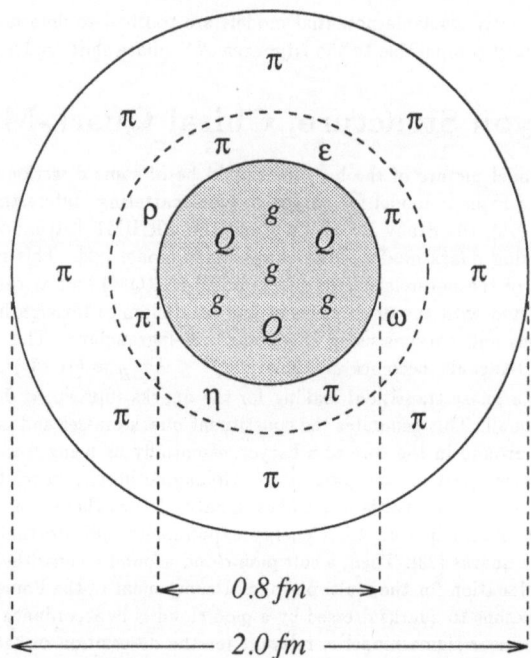

Figure 1: Schematic model of the baryon structure

reliable. Therefore, we prefer to extend the present model later by including the non-adiabatic effects already in the OBE-potentials.

(iii) We extend the OBE-model of [6] further through the inclusion of phenomenological nucleon-nucleon-meson-meson vertices, henceforth referred to as 'pair interactions' or 'pair terms'. The vertices are listed in Table 2.

The motivation for including these 'pair-vertices' is that similar interactions appear in chiral-lagrangians. They can be viewed upon as the result of the out integration of the heavy-meson and resonance degrees of freedom. Moreover, they also represent two-meson exchange potentials. We are less radical than Weinberg, see *e.g.* [25], in that we do not integrate out the degrees of freedom of the mesons with masses below 1 GeV. The techniques to derive the explicit expressions for the potentials corresponding to the meson-pair exchange potentials with soft *i.e.* gaussian form factors, is in essence described in [26]. The new type of graphs that have to be evaluated are those with one pair-vertex and with two pair-vertices.

Fitting this new model to the NN-data, using the 1993 Nijmegen single energy $pp+np$ phase shift analysis [5], leads to an excellent result. We reached for the energies in the range $25 \leq T_{lab} \leq 320$ MeV, which comprises 3709 data, a $\chi^2_{p.d.p.} = 1.16$ [27].

$$J^{PC} = 0^{++} \quad : \quad \mathcal{H}_S \quad = \quad (\bar{\psi}'\psi')\left\{g_{(\pi\pi)_0}\left(\underline{\pi}\cdot\underline{\pi}\right) + g_{\sigma\sigma}\sigma^2\right\}/m_\pi$$

$$J^{PC} = 1^{--} \quad : \quad \mathcal{H}_V \quad = \quad \left[g_{(\pi\pi)_1}\bar{\psi}'\gamma_\mu\underline{\tau}\psi' - \frac{f_{(\pi\pi)_1}}{2M}\bar{\psi}'\sigma_{\mu\nu}\underline{\tau}\psi'\partial^\nu\right]\left(\underline{\pi}\times\partial^\mu\underline{\pi}\right)/m_\pi^2$$

$$J^{PC} = 1^{++} \quad : \quad \mathcal{H}_A \quad = \quad g_{(\pi\rho)_1}\left(\bar{\psi}'\gamma_\mu\gamma_5\underline{\tau}\psi'\right)\left(\underline{\pi}\times\underline{\rho}^\mu\right)/m_\pi$$

$$\mathcal{H}_P \quad = \quad g_{(\pi\sigma)}\left(\bar{\psi}'\gamma_\mu\gamma_5\underline{\tau}\psi'\right)\left(\sigma\partial^\mu\underline{\pi} - \underline{\pi}\partial^\mu\sigma\right)/m_\pi^2$$

$$J^{PC} = 1^{+-} \quad : \quad \mathcal{H}_H \quad = \quad ig_{(\pi\rho)_0}\left(\bar{\psi}'\sigma_{\mu\nu}\gamma_5\psi'\right)\partial^\nu\left(\underline{\pi}\cdot\underline{\rho}^\mu\right)/m_\pi^2$$

$$\mathcal{H}_B \quad = \quad ig_{(\pi\omega)}\left(\bar{\psi}'\sigma_{\mu\nu}\gamma_5\underline{\tau}\psi'\right)\partial^\nu\left(\underline{\pi}\cdot\omega^\mu\right)/m_\pi^2$$

Table 2: Phenomenological Meson-Pair Interactions

The (rationalized) coupling constants and form factor masses are given in Table 3. Here, the $f_\eta$ was not fitted but derived from $f_\pi$ using $\alpha_{pv} = 0.361$. We used for the $\sigma$ a mass of $m_\sigma = 500.4$ MeV, i.e. the lowest mass of the two-pole approximation used in [6]. The use of different form factors for $I_t = 1$ and $I_t = 0$ for vector and scalar exchange did not have much influence on the fit.

The nuclear-bar phase shifts of the new $NN$-model are given in Table 4. In this table, the $I = 1$-phases are $pp$-phases and the $I = 0$-phases are $np$-phases. The $\chi^2$ of the model w.r.t. the PWA-phases is denoted by $\Delta\chi^2$.

The numerical results were obtained by using a coordinate space version of the model. The OPEP treatment was adapted to the PWA by multiplying in momentum space the OPEP of [6] by $\sqrt{M/E(p)}$-factors for the initial and final state.

From Table 4 one notices the great improvement of the new model over the OBE-model [6]. In particular this is obvious for the $^1P_1$-, the $^3D_2$-, and the $^3D_3$-waves. The $^3F_2$-wave however, is bending towards zero too quickly as a function of energy. Here the $2\pi$-potential gives a repulsive tensor force. At this point the inclusion of $\pi \otimes \rho$-, $\pi \otimes \omega$- potentials in the future may be of help in particular. The values reported in Table 3 are very reasonable. The pion coupling was searched and $f_{NN\pi}^2 = 0.072$, which is on the lower side of the determinations listed in Table 1. We have $g_\rho^2 = 0.53$ and $(f/g)_\rho = 4.52$, in reasonable agreement with VDM [29]. The agreement improves if we also take into account the contribution of the $\pi\pi$-pair terms (see remark below). The $\omega$-, $\epsilon$-, and pomeron- couplings are rather similar to those of [6].

Also the meson-pair couplings are accessible to a physical interpretation. The couplings $g_{(\pi\pi)_0}$ and $f_{(\pi\pi)_1}$ are not very small. This, notwithstanding the fact that the $I_t = 0$-channel is dominated by $\epsilon$- and Pomeron-exchange, which tend to cancel each other largely. Similarly, because of the dominance of $\rho$-exchange in the $I_t = 1$-

| ps-pv | | vector | | scalar | | pairs | |
|---|---|---|---|---|---|---|---|
| $f_\pi$ | 0.268 | $g_\rho$ | 0.730 | $g_\delta$ | 1.299 | $g_{(\pi\pi)_0}$ | -0.160 |
| $f_\eta$ | 0.069 | $f_\rho$ | 3.299 | $g_\epsilon$ | 3.573 | $g_{(\pi\pi)_1}$ | -0.001 |
| $f_{\eta'}$ | 0.271 | $g_\omega$ | 3.009 | $g_{A_2}$ | 0.123 | $f_{(\pi\pi)_1}$ | -0.260 |
| | | $f_\omega$ | 0.567 | $g_P$ | 2.346 | $g_{(\pi\rho)_0}$ | 0.073 |
| $\Lambda_{PV}$ | 844.8 | $\Lambda_{V,1}$ | 777.6 | $\Lambda_{S,1}$ | 767.1 | $g_{(\pi\rho)_1}$ | 0.506 |
| | | $\Lambda_{V,0}$ | 744.9 | $\Lambda_{S,0}$ | 835.6 | $g_{(\pi\omega)}$ | -0.001 |
| | | | | $m_P$ | 309.1 | $g_{(\pi\sigma)}$ | -0.170 |
| | | | | | | $g_{(\sigma\sigma)}$ | -0.302 |

Table 3: Form factor masses, meson and meson-pair couplings.

channel one would tend to expect small values for $g_{(\pi\pi)_1}$ and $f_{(\pi\pi)_1}$. However, the pion-pair contribution represents, among other things, the correction to the two-pole approximation used for the description of the broad $\epsilon$ and $\rho$ meson, which is not negligible.

Also, with these $\pi N$-interactions all s- and p-wave pion-nucleon scattering lengths are accounted for very well (see also [2]). In particular, interpreting the $(\pi\pi)_0$-pair contribution as representing in fact the effect of the low mass tail of the broad $\epsilon$-meson, one finds a contribution $\Delta a_{33} \approx 0.10$, which is needed together with the nucleon-pole contribution in order to give the experimental value.

For the $g_{(\pi\rho)_1}$- and $g_{(\pi\sigma)}$-coupling $A_1$-dominance would predict

$$|g_{(\pi\rho)_1}| = \left(\frac{m_\pi}{m_{A_1}}\right)^2 g_{A_1 NN}(0) g_{A_1 \rho\pi}(0) \approx 0.14$$

$$|g_{(\pi\sigma)}| = \left(\frac{m_\pi}{m_{A_1}}\right)^2 g_{A_1 NN}(0) g_{A_1 \sigma\pi}(0) \approx 0.10$$

In obtaining these estimates, we have used the predictions of the chiral-lagrangians in [30] and [31] for $g_{A_1 \pi\rho}(m_{A_1}^2)$ and $g_{A_1 \pi\sigma}(m_{A_1}^2)$. Extrapolation to zero momentum we have done by using a factor $\exp(-m_{A_1}^2/\mathcal{M}^2)$ , where $\mathcal{M} = 1$ GeV. Additional

| $T_{lab}$ | 25 | 50 | 100 | 150 | 215 | 320 |
|-----------|------|------|------|------|------|------|
| ♮ data | 352 | 572 | 399 | 676 | 756 | 954 |
| $\Delta\chi^2$ | 67 | 68 | 30 | 74 | 153 | 335 |
| $^1S_0$ | 49.05 | 38.87 | 24.41 | 13.88 | 3.23 | -9.91 |
| $^3S_1$ | 79.99 | 62.21 | 42.91 | 30.77 | 19.44 | 6.36 |
| $\epsilon_1$ | 1.96 | 2.38 | 2.82 | 3.21 | 3.70 | 4.43 |
| $^3P_0$ | 8.84 | 11.96 | 10.01 | 5.20 | -1.46 | -11.18 |
| $^3P_1$ | -4.85 | -8.24 | -13.22 | -17.32 | -21.94 | -28.19 |
| $^1P_1$ | -6.19 | -9.39 | -13.87 | -17.81 | -22.57 | -29.40 |
| $^3P_2$ | 2.50 | 5.79 | 10.89 | 13.99 | 16.18 | 17.41 |
| $\epsilon_2$ | -0.80 | -1.70 | -2.71 | -3.01 | -2.85 | -2.10 |
| $^3D_1$ | -2.78 | -6.39 | -12.27 | -16.71 | -21.17 | -26.56 |
| $^3D_2$ | 3.67 | 8.91 | 17.37 | 22.48 | 25.44 | 25.54 |
| $^1D_2$ | 0.69 | 1.69 | 3.83 | 5.86 | 7.96 | 9.75 |
| $^3D_3$ | 0.05 | 0.31 | 1.34 | 2.53 | 3.77 | 4.70 |
| $\epsilon_3$ | 0.54 | 1.57 | 3.40 | 4.74 | 5.92 | 7.03 |
| $^3F_2$ | 0.10 | 0.34 | 0.79 | 1.07 | 1.07 | 0.16 |
| $^3F_3$ | -0.22 | -0.66 | -1.47 | -2.11 | -2.82 | -3.97 |
| $^1F_3$ | -0.41 | -1.08 | -2.09 | -2.74 | -3.39 | -4.51 |
| $^3F_4$ | 0.02 | 0.11 | 0.47 | 0.94 | 1.60 | 2.52 |
| $\epsilon_4$ | -0.05 | -0.19 | -0.52 | -0.82 | -1.13 | -1.49 |
| $^3G_3$ | -0.05 | -0.25 | -0.89 | -1.66 | -2.66 | -4.12 |
| $^3G_4$ | 0.17 | 0.70 | 2.09 | 3.49 | 5.15 | 7.33 |
| $^1G_4$ | 0.04 | 0.15 | 0.41 | 0.67 | 1.03 | 1.63 |
| $^3G_5$ | -0.01 | -0.05 | -0.15 | -0.23 | -0.26 | -0.17 |
| $\epsilon_5$ | 0.04 | 0.20 | 0.69 | 1.20 | 1.80 | 2.58 |
| $^3H_4$ | 0.00 | 0.03 | 0.11 | 0.21 | 0.36 | 0.56 |
| $^3H_5$ | -0.01 | -0.08 | -0.29 | -0.52 | -0.77 | -1.10 |
| $^1H_5$ | -0.03 | -0.16 | -0.51 | -0.83 | -1.15 | -1.50 |
| $^3H_6$ | 0.00 | 0.01 | 0.04 | 0.10 | 0.21 | 0.44 |
| $\epsilon_6$ | -0.00 | -0.03 | -0.11 | -0.22 | -0.35 | -0.54 |

Table 4: ESC nuclear bar $pp$ and $np$ phase shifts in degrees.

input in this estimate is that $g_{A_1NN} \approx (m_\pi/m_{A_1})f_{\pi NN} = 2.45$ [32] (see also [33]). Similarly, we find from the chiral-lagrangians the prediction, using $\sigma$-dominance, that roughly $g_{\sigma\sigma} \approx -0.50$, which is not far from $-0.30$ found in the fit. Likewise, assuming that $g_{(\pi\rho)_0}$ and $g_{(\pi\omega)}$ are dominated by respectively the H- and $B_1$-meson, we could estimate from the fitted values the couplings $g_{HNN}$ and $g_{B_1NN}$. Of course, heavy boson dominance is not valid for all these pair couplings. If we would include also the $\pi \otimes \rho$-, $\pi \otimes \omega$- etc. potentials, then the residual interactions are more likely to be boson dominated. Therefore, the present results are preliminary.

# 6  Conclusions and Outlook

The multi-energy Nijmegen PWA poses a nice new challenge to the theory of the low momentum transfer baryon-baryon interactions. The success of our ESC-model indicates that the better quality of the multi-energy Nijmegen PWA with respect to other phase shift analyses, indeed opens the door to a more thorough understanding of the low energy NN-data. To make progress in the problems concerning Few Body Physics, it is imperative that baryon-baryon interactions are used which are based on a very realistic description of nucleon-nucleon scattering. Conclusions about such parameters as the pion-nucleon coupling constant, the relativistic effects, the off-mass-shell effects etc. are otherwise liable to be fallacious.

The chiral-quark-model picture [22] makes it highly implausible that there will be large nucleon-antinucleon-pair effects in the low energy region (see also [34]). Incidentally, a model with large nucleon-antinucleon pair contributions should also include in the intermediate states pion-nucleon resonances up to 3 GeV, nucleon-hyperon-kaon intermediate states etc. Also, the presence of these pairs in nuclear Compton scattering is improbable. In fact, it is likely that the negative energy contributions of the constituents cancel out in the Thomson limit [35].

Multi-soft-pion and multi-meson effects on the other hand are expected, both in chiral-lagrangian models and QCD [24]. However, for reactions dominated by momentum transfers below 1 GeV, interactions based on gluon-exchange are presumably suppressed [22]. Therefore, models based on strong gluon-quark exchanges do not seem very realistic.

The proper theoretical framework for the phenomenological nucleon-nucleon meson-pair vertices seems the non-linear chiral $SU(2) \times SU(2)$- symmetry ( for reference see e.g. [30]. Then, the extension from nucleon-nucleon to baryon-baryon can be tried by employing $SU(3) \times SU(3)$-symmetry (see e.g. [36]). This would introduce only a very restricted set of extra free parameters in for example hyperon-nucleon models.

The extension to higher nucleon-nucleon energies of the ESC-model requires the explicit treatment of the $\Delta_{33}$-resonance degrees of freedom. This can be done immediately and will result in different meson-pair contributions. For low energy scattering this is unnecessary. This follows on the one hand from our successful fit to the low energy data by e.g. the new NN-model described above, and on the other hand this is explained to be possible to a certain degree of accuracy by duality (see the remarks in [2]).

# 7 Acknowledgements

The generous help of C. Terheggen in the implementation of the authors programs on the HP9000-735 computer is gratefully acknowledged. It is also a pleasure to thank the other (former) members of the Nijmegen group, R. Klomp, J.-L. de Kok, M. Rentmeester, V. Stoks, and R. Timmermans for many discussions regarding the various aspects of the multi-energy phase shift analysis. Last but not least, the never fading interest of J.J. de Swart in baryon-baryon interactions should be mentioned.

# References

[1] J.J. de Swart, M.M. Nagels, Th.A. Rijken, and P.A. Verhoeven, Springer Tracts in Modern Physics, Vol. **60**, 138 (1971).

[2] J.J. de Swart, Th.A. Rijken, P.M. Maessen, and R.G. Timmermans Nuov. Cim. **102A**, 203 (1988).

[3] Th.A. Rijken, P.M.M Maessen and J.J. de Swart, in Particles and Fields Series **43**, p.153, AIP New York (1991), and Nucl.Phys. **A547**, 245c (1992).

[4] R. Machleidt, K. Holinde, and Ch. Elster, Physics Reports, **149** 1 (1987); R. Machleidt, Adv. Nucl. Phys. **19**, 189 (1989).
M. Lacombe, B. Loiseau, J.M. Richard, R. Vinh Mau, P. Pirès, and R. de Tourreil, Phys. Rev. **C21**, 861 (1980).

[5] V.G.J. Stoks, R.A.M. Klomp, M.C.M. Rentmeester, and J.J. de Swart, Phys. Rev. **C48**, 792 (1993).

[6] M.M. Nagels, Th.A. Rijken, and J.J. de Swart, Phys. Rev. **D17**, 768 (1978).

[7] J.R. Bergervoet, P.C. van Campen, R.A.M. Klomp, J.-L. de Kok, Th.A. Rijken, V.G.J. Stoks, and J.J. de Swart, Phys. Rev. **C41**, 1435 (1990).

[8] V.G.J. Stoks, R.A.M. Klomp, C.P.F. Terheggen, and J.J. de Swart, submitted to Phys. Rev. C.

[9] J.L. Friar, G.L. Payne, V.G.J. Stoks, and J.J. de Swart, Phys. Lett. B 311, 4 (1993).

[10] J.R. Bergervoet, P.C. van Campen, Th.A. Rijken, and J.J. de Swart, Phys. Rev. Lett. **59**, 2255 (1987).

[11] Th.A. Rijken, V.G.J. Stoks, R.A.M. Klomp, J.-L. de Kok, and J.J. de Swart, Proceedings XIIth Int. Conf. on Few-Body Problems in Physics, Vancouver, Canada, 1989, Nucl. Phys. A508 173c (1990).

[12] G. Höhler and E. Pietarinen, Nucl. Phys. **B95**, 210 (1975); R. Koch and E. Pietarinen, Nucl. Phys. **A336**, 331 (1980).

[13] R.A. Arndt, Z. Li, L.D. Roper, and R.L. Workman, Phys. Rev. Lett. **65**, 157 (1990).

[14] R.A.M. Klomp, V.G.J. Stoks, and J.J. de Swart, Phys. Rev. **C44**, R1258 (1991).

[15] R.G.E. Timmermans, Th.A. Rijken, and J.J. de Swart, Phys. Rev. Lett. **67**, 1074 (1991).

[16] V.G.J. Stoks, R.G.E. Timmermans, and J.J. de Swart, Phys. Rev. C**47**, 512 (1993).

[17] A.W. Thomas and K. Holinde, Phys. Rev. Lett. **63**, 2025 (1989).

[18] T.O.E. Ericson, Proceedings XIIIth Int. Conf. on Few-Body Problems in Physics, Adelaide, Australia, 1992, Nucl. Phys. **A543** 409c (1992); CERN preprint CERN-TH. 6405/92.

[19] A. Chodos, R.L. Jaffe, K. Johnson, C.B. Thorn, and V.F. Weisskopf, Phys. Rev. **D9**, 3471 (1974).

[20] G.E. Brown and M. Rho, Phys. Lett. **82B**, 127 (1979).

[21] S. Théberge, A.W. Thomas, G.A. Miller, Phys. Rev. **D22**, 2838 (1980).

[22] A. Manohar and H. Georgi, Nucl. Phys. **B234**, 189 (1984); S. Weinberg, Physica (Amsterdam) **96A**, 327 (1979).

[23] T. Henkes et al, Phys. Lett. **B283**, 155 (1992); A.M. Smith et al, Phys. Lett. **163B**, 267 (1985).

[24] E. Witten, Nucl. Phys. **B160**, 57 (1979).

[25] S. Weinberg, Phys. Lett. **B251**, 288 (1990), and Nucl. Phys. **B363**, 3 (1991); C. Ordòñez and U. v. Kolck, Phys. Lett. **B291**, 459 (1992).

[26] Th.A. Rijken, Ann. Phys. (N.Y.) **208**, 253 (1990).

[27] In the presentation of this work at the conference only the results for the MAW-X phase shift analysis [28] were available. Here, we present exclusively the results of the ESC-model for the PWA of reference [5].

[28] M.H. MacGregor, R.A. Arndt, and R.M. Wright, Phys. Rev. **182**, 1714 (1969).

[29] J.J. Sakurai, Lectures in Theoretical Physics, Vol. XI-A, p.1 (Gordon & Breach, 1968); Ann.Phys. (N.Y.) **11**, 1 (1960).

[30] S. Weinberg, Phys. Phys. **166**, 1568 (1968); Phys. Phys. **177**, 2604 (1969).

[31] H. Kleinert, Elementary Particle Physics, edited by P.Urban, Acta Physica Austriaca, Supplementum IX, p. 533 (1972).

[32] J. Schwinger, Phys. Lett. **B24**, 473 (1967).

[33] H.E. Haber and G. Kane, Nucl. Phys. **B129**, 429 (1977); W. Grein and P. Kroll, Nucl. Phys. **A377**, 505 (1982).

[34] J.J. de Swart and M.M. Nagels, Fortschr. d. Physik, **28**, 215 (1978).

[35] S.J. Brodsky and J.R. Primack, Ann. Phys. (N.Y.) **52**, 315 (1969).

[36] J. Schechter, Y. Ueda, and G. Venturi, Phys. Rev. **177**, 2311 (1969).

Few-Body Systems, Suppl. 7, 13—21 (1994)

Few-
Body
Systems
© Springer-Verlag 1994
Printed in Austria

# PRECISE MEASUREMENTS OF SPIN OBSERVABLES IN NUCLEON-NUCLEON SCATTERING

J. Sowinski

Indiana University Cyclotron Facility and Department of Physics
Bloomington, IN 47405

## Abstract

A review of the status of nucleon-nucleon scattering at intermediate energies is given. It is found that the tensor component of the N-N force is still poorly determined by the data. This is followed by recent results from high precision experiments at IUCF and descriptions of work in progress. The data are compared with phase shift analyses and potential model predictions. Technological advances that have improved our ability to make precise measurements of spin observables are discussed.

## Introduction

The nucleon-nucleon (N-N) interaction is central to nuclear physics. It has long been commonly assumed that the N-N interaction is well known from low to intermediate energies. Yet, in the past few years it has been found[1,2] that the pion-nucleon coupling constants should be reduced by about 6% from values stable for quite some time. It would also appear that the strength of the the tensor force is still poorly constrained by the data as will be discussed below. I will begin by reviewing the status of N-N elastic scattering. I will then discuss recently completed, as well as ongoing, measurements at IUCF while pointing out new techniques which make high precision measurements of spin observables possible. The data will be compared to phase shift analyses and meson exchange model predictions.

## Status of the Nucleon-Nucleon Interaction

The N-N interaction has been studied via elastic scattering since the earliest days of nuclear physics. Experiments from the early 1950's are still included in the data base. Nevertheless there is still a need for precise N-N measurements. The data are typically compared to predictions from phase shift analyses and meson exchange models. (Above

500 MeV direct amplitude reconstructions[3] are also carried out but I will not discuss those further here.) I will briefly give an overview of these descriptions of the N-N force and then review the data base.

Extensive reviews on the methods used in phase shift analyses (PSA) and current status are available [3,4], so I will present only a short review from an experimentalist's perspective. The basic idea behind a PSA is to parameterize the scattering amplitudes in terms of the phase shift in each of the partial waves. The phase shifts are then determined by fitting the data. However this view of a PSA as a straightforward parameterization of the data is a bit too simple. Considerable theoretical input is necessary. The partial wave expansion is an infinite series for which only the lower partial wave phase shifts can be determined from the data. The higher partial waves are usually fixed to their one pion exchange values, and in some cases heavy boson exchange is also added. Electromagnetic effects from the charges and magnetic moments of the nucleons must also be included properly. Moreover, the energy dependence of the phase shifts must be parameterized in some manner. In some cases the parameterization covers a global energy range fitting all data, for example, from 0-350 MeV or 0-1.6 GeV. "Single energy" fits are also performed, to data not at a single energy but actually to data in a much smaller range of energies than the global fits, e.g. 50-100MeV. A global fit has the advantage that it will filter out "noise" that can occur from experiments at specific energies but could also filter out real energy dependences in the N-N force. Differences between global and single energy fits can point to regions where further measurements might correct problems in the data base or confirm energy dependences not allowed for in the global fits.

There are four primary groups which have long maintained and continue to practice the art of PSA in the intermediate energy region, at Nijmegen[5], at VPI[6], Bugg[7] and at Saclay[8]. They take different approaches on how much to constrain their fits by theoretical input and the level of sophistication of that input. They also make judgments on which data to include and discard. I see two goals of these efforts. One is to arrive at the best description of the N-N force possible from the data for use in calculations. The second is to determine as much of the N-N force as possible from the data. In the first, the best theoretical input can constrain the fits and help filter noise from the data base, hopefully giving the most reliable description. In the second, it is important to differentiate between parts of the force determined by the data and those parts constrained by theoretical input, thus indicating where further measurements can reduce the need for theoretical input. At present the most sophisticated analysis is done by the Nijmegen group[5], in which they achieve a $\chi^2$ per degree of freedom (DOF) of 1 in a global fit, suggesting no loss of important information. As an experimentalist I need to concern myself with how much of the output of a PSA is determined by the data versus the theory. I am interested in making sure the data are of sufficient quality for the task and that, where the theoretical input is a major constraint, this input is tested like any other model. In this context all practitioners take, more or less, reasonable approaches to the problem. Within each PSA the statistical errors on the parameters and predictions of observables are tiny. However, in an individual analysis it is very difficult to judge the systematic errors or model dependence. Differences in predictions of various practitioners can reveal this model dependence in the analyses. Such differences should be considered degrees of freedom allowed by the data and, hence, to point to observables and angle ranges where new data could be of benefit in sharpening our knowledge of the N-N interaction.

There are also serious attempts to generate potential models to fit the N-N data base below pion production threshold. These are based on meson exchange theories and have a goal of predicting the N-N data base in a quantitative way. In the end these models have parameters that must be determined by fits to the data base. With a "good" fit to the data they are then viewed as very accurate descriptions of the N-N force and used in calculations in nuclei. The potentials most frequently compared to the N-N data base directly are the Bonn[9] and Paris[10] potentials. Differences in the predictions between the potentials can arise because of the theoretical input, but can also arise from the fact that they are constructed at different times meaning they have been fit to different data bases. In particular the Paris potential is now 13 years old. The overall quality of the fits of a wide range of potentials to the N-N data base was recently investigated [11] and it was found that both of these potentials do an acceptable job with a $\chi^2$ per DOF of ~ 2. A potential, NijmRdl, recently developed by the Nijmegen group is found to reproduce the data base as well as the groups PSA, i.e. $\chi^2$ per DOF of 1, while an old style potential, Nijm78, fits the data base about as well as Bonn or Paris.

Let me turn now to describe the current situation in the data base. Much of the data are quite old. It is often of not particularly good statistical precision and has poorly understood systematic errors. The data primarily consist of cross sections, analyzing powers, spin transfer and spin correlation data. (No one convention for the naming of observables has caught on. (See ref. 12 for definitions and cross references.) The spin transfer and spin correlation data are often sparse in energy and angle and often incomplete in the combinations of spin directions of interest. The n-p data base for these observables is represented in Fig. 1. We see that n-p spin transfer data are particularly scarce and often quite old. We also see the effect of the first of

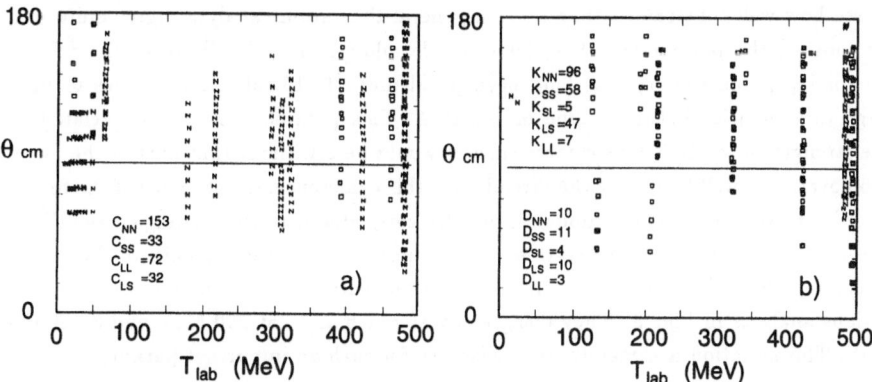

Fig. 1. Data base for n-p scattering in Arndt's program SAID[13]. Points are plotted (N=post 1987, □=pre 1988) for a datum of the given type at the energy and angle measured. a) Spin correlations. b) Spin transfer parameters.

the new technologies I will mention, the common use of polarized neutron beams and cryogenic polarized proton targets. This has resulted in a quintupling of the n-p spin correlation data over the past 5 years, making these data more numerous than the spin transfer observables. Nevertheless, it is still not possible to do an analysis for the n-p channel without including knowledge of the isovector phase shifts from the p-p channel.

Statistical data as shown in Fig. 1 do not provide any information on the quality of the measurements. In Fig. 2 I show measurements and predictions for spin transfer observables in n-p and p-p scattering. The quality of the data is typical for these types of observables in the data base. The differences in the predictions are among the largest I have found in making such comparisons near 200 MeV.

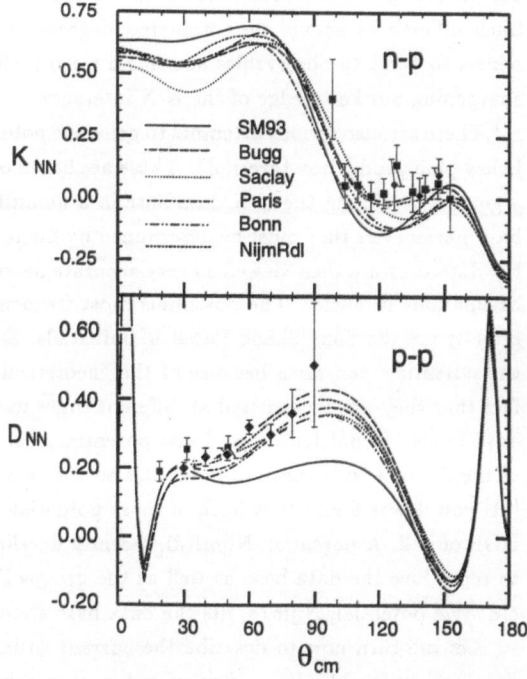

Fig. 2. Spin transfer parameters for both spins normal to the scattering plane near 200 MeV. For n-p the neutron beam is polarized and the polarization of the proton is measured. SM93 refers to the global fit by the VPI group[13] in the summer of 1993.

We can also look at the extracted phase shift parameters from the various groups to see how well constrained they are. In general the various analyses agree quite well for most of the parameters. See, for example, plots in ref. 5. However the $^3S_1 - ^3D_1$ mixing parameter, $\epsilon_1$, has long been problematic[14,15,16] and continues to have large discrepancies between some of the PSA. This parameter is directly related to the strength of the N-N tensor force since it would be 0 without that part of the force. Moreover, as is well known by the attendees of this conference, the strength of the tensor force in the various potential models is directly correlated with the binding energy of the triton. This and other manifestations of the uncertainty in the strength of the tensor force are summarized in ref. 16. I have collected current values of $\epsilon_1$ extracted from various analyses in Fig. 3. They disagree by as much as 1° at 200 MeV and 3° at 100 MeV. This situation is certainly unsatisfactory for such an important parameter.

Is there a need for more N-N data? In general I would say the people using the N-N interaction for calculations in nuclei seem to be happy with the quality of the existing information. The PSA do a very good job of taking all the data and assembling it into quite accurate predictions of observables. One exception which needs to be addressed is the problem with $\epsilon_1$ just mentioned. Other special situations arise from time to time such as the question of the $\pi N$ coupling constants which I shall return to later. And there are always individual energy regions that could be tidied up. For example the data base near 100 MeV is particularly thin. An ambitious goal would be to measure enough n-p observables to do an analysis independent of p-p scattering thus allowing an investigation of charge independence violation in the N-N force. However, a clear indication of the number of observables needed for this effort is required to evaluate its feasibility. I will close this section with a quote from the Nijmegen group's recent paper[5]:

> "The fact that the multi-energy phase shifts can be determined very accurately also implies that angular dependence of the cross sections, analyzing powers, spin correlation parameters, etc are fixed rather well.... In almost all cases our determination of the normalization is (much) more accurate than the uncertainty quoted by the experimentalists."

This certainly is a challenge to experimentalists and a number of measurements should be made to test the predictive power of phase shift analyses at the claimed level of precision.

## Precise N-N Measurements from IUCF

I will now describe some recent and ongoing measurements. I will try to point out how new techniques are being used to achieve high precision in the measurements. Many of these techniques have been or are being used to study the N-N interaction at labs such as TRIUMF, PSI, LAMPF and SACLAY. Here the examples will come from experiments at IUCF because these are what I know best.

Experimentalists have learned to measure spin dependent observables to a very high degree of precision. Examples such as tests of parity and charge symmetry are described in the talk at this conference by van Oers. One such charge symmetry violation measurement[18] was performed at IUCF. In this experiment we measured a difference in the n-p scattering analyzing power measured once with the neutron polarized and again with the proton polarized. This difference was found to be $\Delta A = (33.1 \pm 5.9 \pm 4.3) \times 10^{-4}$. This experiment used a polarized neu-

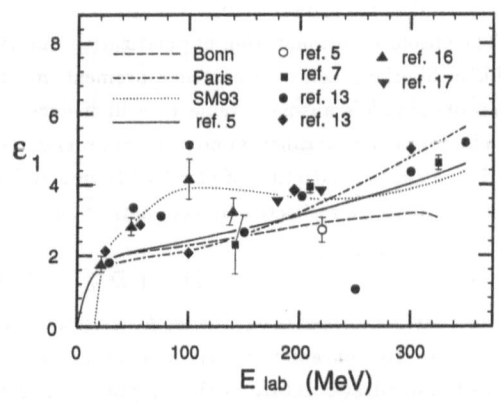

Fig. 3. The $^3S_1 - {}^3D_1$ mixing parameter, $\epsilon_1$, from various analyses vs. energy. Curves are potentials or global PSA solutions. Points are single energy analyses with error bars when available. ($\bullet$=1993, $\blacklozenge$=1989).

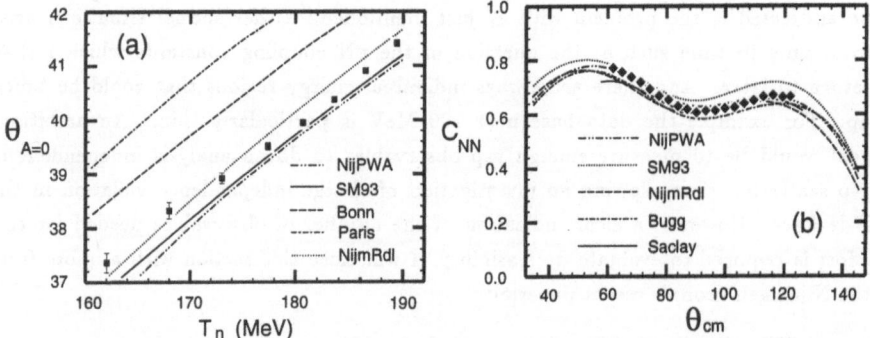

Fig. 4. Measured observables for n-p scattering. a) Lab angle where the analyzing power goes through 0 vs incident energy. b) Spin correlation parameter with both the neutron beam spin and proton target spin normal to the scattering plane at 183 MeV.

tron beam and cryogenic polarized target, the technology mentioned previously which has resulted in a huge increase in the amount of spin correlation data for n-p scattering. In making the charge symmetry measurements we also measured the conventional analyzing power and the spin correlation parameter with both spins normal, here called $C_{NN}$, to very high statistical precision. Preliminary measurements of lower precision have already been published [19]. The neutron's beam energy has a significant spread. By measuring the time of flight of beam neutrons with respect to the cyclotron rf we separate the data into energy bins. This allows us to extract the angle at which the analyzing power crosses zero as a function of energy as is shown in Fig. 4 along with $C_{NN}$. Within the uncertainty in the beam energy ($\pm 2$ MeV) the VPI (SM93) and Nijmegen phase shift analyses along with the NijmRdl potential agree with the data while the Paris and Bonn potential miss significantly. The $C_{NN}$ measurements are still unnormalized but shape differences with many of the curves are apparent. Others are similar in shape but differ primarily in normalization.

The absolute normalization of polarization observables has always proved a difficult problem. Special reactions[20] or measurements must be found which allow one to know a spin observable's absolute value to a high degree of precision, say 1% or better, to provide a proper experimental normalization and test the phase shift claims. One such case is the elastic scattering of spin 1/2 from a spin 0 target. With this spin structure the amplitudes are restricted by symmetries forcing relationships between observables. For example here,

$$D_{LS} + D_{LL} + A \equiv 1.$$

We see that if the analyzing power, A, is close to 1 relatively crude measurements of the spin transfer parameters, $D_{LS}$ and $D_{LL}$, will determine A to very high accuracy due to the quadratic nature of the equation. (There are other relationships useful for calibrating polarimeters used in spin transfer measurements.) Large elastic scattering analyzing powers and cross sections on light spin 0 targets are a common feature near 200 MeV. We have recently measured[21] the analyzing power in p-$^{12}$C elastic scattering to be $0.99963^{+0.00021}_{-0.00030}$ at $\theta_{lab} = 17.3°$ and 189 MeV. This is a very convenient standard because it can be measured relatively easily in many situations.

This p-$^{12}$C standard has been transferred to the p-p elastic scattering analyzing powers at IUCF[22] and then used to normalize small angle p-p scattering data[23]. These data (corrected for small energy differences) along with recently analyzed data, which complete the angular distribution of the analyzing power, are shown in Fig. 5. These new measurements have statistical precision of about 0.002 or about 1% of the peak value of the analyzing power and a normalization uncertainty of ~1%. The predictions of the Nijmegen PSA and

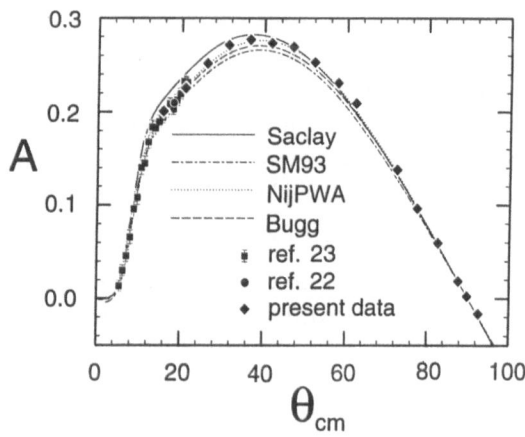

Fig. 5. Analyzing power for p-p scattering at 181 MeV.

the VPI multi-energy solution, VL40, are essentially identical and agree very well with the data at small angles. At the central angles there is about a 4% discrepancy. We are now in the process of trying to use these precisely normalized p-p data to normalize the n-p data discussed above. This is being done by making use of n-p and p-p data taken simultaneously on a polarized proton target. Comparison of the two p-p data sets determines the target polarization and hence the normalization for the n-p data.

The small angle p-p measurements (note that the smallest lab angle is about 3°) discussed above illustrate another new technique, that of internal targets in storage rings. These targets are pure, windowless and very thin. This results in low backgrounds and the ability to detect very low energy recoils, ~1MeV, allowing coincident detection of both nucleons at the extremes of the angle range. Moreover there are techniques to produce polarized targets with the same advantages[24]. Let me illustrate one advantage in practical terms. In the n-p and p-p measurements with a cryogenic polarized target discussed above, the elastic scattering signal rides on a quasi-free scattering background of about 40% (from nuclei like C and O in the target material) with no cuts, which is reduced to about 8% using kinematic cuts and finally eliminated with subtraction of data from a non-hydrogenous "dummy" target. In recent tests with an internal target in the IUCF Cooler the background under the peak was less than 1% with no cuts. A polarized proton target developed at the University of Wisconsin[25] has just been installed in our ring and measurements are beginning. A complete set of spin correlations in p-p scattering will be made at 185 MeV. The normal and sideways spin correlations, $C_{NN}$ and $C_{SS}$, will be measured between 100 and 500 MeV. Spin dependent total cross section measurements in pp→pp$\pi^0$ are also planned.

We are currently performing one other N-N experiment at IUCF. The purpose is to make precise measurements of an observable sensitive to the value of the pion coupling constant $g_0^2$. As discussed in the talk at this conference by Rijken, the pion coupling constants have had their values revised downward in the past few years[1,2]. There may still be open questions as to whether these new values are consistent with the properties of the deuteron[26]. Bugg has previously extracted values for $g_0^2$ in his single energy analyses[27]. He finds that the normal to normal spin transfer parameter $D_{NN}$ is sensitive to the coupling constant. The fact that sensitivity is observed in this type of spin observable is not surprising if one looks at the form of the N-N scattering amplitudes and the form of the amplitude for one pion exchange

$$f = \alpha + i\gamma(\sigma_{1n} + \sigma_{2n}) + \beta\sigma_{1n}\sigma_{2n} + \delta\sigma_{1q}\sigma_{2q} + \epsilon\sigma_{1\ell}\sigma_{2\ell} \quad and \quad f_\pi = g_0^2\frac{\sigma_{1q}\sigma_{2q}}{t - m_\pi^2}.$$

We see that the $\delta$ term has the same spin structure as the one pion exchange. If the observable $D_{NN}$ is written in terms of the amplitudes one finds that

$$1 - D_{NN} = 2\frac{\delta^2 + \epsilon^2}{d\sigma/d\Omega}$$

so that when $\epsilon$ is small at forward angles there is good sensitivity to the pion exchange term. Fig. 6 shows $D_{NN}$ for two different values of $g_0^2$ as calculated in Bugg's PSA. We see that measurements with a precision of about 0.01 are needed to differentiate between the curves. Techniques related to the use of spin 0 targets for calibrating our polarimeters should allow us to achieve this precision. The measurements will also continue out to about 90° in the center of mass where the predictions, as shown in Fig. 2, vary substantially.

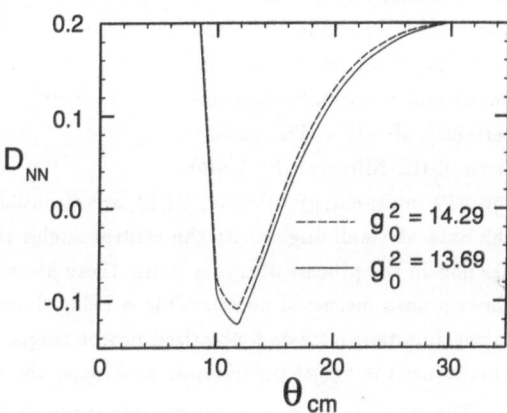

Fig 6. $D_{NN}$ in p-p scattering at 200 MeV as calculated in Bugg's PSA for two different values of the pion coupling constant $g_0^2$.

## Conclusion

I have reviewed the status of N-N elastic scattering at intermediate energies. I would characterize the data base as less than ideal considering the new techniques available today. Nevertheless, it appears more or less adequate for describing the N-N force in the framework of phase shift analyses and judging the quality of potential models. However, in some specific cases there are still open questions. The tensor force, as represented by $\epsilon_1$, is still poorly constrained by existing measurements. It is also found that due to the lack of data the n-p channel must be fit taking isovector phases from the p-p channel, preventing the investigation of charge independence violation in the N-N interaction. I have also described how new techniques, such as high quality polarized beams and targets, new calibration techniques and, in the near future, polarized targets internal to storage rings, are allowing experimentalists to make measurements of high precision. These open questions and new techniques should continue to drive the field in the years to come.

## References

[1] V. Stoks, R. Timmermans, J.J. de Swart, Phys. Rev. C **47**, 512, (1993); and references therein.

[2] R.A. Arndt, et al., Phys. Rev. D. **44**, 289 (1991); Phys. Rev. Lett. **65**, 157 (1990).

[3] C. Lachanoine-Leluc, et al., J. Physique **48**, 985 (1987); Rev. Mod. Phys. **65**, 47 (1993).

[4] D.V. Bugg, Ann. Rev. Nucl. Part. Sci. **35**, 295 (1985).

[5] V.G.J. Stoks, et al., Phys. Rev. C **48**, 792 (1993).

[6] R.A. Arndt, et al., Phys. Rev. D **45**, 3995 (1992).

[7] D.V. Bugg, Phys. Rev. C **41**, 2708 (1990).

[8] J. Bystricky, F. Lehar, C. Lechanoine-Leluc, J. Phys. **48**, 199 (1987).

[9] There are many versions of the Bonn potential: R. Machleidt, K. Holinde C. Elster, Phys. Rep. **149**, 1 (1987); R. Machleidt, Adv. Nucl. Phys. **19**, 189 (1989); J. Haidenbauer and K. Holinde, Phys. Rev. C **40**, 2465 (1989). Bonn curves herein are generated form SAID [13].

[10] M.Lacombe, et al., Phys. Rev. C **21**, 861 (1980).

[11] V. Stoks and J.J. de Swart, Phys. Rev. C **47**, 761 (1993).

[12] I use A for the analyzing power, C for spin correlations, D for spin transfer to the scattered particle and K for spin transfer to the recoil particle. Subscripts N,L,S refer to measured particle polarizations in the direction normal to the scattering plane, or longitudinal and sideways with respect to particle momentum, respectively. A translation table between various notations can be found in F. Arash, M. Moravcsik and G. Goldstein, Phys. Rev. D. **32**, 74 (1985). Detailed definitions can be found in J. Bystricky, F. Lehar, and P. Winternitz, J. Phys. **39**, 1 (1978).

[13] Much information on the data base, the VPI and other PSA and potential model predictions can be extracted from the program SAID at VPI. See ref. 6 for details on accessing this program via the network. This program was used extensively in preparing figures for this talk.

[14] D.V. Bugg, in *Progress in Particle and Nuclear Physics*, ed. D.H. Wilkinson, (Pergamon, Oxford, 1981), Vol. 7, p. 47.

[15] G.S. Chulick, et al., Phys. Rev. C **37**, 1549 (1988).

[16] R. Henneck, Phys. Rev. C **47**, 1859 (1993).

[17] C. Leluc, private communication.

[18] S.E. Vigdor, et al., Phys. Rev. C **46**, 410 (1992); L.D. Knutson, et al., Phys. Rev. Lett. **66**, 1410 (1991).

[19] J. Sowinski, et al., Phys. Lett. B **199**, 341 (1987).

[20] G.G. Ohlsen, Rep. Prog. Phys. **35**, 717 (1972).

[21] S.W. Wissink, et al., Phys. Rev. C **45**, R504 (1992).

[22] B. von Przewoski, et al., Phys. Rev. C **44**, 44 (1991).

[23] W.K. Pitts, et al., Phys. Rev. C **45**, R121 (1992).

[24] Proceedings of the Conference on Polarized Ion Sources and Polarized Gas Targets, Madison, May 1993.

[25] A. Roberts in Proceedings of the Conference on Polarized Ion Sources and Polarized Gas Targets, Madison, May 1993.

[26] R. Machleidt and F. Sammurruca, Phys. Rev. Lett. **66**, 564 (1991).

[27] D.V. Bugg, et al., J. Phys. G **4**, 1025 (1978).

Few-Body Systems, Suppl. 7, 22—33 (1994)

# PRODUCTION OF ANTIHYPERON-HYPERON PAIRS AT LEAR

K. Kilian for the PS185 Collaboration [1]
Institut für Kernphysik, Forschungszentrum Jülich GmbH, D-52425 Jülich

## Abstract

The reactions $\bar{p}p \to \overline{Y}Y$ and $\bar{p}p \to \overline{Y}Y + X^\circ$ are investigated at LEAR in the momentum range from theshold up to 2 GeV/c. Integrated and differential cross sections as well as polarizations and spin correlations have been measured for the $\overline{\Lambda}\Lambda$ channel. Total and differential cross sections were also obtained for the charged $\overline{\Sigma}\Sigma$, the $\overline{\Lambda}\Sigma + \mathbf{cc}$ and the $\overline{\Lambda}\Lambda\pi^\circ$ channels. The data allow to test theoretical predictions based on boson exchange models or models inspired by effective quarks.

## 1. Introduction

The low energy antiproton ring LEAR [2] at CERN produces phasespacecooled antiproton beams of unequaled intensity, purity and phasespacedensity. Its beam of only $\leq 1$ mm$^2$ size is well suited for a very effective experimental study of hyperon production in $\bar{p}p$ interactions. Especially the two body reaction $\bar{p}p \to \overline{Y}Y$ close to theshold has attractive features: The delayed decays of $\overline{\Lambda} \to \bar{p}\pi^+$ and $\Lambda \to p\pi^-$ are spatially separated from the production vertex. This provides both a clean experimental trigger pattern and a means to reconstruct fully the kinematical parameters of each event. Since the weak $\overline{\Lambda}$ and $\Lambda$ decays are parity violating, their asymmetry also allows to determine the $\overline{\Lambda}$ and $\Lambda$ polarization and spin correlation. On the other hand the reaction mechanism should be relatively simple at least close to threshold where only few partial waves contribute. Fig. 1 shows as a function of the $\bar{p}$ laboratory momentum the center of mass momenta of the entrance $p_{\bar{p}}^*$ and exit particles $p_{\overline{\Lambda}}^*$ and their typical angular momentum range. Of course very close to the $\overline{\Lambda}\Lambda$ threshold at 1435.3 MeV/c $\bar{p}$ momentum there should be nearly pure S-wave $\overline{\Lambda}\Lambda$ since $p_{\overline{\Lambda}}^* \approx 0$. The entrance channel however supplies (for kinematical reasons) partial waves with high angular momentum already right from $\overline{\Lambda}\Lambda$ threshold on. This may enhance early onset of $\ell > 0$ partial waves in the $\overline{\Lambda}\Lambda$ channel. It may also lead to an enhancement of tensor transitions and, connected with that to a preference of $\bar{p}p$ and $\overline{\Lambda}\Lambda$ in a spintriplet state. The spin flip of $\Delta s = +2$ allows to couple with $\Delta L = -2$ from the large supply of D (and F) waves in $\bar{p}p$ to kinematically favoured lower angular momentum S (and P) waves in $\overline{\Lambda}\Lambda$.

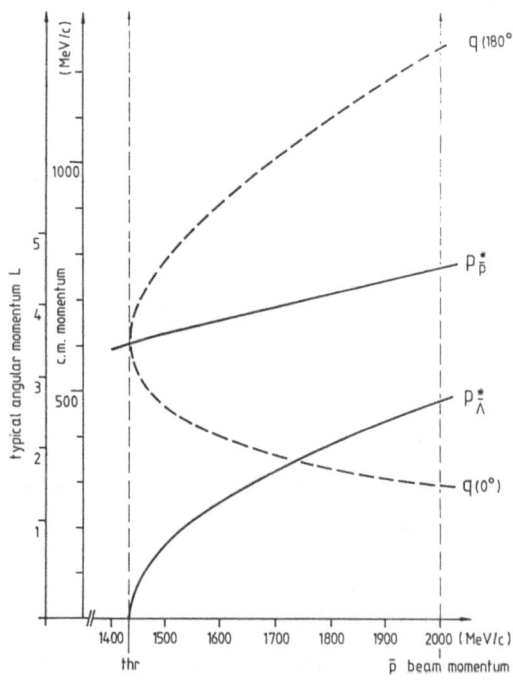

Fig. 1: For the reaction $\bar{p}p \to \bar{\Lambda}\Lambda$ are shown as function of $\bar{p}$ momentum

- the center of mass momenta of incoming $\bar{p}$ and outgoing $\bar{\Lambda}$ (solid curves) and their relation to typical angular momenta $\ell$ for a baryon radius of 1.3 fm

- minimal (at 0°) and maximal (180°) momentum transfers q from $\bar{p}$ to $\bar{\Lambda}$ (dashed curves).

Also shown in Fig. 1 is the range of momentum transfers q in the $\bar{p}p \to \bar{\Lambda}\Lambda$ reaction at LEAR. Right at threshold q is already 600 MeV/c. At maximum LEAR momentum $\bar{\Lambda}$ forward production needs only q ≈ 300 MeV/c while backward production proceeds with q ~ 1250 MeV/c or a wave number $k \geq 6$ fm$^{-1}$. This corresponds to a spatial resolution of $(2\pi/k) \sim 1$ fm). This is certainly not enough to reach a perturbative view of quark gluon dynamics. The quark-gluon degrees of freedom are still partly frozen. Hadronic degrees of freedom - baryons and mesons, K and K* exchanges - are still useful. Nevertheless new flavours - $\bar{s}s$ quark pairs - are created and it may already be useful to apply "effective" quarks.

In the additive quark model, the $\bar{\Lambda}$ and $\Lambda$ hyperons are composed of 3 quarks in a relative S-state. The $\bar{s}$ and s quarks, respectively, are coupled to a spin- and isospin zero $(\bar{u}\bar{d})$ and (ud) quark pair. Therefore the measured polarization parameters of the $\bar{\Lambda}$ and $\Lambda$ are the same as those of the $\bar{s}$ and s quarks. If the light quark pairs (diquarks) are spectators in the $\bar{s}s$ creation process, then the $\bar{p}p \to \bar{\Lambda}\Lambda$ reaction may provide a maximum of information on the $\bar{s}s$ loop, for example on the $\bar{s}s$ quark spin correlation and angular momentum. This is related to the question, if an effective quark-antiquark pair is created with $^3S_1$- or $^3P_0$. quantum numbers. On the other hand, the $\bar{p}p \to \bar{\Lambda}\Lambda$ reaction might be viewed as a t-channel exchange of K-mesons (Fig. 2a). Cross sections and polarizations are sensitive to the transition potential and, to a large extent, to the $\bar{p}p$ initial state and $\bar{\Lambda}\Lambda$ final state interaction. These distortions are strongly inelastic and depend on the partial waves involved.

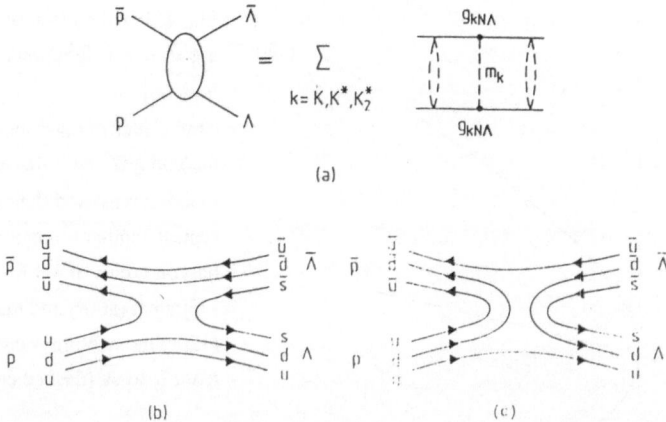

(a)

(b)                                                    (c)

Fig. 2: The reaction $\bar{p}p \to \bar{\Lambda}\Lambda$, viewed (a) as a t-channel meson exchange, (b) as a $\bar{u}u$ quark-pair annihilation and $\bar{s}s$ creation process, (c) as diquark pair annihilation and creation process.

The strange meson exchange in the t-channel corresponds, in the simplest case to a $\bar{u}u$ quark pair annihilation followed by a $\bar{s}s$ creation in the direct channel (Fig. 2b). Also the more complicated diquark-antidiquark annihilation and creation process (Fig. 2c) can be considered which corresponds to a two meson exchange. Certainly also here additional distortion effects are important. There is a duality - not a contradiction - between quark- [3-5] and boson exchange [6-8] models. Due to the rich information a phenomenological parameterization in partial waves is very instructive [9].

## 2. Experimantal Setup

The detector arrangement is shown in fig. 3. Its central part is a track imaging stack of 10 MWPC and 13 drift chamber planes which can be considered as a non magnetic decay spectrometer and decay polarimeter. From all $\bar{\Lambda}\Lambda$ events which decay typically after 3 to 7 cm, about 41 % go into four charged particles: $\bar{\Lambda}\Lambda \to \bar{p}\pi^+p\pi^-$. These tracks can be reconstructed and their kinematics fully determined. From their decaypoint distribution and momenta we can determine the $\bar{\Lambda}$ and $\Lambda$ lifetime separately. The equality of $\tau_{\bar{\Lambda}}$ and $\tau_{\Lambda}$ is required from CPT invariance. The extent to which $\tau_{\bar{\Lambda}} = \tau_{\Lambda}$ thus shows the statistical and systematical precision of the experiment. We obtained e.g. with $7.2 \cdot 10^{10}$ antiprotons at 1.64 GeV/c from about 42000 $\bar{\Lambda}\Lambda$ events [10]

$$\tau_{\bar{\Lambda}} = (258.4 \pm 4.7 \pm 5.3) \text{ ps} \qquad \tau_{\Lambda} = (265.2 \pm 4.3 \pm 5.3) \text{ ps}.$$

The PS185 experiment gives so far the most precise comparison of $\bar{\Lambda}$ and $\Lambda$ lifetimes. The cone, where $\bar{\Lambda}$ and $\Lambda$ can decay and the superimposed cone for the forward confined decay baryons, is fully covered by the chamber stack, the trigger hodoscope and the baryon number identifier, as shown in fig. 3. Typical $\bar{p}$ currents in the 1 hour spills from LEAR were 1 to $6 \cdot 10^5 \bar{p}/s$. The $\bar{p}$ beam passes through beam defining scintillators (S1) and then 5 target slices of 2.5 mm thickness and 2.5 mm diameter (4 polyethylene cells as proton target and one graphite cell for background checks). Close to threshold we use the energy degradation in the cells to measure excitation functions [11]. These slices are surrounded by anticounter cylinders (S2) and sandwiched between 0.2 mm Veto Scintillators (S3). As trigger for the delayed $\overline{\Lambda}$ and $\Lambda$ decays, we require a beam $\bar{p}$ in S1, no hit in S2 and S3 and a hit in the scintillator hodoscope behind the chamber stack. Due to the low mass of the detector system, the number of background triggers (mainly from $\pi^0 \rightarrow 2\gamma$ and $\gamma \rightarrow e^+e^-$ conversion and $\bar{n}n$ production and subsequent reactions) can be kept small ($\leq 3 \cdot 10^{-4}$ per $\bar{p}$ in the beam). The $\overline{\Lambda}$ and $\Lambda$ decay products passing through the hodoscope enter a solenoid with a 0.1 T vertical magnetic field by penetrating through the Aluminium-coil of the solenoid. Three drift chambers inside the solenoid allow to distinguish the baryon number ($\pm1$) of the two identified decay vertices through sign measurements of the particle charges.

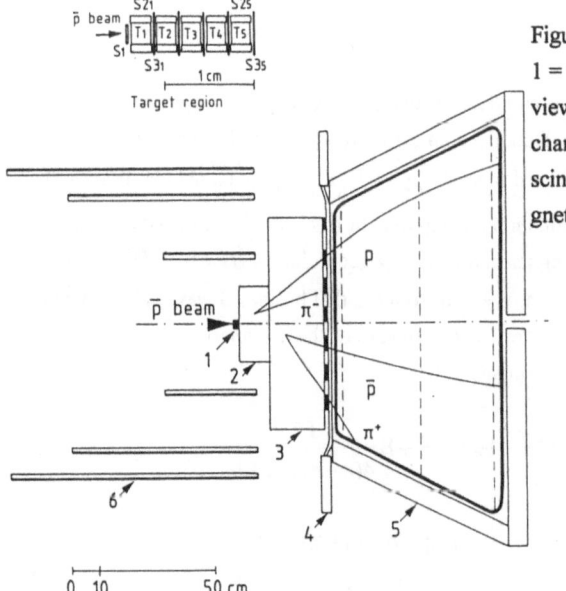

Figure 3: Experimental set-up with: 1 = target region (see magnified view), 2 = multiwire proportional chambers, 3 = drift chambers, 4 = scintillator hodoscope, and 5 = magnetic solenoid with drift chambers.

From the trigger events, $\overline{\Lambda}\Lambda$ candidates were selected off line by geometrical and kinematical checks of the V shaped decay patterns. A subsequent, full kinematic fitting of pairs of decay patterns, assuming a $\bar{p}p \rightarrow \overline{\Lambda}\Lambda \rightarrow \bar{p}\pi^+p\pi^-$ event, gave optimized information about all kinematical details and excluded background. Acceptance modulations were corrected with the help of Monte Carlo calculations.

## 3. Results

The measured $\overline{\Lambda}\Lambda$ cross sections are shown in fig. 4a.

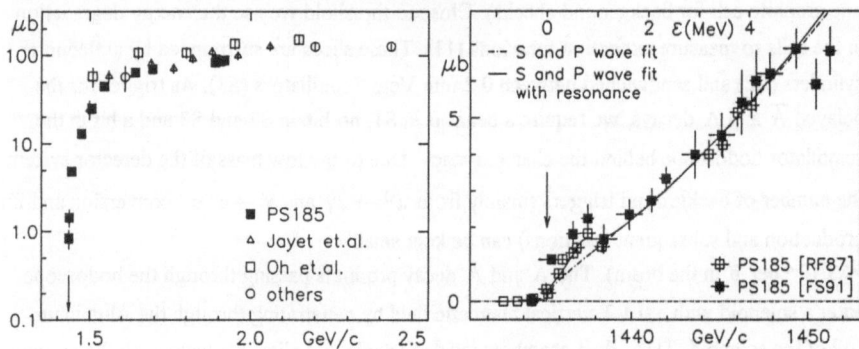

Figure 4: Momentum dependence of the $\overline{p}p \rightarrow \overline{\Lambda}\Lambda$ production cross section (a) in comparison with earlier data, (b) results very close to threshold (1435.3 MeV/c). For the curve see text.

The data from PS185 are consistent and have small errors. An expanded view of the threshold region is given in fig. 4b. Data from an earlier run [11] and from a new measurement [12] which was dedicated to study $\overline{p}p \rightarrow K_s K_s$ [9] are combined. In order to describe this threshold excitation function an empirical fit has been done. Based on phase space arguments $\sigma_{tot}$ depends on angular momentum $\ell$ according to $\sigma_\ell \approx b_\ell \cdot \varepsilon^{\ell+1/2}$ where $\varepsilon$ is the excess energy in the $\overline{\Lambda}\Lambda$ system. We find from $\sigma_{tot}$ that besides S waves also P waves are needed even below $\varepsilon = 6$ MeV. In order to take care of the irregularity at $\sim 1$ MeV we included a Breit-Wigner resonance term and get:

$$\sigma_{tot} = b_1 \varepsilon^{1/2} + b_2 \varepsilon^{3/2} + B \frac{\Gamma^2}{4(\varepsilon - \varepsilon_r)^2 + \Gamma^2}$$

with

$b_1$ = $0.94 \pm 0.08 \ \mu b/MeV^{1/2}$    $b_2$ = $0.44 \pm 0.03 \ \mu b/MeV^{3/2}$

$B$ = $0.6 \pm 0.1 \ \mu b$    $\varepsilon_r$ = $0.7 \pm 0.2$ MeV    $\Gamma$ = $0.6 \pm 0.1$ MeV.

A new measurement has been approved at LEAR to clarify this situation. By collecting enough statistics for precise angular distributions of cross section and polarization, quantum numbers can be determined.

The manifestation of a strong P wave component down to lowest momenta is experimentally further supported by finding anisotropic differential cross sections and finite $\overline{\Lambda}$ polarization. There are two explanations for that: In models describing the reaction with optical potentials the strong short range absorptive part due to annihilation causes a suppression of the S-wave

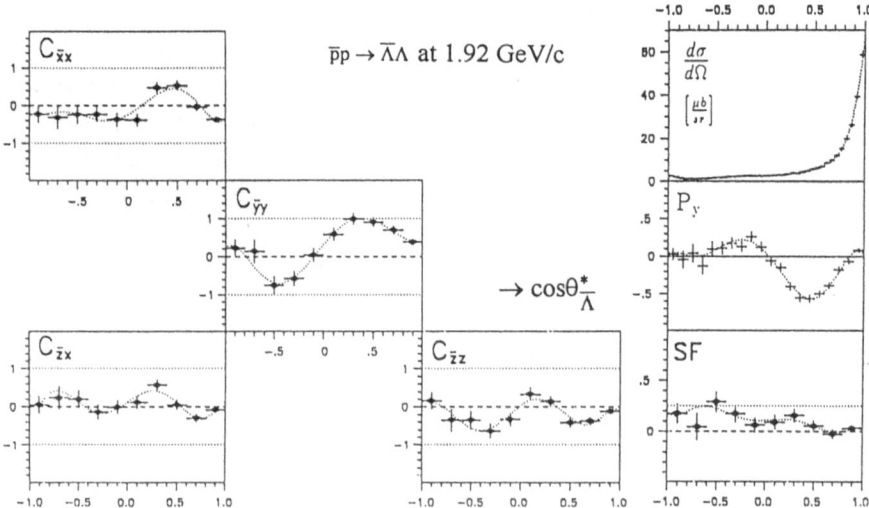

Fig. 5: Center of mass angular distributions of observables in the $\bar{p}p \to \bar{\Lambda}\Lambda$ reaction at 1.92 GeV/c. $P_y$ is the average of $\bar{\Lambda}$ and $\Lambda$ polarization. $C_{ij}$ are the spin-correlation coefficients and SF is the singlet fraction.

contribution [6-8]. Alternatively an enhancement of the P-wave [14] or D-wave [15] amplitude may be produced by resonant states between antibaryon and baryon.

At higher energies due to the higher cross sections we have now precise data at several energies [10,16-19]. Fig. 5 shows as an example results at 1.92 GeV/c [17]. Here 34.000 $\bar{p}p \to \bar{\Lambda}\Lambda$ events were collected with a total of $5.4 \cdot 10^{10}$ beam antiprotons. Shown are center of mass distributions of the differential cross section $d\sigma/d\cos\Theta$, $\bar{\Lambda}$ polarization P, $\bar{\Lambda}\Lambda$ spin correlation coefficients $C_{ij}$ and the $\bar{\Lambda}\Lambda$ singlet fraction SF. The differential cross section has a characteristic steep increase towards forward angles. Plotted over the reduced four momentum transfer $t' = t - t_{min} = 2p_{\bar{p}}^* \cdot p_{\bar{\Lambda}}^* (\cos\Theta^* - 1)$ one finds [19] that $(d\sigma/dt')$ can be parametrized like $d\sigma/dt \approx d\sigma/dt'|_o \cdot \exp(-b \cdot |t'|)$ with an energy independent slope parameter $b \simeq$ 8.5 $(GeV/c)^{-2}$. At $t' \sim 0.15$ GeV/c² there is a transition from the steep slope to a flat angular distribution.

An expansion of the differential cross sections in Legendre polynomials and of the polarization P in adjoint Legendre polynomials was made

$$d\sigma/d\cos\Theta = \sum_{m=0}^{M} A_m P_m(\cos\Theta) \quad \text{and} \quad P = \sum_{n=0}^{N} B_n \cdot P_n^1(\cos\Theta).$$

The results are summarized in Fig. 6a and b. The highest necessary index M or N gives a limit for the highest angular momentum $\ell$ playing a role in the reaction since M or $N \leq 2\,\ell_{max}$.

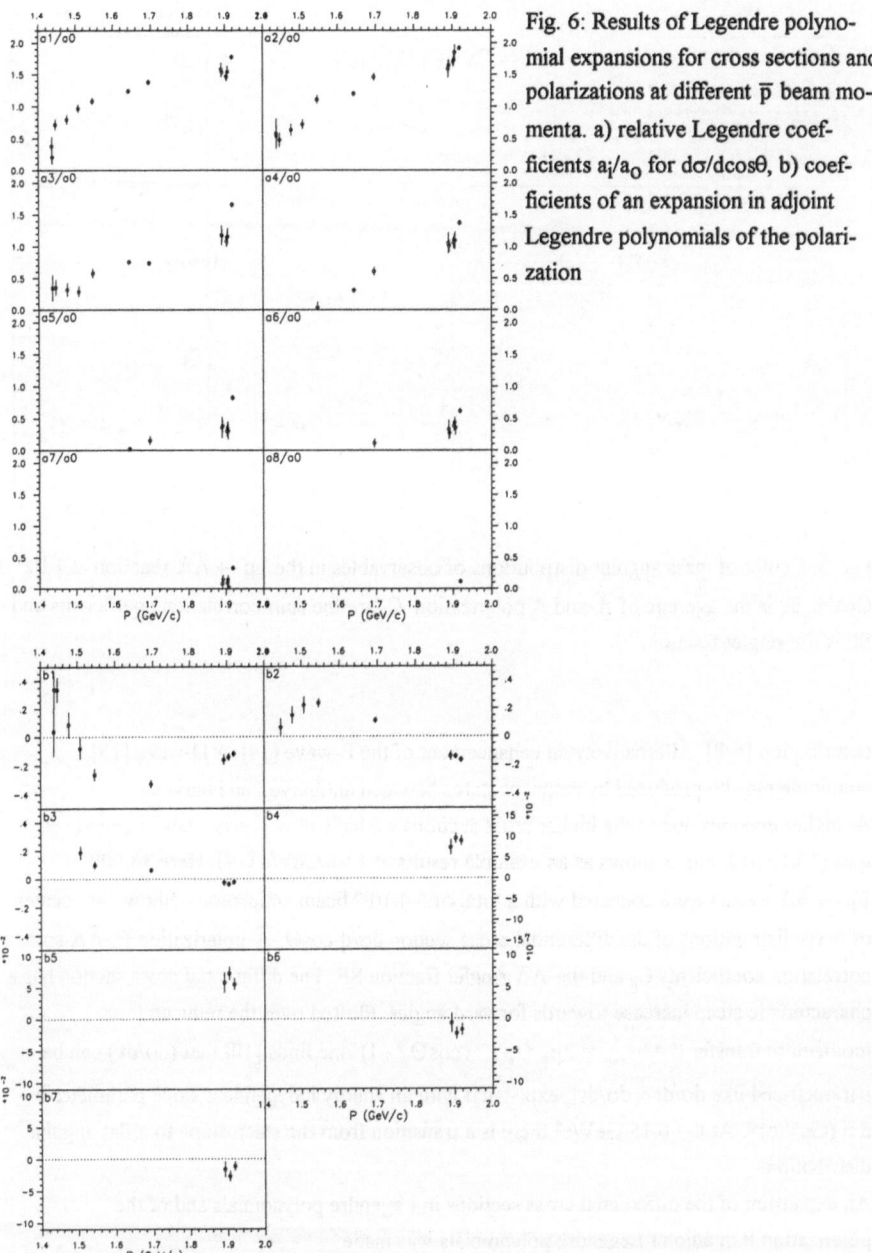

Fig. 6: Results of Legendre polynomial expansions for cross sections and polarizations at different $\bar{p}$ beam momenta. a) relative Legendre coefficients $a_i/a_0$ for $d\sigma/d\cos\theta$, b) coefficients of an expansion in adjoint Legendre polynomials of the polarization

From the steep rise of $A_2/A_0$ in Fig. 6a for example one can see again that $\ell \geq 1$ is important already close to threshold.

The $\bar{\Lambda}$ and $\Lambda$ polarization was explored separately using the weak decays $\bar{\Lambda} \to \bar{p}\pi^+$ and $\Lambda \to p\pi^-$. In the restframe of the hyperon, the outgoing nucleon is preferentially emitted along the $\Lambda$ spin direction, yielding the distribution $I(\beta) = I_o(1 + \alpha_\Lambda P \cos\beta)$ for a sample with polarization P. The $\cos\beta$ is the projection of the nucleon decay direction in the $\Lambda$ rest frame onto the normal of the reaction plane, and $\alpha_\Lambda = 0.642 \pm 0.013$ is the $\Lambda \to p\pi$ decay asymmetry parameter. One has $\alpha_\Lambda = -\alpha_{\bar{\Lambda}}$ if CP conservation holds. The polarizations are transverse to the $\bar{\Lambda}\Lambda$ production plane due to parity conservation and equal $(p_{\bar{\Lambda}} = p_\Lambda)$ due to charge conjugation invariance. An apparent difference in $P_{\bar{\Lambda}}$ and $P_\Lambda$ would signal a CP violating effect in the weak decay parameters $\alpha_{\bar{\Lambda}}$ and $\alpha_\Lambda$. The relevant quantity which measures CP violation is $A = (\alpha_\Lambda P_\Lambda + \alpha_{\bar{\Lambda}} P_{\bar{\Lambda}})/(\alpha_\Lambda P_\Lambda - \alpha_{\bar{\Lambda}} P_{\bar{\Lambda}}) = (\alpha_\Lambda + \alpha_{\bar{\Lambda}})/(\alpha_\Lambda - \alpha_{\bar{\Lambda}})$. CP conservation requires A=0. In Fig. 7 this parameter A is shown as determined in our experiment for different $\bar{p}$ beam momenta including already published data [20]. The average amounts to A=0.022 $\pm$ 0.019 and is consistent with zero.

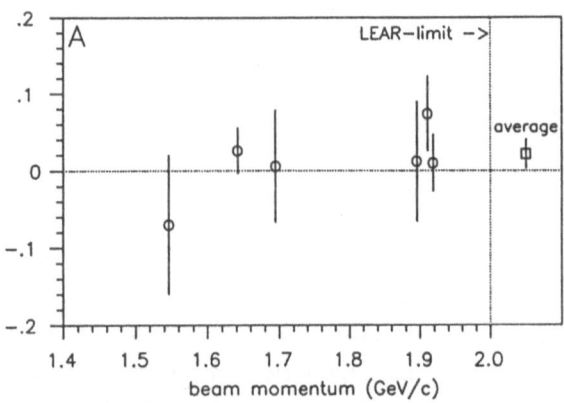

Fig. 7: Results for the quantity $A = (\alpha + \bar{\alpha})/(\alpha - \bar{\alpha})$ which tests the equality of the $\bar{\Lambda}$ and $\Lambda$ decay asymmetry, as obtained at different beam momenta. The average is $A = 0.022 \pm 0.019$.

First good results were obtained from our data for the spin correlation coefficients $C_{ij}$ in the exit system and for the singlet fraction SF = $1/4(1+C_{xx}-C_{yy}+C_{zz})$. The $\bar{\Lambda}\Lambda$ system is in a pure singlet state for SF = 1 and in a pure triplet state for SF = 0. For a statistical mixture of 1 singlet to 3 triplet one expects SF = 0.25. The experimental results are shown for one energy in fig. 5. Averages integrated over all angles are summarized in Fig. 8 as function of $\bar{p}$ momentum. We find that nearly all $\bar{\Lambda}\Lambda$ events are in a triplet state. There are reasons to expect this as mentioned in the introduction. However if one considers a direct channel meson exchange then $\eta$ meson type pseudoscalars which lead to singlet transition should be important due to their large $\bar{s}s$ quark content. This reaction $\bar{p}p \to "\eta" \to \bar{\Lambda}\Lambda$ was studied [21] and a strong cancellation between $\eta$ and $\eta'$ was found which suppresses a singlet contribution.

It is interesting to study also $\bar{p}p \to \bar{\Sigma}^\circ\Lambda + cc$ and $\bar{p}p \to \bar{\Sigma}\Sigma$, where the $\bar{s}s$ quark pair creation is embedded in a hadronic surrounding which differs from the $\bar{\Lambda}\Lambda$ case in a known way. The light diquarks in $\Sigma$ have isospin and spin 1. Of special interest will be a measurement of the

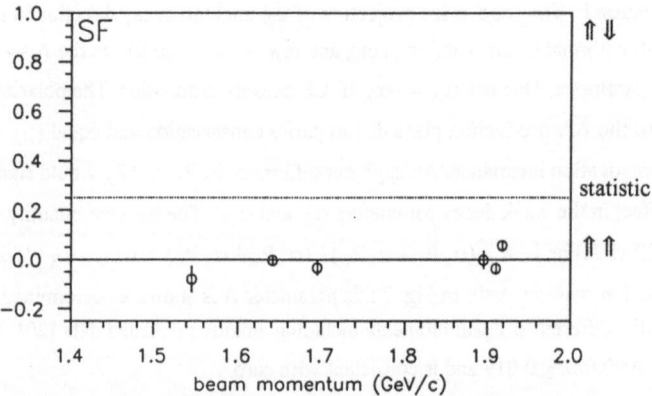

Fig. 8: Probability of $\overline{\Lambda}\Lambda$ singlet production in $\overline{p}p \rightarrow \overline{\Lambda}\Lambda$ at different $\overline{p}$ beam momenta. Averages over the whole $\overline{\Lambda}$ center of mass angular ranges are shown.

channel $\overline{\Sigma}^-\Sigma^-$, where one K-exchange and the corresponding simplest quark line diagram of fig. 1b is forbidden. Charge conservation on diquark annihilation and creation (fig. 1c) is the leading process whose importance and dynamics might be determined from those experiments. First data have been taken.

Already published are first results on $\overline{p}p \rightarrow \overline{\Sigma}^0\Lambda + cc$ [22]. Here one might see the short range behaviour since the long range K exchange is much suppressed with respect to the heavier K* exchange. Our low energy data were also used to determine the $KN\Sigma$ coupling constant $f^2_{\Lambda pk} = 0.071 \pm 0.007$ [23].

The high precision and kinematically complete information on $\overline{\Lambda}$ and $\Lambda$ of the PS185 detector allows to evaluate the four vector of a possibly missing third particle $X^0$ in the reaction $\overline{p}p \rightarrow \overline{\Lambda}\Lambda X^0$ [12]. In the case of $\overline{p}p \rightarrow \overline{\Sigma}^0\Lambda$ the prompt decay $\Sigma^0 \rightarrow \Lambda\gamma$ leads to a missing photon and the four vector squared of $X^0$ gives $m^2_{X^0} = 0$. We also found events with $m^2_{X^0} = m^2_{\pi^0}$ and thus were able to produce a sample of fully determined three body events $\overline{p}p \rightarrow \overline{\Lambda}\Lambda\pi^0$ [24]. Fig. 9 shows the total cross section for this reaction. In Fig. 10 is shown the mass spectrum of the $\overline{\Lambda}\Lambda$ subsystem at one beam energy in comparison to phase space expectations.

It is visible that there is an enhancement of low energy $\overline{\Lambda}\Lambda$ events. With higher resolution it might become possible to find out the reason for this enhancement. It could be related to the anomaly in the low energy $\overline{p}p \rightarrow \overline{\Lambda}\Lambda$ cross section (Fig. 4).

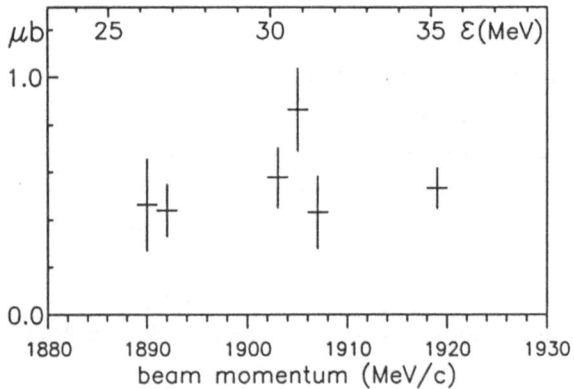

Fig. 9: Total cross section for the 3body reaction $\bar{p}p \to \bar{\Lambda}\Lambda\pi^\circ$ at various beam momenta and (top scale) center of mass excess energies.

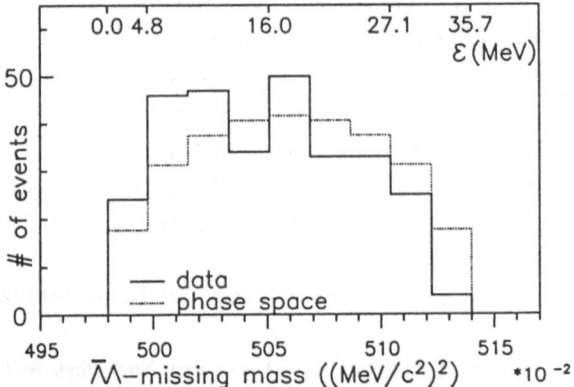

Fig. 10: Spectrum of the $\bar{\Lambda}\Lambda$ subsystem obtained in the 3body reaction $\bar{p}p \to \bar{\Lambda}\Lambda\pi^\circ$ at 1919 MeV/c beam momentum (solid line) and expected phase space behaviour (dotted line). There is an excess of low energy events.

For the future it is planned to study $\bar{p}p \to \bar{\Lambda}\Lambda$ with a polarized frozen spin target. This will allow to determine the depolarization parameter $D_{NN}$ which measures the spintransfer from proton to $\Lambda$ (and $\bar{p}$ to $\bar{\Lambda}$). A tensor interaction which needs spinflips could be distinguished from purely central interaction thus allowing to distinguish clearly between the actually used quark line inspired models (no spinflip) against boson exchange models (strong spinflip) [25].

We wish to thank the LEAR accelerator team for the development and excellent operation of the antiproton beams required for this experiment. Support is gratefully acknowledged from the Austrian Science Foundation, the German Bundesministerium für Forschung und Technologie, the Swedish Natural Science Research Council, and the United States Department of Energy and the United States National Science Foundation.

References:

[ 1]  PS185 collaboration: P.D. Barnes, R. Besold, P. Birien, B. Bonner, W.H. Breunlich,
      G. Decker, G. Diebold, W. Dutty, R.A. Eisenstein, G. Ericsson, W. Eyrich, H. Fischer,
      R. von Frankenberg, G. Franklin, J. Franz, N. Hamann, D. Hertzog, A. Hofmann,
      T. Johansson, K. Kilian, C.J. Maher, D. Malz, N. Naegele, W. Oelert, S. Ohlsson,
      H. Ortner, P. Pawlek, B. Quinn, P. Reimer, E. Rössle, H. Schledermann, H. Schmitt,
      T. Sefzick, G. Sehl, J. Seydoux, F. Stinzing and J. Szymanski, R. Tayloe, P. Woldt,
      V. Zeps, M. Ziolkowski

[ 2]  P. Lefevre, in "Physics at LEAR with Low Energy Antiprotons", C. Amsler,
      G. Backenstoss, R. Klapisch, C. Leluc, D. Simon, and L. Tauscher, eds., Harwood
      Academic Puslishers, Chur, Switzerland (1988).

[ 3]  H. Genz and S. Tatur, Phys. Rev. D30 (1984) 63.
      G. Brix, H. Genz, and S. Satur, Phys. Rev. D39 (1989) 2054.

[ 4]  H.R. Rubinstein and H. Snellman, Phys. Lett. 165B (1985) 187.
      S. Furui and A. Fäßler, Nucl. Phys. A468 (1987) 669.
      P. Kroll and W. Schweiger, Nucl. Phys. A474 (1987) 608
      P. Kroll, B. Quadder, and W. Schweiger, Nucl. Phys. B316 (1989) 373.
      M. Burkart and M. Dillig, Phys. Rev. C37 (1988) 1362.

[ 5]  M.A. Alberg, E.M. Henley, L. Wilets, and P.D. Kunz, Nucl. Phys. A508 (1990) 323c.

[ 6]  M. Kohno and W. Weise, Phys. Lett. 179 B (1986) 15; Phys. Lett. 206 B (1988) 584;
      Nucl. Phys. A479 (1988) 433c.

[ 7]  P. LaFrance, B. Loiseau, and R. Vinh Mau, Phys. Lett. B214 (1988) 317; Nucl. Phys.
      A528 (1991) 557.
      R.G.E. Timmermanns, T.A. Rijken, and J.J. de Swart, Nucl. Phys. A479 (1988) 383c.
      F. Tabakin and R.A. Eisenstein, Phys. Rev. C31 (1985) 1857.

[ 8]  J. Haidenbauer, T. Hippchen, K. Holinde, B. Holzenkamp, V. Mull and J. Speth, Phys.
      Rev. C45 (1992) 931.
      J. Haidenbauer, K. Holinde, V. Mull and J. Speth, Phys. Rev. C46 (1992) 2158.

[ 9]  F. Tabakin, R.A. Eisenstein and Yang Lu, Phys. Rev. C44 (1991) 1749.

[10]  Thesis H. Fischer, University of Freiburg 1992.

[11]  P.D. Barnes et al., Phys. Lett. B229 (1989) 432.

[12]  Thesis F. Stinzing, University of Erlangen-Nürnberg 1991.

[13]  P.D. Barnes et al., Phys. Lett. B309 (1993) 469.

[14]  O.D. Dalkarov, K.V. Protasov, I.S. Shapiro, Nucl. Phys. B (Proc. Suppl.) 8 (1989),
      110.
      J. Carbonell et al., Nucl. Phys. A5355 (1991) 651.
      J. Carbonell, O.D. Dalkarov, K.V. Protasov and I.S. Shapiro, CERN-TH 6096/91.

[15]  J. Carbonell et al., Nucl. Phys. A558 (1993) 353c.

[16]  P.D. Barnes et al., Nucl. Phys. A526 (1991) 575

[17]  Thesis M. Ziolkowski, University of Krakau 1992, also Jülich report Jül-2703, Dec.
      1992.

[18]  Thesis T. Sefzick, University of Bonn 1991, also Jülich report Jül-2584, Feb. 1992.

[19]  H. Schmitt et al., Nucl. Phys. B (Proc. Suppl.) 8 (1989) 162.

[20]  P.D. Barnes, et al. Phys. Lett. 199B (1987) 147.

[21]  M.A. Alberg, E.M. Henley, W. Weise, Phys. Lett. B255 (1991) 498.

[22]  P.D. Barnes et al., Phys. Lett. B246 (1990) 273.

[23]  R. Timmermans, T. Rijken, J. de Swart, Phys. Lett. B257 (1991) 227.

[24]  Thesis G. Decker, University of Bonn 1992, also Jülich report Jül-2645, July 1992

[25]  J. Haidenbauer et al., Phys. Rev. Lett. B291 (1992) 223.

Few-Body Systems, Suppl. 7, 34—41 (1994)

Few-
Body
Systems
© Springer-Verlag 1994
Printed in Austria

# Antiproton-proton partial-wave analysis below 925 MeV/c

R. Timmermans

*Theory Division, Los Alamos National Laboratory, Los Alamos, NM 87545, USA*

Th.A. Rijken and J.J. de Swart

*Institute for Theoretical Physics, University of Nijmegen, Nijmegen, The Netherlands*

## I. Introduction

Partial-wave analyses (PWAs) have a long history in the fields of $\pi N$ and $NN$ scattering. Due to the poor quality of low-energy antiproton beams and the resulting absence of accurate experimental data, analogous model-independent studies of the much more complex $\bar{p}p$ system have in the past always been impossible. In recent years, however, experimental progress has been very significant, in particular due to the coming in 1983 of the Low-Energy Antiproton Ring (LEAR) facility at CERN. While in the pre-LEAR era spin-dependent observables and charge-exchange ($\bar{p}p \to \bar{n}n$) data were almost nonexistent, the situation between 400 and 925 MeV/c is now quite good: the LEAR collaborations PS172, PS173, PS198, and PS199 have measured a variety of observables with impressive accuracy. High-quality analyzing-power data have been obtained for the elastic [1] and charge-exchange [2] reactions. Very recently, even charge-exchange depolarization data have become available [3]. Unfortunately, the practical difficulties involved in constructing a high-quality "cooled" antiproton beam of lower momentum are large. Consequently, the $\bar{p}p$ database below about 400 MeV/c is still by far not as good as one would like, in striking contrast to the $pp$ case where very accurate data exist as low as $T_L = 0.35$ MeV ($p_L = 25$ MeV/c). It also remains an outstanding experimental challenge to construct a polarized antiproton beam to further probe the spin structure of the interaction.

During the last 10 years a new method has been developed by the Nijmegen group to perform PWAs of the abundant and accurate $NN$ ($pp$ and $np$) scattering data below $T_L = 350$ MeV [4, 5]. With the now available high-quality data from LEAR and KEK, we have been able to extend the methods used in these $NN$ PWAs to perform an energy-dependent PWA of all $\bar{p}p$ scattering data below $p_L = 925$ MeV/c ($T_L = 379$ MeV). This work was started in 1987 [6] and has only recently been finished [7]. The same methods of PWA have also been applied [8] to the strangeness-exchange reaction $\bar{p}p \to \bar{\Lambda}\Lambda$, for which the PS185 group at LEAR has obtained beautiful data. In the next section we review the theoretical ideas behind these Nijmegen PWAs, and in section III we apply these ideas and methods to the case of $\bar{p}p$ scattering. In section IV some results of this $\bar{p}p$ PWA are presented and discussed.

After almost a decade of LEAR, it is fair to say that in this field theory has some catching up to do with respect to experiment. Since the partial-wave amplitudes or the phase-shift parameters are in a sense the meeting ground between theory and experiment, the results of the present PWA should be very useful in many ways. They can be used to improve models [9, 10] for the $\overline{N}N$ interaction. Apart from the fact that this provides independent and complementary [7] information about the spin- and isospin structure of the $NN$ force, the $\overline{N}N$ interaction is needed as input in many other $\overline{p}p$ subfields. Studies of for instance protonium (the $\overline{p}p$ atom) or specific annihilation processes like $\overline{p}p \to \pi^+\pi^-$, $K^+K^-$ require a realistic treatment of the initial $\overline{p}p$ interaction. At the same time, this PWA could be helpful in planning new experiments at LEAR, the future of which is of course crucial to this field.

## II. Methods of partial-wave analysis

The hallmark of the Nijmegen energy-dependent PWAs is the sophisticated manner in which the energy dependence of the partial-wave amplitudes is parametrized. At the basis of the PWA is the trivial observation that in the low-energy region (long wave lengths) the long-range interaction is very important. It is this long-range interaction that is responsible for the rapid variations with energy of the scattering amplitudes. Short-range interactions lead to much slower energy variations of the amplitudes. One usually looks for a function in the problem that one can parametrize as easily as possible, i.e. one that does not contain the contributions from these long-range processes. Because these long-range interactions are at the same time model independent (in the sense that they are or at least should be the same in all $NN$ and $\overline{N}N$ models), they can then be taken into account separately and exactly.

It is, of course, not a good idea to try to parametrize the partial-wave $S$ matrix itself, since it does contain all of these long-range effects. As a function of complex energy, the $S$ matrix has a (kinematical) right-hand unitarity cut, other right-hand cuts due to the coupling to inelastic channels, and (dynamical) left-hand cuts due to particle exchanges. The left-hand cuts that are the closest to the origin $T_L = 0$ correspond to the longest-range processes. The left-hand cuts that start far away from the origin are due to the short-range interactions. For instance, the infinite-range Coulomb potential ($V \sim 1/r$) produces an essential singularity and a branch point at $T_L = 0$, vacuum polarization ($V \sim \exp(-2m_e r)/r^{5/2}$) produces a cut at $T_L = -0.6$ keV, and one-pion exchange ($V \sim \exp(-m_\pi r)/r$) leads to a cut starting at $T_L = -9.7$ MeV. One sees that the crux is to find a quantity in which the cuts nearest to the origin are not present. This quantity then allows an *analytical* parametrization in energy or $k^2$ in an enlarged domain up to the next left- or right-hand cut present.

A familiar example of such a quantity with improved analyticity properties is the modified effective-range function [4, 11]. The Coulomb-modified effective-range function for the $pp$ $^1S_0$ state was originally derived (in a rather intuitive way) by Landau and Smorodinsky [12]. When only the Coulomb potential is present the boundary condition for the radial wave function $\Phi(0) = 0$ is of course satisfied by $F$, the regular Coulomb wave function (for $\ell = 0$). When, however, there is an additional short-range interaction this condition is satisfied by a linear combination of $F$ and $G$, the irregular Coulomb wave function, so up to normalization

$$\Phi(r) = F(r)\cot\delta_0 + G(r) , \qquad (1)$$

where $\delta_0$ is the nuclear phase shift in the presence of the Coulomb interaction ($\delta(^1S_0) = \delta_0 + \sigma_0$), as can be seen from the asymptotic behavior of $\Phi$. When the wave length is very large (very low energy), one can take the limit in which the range of the additional (strong) interaction goes to zero and its presence is *only* revealed by a modified boundary condition at $r = 0$. An equation for $\cot \delta_0$ can then be obtained by evaluating the logarithmic derivative of the wave function, which we call $P(k, r)$, for $k \to 0$. In the $np$ case this quantity $P(k, 0) = k \cot \delta_0$ approaches a constant:

$$P(k, \varepsilon) = \left( \frac{d\Phi}{dr} / \Phi \right)_{r=\varepsilon} \to -\frac{1}{a} . \tag{2}$$

In the $pp$ case, the evaluation has to be done at $r = \varepsilon$ because of a term $\ln \varepsilon$ that appears due to the singular behavior of $G$. This term one absorbes in the constant $-1/a$, along with some further constant terms. Then one lets $\varepsilon \to 0$ and immediately obtains the Coulomb-modified ("zero-range") effective-range function. It can be shown that after these manipulations the resulting left-hand side of Eqn. (2) is an analytical (actually meromorphic) function of the energy, so that the right-hand side can be written as a power series in $k^2$ (this means dropping the zero-range approximation).

The analytical expansion of the Coulomb-modified effective-range function breaks down already at $T_L = -9.7$ MeV, where the one-pion–exchange cut starts. It is possible to derive a new "pion-modified" effective-range function from which also this cut has been removed [4]. Let the regular and irregular wave functions for the case where only the Coulomb and pion-exchange potentials are present be called $F_\pi$ and $G_\pi$. (For the purpose of the present discussion, we ignore vacuum polarization.) The wave function can then be written as

$$\Phi(r) = F_\pi(r) \cot \delta_0 + G_\pi(r) , \tag{3}$$

where $\delta_0$ is now the phase shift due to the short-range remainder of the strong interaction ($\delta(^1S_0) = \delta_0 + \pi_0 + \sigma_0$, where $\pi_0$ is the one-pion–exchange phase shift in the presence of the Coulomb potential). However, proceeding in similar fashion as above, one encounters an important problem here. The evaluation of $P(k, \varepsilon)$ has to be done numerically, since $F_\pi$ and $G_\pi$ are not known in analytical form. Due to the singular behavior of $G_\pi$ when $\varepsilon \to 0$, it is very hard to maintain sufficient numerical accurary, especially for higher orbital angular momenta.

At this point one has to realize that this numerical problem of the modified effective-range function is really an *artificial* problem: it crops up due to the singular behavior of the irregular function of the *long-range* potential *near the origin*. However, it is precisely this short-range interaction that one wants to parametrize, since it is essentially unknown, very complicated, and leads to only slow energy variations of the scattering amplitudes. Looking at Eqn. (3), one observes that it is valid for *any* $r$, so why not evaluate $P(k, r)$ at a *finite* value $r = b$, instead of at $r = \varepsilon$?

This is essentially what is done in the Nijmegen PWAs. The wave functions are obtained by solving the (relativistic) Schrödinger equation. Suppose one starts at a point $r_\infty$ where only the Coulomb potential is present. Integrating inwards, one picks up sequentially the contributions (varying rapidly with energy) from the electromagnetic potentials, one-pion exchange, and contributions (varying slower with energy) from other meson exchanges. One then stops at a point $r = b$. If there are no additional interactions for $r < b$, the boundary condition $P(k, b)$ at $r = b$ is obviously satisfied by the regular wave function corresponding to precisely this potential tail. For small enough $b$ the model used for $r > b$ will of course not be correct, and the boundary condition has to be modified, as in the above examples. In practice, it works the other way: one starts integrating at $r = b$ and $P(k, b)$ is parametrized as a function of energy. Also the best value for $b$ is determined by fitting the data. In general multichannel problems $P(k, b)$ becomes a matrix. This $P$ matrix has the required improved analyticity properties. When there are no nearby right-hand cuts, it is an analytical (again: actually meromorphic) function of $k^2$ in a domain bounded by the nearest left-hand cut not removed by including (or including incorrectly) the corresponding exchange in the potential tail for $r > b$. It can happen, of course, that short-range dynamics gives rise to a rapid energy variation of the amplitudes, as in the case of a resonance. This would have to be taken into account in the $P$ matrix, for instance by including a pole in the parametrization. It is seen that the formalism used in the Nijmegen PWAs is similar to the boundary-condition approach to the strong interactions that goes back to the work of Feshbach and Lomon [13] and earlier. The philosophy, however, is very different. The term $P$ matrix (for "pole" matrix) was introduced by Jaffe and Low [14] in the framework of the bag model.

## III. An antiproton-proton partial-wave analysis

Let us now be more specific and apply the foregoing ideas to the case of $\bar{p}p$ scattering. In all the Nijmegen PWAs, the two-body scattering process is described with the relativistic Schrödinger equation [15, 16], which is essentially a coordinate-space version of the Blankenbecler-Sugar equation. It reads the same as the ordinary Schrödinger equation

$$\left(\Delta + k^2 - 2mV\right)\psi(\mathbf{r}) = 0 , \tag{4}$$

except that the proper relativistic relation between energy and momentum is used. It is well known how to derive the potentials for use in this equation [15, 16]. In this relativistic framework, there is no known quantum-mechanical interpretation for the "wave function" $\psi(\mathbf{r})$. It is perhaps best to regard it as just a tool that allows one to compute the correct relativistic scattering amplitude (e.g. the poles are the correct bound states). We solve Eqn. (4) for the coupled $\bar{p}p$ and $\bar{n}n$ channels. The mass difference between proton and neutron is included in order to account for the $\bar{n}n$ threshold at $p_L = 99$ MeV/c.

The interaction in the region $r > b$ is described by a theoretically well-founded $\overline{NN}$ potential. This potential is given by

$$V = V_C + V_{MM} + V_N , \tag{5}$$

where $V_C$ and $V_{MM}$ are the relativistic Coulomb and magnetic-moment interaction respectively. $V_N$ is the $\overline{NN}$ meson-exchange potential. It consists of one-pion exchange and the (charge-conjugated) heavy-meson and pomeron exchanges from the 1978 Nijmegen $NN$ potential [17]. As argued in the previous section, the rapid energy variations of the amplitudes due to the long-range electromagnetic interactions and one-pion exchange are now included *exactly*.

Let us next turn to the parametrization of the short-range interactions for $r < b$ by way of the $P$-matrix boundary condition at $r = b = 1.3$ fm. Due to the coupling to the annihilation channels, the $S$ matrix has a right-hand cut starting at $T_L = 0$. (In the $pp$ case this cut starts only at the $pp \rightarrow pp\pi^0$ threshold at $T_L = 280$ MeV.) As these annihilation processes are of short range (and so give rise to slow energy variations of the amplitudes), this right-hand cut has to be present in the $P$ matrix, which we therefore take to be complex. (Similarly, the effective-range parameters for the $\overline{NN}$ case are complex.) The choice of the value for $b$ is rather critical, more so than in the $NN$ case (where it was taken to be $b = 1.4$ fm). The best results are obtained for $b = 1.3$ fm. Since for $r > b$ we use only a real potential, the coupling to the annihilation channels is completely represented by the boundary condition. We conclude therefore that the range of the annihilation process is in fact about 1.3 fm [7].

The electromagnetic interactions that we use are adapted from the improved Coulomb potential [16]. This potential, designed specifically for use in the relativistic Schrödinger equation, contains relativistic corrections to the static Coulomb potential and (in its off-shell behavior) the main contributions from the two-photon–exchange diagrams. All these effects are included in the Nijmegen $pp$ PWA [4, 5], as well as the vacuum-polarization potential. In our case it suffices to use the following spin-dependent one-photon–exchange potentials

$$V_\gamma(r) = -\frac{\alpha'}{r} + \frac{\mu_p^2}{4M_p^2}\frac{\alpha}{r^3}S_{12} + \frac{8\mu_p - 2}{4M_p^2}\frac{\alpha}{r^3}\mathbf{L}\cdot\mathbf{S} \quad \text{for } \bar{p}p \to \bar{p}p , \tag{6}$$

and

$$V_\gamma(r) = \frac{\mu_n^2}{4M_n^2}\frac{\alpha}{r^3}S_{12} \quad \text{for } \bar{n}n \to \bar{n}n . \tag{7}$$

The magnetic moments of the proton and neutron are $\mu_p = 1 + \kappa_p = 2.793$ and $\mu_n = \kappa_n = -1.913$, respectively. The use of $\alpha'$ in the central potential for $\bar{p}p \to \bar{p}p$ takes care of the main relativistic corrections to the Coulomb potential. It is given by $\alpha'/\alpha = 2k/Mv_L$ where $v_L$ is the velocity of the antiproton in the laboratory system. At 600 MeV/c one has for instance $v_L = 0.54$ and $\alpha'/\alpha = 1.135$, a correction of 13.5% to the static Coulomb potential. The spin-orbit potential comes from the interaction of the magnetic moment of one particle with the Coulomb field of the other particle (and is consequently absent in $\bar{n}n \to \bar{n}n$). It includes a relativistic correction due to the Thomas precession. The tensor potential comes from the interaction of the two magnetic moments. Vacuum polarization and two-photon–exchange effects are negligible in our case. The proper treatment of these electromagnetic effects in the evaluation of the scattering amplitudes is a nontrivial matter [7]. The following simple one-pion–exchange potential without a form factor is used

$$V_\pi(r) = f_{NN\pi}^2\frac{M}{\sqrt{k^2 + M^2}}\frac{m^2}{m_{\pi\pm}^2}\frac{1}{3}\left[\sigma_1\cdot\sigma_2 + S_{12}\left(1 + \frac{3}{(mr)} + \frac{3}{(mr)^2}\right)\right]\frac{e^{-mr}}{r} , \tag{8}$$

where $m$ is the mass of the pion and $f_{NN\pi}^2 = 0.0745$ is the rationalized pion-nucleon coupling constant [18]. The mass difference between the $\pi^0$ and $\pi^\pm$ is included.

Let us finish this section with some more general remarks about PWAs. Even for the $pp$ case, where the database is of high quality and the observables are very well mapped out, a PWA is impossible without a substantial amount of theoretical input or constraints. For the $np$ and $\bar{p}p$ PWAs, this is true a fortiori. For instance, one has to make some assumptions about the validity of symmetries like charge independence or (as in our case) charge conjugation. Obviously, one has to careful here: sometimes general physical principles are inspired by local renormalizable field theories and not strictly valid for extended objects like hadrons. A good example can be found in $\pi N$ PWAs, where one usually implements full Mandelstam analyticity [19]. The amplitudes are assumed to be analytic functions of the two complex variables $s$ and $t$ except for singularities from the mass spectrum and unitarity. These amplitudes then exhibit crossing symmetry. It is not clear at all to what extent low-energy hadron dynamics actually satisfies this symmetry.

Using strong and mostly model-independent theoretical constraints it has turned out to be possible to perform an *energy-dependent* or *multienergy* PWA of the $\bar{p}p$ data. However, it is quite a different ballgame to perform *energy-independent* or *single-energy* $\bar{p}p$ PWAs. In a single-energy $\bar{p}p$ PWA one has to determine in principle 20 phase-shift parameters for each $J \neq 0$ (8 for $J = 0$), which is four times as many as in a single-energy $np$ PWA [7]! Almost certainly the present $\bar{p}p$ database does not allow satisfactory energy-independent PWAs. One has to realize, however, that even in the $NN$ field the usefulness of energy-independent PWAs is more limited than is perhaps generally thought. When one has a *good* energy-dependent PWA, the best value for a phase shift (or the pion-nucleon coupling constant!) is definitely the one determined in the energy-dependent PWA, and not the one from an energy-independent PWA. One reason is that an energy-independent PWA contains no information about the energy dependence of the amplitudes. This makes it for instance less stable than an energy-dependent PWA with respect to the addition of new data to the database. Also, a set of energy-independent PWAs is usually overparametrized compared to a good energy-dependent PWA in the same energy region. It thus almost certainly contains noise. For an extensive discussion of this important point, see Ref. [5].

## IV. Some results of the analysis

While in $NN$ PWAs there is essentially agreement on the correct database (especially for $pp$), we had to spend a lot of time and effort into collecting, scrutinizing, and cleaning up the world set of $\bar{p}p$ scattering data, which contains a lot of flaws and contradictory data. Exactly the same statistical arguments were used in this process as were used in the set-up of the Nijmegen $NN$ database [4, 5]. This means for instance that data with a very improbable high *or low* $\chi^2$ are rejected on statistical grounds. The resulting Nijmegen 1993 $\bar{p}p$ database in the momentum interval 119–923 MeV/c is unique in the world and consists of $N_{\text{data}} = 3646$ $\bar{p}p$ data. It is extensively discussed in Ref. [7]. In the final fit to this database we reach $\chi^2 = 3801.0$ or $\chi^2/N_{\text{data}} = 1.04$. The number of boundary-condition parameters needed is 30, which is a reasonable number, in view of the fact that 21 parameters were needed in the Nijmegen $pp$ PWA and an additional 18 in the $np$ PWA. The total number of degrees of freedom is $N_{\text{df}} = 3503$, which means that $\chi^2/N_{\text{df}} = 1.09$.

If the database is a correct statistical ensemble and if the theoretical model is correct, one expects that $\langle \chi^2 \rangle = N_{\text{df}} = 3503$ with an error of $\sqrt{2N_{\text{df}}} = 84$. This means that in our PWA we are 298 or only 3.5 standard deviations away from the expectation value for $\chi^2$. We conclude that although there is still room for improvement, our *energy-dependent solution is essentially correct statistically*. As a consequence, the values for the phase-shift parameters (and also for the pion-nucleon coupling constant) and the statistical errors (obtained in the standard manner from the error matrix) are essentially correct.

In our 1991 preliminary PWA [6] we were able to determine the charged-pion–nucleon coupling constant $f_c^2 \equiv f_{pn\pi^+} f_{np\pi^-}/2$ from the data on the charge-exchange reaction $\bar{p}p \rightarrow \bar{n}n$, in which only isovector mesons can be exchanged. The result found was $f_c^2 = 0.0751(17)$, at the pion pole. The error is purely statistical. In our final analysis, we have repeated the determination of $f_c^2$, but this time from the complete 1993 Nijmegen database. The coupling constants of the neutral pion were kept at the value of $f_{pp\pi^0}^2 = f_{nn\pi^0}^2 = 0.0745$ [18]. We now find $f_c^2 = 0.0732(11)$, at the pion pole. This result supersedes our previous value from Ref. [6]. Again, the error is of statistical origin only. In view of the enormous amount of work involved, it is very hard to estimate possible systematic errors on this result. We have checked that there are no systematic errors due to form-factor effects or due to uncertainties in $\rho^{\pm}(770)$ exchange. In the Nijmegen $pp$ PWA systematic errors could be more thoroughly investigated and they were found to be small [18]. In our case the systematic errors are probably larger than for the $pp$ case, but it is very encouraging that the result for $f_c^2$ is in good agreement with recent determinations $f_c^2 = 0.0735(15)$ from $\pi^{\pm}p$ [20] scattering and $f_{pp\pi^0}^2 = 0.0745(6)$ and $f_c^2 = 0.0748(3)$ from $NN$ scattering [18]. Very probably the new LEAR experiment PS206 on $\bar{p}p \rightarrow \bar{n}n$ will further constrain the charged-pion–nucleon coupling constant.

In Fig. 1 the differential cross section at 693 MeV/c and the analyzing power at 875 MeV/c are shown for $\bar{p}p \rightarrow \bar{n}n$. The data are from PS199 [2]. One can see the truly remarkable accuracy of the cross-section data and the analyzing-power data in the forward region. The "dip-bump" structure in $d\sigma/d\Omega$ at forward angles is due to the interference of one-pion exchange and a smooth background.

The fact that the available charge-exchange data already pin down the charged-pion coupling constant with a remarkable small statistical error is only one example of how at present *quantitative* information can be extracted from the $\bar{p}p$ system. We can mention

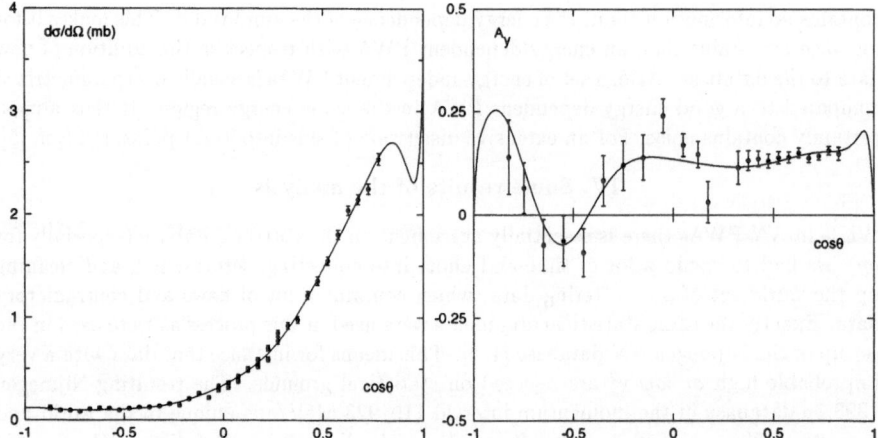

Figure 1: Differential cross section at 693 MeV/c and analyzing power at 875 MeV/c for the charge-exchange reaction $\bar{p}p \rightarrow \bar{n}n$. The data are from PS199 [2]. The curves are from the Nijmegen PWA [7].

some subtle effects that are also visible in the data. The use of $\alpha'$ instead of $\alpha$, i.e. the main relativistic correction to the static Coulomb potential, gives a drop of $\Delta\chi^2 = 30$, or 5.5 standard deviations. The inclusion of the magnetic-moment interaction gives a drop of $\Delta\chi^2 = 14$, or 3.7 standard deviations. Even the use of the correct pion masses instead of an average mass of 138 MeV is a 3 standard-deviation effect.

Since the present $\bar{p}p$ PWA is the first of its kind, we have also proposed a convention for extracting phase-shift and inelasticity parameters from the $S$ matrix. In the presence of coupling to annihilation channels the $S$ matrix describing $\overline{N}N$ scattering is only a submatrix of the much larger multichannel $S$ matrix. It is therefore still symmetric, but no longer unitary. This doubles the number of parameters needed. For the partial waves with $\ell = J$, $s = 0,1$ one obviously writes $S = \eta \exp(2i\delta)$. For the states with $\ell = J \pm 1$, $s = 1$, coupled by the tensor force, six parameters are needed to parametrize the $2 \times 2$ $S$ matrix. In this case it is not so easy to think of a convenient parametrization which satisfies all constraints from unitarity, is completely general, and free from nontrivial ambiguities. We have used a generalization [21] of the "bar-phase" convention commonly used in $NN$ scattering. One writes (with the notation $\bar{\delta}_{\ell J}$ for the phase shift)

$$S^J = \exp(i\bar{\delta}) \begin{pmatrix} \cos\bar{\epsilon}_J & i\sin\bar{\epsilon}_J \\ i\sin\bar{\epsilon}_J & \cos\bar{\epsilon}_J \end{pmatrix} H^J \begin{pmatrix} \cos\bar{\epsilon}_J & i\sin\bar{\epsilon}_J \\ i\sin\bar{\epsilon}_J & \cos\bar{\epsilon}_J \end{pmatrix} \exp(i\bar{\delta}) , \qquad (9)$$

where $\bar{\delta} = \mathrm{diag}(\bar{\delta}_{J-1,J}, \bar{\delta}_{J+1,J})$ and $\bar{\epsilon}_J$ is the mixing parameter. $H^J$ is a three-parameter real and symmetric matrix representing inelasticity. It can be diagonalized in Blatt-Biedenharn fashion

$$H^J = \begin{pmatrix} \cos\omega_J & -\sin\omega_J \\ \sin\omega_J & \cos\omega_J \end{pmatrix} \begin{pmatrix} \eta_{J-1,J} & 0 \\ 0 & \eta_{J+1,J} \end{pmatrix} \begin{pmatrix} \cos\omega_J & \sin\omega_J \\ -\sin\omega_J & \cos\omega_J \end{pmatrix} , \qquad (10)$$

where the diagonal matrix contains the "eigeninelasticities" $\eta_{J-1,J}$ and $\eta_{J+1,J}$, and $\omega_J$ is again a mixing parameter. We are presently in the process of doing a careful evaluation of these phase-shift and inelasticity parameters and their errors. Unfortunately, this involves a large amount of work. These and other issues will be the subject of future publications.

# References

[1] R.A. Kunne *et al.*, Phys. Lett. B **206**, 557 (1988); Nucl. Phys. **B323**, 1 (1989); R. Bertini *et al.*, Phys. Lett. B **228**, 531 (1989); F. Perrot *et al.*, *ibid.* **261**, 188 (1991).

[2] R. Birsa *et al.*, Phys. Lett. B **246**, 267 (1990); *ibid.* **273**, 533 (1991).

[3] R. Birsa *et al.*, Phys. Lett. B **302**, 517 (1993).

[4] J.R. Bergervoet, P.C. van Campen, W.A. van der Sanden, and J.J. de Swart, Phys. Rev. C **38**, 15 (1988); J.R. Bergervoet, P.C. van Campen, R.A.M. Klomp, J.-L. de Kok, T.A. Rijken, V.G.J. Stoks, and J.J. de Swart, *ibid.* **41**, 1435 (1990).

[5] V.G.J. Stoks, R.A.M. Klomp, M.C.M. Rentmeester, and J.J. de Swart, Phys. Rev. C, to appear (1993).

[6] R. Timmermans, Ph.D. thesis, University of Nijmegen, The Netherlands (1991); R. Timmermans, Th.A. Rijken, and J.J. de Swart, Phys. Rev. Lett. **67**, 1074 (1991).

[7] R. Timmermans, Th.A. Rijken, and J.J. de Swart, submitted to Phys. Rev. D.

[8] R. Timmermans, Th.A. Rijken, and J.J. de Swart, Phys. Lett. B **257**, 227 (1991); Phys. Rev. D **45**, 2288 (1992).

[9] J. Côté, M. Lacombe, B. Loiseau, B. Moussallam, and R. Vinh Mau, Phys. Rev. Lett. **48**, 1319 (1982); M. Pignone, M. Lacombe, B. Loiseau, and R. Vinh Mau, *ibid.* **67**, 2423 (1991).

[10] P.H. Timmers, W.A. van der Sanden, and J.J. de Swart, Phys. Rev. D **29**, 1928 (1984); *ibid.* **31**, 99 (1985).

[11] H. van Haeringen and L.P. Kok, Phys. Rev. A **26**, 1218 (1982).

[12] L. Landau and J. Smorodinsky, J. Phys. USSR **8**, 154 (1944).

[13] H. Feshbach and E.L. Lomon, Ann. Phys. (NY) **29**, 19 (1964).

[14] R.L. Jaffe and F.E. Low, Phys. Rev. D **19**, 2105 (1979).

[15] J.J. de Swart and M.M. Nagels, Fortschr. Phys. **26**, 215 (1978).

[16] G.J.M. Austen and J.J. de Swart, Phys. Rev. Lett. **50**, 2039 (1983).

[17] M.M. Nagels, T.A. Rijken, and J.J. de Swart, Phys. Rev. D **17**, 768 (1978).

[18] V. Stoks, R. Timmermans, and J.J. de Swart, Phys. Rev. C **47**, 512 (1993).

[19] S. Mandelstam, Phys. Rev. **112**, 1344 (1958).

[20] R.A. Arndt, Z. Li, L.D. Roper, and R.L. Workman, Phys. Rev. Lett. **65**, 157 (1990).

[21] R.A. Bryan, Phys. Rev. C **24**, 2659 (1981); *ibid.* **30**, 305 (1984); **39**, 783 (1989).

Few-Body Systems, Suppl. 7, 42–49 (1994)

Few-
Body
Systems

# SPIN OBSERVABLES AND ANNIHILATION IN ANTIPROTON–PROTON REACTIONS

Frank Tabakin

Dept. of Physics and Astronomy

University of Pittsburgh

Pittsburgh, PA 15260 U.S.A.

## ABSTRACT

The reaction $\bar{p} + p \to \bar{\Lambda}\Lambda$ is examined with emphasis on comparing amplitudes extracted from recent LEAR data using effective range ideas to those based on using mesonic or quark degrees of freedom.

## INTRODUCTION

Antiproton-proton reactions are dominated by annihilation to multimeson final states. Impressive experiments at the low energy antiproton ring (LEAR) facility provide precise and selective information about exclusive reactions. A major issue is to learn if these are mesonic and/or quark and gluonic degrees of freedom in a few body system.

The selective reaction $\bar{p} + p \to \bar{\Lambda}\Lambda$ is of particular interest for several reasons. The PS185 experiment, as discussed by K. Kilian [1], provides precision data on this associated strangeness production reaction. Measurement is made not only of the total and differential cross-sections, but also $\Lambda$ and $\bar{\Lambda}$ polarizations and spin correlations, which are observed using the $\Lambda$'s "self-analyzing" anisotropic, weak, $\Lambda \to p + \pi^-$ decay. In addition, polarized targets and beams will make "complete experiments" possible. A new approach for deducing experiments needed to provide a complete set will appear in a separate paper. Another important feature of this reaction is that possibly the same quark or meson exchange and annihilation ideas can be applied to production of $\overline{\Lambda}\Sigma$, $\overline{\Sigma}\Sigma$ and $\phi\phi$ (the JETSET experiment at LEAR).

Can such exclusive experiments yield understanding of annihilation, of spin observables, and of explicit quark-gluon and/or meson-isobar dynamics[2]? Many authors [2,3] have studied this reaction from a quark model viewpoint wherein the final $\bar{s}s$ pair couples with "spectator" quarks to form the final $\Lambda\overline{\Lambda}$ (or $\Sigma$'s). A basic part of this picture is the s-channel exchange of one or more $1^-$ gluons. Hence, the reaction is thought to proceed via $^3S_1$- and $^3P_{0+}$ states, with no singlet terms. One possible mechanism for generating singlet amplitudes is by s-channel production of $\eta$, $\eta'$ mesons, which have sizable $\bar{s}s$ content. However, that mechanism for creating singlet terms has been shown [3] to be small in a dynamical model. Thus, quark-model inspired ideas suggest small singlet terms in the $\overline{\Lambda}\Lambda$ production. In contrast, the competing mesonic description [4-7] by kaon exchange, yields strong tensor (and also strong L·S) interactions and then the triplet interactions simply overwhelm the nonzero, but small, singlet terms.

In the mesonic description, the $KN\Lambda$ form factor and associated momentum transfers correspond to sizable hadron overlap, which involves the nonperturbative quark realm. Also annihilation into mesons involves baryon exchange and hence a short-ranged ( $\lambda_N \sim .2$ fm ) transition operator, $\mathcal{V}$. One could object that this is too small a range of annihilation. A reply made by Holinde [5] and independently by Liu and Tabakin [6], is that in coupled–channels(CC) dynamics the range of the transition operator $\mathcal{V}$ is of short range, but the CC effective annihilation operator $V = \mathcal{V}$ $(1/e)$ $\mathcal{V}$, has a reasonable range of about one fm. Here $1/e$ represent the intermediate meson state propagator. In addition, the major effect of CC dynamics is that flux feeds in as well as out of the $\bar{p}p$ channel via intermediate mesons, which influences not only the range of annihilation, but also the associated short distance elastic channel wave functions.

Objections to the quark model approach are that we are in a nonperturbative region where the constituent quark model is most difficult and that the treatment of annihilation is just as complicated as the mesonic approach, which at least can make contact in some charge conjugation sense with the large body of two-nucleon information.

A major difference between models is their treatment of annihilation. Some models include annihilation by local, state and energy dependent optical potentials, while others use the more physical CC "Nijmegen philosophy" of introducing effective two-body channels to provide for net loss of flux [4]. The basic annihilation process $\bar{p}p \rightarrow$ mesons yields a highly momentum dependent transition operator $\mathcal{V}$, because nonstatic ( e.g. $\sigma \cdot \nabla$)

terms occur at meson vertices [6,7]. Despite the basic requirement that annihilation must be nonstatic, local descriptions of annihilation are often used to simplify computation, even in more sophisticated CC descriptions of annihilation [4]. At short-distances, the $\bar{p}p$ elastic channel wave function depends considerably on the annihilation model adopted. For example, a local optical potential leads to a strong attenuation of the short distance wave function, especially for low partial waves. In contrast, a description of annihilation solely by a coupled channels description (which corresponds to the actual dynamics) leads to a $\bar{p}p$ elastic channel wave function which dampens, but also oscillates at short distance, thus offering some hope of revealing short distance dynamics. Hence, the annihilation model adopted influences the role of short distances and of low partial waves and hence also affects spin observables. Let us examine the data and associated amplitudes.

## DATA and PROFILE FUNCTIONS

To extract amplitude information, consider the products $\mathcal{I}(\theta)P(\theta)$ and $\mathcal{I}(\theta)C_{2M}(\theta)$, where $\mathcal{I}(\theta)$ is related to the differential cross section by $\sigma(\theta) = \rho_0 \mathcal{I}(\theta)$, where $\rho_0 = (q/2ps)(m_p m_\Lambda)^2$. Examination of these "profile functions" by Tabakin, Eisenstein, and Lu (TEL) [8] revealed amplitude constraints, especially near threshold. Some of these rules and their relation to quark and meson based theories will be discussed later. Spherical tensor combinations of the Cartesian spin correlations are formed using $C_{2M} \sim [\sigma_{\overline{\Lambda}} \otimes \sigma_\Lambda]_{2M}$. The basic PS185 data [1], organized in profile function form [8], and fit with associated Legendre expansions, are shown in Fig. 1 . There are nine momenta for which data exists, with more expected soon. Data at $p_{lab} = 1.43695$ GeV/c is not shown, but was included in the fits.

Data for the total and differential cross-sections and the $\Lambda$ polarization exist for all nine incident $\bar{p}$ momenta. However, it is only for the four momenta above 1.507 GeV/c that spin correlations are available. The polarization profiles and associated Legendre fits are also shown in Fig. 1, where a polarization node occurs at a fixed momentum transfer[1]. Plots of some $C_{2M}$'s are provided in Fig. 1, which also include interesting nodal evolution behavior[8].

From the spin correlations, one can form the singlet fraction, which indicates the role of singlet processes. Note that the singlet fraction and the singlet amplitude hover near zero - at least for this range of momenta. The small singlet fraction is either zero for a dynamical reason (dominance of strong tensor forces [4-7] and/or the suppression of $\eta$ $\eta'$ s-channel exchange [3]) or there could be a cancellation or suppression which occurs just in the 1.546 to 1.919 GeV/c region. The value of the singlet fraction away from this region is unknown.

## AMPLITUDES COMPARED

Meson- and quark-inspired models are consistent with these data. It is of interest to compare amplitudes generated by different dynamical models. Fortunately, the Washington-Colorado group[3] has kindly provided their quark-model amplitude results. Timmermans et al. [4] provide amplitudes fit to data using a one kaon exchange at large distances plus a P-matrix approach for the short distance dynamics. Amplitudes for the Nijmegen or Bonn CC meson dynamical calculations are not available, although their partial cross sections[5] are in general agreement with the Timmermans et al. [4] amplitudes. Thus we can compare the available amplitudes with general threshold rules discussed by TEL [8].

The basic idea used in TEL is that a truncation to a few amplitudes near threshold $(^3S_1, ^3D_1, ^3P_{0,1,2}, T_{SD}, T_{DS}$ ) suffices to yield a good fit to data. Also, near threshold data indicate the need not only for P-waves, but for spin-orbit type P-wave splitting, e. g. a matrix element dependence proportional to $J(J + 1) - S(S + 1) - L(L + 1)$. Furthermore, they concluded that near threshold the P-wave combination $\phi_P = (3T_{3P1} - T_{3P2} - 2T_{3P0})$, should be within 90° of the $T_{3S1}$ amplitude to provide forward angle peaking. No adjustment was made for particularly strong S-wave suppression or for F to P noncentral transitions, nor did the direct amplitude fit to the threshold data require such assumptions. The term "near threshold" is vague and the range of applicability of their truncations was analyzed, but not pinned down.

The full Nijmegen mesonic model [4] uses static $\bar{p}p$ interactions and local transition operators in a coupled-channels approach; their amplitude fits are however based on a direct fit to data using a one kaon exchange potential. In the quark-inspired model [3] a cluster model with phenomenological annihilation is adopted. These results can now be compared to the earlier TEL work.

First the Nijmegen and Washington-Colorado amplitudes, displayed as Argand diagrams in Fig. 2 for S, P and D-waves, will be discussed. Note in all cases we have introduced arbitrary scaling factors in order to compare amplitudes trends on a common basis. (The notation $T_{SD}$ etc. refers to the $^3D_1 \rightarrow ^3S_1$ noncentral transition). Note that the Nijmegen S-waves show a sizable tensor force since their $T_{SD}$ amplitude is large and $T_{SD} > T_{3S_1}$, at all momenta. For the Washington case $T_{SD}$ is relatively smaller than

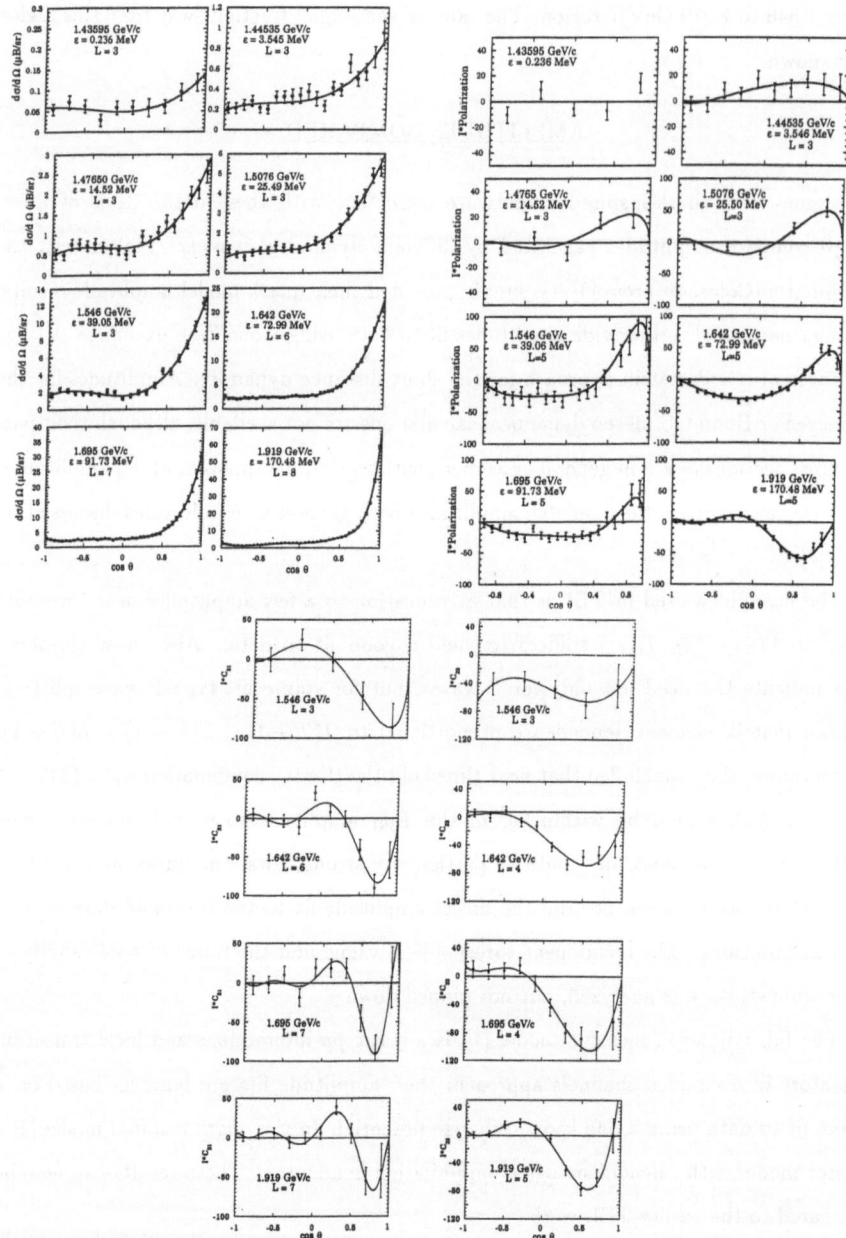

Figure 1: The PS185 $p_{lab} = 1.436$ to $1.919$ GeV/c data[1] for the $\bar{p}p \to \bar{\Lambda}\Lambda$ differential cross-section, polarization, spin correlations(in spherical tensor form) and the singlet fraction. Here profile forms are presented. The solid curves are Legendre fits. The latest data are at 1.642 and 1.919 GeV/c; the 1.437 data are available but not presented here.

for Nijmegen and at all momenta $T_{SD} < T_{3S_1}$. Another feature for both cases is that the S-wave amplitudes exhibit large curvature in the Argand diagrams as a function of $p_{lab}$ (this curvature in the real versus imaginary amplitude plots(Argand diagrams) is called "looping"). This S-wave looping is in contrast to the straight line evolution of the simplest scattering length approximation $T_\ell \to a_\ell\, q_{cm}^\ell$ adopted by TEL. Both the Nijmegen and the Washington P-waves are sizable and split, with only slight looping until the higher momenta. These amplitudes obey the TEL rules for P-wave splitting and cross-section peaking rules; however, the order of the P-wave splittings are quite different. The Washington P- and D-waves follow a spin-orbit order while the Nijmegen P-waves are split by a combination of tensor and spin-orbit interactions. Both find important $T_{PF}$ contributions, with the Nijmegen values being quite large. Similar remarks hold for the D-waves. Clearly, the Nijmegen fits, which are guided by mesonic dynamics at long distances, involve much stronger noncentral terms than the quark-based Washington amplitudes. These features are consequences of their dynamics and annihilation mechanisms.

In the absence of annihilation effects, one might expect a natural ordering of $T_{DG} \approx T_{DS} < T_{PF} < T_{SD}$. However, with large S-wave attenuation, arising from local annihilation models, the S-wave final state is greatly reduced, leading to the large relative $T_{PF}$ contribution seen in Fig. 2. This feature was absent from TEL. It appears in the Bonn [5] model as well as in the quark-model[3] to varying degrees. Another significant difference between the Nijmegen and Washington amplitudes is that the Nijmegen model has nonzero, albeit small, singlet amplitudes. As the momentum increases these singlet amplitudes might increase and not cancel, which might be detected in higher $p_{lab}$ spin correlation data.

From the above results, it appears that the simplest scattering length approximation, $T_\ell = a_\ell\, q^\ell$, is not sufficient for S-waves and one should include higher powers of $q_{cm}$ via an effective range expansion. We begin by using the form: $T_\ell = q^\ell a_\ell/(1 + b_\ell q^2)$, where the $b_\ell$ factor produces looping of the amplitude in the Argand diagram. The use of such an "effective range" approximation has been discussed more generally by several authors [9,10]. A simple $T_\ell = a_\ell\, q^\ell$ approximation seems adequate for P-waves since the above P-waves exhibit only slight looping. The role of the F to P transition also needs to be examined[5]. The TEL model, which is a simple scattering length, truncated basis analysis, clearly needs to be extended. Thus, the $p_{lab} \leq 1.546$ GeV/c data has been refit[8] with an extended state basis: $^1S_0$, $^3P_{0,1,2}$, $^1P_1$, $^3S_1$, $^3D_1$, $T_{SD}$, $T_{DS}$, $^1D_2$, $T_{PF}$, $^3P_2$, $^3D_2$, $T_{DG}$, $^3D_3$ and with the ansatz: $T_\ell = q^\ell a_\ell/(1 + b_\ell q^2)$. At this stage, we have taken $b_\ell$ to be nonzero only for the $^3S_1$ and $T_{SD}$ amplitudes. With these alterations, a new fit to the cross-section and polarization data for $p_{lab} \leq 1.546$ GeV/c was achieved with $\chi^2/N_{freedom}$ of 1.3. The motivation here is to see what the data require concerning the basic amplitudes, with minimal assumptions about dynamics. The new TE amplitudes are seen on the right side of Fig. 2, where a $^3F_2 \rightarrow ^3P_2$ amplitude now arises along with a suppression of the $^3D_1 \rightarrow^3 S_1$ amplitude. There is a clear need for a sizable $T_{PF}$ contribution.

Of course, a full determination of the amplitudes requires a theoretical description. At this stage, a fully nonstatic coupled channels approach to antiproton-proton reactions which has the flexibility to include both long-range mesonic effects and short distance quark information is needed. In such an effort[7] the specific annihilation channel information should be incorporated by nonstatic effective two-body channels.

## ACKNOWLEDGMENTS

The author is deeply appreciative of the kind help and encouragement provided by Professors R. A. Eisenstein, M. A. Alberg and P. D. Kunz.

## REFERENCES

1. Kilian, K. K., "Production of Hyperon-Antihyperon Pairs at LEAR." these proceedings. Also see: Barnes, P. D. et al.: Phys.Lett. **B229** 432(1989), Phys. Lett. **189**, 249(1987), Nucl. Phys. **A526** 575(1991).

2. Dover, C. B. et al.: Prog. Part. Nucl. Phys. **29**, 87(1992); Amsler. C., and Myhrer, F.: Ann. Rev. Nucl. Part. Sci. **41**,219(1991).

3. Alberg, M.A., Henley, E .M., Wilets, L., Kunz. P. D.: Nucl. Phys. **A560**,365(1993). Alberg, M.A., Henley, E .M., Weise, W.: Physics Letters **B255**,498(1991).

4. Timmermans, R. G. E.: "Antiproton-Proton Partial Wave Analysis" these proceedings, and Ph. D. Nijmegen(1991). Timmermans, R. G. E., Rijken, T. A. , deSwart, J. J.: Nucl. Phys. **A479**, 383C(1988), Phys. Lett. **B257**, 227(1991), and Phys. Rev. **D45**, 2288(1992).

5. Haidenbauer, J., Holinde, K., Mull, V., Speth, J.: Phys.Lett. **B291**, 223(1992). Haidenbauer, J., Hippchen, T., Holinde, K., Holzenkamp, B., Mull, V., Speth, J.: Phys. Rev. **C45**, 931 (1992).

6. Liu, G. Q.: Ph. D. Thesis, University of Pittsburgh (1987), and Liu, G. Q. et al.: Phys. Rev. **C41**, 655(1990).

7. Lu, Y.: Ph. D. Thesis, University of Pittsburgh (1991).

8. Tabakin, F., Eisenstein, R. A., Lu, Y.: Phys. Rev. **C44**, 1749(1991). Also see: Eisenstein, R. A.:"Hyperon-antihyperon Production" to be published in the proceedings of the NAN'93 Conference, Moscow, Russia (1993).

9. Pirner, H.J., Kerbikov, B., Mahalanabis, J.: Z. Phys. **A338**,111(1991). Grach, I.L., Kerbikov, B.O., Simonov, Yu.A.: Phys.Lett. **208B**,309(1988).

10. Carbonell, J., Protasov, K.: Hyperfine Interactions **76**, 327(1993).

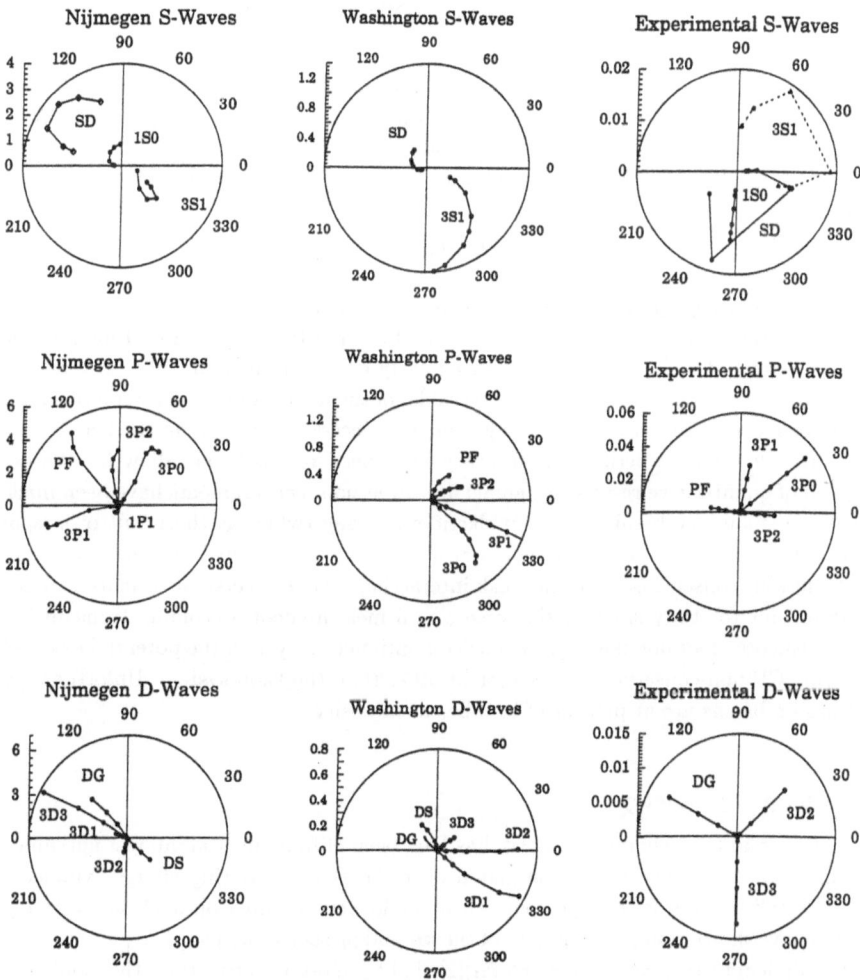

Figure 2: Argand Diagrams for the S, P and D-wave amplitudes for the NIJMEGEN[4], WASHINGTON[3], and our new fits to data(EXPERIMENT) which extend TEL[8] to include S-wave looping.

Few-Body Systems, Suppl. 7, 50—63 (1994)

# THE NUCLEON-NUCLEON INTERACTION AND VIOLATION OF FUNDAMENTAL SYMMETRIES

W.T.H. van Oers

Department of Physics, University of Manitoba, Winnipeg, MB, Canada R3T 2N2
and
TRIUMF, 4004 Wesbrook Mall, Vancouver, B.C., V6T 2A3

## Abstract

The interplay of the nucleon-nucleon interaction and its observables with the "fundamental" symmetries of isospin conservation, parity conservation, time-reversal invariance or CP conservation was realized early on. Many tests of these symmetries through measurements of particular observables of the nucleon-nucleon interaction have been made over a time frame spanning some five decades. It is only in the last decade or so that levels of experimental accuracy have been reached that allow for the deduction of quantitative results of significance. Precision measurements have been made of charge symmetry breaking in $n$-$p$ elastic scattering (which is the result of isospin non-conservation) and of parity violation in $p$-$p$ scattering (which is a manifestation of the flavour conserving hadronic weak interaction). Time reversal invariance is much more difficult to study since in this case a null measurement, excluding transmission measurements, does not exist. In the nucleon-antinucleon system the potential exists of studying CP non-conservation in a system other than the kaon system. Unfortunately antiproton beams are at present of insufficient intensity.

## 1. Charge Symmetry

The validity of charge symmetry has long been of fundamental interest and much circumstantial evidence has accumulated over the years favouring charge symmetry breaking (CSB) in order of a percent. Although low energy nucleon-nucleon scattering studies have shown a slight inequality of the $nn$ and $pp$ scattering lengths[1] with the $nn$ scattering length ($a_{nn} = -18.6 \pm 0.4$ fm)[2] slightly more negative than the Coulomb-corrected $pp$ scattering length ($a_{pp} = -17.3 \pm 0.4$ fm)[3], it has proved very difficult to remove experimental and theoretical uncertainties to isolate charge symmetry breaking unequivocally.

A new experimental effort to measure the low-energy $n$-$n$ scattering parameters is underway at TRIUMF. The experiment is a triple coincidence measurement of the $\pi^-$-$d$

$\rightarrow \gamma$-$n$-$n$ reaction with stopped pions. It is anticipated that the $n$-$n$ scattering length $a_{nn}$ and effective range parameter $r_{nn}$ can be determined to an accuracy comparable to or better than obtained previously in the PSI measurements.

The inequality of $a_{nn}$ and $a_{pp}$ reflects the presence of a charge-asymmetric/charge-dependent interaction, which preserves symmetry under the interchange of the two nucleons in isotopic spin space. The interaction can be written $V_{12} = D(\tau_3(1) + \tau_3(2))$, with $\tau_3(1)$ and $\tau_3(2)$ the third components of the isotopic spin of nucleons 1 and 2, and $D$ a function of space and spin coordinates[4], – the class III interaction.

Charge symmetry also leads to the complete separation of the isoscalar and isovector components of the $n$-$p$ interaction. This in turn leads to the equality of the differential cross sections for polarized neutrons scattering from unpolarized protons and vice versa. As a result $A_n(\theta) \equiv A_p(\theta)$ where $A$ denotes the analyzing power and where the subscript represents the polarized nucleon. A nonvanishing asymmetry difference is directly proportional to the isotopic spin singlet-triplet, spin triplet-singlet mixing amplitude and therefore direct evidence of a charge-asymmetric/charge-dependent interaction, which is asymmetric under the interchange of nucleons 1 and 2 in isotopic spin space. The interaction is of the form $V_{12} = E(\tau_3(1) - \tau_3(2)) + F(\vec{\tau}(1) \times \vec{\tau}(2))_3$, with $E$ and $F$ functions of space and spin coordinates,- the class IV interaction. Following the scattering formalism as defined by LaFrance and Winternitz[6] the scattering matrix for $n$-$p$ elastic scattering can be written

$$M(\vec{k}_f, \vec{k}_i) = \frac{1}{2}\Big\{(a+b) + (a-b)(\vec{\sigma}_1 \cdot \hat{n})(\vec{\sigma}_2 \cdot \hat{n}) + (c+d)(\vec{\sigma}_1 \cdot \hat{m})(\vec{\sigma}_2 \cdot \hat{m})$$
$$+ (c-d)(\vec{\sigma}_1 \cdot \hat{\ell})(\vec{\sigma}_2 \cdot \hat{\ell}) + e(\vec{\sigma}_1 + \vec{\sigma}_2) \cdot \hat{n} + f(\vec{\sigma}_1 - \vec{\sigma}_2) \cdot \hat{n}\Big\}. \qquad (1)$$

Here $\hat{\ell}$, $\hat{m}$ and $\hat{n}$ are unit vectors given as

$$\hat{\ell} = \frac{\vec{k}_i + \vec{k}_f}{|\vec{k}_i + \vec{k}_f|}; \quad \hat{m} = \frac{\vec{k}_f - \vec{k}_i}{|\vec{k}_f - \vec{k}_i|}; \quad \hat{n} = \frac{\vec{k}_i \times \vec{k}_f}{|\vec{k}_i \times \vec{k}_f|}; \qquad (2)$$

with $\vec{k}_i$ and $\vec{k}_f$ the initial and final state centre-of-mass nucleon momenta. The amplitudes $a, b, c, d, e$, and $f$ are functions of centre-of-mass energy $E$ and scattering angle $\theta$. Written explicitly, the difference in the analyzing powers

$$\Delta A(\theta) \equiv A_n(\theta) - A_p(\theta) = \frac{2}{\sigma_0}Re(b^*f), \qquad (3)$$

showing the proportionality to $f$, with $\sigma_0$ the differential cross section for the scattering of unpolarized neutrons from unpolarized protons. Experimental considerations show that the quantity next in order of difficulty of measuring is the difference in the spin-correlation parameters $C_{xz}(\theta)$ and $C_{zx}(\theta)$. The correlation parameter $C_{xz}$ has the projectile spin transverse to the beam direction in the scattering plane and the target spin longitudinal with the incident beam direction, while for $C_{zx}$ the reverse holds. Again charge symmetry leads to the equality of $C_{xz}(\theta)$ and $C_{zx}(\theta)$, but if charge symmetry is broken then one will be able to measure a difference:

$$\Delta C(\theta) \equiv C_{xz}(\theta) - C_{zx}(\theta) = \frac{2}{\sigma_0}Im(c^*f). \qquad (4)$$

Charge symmetry is a less restrictive form of isotopic spin conservation, according to which interactions must be invariant under a specific rotation in isotopic spin space that interchanges every particle with its isotopic spin mirror (rotation by 180° about the $I_2$ axis, which reverses the sign of the third component of isotopic spin ($I_3 \leftrightarrow -I_3$)). Consequently on the quark level $\mu \leftrightarrow d$, on the meson level $\pi^+ \leftrightarrow \pi^-$, and on the nucleon level $p \leftrightarrow n$. Since the nucleon-nucleon interaction in the meson exchange model description is mediated manifestly by pion exchanges, and since the pion masses are invariant under the above rotation, experiments that measure charge symmetry breaking potentially probe smaller isotopic spin non-conservation effects that arise fundamentally at the quark level. In this context, isotopic spin non-conservation stems from the inequality of the $u$ and $d$ quark masses and from the Coulomb interaction among the quarks.

The first measurement ever of CSB in $n$-$p$ elastic scattering was performed at TRIUMF[7]. The measurement of $\Delta A \equiv A_n - A_p$, at the zero-crossing angle of the analyzing power, at an incident neutron energy of 477 MeV, has yielded $\Delta A = (47 \pm 22 \pm 8) \times 10^{-4}$, a little over two standard deviations effect. More recently the results of a similar experiment at a neutron energy of 183 MeV performed at IUCF have been reported[8]. The measured value of $\Delta A \equiv A_n - A_p$, averaged over the angular range $82.2° \leq \theta_{cm} \leq 116.1°$ over which $< A(\theta) >$ averages to zero, is $(33.1 \pm 5.9 \pm 4.3) \times 10^{-4}$, where again the first error represents mainly the statistical uncertainty and the second error the systematic uncertainty. The latter result differs from zero by 4.5 standard deviations (see Fig. 1). It differs from the value expected from the electromagnetic spin-orbit interaction by 3.4 standard deviations. This difference represents the strongest experimental evidence to date of CSB in the nuclear interaction. Due to the intrinsic difficulties present, measurements of $\Delta C \equiv C_{xz} - C_{zx}$ have not yet been attempted.

There are difficulties in extracting an angular distribution of $\Delta A(\theta)$. This follows directly from the expression for the difference in the asymmetries for beam and target polarized, respectively, or

$$\epsilon_b - \epsilon_t = \Delta A (P_b + P_t)/2+ < A > (P_b - P_t), \tag{5}$$

pointing to the need for calibration of the beam and target polarizations ($P_b$ and $P_t$) with an accuracy unattainable at present. In the analysis of the IUCF experiment this difficulty was overcome by adjusting the ratio of ($P_b/P_t$) until the error-weighted rms value of $\Delta A(\theta)$ over the angular range of the experiment reached minimal variance. Following this procedure a twelve point angular distribution was obtained. The procedure does not work at 477 MeV where $\Delta A(\theta)$ and $< A(\theta) >$ have zero-crossing angles in close proximity and consequently the angular dependencies are no longer orthogonal. If the theoretical calculations were consistent in their predictions of the zero-crossing angle of $\Delta A(\theta)$ one could in principle also determine $\Delta P = P_b - P_t$ and consequently the angular distribution of $\Delta A(\theta)$ would follow.

In general the measured analyzing power differences of the IUCF and TRIUMF experiments are well reproduced by theoretical predictions based on meson exchange potential models, which indirectly incorporate quark level effects. The calculations include contributions from one photon exchange (the magnetic moment of the neutron interacting with the current of the proton), from the neutron-proton mass difference affecting charged one $\pi$- and $\rho$-exchange, and from the more interesting isospin mixing $\rho^0$-$\omega^0$ meson exchange. Some other smaller effects (like two $\pi$-exchanges not included in $\rho$-exchange) have also been evaluated. The effects of $\pi^0$-$\gamma$ exchanges have not yet been calculated. The theoretical results indicated by the solid and dashed lines in Fig. 1 are based on a momentum space version of the Bonn nucleon-nucleon potential[9]. Note that the first two contributions mentioned suffice to give a theoretical prediction in agreement with the TRIUMF result. This is because at 477 MeV the effect arising from $\rho^0$-$\omega^0$ mixing crosses zero close to the zero-crossing angle of $< A(\theta) >$. It is clear that the IUCF result at 183 MeV requires inclusion of the $\rho^0$-$\omega^0$ meson mixing contribution, an approximately two standard deviation effect.

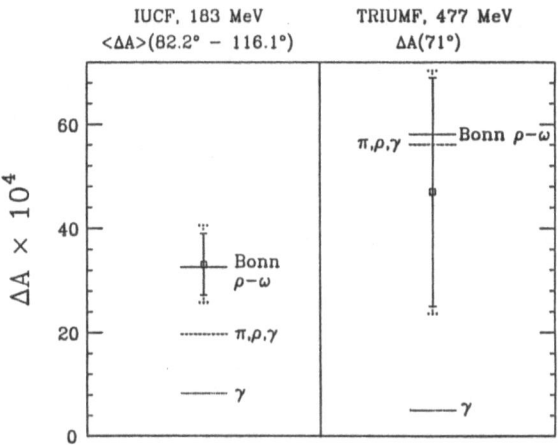

Fig. 1. Measured values of $\Delta A \equiv A_n - A_p$ for $n$-$p$ elastic scattering at 183 MeV(IUCF) and 477 MeV(TRIUMF). The horizontal lines represent theoretical predictions of HHT(Ref.[9]).

The differences in the calculations of Holzenkamp, Holinde and Thomas (HHT)[9] and of Beyer and Williams (BW)[10] reflect the poorly known $\rho NN$ and $\omega NN$ coupling constants. The better agreement with the data provided by calculations employing the stronger coupling constants measures internal consistence among the ingredients of the HHT calculations. A calculation self-consistent in the isotopic spin conserving neutron-proton interaction would be most welcome. However, the theoretical picture may be far from settled as argued in several recent papers about off-energy-shell effects[11].

In order to delineate the various contributions to CSB, a new measurement at 350 MeV has been made at TRIUMF (E369) (Ref.[12]), with data taking completed in the spring of 1993. This new measurement is very similar to the earlier TRIUMF measurement at 477 MeV[7]. Again designed as a null experiment, its goal is to achieve an accuracy in $\Delta A \equiv A_n - A_p$ of $\pm 0.0006$(or $\pm 0.015°$ in the zero crossing angle), which is a factor four improvement over the earlier measurement. The 350 MeV neutron beam was produced using the $(p,n)$ reaction on deuterium. The proton beam had an intensity of about 2 $\mu A$ and a polarization of about 0.70 and was incident on a 0.21 m long LD$_2$ target. The energy, the polarization, the position and direction of the proton beam were monitored throughout the experiment and controlled (in the case of position and direction) using a feedback system coupling two sets of split-plate secondary emission monitors (which determined the median of the intensity distribution) with steering magnets upstream in the beam transport line. At the two sets of split-plate secondary emission monitors the beam position was kept fixed with a $\sigma \leq 0.1$mm in both $x$ and $y$ intensity profiles). The beam energy monitor, based on range determinations, allowed the beam energy to be kept constant with a $\sigma$ of less than 30 keV (through minute changes in rf of the cyclotron and stripper foil position). The polarization was transferred from the proton to the neutron by making use of the large sideways to sideways polarization transfer coefficient $r_t$ (-0.88 at 364 MeV). Required rotations of the polarization directions are obtained by a solenoid magnet (for the proton polarization direction) and a combination of two dipole magnets (for the neutron polarization direction). The 9° neutron beam passed a 3.3 m long, tapered steel collimator before impinging on a frozen spin type polarized proton target positioned at 12.85 m from the centre of the LD$_2$ target. The frozen spin target contains butanol beads of $\sim$1.5 mm diameter contained in a target cell of dimensions 20 mm wide, by 35 mm long, by 55 mm high. Typical polarizations were 0.80 or higher. Scattered neutrons were detected in the angle range 24.0° to 42.4° in large area scintillation counters, while the recoil protons were observed in scintillation counter/wire chamber telescopes nominally centred at 53°. The detection apparatus had reflection symmetry about the neutron beam axis to increase the event rate and allow certain systematic errors to be cancelled (a three dimensional picture of the experimental setup is shown in Fig. 2). At the zero-crossing angle all systematic errors, except those due to background corrections, are eliminated to second order. With a solid angle considerably larger than in the previous experiment (at 477 MeV) it will be possible to obtain an angular distribution for $\Delta A(\theta)$ which includes a region where the interesting $\rho^0$-$\omega^0$ mixing is relatively large. But note that at most the shape of the angular distribution of $\Delta A$ is accessible in these experiments unless one is able to determine $\Delta P/P$ with high precision, where $\Delta P = P_b - P_t$ and $P = (P_b + P_t)/2$. In order to arrive at the desired precision in $\Delta A(\theta)$ of about $\pm 0.0020$ in bins of a few degrees, the experiment made use of the high intensity provided by the recently developed optically pumped polarized H$^-$ ion source. Using various slits inside the cyclotron it was possible to reduce the time spread of the neutron beam to less than 1 ns (FWHM), thus improving time-of-flight measurements considerably.

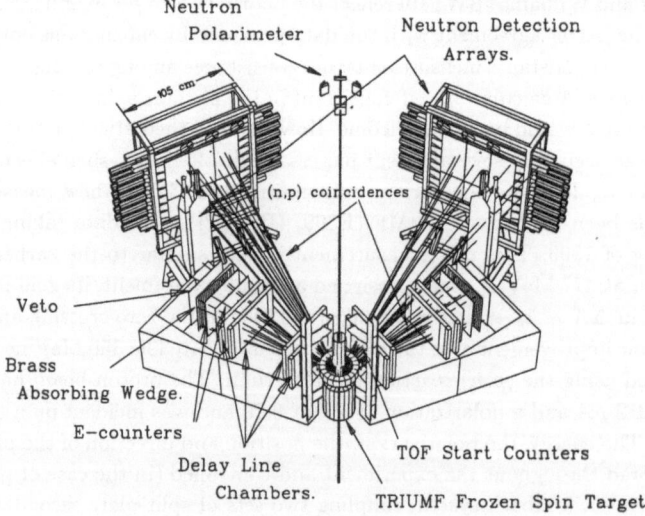

Fig. 2. Three dimensional view of the detection apparatus.

Neutron-proton scattering events are identified using vertex reconstruction and cuts on opening angle, on coplanarity, on the neutron energy, the proton energy and their energy sum, and on the transverse momentum sum. Alternatively, constraints are applied to the sum of the $\chi^2$ variables. Data analysis is progressing well with a final result expected in a year's time.

Another experiment that tests the effects of class IV interactions is the measurement of the forward-backward asymmetry in the reaction $n\text{-}p \to d\text{-}\pi^0$. In the absence of such interactions isospin is conserved and consequently the angular distribution is symmetric about $90°$ in the centre-of-mass. A nonvanishing forward-backward asymmetry difference in the deuteron angular distribution

$$A_{fb} \equiv \frac{\sigma(\theta) - \sigma(\pi - \theta)}{\sigma(\theta) + \sigma(\pi - \theta)} \tag{6}$$

is direct evidence of a charge-asymmetric/charge-dependent interaction, which is asymmetric under the interchange of nucleons 1 and 2 in isotopic spin space.

The first calculations of $A_{fb}$ by Cheung et al.[13] focussed on energies near the $\Delta$ resonance and found that the dominant contribution to $A_{fb}$ at these energies was $\pi^0\text{-}\eta$ mixing. Later more complete calculations by Niskanen et al.[14] show an energy dependence of $A_{fb}$ that suggests a large negative value below 300 MeV. Recent calculations[14] give an angle integrated value of approximately $-50 * 10^{-4}$ for $A_{fb}$ near 280 MeV with $\pi^0\text{-}\eta$ and $\pi^0\text{-}\eta'$ mixing contributing almost an order of magnitude more than the $n\text{-}p$ mass difference affecting charged one $\pi$-and $\rho$-exchange and $\rho^0\text{-}\omega^0$ mixing. These contributions are strongly dependent on the meson mixing matrix elements and on the $\eta NN$ coupling constant. While $A_{fb}$ has been measured for a range of energies, the large uncertainties of these measurements make them rather inconclusive[15–18].

A new measurement of $A_{fb}$ at 281 MeV is in a preparatory stage at TRIUMF[19]. Using a broad range magnetic spectrometer set at zero degrees, the full angular distribution of the recoil deuterons will be measured in a single momentum setting, eliminating many systematic uncertainties. Remaining systematic uncertainties will be suppressed by measuring the response of deuterons with the same momentum and trajectory from $p\text{-}p \to d\text{-}\pi^+$, which must be symmetric about $90°$ in the centre-of-mass. It is intended to measure $A_{fb}$ with a precision of approximately $\pm 7 * 10^{-4}$.

## 2. The flavour conserving hadronic weak interaction: parity violation experiments

Because flavour changing neutral currents are almost completely suppressed by the G.I.M. mechanism, the study of hadronic neutral currents in nuclear systems provides a unique window of hadronic weak neutral currents. Parity violation in nuclear systems is the only flavour conserving process in which hadronic weak neutral currents can be observed. At low and intermediate energies, the parity violating weak $N\text{-}N$ interaction is described in terms of a meson exchange model involving a strong interaction vertex, and a weak interaction vertex. The strong interaction vertex is thought to be understood and is represented by the conventional parameterization of the $N\text{-}N$ interaction. The weak interaction vertex is calculated from the Weinberg-Salam model assuming the $W$- and $Z$-bosons are exchanged between the intermediate mesons ($\pi$, $\rho$, and $\omega$) and constituent quarks of the nucleon. The interaction can then be described in terms of six weak meson-nucleon coupling constants. The exchange of neutral scalar mesons is suppressed by CP conservation by a factor of a few times $10^{-3}$. The six weak meson-nucleon coupling constants $(f_\pi^1, h_\rho^0, h_\rho^1, h_\rho^2, h_\omega^0, h_\omega^1$, with the superscripts indicating isospin changes) have been calculated by Desplanques, Donoghue, and Holstein (DDH)[20] synthesizing the quark model and SU(6) and treating strong interaction effects in renormalization group theory. These authors tabulated "best guess values" and "reasonable ranges". Similar calculations have been made more recently by Dubovik and Zenkin (DZ)[21]. Extending the earlier work in the nucleon sector, Feldman, Crawford, Dubach and Holstein (FCDH)[22] included the weak $\Delta$-nucleon-meson and $\Delta$-$\Delta$-meson parity violating vertices for $\pi, \rho$, and $\omega$ mesons. The latter authors also present both theoretical "best guess" values and a theoretical "reasonable range" for the weak meson-nucleon coupling constants. Using the expressions of an earlier paper by Desplanques (D)[23] the same authors present a third set of weak meson-nucleon coupling constants. It is apparent that these coupling constants carry considerable ranges of uncertainty (see Table 1). A complete determination of the six weak meson-nucleon coupling constants requires at least six linearly independent pieces of experimental information. As of to date essentially only four experimental constraints of significance exist. As a consequence one needs several new precision parity violation measurements.

Measurements of parity violation have dealt with nuclei where nuclear matrix elements amplify the parity violating part of the hadronic interaction, as in measurements of the circular polarization of gamma rays emitted from parity mixed doublets of nuclei in the $2s$-$1d$ shell[24] and in measurements of the helicity dependence of the scattering cross section or capture cross section of epithermal neutrons on heavy nuclei, like $^{238}$U and $^{232}$Th (Ref.[24]). For these measurements there is a dynamical enhancement factor consisting of the product of the reciprocal of the energy difference of the parity mixed states and the ratio of the particle decay widths of the states in question. The total enhancement factor may be as large as $10^5$ giving parity violating effects in the order of a percent or larger. Other measurements have dealt with the circular polarization of $\gamma$ rays in $n$-$p$ capture[26] or have dealt with the angle-averaged

Table 1. Weak meson-nucleon couplings constants

| Coupling | Theoretical | | | | | | Experimental | |
| --- | --- | --- | --- | --- | --- | --- | --- | --- |
| | range ×10⁷ (DDH) | 'best value' value ×10⁷ (DDH) | value ×10⁷ (DZ) | range ×10⁷ (FCDH) | 'best value' ×10⁷ (FCDH) | value ×10⁷ (D) | best fit ×10⁷ | range ×10⁷ |
| $f_\pi^1$ | 0 → 11.4 | 4.6 | 1.3 | 0 → 6.5 | 2.7 | 2.7 | 4.3 | 0.3 → 8.4 |
| $h_\rho^0$ | -31 → 11.4 | -11.4 | -8.3 | -31 → 11 | -3.8 | -6.1 | -3.9 | -6.4 → -1.3 |
| $h_\rho^1$ | -0.38 → 0 | -0.19 | 0.39 | -1.1 → 0.4 | -0.4 | -0.4 | -0.15 | -0.21 → -0.10 |
| $h_\rho^2$ | -11.0 → -7.6 | -9.5 | -6.7 | -9.5 → -6.1 | -6.8 | -6.8 | -7.3 | |
| $h_\omega^0$ | -10.3 → 5.7 | -1.9 | -3.9 | -10.6 → 2.7 | -4.9 | -6.5 | -5.9 | -6.7 → -5.1 |
| $h_\omega^1$ | -1.9 → -0.8 | -1.1 | -2.2 | -3.8 → -1.1 | -2.3 | -2.3 | -2.2 | -2.3 → -2.1 |

longitudinal analyzing power $A_z$ in $p$-$^4$He [27], in $p$-$d$ [28], or in $p$-$p$ scattering [29]. Here $A_z$ is defined as $A_z = (\sigma^+ - \sigma^-)/(\sigma^+ + \sigma^-)$, where $\sigma^+$ and $\sigma^-$ represent the scattering cross sections for polarized incident protons of positive and negative helicity, respectively, integrated over some appropriate angle range. A non-zero value of $A_z$ implies parity violation due to the non-zero pseudoscalar observable $\vec{\sigma} \cdot \vec{p}$, with $\vec{\sigma}$ the spin and $\vec{p}$ the momentum of the incident particle. It is to be noted that the measurement of $A_z$ in $p$-$p$ scattering is sensitive only to the short-range part of the parity violating interaction. In practice, it is almost impossible to measure the angular dependence of the longitudinal analyzing power $A_z(\theta)$ and consequently the measurements have to resort to an angle-averaged longitudinal analyzing power $A_z$. This can be accomplished through a scattering measurement, restricted to the lower energies, or a transmission (attenuation) measurement.

Impressively precise measurements of $A_z$ in $p$-$p$ scattering have been made at 13.6 MeV [$A_z = (-0.93\pm0.20\pm0.05)\times10^{-7}$] at the University of Bonn[30] and 45 MeV [$A_z = (-1.50\pm0.22)\times10^{-7}$], at the Paul Scherrer Institute (PSI) [31]. These measurements involve essentially only the first parity violating transition amplitude ($^1S_0$–$^3P_0$) in a partial wave decomposition of $A_z$. Note that from the PSI measurement at 45 MeV and the $\sqrt{E}$ energy dependence of $A_z$ at low energies one can extrapolate $A_z$ at 13.6 MeV to be $A_z = (-0.8\pm0.1)\times10^{-7}$. Both results allow pinning down the combination of effective $\rho$ and $\omega$ weak meson-nucleon coupling constants $h_\rho^{pp}$ and $h_\omega^{pp}$ with $h_\rho^{pp} = h_\rho^0 + h_\rho^1 + h_\rho^2/\sqrt{6}$ and $h_\omega^{pp} = h_\omega^0 + h_\omega^1$. Following the approach of Adelberger and Haxton[32], the most significant nuclear parity violation data were used in fitting a two-parameter expression based on the formalism of DDH. It is apparent that the precise measurements of $A_z$ in $p$-$p$ scattering have helped in constraining the range of values for $h_\rho$ and $h_\omega$ (see table 1). However the experimental value for $f_\pi^1 = \left[\left(0.28^{+0.89}_{-0.28}\right)*10^{-7}\right]$ is still at the border of the deduced range[24]. The energy dependence of the first two partial wave contributions to $A_z$ corresponding to the transition amplitudes ($^1S_0$–$^3P_0$) and ($^3P_2$–$^1D_2$) are shown in Fig. 3 (Ref.[33]). In the low energy region there is good agreement between these theoretical predictions and experiment. There exists one measurement of $A_z$ in $p$-$p$ scattering at the higher energy of 800 MeV [$A_z = (2.4\pm1.1)*10^{-7}$] (Ref. [34]).

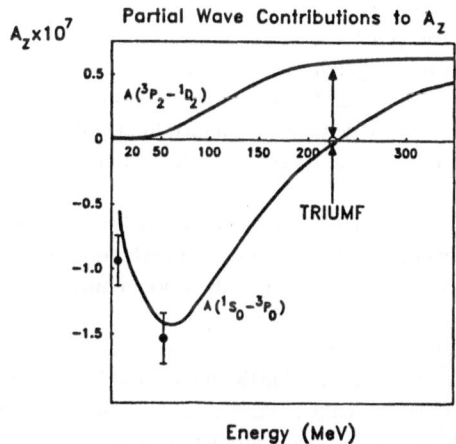

Fig. 3. Partial wave contributions to $A_z$ in $\vec{p}$—$p$ scattering as a function of incident energy.

As shown in Fig. 3 a unique feature is present at 230 MeV: The ($^1S_0$–$^3P_0$) partial wave contribution to $A_z$ integrates to zero. This reflects the direct cancellation of $^1S_0$ and $^3P_0$ strong interaction phases and is completely independent of the weak meson-nucleon coupling constants. Thus, with the exception of a small contribution ($\sim 5\%$) from the ($^1D_2$–$^3F_2$) transition amplitude, a measurement of $A_z$ at this energy constitutes a measurement of the ($^3P_2$–$^1D_2$) partial wave contribution alone. It has been shown by Simonius [35] that the ($^3P_2$–$^1D_2$) partial wave contribution depends only weakly on $\omega$-exchange whereas $\rho$-exchange and $\omega$-exchange contribute to the ($^1S_0$–$^3P_0$) transition amplitude with equal weight. Consequently, a measurement of $A_z$ at 230 MeV constitutes a determination of $h_\rho^{pp}$. The energy dependence of the real parts of the $p$-$p$ phase shifts predicts that, neglecting inelasticity, the ($^1D_2$–$^3F_2$) transition amplitude changes sign at about 650 MeV, and that the ($^3P_2$–$^1D_2$) transition amplitude changes sign at about 950 MeV. Consequently a precision measurement at an energy of 650 MeV as well as at 230 MeV would then provide determinations of both $h_\rho^{pp}$ and $h_\omega^{pp}$ and would be complementary to the low energy measurements of $A_z$ in $p$-$p$ scattering.

Various theoretical predictions of the longitudinal analyzing power $A_z$ have been reported; at 230 MeV the values for $A_z$ predicted are $+0.6 \times 10^{-7}$ (Ref. [33]), $+0.7 \times 10^{-7}$ (Ref. [36]), and $+0.6 \times 10^{-7}$ (Ref. [37]). Extensions of the single meson exchange model have been made to include $2\pi$ and $\rho a$ exchange via $N - \Delta$ intermediate states to which the $(^3P_2 - ^1D_2)$ transition amplitude is particularly sensitive. For example, Iqbal and Niskanen [36] find that the $\Delta$ isobar contribution at 230 MeV may be as large as the single $\rho$ exchange contribution, enhancing the value of $A_z$ by a factor of two. This indicates that an accuracy of $\pm 2 \times 10^{-8}$ in $A_z$ at 230 MeV would provide a significant determination of parity violation in $p$-$p$ scattering. Such an experiment is in an advanced state of preparation at TRIUMF [38].

In the TRIUMF experiment a beam of 500 nA with a polarization of $\sim 0.80$, extracted from the optically pumped polarized ion source, is accelerated through the cyclotron to an energy of 230 MeV. A combination of solenoid-dipole-solenoid-dipole magnets in the beam line provides a longitudinally polarized beam with positive or negative helicity. Beam transport is arranged in such a manner that an achromatic waist is produced at the detection apparatus with the horizontal and vertical beam parameters decoupled from each other. The quantity $A_z$ then follows from the helicity dependence of the $p$-$p$ total cross section as determined in precise measurements of the transmission through or the scattering from a 0.40 m long $LH_2$ target: $A_z = (-1/P)(T/S)(T^+ - T^-)/(T^+ + T^-)$, and $A_z = (1/P)(S^+ - S^-)/(S^+ + S^-)$, respectively, where $P$ is the incident beam longitudinal polarization, $T = 1 - S$ is the average transmission through the target, and the $+$ and $-$ signs indicate the helicity state.

There are many other effects which mimic such a helicity correlated change in transmission or scattering. Very strict constraints are imposed on the incident longitudinally polarized beam in terms of position, direction, emittance, intensity, polarization and energy together with deviations of the detection apparatus from cylindrical symmetry. Systematic errors, arising from the imperfections are individually not to exceed one tenth of the total error on $A_z$. Particularly troublesome are residual transverse polarizations. Of these, the so-called "circulating" or first moments of transverse polarization ($< xP_y >$ and $< yP_x >$) are predicted from simulations to give systematic errors of $A_z$ of about $4 \times 10^{-5}$ mm$^{-1}$. In addition to the strict constraints on incident beam and detection apparatus, required to minimize systematic errors, the approach which will be followed is to further measure the sensitivity to residual errors, to monitor these errors and to make corrections as necessary.

All the required diagnostic and measurement apparatae have been designed, constructed and tested in a large number of engineering runs on several beam lines. The only exception is the $LH_2$ target which is still under construction. The beam line dedicated to the parity violation experiment will be completed this fall so that control measurements and data taking can commence in 1994.

Figure 4 provides a diagram of the experimental setup (transmission measurement). The beam, incident from the lower right, passes first a series of diagnostic devices - a pair of beam intensity profile monitors (IPM's), which also determine the beam current median, and a pair of transverse polarization profile monitors (PPM's) - before reaching the $LH_2$ target which is preceded and followed by transverse electric

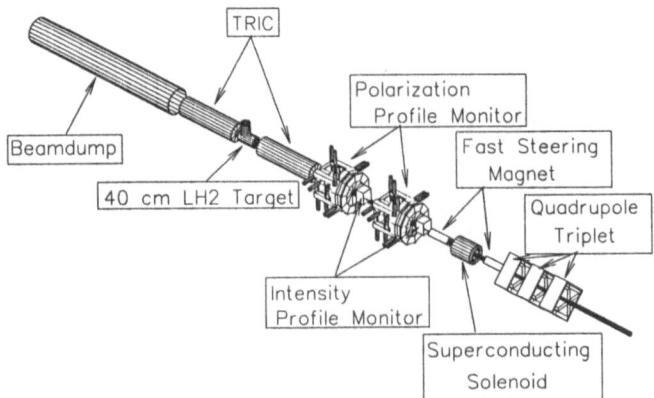

Fig. 4. Three dimensional view of the parity violation experiment at 230 MeV at TRIUMF.

field ionization chambers (TRIC's) which measure the beam current. Fast steering, ferrite core, magnets acting through a feed back loop keep the beam current median fixed to ±10 μm at the target. The superconducting solenoid between the fast steering magnets is for control measurements only and is normally turned off. The incident beam is stopped in a well shielded beam dump. Table 2 presents the requirements on beam and detector properties and what has been achieved so far. Clearly the parity apparatus is meeting the design specifications.

Further measurements of $A_z$ in $\vec{p}$-$p$ scattering are being planned at COSY [39]. The incident proton energies are 230 MeV, for reasons outlined above, and 1.5 GeV. The choice of the latter energy or even higher energy available at COSY ($< 2.5$ GeV)

Table 2. Beam property and detector requirements

| Beam property | Nominal dc value | Random fluct. | Helicity corr. | Detector | Required error (1 sec) | Achieved error (1 sec) |
|---|---|---|---|---|---|---|
| I | 500 nA | $\frac{\delta I}{I} = 0.01$ | $\frac{\delta I}{I} = 10^{-5}$ | TRIC | $\frac{\delta I}{I} = 3 \times 10^{-7}$† | $0.4 \le \alpha \le 1.0^*$ |
| E | 230 MeV | 42 keV | 5 eV | BEM | | 27 keV($\sigma$) |
| $x, y$ | 0 | 0.1 mm | 0.001 mm | IPM | 1 mm | 0.1 mm |
| $\sigma_x, \sigma_y$ | 5 mm | 0.1 mm | 0.001 mm | IPM | 0.01 mm | 0.01 mm |
| $P_{trans}$ | 0 | 0.01 | 0.001 | PPM | 0.1 | 0.02 |
| $< xP_y >$, $< yP_x >$ | 0 | 0.1 mm | $10^{-4}$ mm | PPM | 0.1 mm | 0.1 mm |

*($\alpha$ is a function of beam current, gas pressure, high voltage, collector plate configuration and digitizing frequency).
†(for $\alpha = 0.5$)

is in part motivated by the earlier 5.13 GeV measurement of $A_z$ (on a water target) at the ZGS which resulted in $A_z = (26.5 \pm 6.0) * 10^{-7}$ (Ref. [40]). The theoretical prediction shown in Fig. 5 is due to Goldman and Preston [42]. It is based on a diquark model giving a parity violating component of the nucleon wave function. The theoretical interpretation of the unexpectedly large result at 5.13 GeV has created a great deal of controversy [43] and has been criticised. It was followed more recently with a rebuttal by one of the original authors[44]. Clearly, the 5.13 GeV result presents a great challenge both in terms of experimental confirmation through a new measurement (at an energy of tens of GeV) and theoretical explanation. Such a measurement may be performed

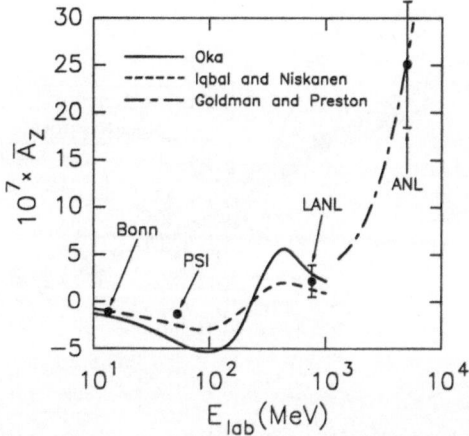

Fig. 5. Energy dependence of $A_z$. Various theoretical curves are compared to the data. The predictions of Iqbal and Niskanen (Ref. [36]) and Oka (Ref. [41]) are based on the meson exchange model, the one of Goldman and Preston (Ref. [42]) on a diquark model.

at an advanced hadron facility with its intense longitudinally polarized proton beam ($I \geq 1\,\mu A$, $P \approx 0.7$) [45].

## 3. CP

CP non-conservation has been established exclusively through measurements of the parameter $\epsilon$ in $K^0$ decays. Although CP non-conservation can be accommodated within the framework of the standard model, its origin and magnitude remain obscure. One of the more important questions is whether CP non-conservation is unique to the neutral kaon system or whether CP non-conservation is also manifested by other particles as predicted by the standard model, e.g., in the decays of hyperons and neutral $B$-mesons. Another important question is whether CP non-conservation occurs only in $|\Delta S| = 2$ weak decays, as formulated by the superweak model of Wolfenstein[46], or is also demonstrated in direct $|\Delta S| = 1$ transitions, as predicted by the standard model. However, the observation of direct CP non-conservation remains elusive, stemming from the discrepancy between the latest CERN (NA31) and FNAL (E731) results on $\epsilon'/\epsilon$: $\mathcal{R}e(\epsilon'/\epsilon) = (2.3 \pm 0.7) \times 10^{-3}$ (Ref. [47]) and $\mathcal{R}e(\epsilon'/\epsilon) = (7.4 \pm 5.2 \pm 2.9) \times 10^{-4}$ (Ref. [48]), respectively. Consequently, any non-zero result from any other system sensitive to direct CP non-conservation would be very important. Limits on the magnitude of CP non-conservation are less useful, especially in nuclear systems where there are uncertainties in interpretation. There exists no direct evidence for time-reversal non-invariance but it is strongly implied by CPT invariance together with CP non-conservation.

Straight forward in interpretation are comparisons of decay asymmetries in hyperon non-leptonic decays, as in $\overline{\Lambda} \to \overline{p}\pi^+$ and $\Lambda \to p\pi^-$, from the analysis of $J/\psi \to \overline{\Lambda}\Lambda$ decays and from the exclusive process $\overline{p}p \to \overline{\Lambda}\Lambda$. For the reaction $\overline{p}p \to \overline{\Lambda}\Lambda$ and also $\overline{p}p \to \overline{\Xi}^+\Xi^-$ four quantities may be considered for which non-zero values would indicate CP non-conservation:

$$\Delta = (\Gamma - \overline{\Gamma})/(\Gamma + \overline{\Gamma}),$$
$$A = (\alpha + \overline{\alpha})/(\alpha - \overline{\alpha}),$$
$$B = (\beta + \overline{\beta})/(\beta - \overline{\beta}),$$
$$B' = (\beta + \overline{\beta})/(\alpha - \overline{\alpha}),$$

where $\overline{\Gamma}$, $\overline{\alpha}$, and $\overline{\beta}$ refer to the antihyperon. In these ratios, $\Gamma$, $\alpha$, and $\beta$ are the partial decay width and the decay parameters for a given non-leptonic decay mode of the hyperon. The latter are given in Table 3 [49].

Table 3: $\Xi^-$ and $\Lambda^0$ hyperon decay parameters.

| Mode | $\alpha$ | $\beta$ |
|---|---|---|
| $\Xi^- \to \Lambda^0\pi^-$ | -0.456 ± 0.014 | 0.062 ± 0.062 |
| $\Lambda^0 \to p\pi^-$ | 0.642 ± 0.013 | -0.087 ± 0.047 |

The ratios can be estimated to be related by roughly $0.1B \approx B' \approx 10A \approx 100\Delta$. Note that the source of $\Delta$ is an interference between $|\Delta I| = \frac{1}{2}$ and $\frac{3}{2}$ transitions, whereas the quantities A, B, and B' are stemming chiefly from an interference between $S$ and $P$-wave amplitudes in the $|\Delta I| = \frac{1}{2}$ transitions. Estimates of $A(\Lambda^0)$ are of the order $10^{-4}$ to $10^{-5}$ [50].

From an experimental point of view it is most promising to measure the ratio $A(\Lambda^0)$ which reflects the difference in the $\Lambda$ and $\overline{\Lambda}$ decay asymmetry parameters. The ratio $B'(\Lambda^0)$ has a greater sensitivity to a CP non-conserving effect. It depends on a measurement of the difference on the polarizations of the $\Lambda$ and $\overline{\Lambda}$ decay products with the decay product (proton) polarization given by

$$\overline{P}_p = \frac{(\alpha + \vec{P}_\Lambda \cdot \hat{p}_p)\hat{p}_p + \beta(\vec{P}_\Lambda \times \hat{p}_p) + \sqrt{1 - \alpha^2 - \beta^2}[\hat{p}_p \times (\vec{P}_\Lambda \times \hat{p}_p)]}{(1 + \alpha \vec{P}_\Lambda \cdot \hat{p}_p)} \qquad (7)$$

where $\vec{P}_\Lambda$ is the $\Lambda$ hyperon polarization and $\hat{p}_p$ is the proton momentum direction in the rest frame of the $\Lambda$. Such a measurement requires knowledge of analyzing powers to a level which cannot be obtained at present (certainly not for antiprotons). The asymmetry parameter

$$A = (\alpha P_\Lambda + \overline{\alpha} P_{\overline{\Lambda}})/(\alpha P_\Lambda - \overline{\alpha} P_{\overline{\Lambda}})$$

has been determined at CERN-LEAR (PS185), $A = -0.022 \pm 0.019$ [51]. It represents the average of various incident $\overline{p}$ momenta. It might be possible with improvements in the incident $\overline{p}$ beam and in the detection system to improve on the above results. It is to be noted that if $A(\Lambda^0) \approx 10^{-4}$ is to be measured with 3 $\sigma$ accuracy, about $5 * 10^9$ events would be needed.

For a measurement of the ratio $B'$, the exclusive production process $\overline{p}p \to \overline{\Xi}^+\Xi^- \to \overline{\Lambda}\pi^+\Lambda\pi^-$ presents a promising, unique case. It would allow for measuring the hyperon decay products polarization differences in addition to the decay products asymmetries. The exclusive production process $\overline{p}p \to \overline{\Xi}^+\Xi^-$ cannot be reached at CERN-LEAR. With very little prospect for constructing a SuperLEAR facility at CERN, the study of the exclusive production process $\overline{p}p \to \overline{\Xi}^+\Xi^-$, to search for $\Delta S = 1$ direct CP non-conservation, awaits the realization of an advanced hadron facility.

Another possibility is to produce secondary beams of $\overline{\Xi}^+$ and $\Xi^-$ hyperons and to study their respective decays (FNAL(P871)) [52]. In a narrow forward cone the $\overline{\Xi}^+$ and $\Xi^-$ hyperons will be unpolarized and consequently their decay products $\overline{\Lambda}$ and $\Lambda$ will be in a helicity state with polarization given by $\overline{\alpha}_{\overline{\Xi}}$ and $\alpha_\Xi$, respectively. One can then determine the decay products asymmetry parameter

$$A = \left(\alpha_\Lambda \alpha_\Xi - \overline{\alpha}_{\overline{\Lambda}}\overline{\alpha}_{\overline{\Xi}}\right)\Big/\left(\alpha_\Lambda \alpha_\Xi + \overline{\alpha}_{\overline{\Lambda}}\overline{\alpha}_{\overline{\Xi}}\right).$$

If $4 \times 10^9$ $\Xi^-$ and $\overline{\Xi}^+$ decays can be collected in fixed target runs, a measurement of the decay parameter asymmetry $A$ to a sensitivity of $10^{-4}$ may be reached.

## 4. Summary

The intermediate energy physics facilities have made it possible to measure the violation of symmetry relations and conservation laws to very high precision, probing in many cases previously inaccessible features of strongly interacting particles. Spin is an important commodity in studying such violations in $N - N$ and $\overline{N} - N$ systems. These studies have given a perspective of a broad range of future precision experiments which strongly depend on the realization of an advanced hadron facility.

## References

[1] Gabioud, B. et al., Phys. Rev. Lett. **42**, 1508 (1979); Nucl. Phys. **A420**, 496 (1984).

[2] Schori, O. et al., Phys. Rev. **C35**, 2252 (1987).

[3] Miller, G.A., Nefkens, B.M.K. and Slaus, I., Phys. Rep. **194**, 1 (1990).

[4] TRIUMF Experimental Proposal (E661), spokesperson Kovash, M.A.

[5] Henley, E.M. and Miller, G.A., Mesons in Nuclei, ed. M. Rho and D.H. Wilkinson (North Holland, Amsterdam, 1979) p. 405.

[6] LaFrance, P. and Winternitz, P., J. Physique, **41**, 1391 (1980).

[7] Abegg, R. et al., Phys. Rev. **D39**, 2464 (1989).

[8] Vigdor, S.E. et al., Phys. Rev. **C46**, 410 (1992).

[9] Holzenkamp, B., Holinde, K. and Thomas, A.W., Phys. Lett. **195B**, 121 (1987).

[10] Beyer, M. and Williams, A.G., Phys. Rev. **C38**, 779 (1988).

[11] Goldman, T., Henderson, J.A. and Thomas, A.W., Few Body Systems **12**, 123 (1992); Hatsuda, T., Henley, E.M., Meissner, T. and Krein, G., (private communication).

[12] TRIUMF Experimental Proposal (E369), spokespersons Greeniaus, L.G. and van Oers, W.T.H.

[13] Cheung, C.Y., Henley, E.M. and Miller,G.A., Phys. Rev. Lett. **43**, 1215 (1979); Nucl. Phys. **A348**, 365 (1980).

[14] Niskanen, J.A., Sebestyen, M. and Thomas, A.W., Phys. Rev. **C38**, 838 (1988); Niskanen, J.A., (private communication).

[15] Hollas, C.L. et al., Phys. Rev. **C24**, 1561 (1981).

[16] Wilson, S.S. et al., Nucl. Phys. **B33**, 253 (1971).

[17] Bartlett, D.F. et al., Phys. Rev. **D1**, 1984 (1970).

[18] Hutcheon, D.A. et al., Nucl. Phys. **A535**, 618 (1991).

[19] TRIUMF Experimental Proposal (E704), spokespersons Korkmaz, E. and Opper, A.K.

[20] Desplanques, B., Donoghue, J.F. and Holstein, B.R., Ann. Phys. (N.Y.) **124**, 449 (1980).

[21] Dubovik, V.M. and Zenkin, S.V., Ann. Phys. (N.Y.) **172**, 100 (1986).

[22] Feldman, G.B., Crawford, G.A., Dubach, J. and Holstein, B.R., Phys. Rev. **C43**, 863 (1991).

[23] Desplanques, B., Nucl. Phys. **A335**, 147 (1980).

[24] Evans, H.C. et al., Phys. Rev. Lett. **55**, 791 (1985); Page, S.A. et al., Phys. Rev., **C35**, 1119 (1987); Bini, M. et al., Phys. Rev. Lett. **55**, 795 (1985); and references therein.

[25] Zhu, X. et al., Phys. Rev. **C46**, 768 (1992); Frankle, C.M., et al., Phys. Rev. **C46**, 778 (1992).

[26] Knyazkov, V.A. *et al.*, Nucl. Phys. **A417**, 209 (1984).

[27] Lang, J. *et al.*, Phys. Rev. **C34**, 1545 (1986).

[28] Nagle, D.E. *et al.*, *High energy physics with polarized beams and targets*, ed. Thomas, G.H., AIP Conference Proceedings No. 51 (AIP, New York, 1979) 224; Kistryn, S. *et al.*, *High energy spin physics*, 1988, ed. Heller, K.J., AIP Conference Proceedings, No. 187, (AIP, New York, 1989), 468; Mischke, R.E. *ibid*, 463.

[29] Simonius, M., Phys. Lett. **41B**, 415 (1972).

[30] Eversheim, P.D. *et al.*, Phys. Lett. **B256**, 11 (1991); Eversheim, P.D., (private communication).

[31] Kistryn, S. *et al.*, Phys. Rev. Lett. **58**, 1616 (1987).

[32] Adelberger, E.G. and Haxton, W.C., Ann. Rev. Nucl. Part. Sci. **35**, 510 (1985); Hamian, A.A., (private communication).

[33] Simonius, M., *Proc. of the symposium/workshop on spin and symmetries*, ed. Ramsay, W.D. and van Oers, W.T.H., Can. J. Phys. **66**, 245 (1988).

[34] Yuan, V. *et al.*, Phys. Rev. Lett. **57**, 1680 (1986).

[35] Simonius, M., *Interaction studies in nuclei*, ed. Jochim, H. and Ziegler, B. (North Holland, Amsterdam, 1975) 3; *High energy physics with polarized beams and polarized targets*, ed. Joseph, C. and Soffer, J. (Birkhauser, Basel, (1981)) 355.

[36] Iqbal, M.J. and Niskanen, J.A., Phys. Rev. **C42**, 1872 (1990); ibid; **C45**, 2648(1992); **C45**, 3021(1992).

[37] Driscoll, D.E. and Miller, G.A., Phys. Rev. **C39**, 1951 (1989).

[38] TRIUMF Experimental Proposal (Expt. 497), Spokespersons Birchall, J., Page, S.A. and van Oers, W.T.H.

[39] Eversheim, P.D., Hinterberger, F., Paetz gen Schieck, H. and Kretschmer, W., *High energy spin physics*, ed. Althoff, K-H. and Meyer, W. (Springer Verlag, Berlin, (1991)) 573.

[40] Lockyer, N. *et al.*, Phys. Rev. **D30**, 860 (1984).

[41] Oka, T., Prog. Theor. Phys. **66**, 977 (1981).

[42] Goldman, T. and Preston, D., Phys. Lett. **168B**, 415 (1986).

[43] Simonius, M. and Unger, L., Phys. Lett. **B198**, 547 (1987).

[44] Goldman, T., *Future Directions in Particle and Nuclear Physics at Multi-GeV Hadron Beam Facilities*, ed. Geesaman, D.F. (in press).

[45] van Oers, W.T.H., *High energy spin physics*, ed. Althoff, K-H. and Meyer, W. (Springer Verlag, Berlin, (1991)) 335.

[46] Wolfenstein, L., Phys. Rev. Lett. **13**, 562 (1964).

[47] Burkhardt, H. *et al.*, Phys. Lett. **B206**, 169 (1988); Perdereau, O., *'92 Electroweak Interactions and Unified Theories*, ed. Tran Thanh Van, J., (Editions Frontieres, Gif-sur-Yvette, 1992)365.

[48] Gibbons, L.K. *et al.*, Phys. Rev. Lett. **70**, 1203 (1993).

[49] Review of Particle Properties, Phys. Rev. **D45**, 1 (1992).

[50] He, X.-G., Steger, H. and Valencia, G., Phys. Lett. **B272**, 411 (1991).

[51] Kilian, K.K., this conference.

[52] FNAL Experimental Proposal P871, spokesperson Luk, K.B.

Few-Body Systems, Suppl. 7, 64—67 (1994)

Few-
Body
Systems
© Springer-Verlag 1994
Printed in Austria

# STATUS OF THE VIRGINIA TECH PARTIAL-WAVE ANALYSES

Richard A. Arndt and Ron L. Workman

Department of Physics
Virginia Polytechnic Institute and State University
Blacksburg, Virginia 24061

### Abstract

We review the status of partial-wave analyses performed by the Virginia Tech group. The pion-induced reactions, specifically $\pi N \to \pi N$ and $\pi d \to pp$, are described in more detail. We discuss the effect of dispersion relation constraints in our $\pi N$ analyses. The appearance of loops in the $\pi d \to pp$ Argand diagram is also indicated.

## I. Introduction

A number of medium-energy scattering processes have been analyzed by the Virginia Tech group. Our results and databases are available on the SAID system, which is an interactive graphics program available to the nuclear/particle physics community[1]. The SAID program currently contains the reactions: $\pi N \to \pi N$, $NN \to NN$, $K^+N \to K^+N$, $\gamma N \to \pi N$, and $\pi d \to pp$. The last reaction[2] was added this summer and further additions are planned. Databases are currently being assembled for the reactions $\pi d \to \pi d$ and $eN \to e'N\pi$.

The above analyses are regularly updated as new data are added. Users who 'dial-in' to the system can compare a number of partial-wave solutions against the current databases. We are presently updating the $\pi N$ and $NN$ solutions. The $np$ elastic scattering database has recently been augmented by a series of SATURN measurements[3]. These have been included in the most recent $NN$ analysis. At low energies, our database has been made compatible with the Nijmegen set[4].

Following a suggestion from the organizers, we will restrict our discussion to the processes: $\pi N \to \pi N$ and $\pi d \to pp$. The most recent set of $\pi N$ analyses[5] are interesting as they have been constrained via fixed-t and forward dispersion relations. We will compare these to previous constrained and unconstrained analyses. The $\pi d$ analysis is also interesting. The dominant partial-wave amplitudes produce correlated loops in an Argand plot. We compare with the previous Bugg[6] and Hiroshima[7] analyses.

## II. The Pion-Nucleon Partial-Wave Analyses

We have carried out a number of analyses using different subsets of dispersion relation (DR) constraints. Three main types of analysis were performed: (a) unconstrained, (b) constrained only via the Hüper fixed-$t$ DR, and (c) constrained through the Hüper and $C^{\pm}$ forward DR. The Hüper plot, which was used in our initial extraction of the $\pi NN$ coupling[8], has been used to constrain our analyses at several fixed values of $t$. From the relations for the $\pi^{\pm}p$ invariant amplitudes $B_{\pm}(\nu, t) = B^{+}(\nu, t) \mp B^{-}(\nu, t)$ the following dispersion relation can be constructed

$$(\nu_B \pm \nu)\left\{ \mp ReB_{\pm}(\nu, t) \pm \frac{\nu}{\pi}\int_{\nu_1}^{\infty}\left[\frac{ImB_+}{\nu' \mp \nu} + \frac{ImB_-}{\nu' \pm \nu}\right]\frac{d\nu'}{\nu'}\right\} = \frac{g^2}{M} + \tilde{B}(0, t)(\nu_B \pm \nu), \quad (1)$$

with

$$\tilde{B}(0, t) = \frac{2}{\pi}\int_{\nu_1}^{\infty}\frac{ImB^-(\nu', t)}{\nu'}d\nu' \qquad (2)$$

where $s$, $t$, and $u$ are the usual Mandelstam variables, $\nu = (s-u)/4M$, $\nu_B = (t-2\mu^2)/4M$ and $\nu_1 = \mu + t/4M$, $\mu$ and $M$ being the charged-pion and proton masses respectively. From the right-hand-side of Eq.(1), it is evident that this relation should be linear in $(\nu_B \pm \nu)$ with an intercept corresponding to $g^2/M$. Deviations from linearity were corrected iteratively through constraints on $ReB_{\pm}$.

The forward isospin-odd amplitude $C^-(\omega)$ was used to link the $a^{(-)}$ scattering length with $g^2/4\pi$. As the GMO integral[9] is slowly convergent, we chose to use the subtracted form

$$a^{(-)} = \frac{1}{4\pi}\frac{\mu}{\omega}\frac{M}{M + \mu}\left\{ReC^-(\omega) + \frac{\mu^2}{2M^2}\frac{g^2k^2\omega}{(\omega^2 - \omega_B^2)(\mu^2 - \omega_B^2)} - I^{(-)}(k)\right\}, \qquad (3)$$

with $k$ being the pion lab momentum, $\omega_B = \nu_B(t = 0)$, and

$$I^{(-)}(k) = k^2\omega\frac{2}{\pi}\int_0^{\infty}\frac{\sigma^-(k)}{k'^2 - k^2}\frac{dk'}{\omega'}, \qquad (4)$$

to constrain our solutions. This reduced our reliance on high energy contributions.

The forward $C^+$ DR was used in the following form

$$a^{(+)} = \frac{1}{4\pi}\frac{M}{M + \mu}\left\{ReC^+(\omega) + \frac{g^2}{M}\frac{k^2\omega_B^2}{(\omega_B^2 - \omega^2)(\mu^2 - \omega_B^2)} - I^{(+)}(k)\right\}, \qquad (5)$$

with

$$I^{(+)}(k) = \frac{2k^2}{\pi}\int_0^{\infty}\frac{\sigma^+(k')}{k'^2 - k^2}dk'. \qquad (6)$$

These expressions demonstrate that a constant value is expected when the right-hand-sides of Eqs. (3) and (5) are evaluated. This constant was used to constrain our analyses. Deviations were corrected through constraints imposed on $ReC^{\pm}$ in our fits.

In practice the above integrals were evaluated up to some $k_{max}$ ensuring sufficient accuracy for the analyses. In the case of the forward $C^+$ DR, two different high energy parameterizations were tested[10] for values of pion laboratory kinetic energy above 4 GeV. The results were found to be insensitive to this choice. The Hüper and subtracted

$C^-$ DR were evaluated up to a pion laboratory kinetic energy of 4 GeV, the contribution from 2-4 GeV coming from the Karlsruhe solution[11]. The $C^\pm$ DR constraints were imposed at 25 MeV intervals, from a $T_{Lab}$ of 25 MeV to 600 MeV. The Hüper-plot constraints covered different ranges of $T_{Lab}$ depending upon the value of $t$ - which extended from 0 to $-0.3$ GeV$^2$.

Analyses were performed with the value of $g^2/4\pi$ constrained to 13.0, 13.5, 14.0, and 14.5. The value of $a^{(-)}$ was determined from the GMO sum rule[9]. For each coupling choice, the value of $a^{(+)}$ was constrained to be either $-0.025$, $-0.050$, or $-0.075$ in inverse pion mass units. This resulted in a grid of 12 solutions. We generally found a $\pi NN$ coupling below the Karlsruhe value. This lower value was found when one examined the $\chi^2$ due to either the experimental data or the constraints. A simple quadratic fit was made to the 4 $\chi^2$ values spanning our range of choices for $g^2/4\pi$. An average value near 13.6 was found when we computed the $\chi^2$ from data below 2.1 GeV. Here the preferred value of $g^2/4\pi$ was fairly constant, independent of the charge channel or choice of $a^{(+)}$. Concentrating on the region from 100 to 600 MeV, we also found an average near 13.6. Over this region, however, there was a greater spread of values from the different charge channels.

As one might expect, the unconstrained fit had the lowest $\chi^2$/datum, and the overall $\chi^2$ increased as more constraints were added. In general we can say that the $C^-$ DR constraint was easily satisfied (generally to the 1% level) in all of our fits. The $C^+$ DR relation was much more difficult to satisfy precisely[12].

The generation of these fits, with various combinations of constraints, was a very lengthy procedure. These tests would not have been feasible without access to a number of fast UNIX workstations. In some cases, the convergence of our iterative procedure was very slow. This was particularly true for unfavorable combinations of the coupling constant and scattering lengths.

### III. Analysis of the Reaction: $\pi d \to pp$

We have recently completed[2] an energy-dependent analysis (along with a set of energy-independent analyses) of the reaction $\pi d \to pp$. These analyses extend from threshold to 500 MeV in the laboratory kinetic energy of the pion. The value of $\sqrt{s}$ varies from 2.015 GeV to 2.437 GeV and spans the range of energies typically associated with dibaryon candidates.

The present solution resulted in a $\chi^2$ of 7065 for 4440 data. The stability of the fit was tested by 'pruning' data more than 3 standard deviations from the solution. This resulted in the elimination of 135 data scattered throughout the database, with a consequent decrease in $\chi^2$ to 5297. Upon searching, the $\chi^2$ reduced to 5265. No significant change in the solution was found.

The $\pi d \to pp$ reaction is dominated by only a few partial-wave amplitudes - the $^1D_2P$, $^3F_3D$, $^3P_2D$, and $^3P_1S$. [ We use the notation: $^{2S+1}L_JL^\pi$ to connect the $\pi d \to pp$ partial-wave with the respective $pp$ state, $^{2S+1}L_J$. $L^\pi$ labels the $\pi d$ state. ] While different phase conventions have been used in previous analyses, we have rotated the recent analyses from

Bugg and the Hiroshima group to our phase convention for the purpose of comparison. Agreement with the solution of Bugg is quite reasonable. Comparison with the Hiroshima analysis generally reveals much larger discrepancies.

The dominant amplitudes show a striking 'resonancelike' behavior when plotted in an Argand diagram. The loops also show a correlated energy dependence. The mechanism generating these structures is not generally agreed upon (Bugg and Strakovsky hold disparate views regarding the influence of dibaryons in this reaction).

## Acknowledgments

We acknowledge the contributions of D.V.Bugg, J.S.Hyslop, Z.Li, M.W.McNaughton, M.M.Pavan, L.D.Roper, and I.I.Strakovsky in the analyses. This work was supported in part by a U.S. Department of Energy Grant No. DE-FG05-88ER40454.

# References

[1] The SAID databases and partial-wave solutions can be accessed via Telnet. Call vtinte.phys.vt.edu with the user/password: physics/quantum. Help is available from phys0 (Arndt) or workman @ vtvm1.bitnet.

[2] R.A. Arndt, I.I. Strakovsky, R.L. Workman and D.V. Bugg, (submitted).

[3] J. Ball et al., LNS/Ph/93-05 to 12; R. Binz, Ph.D. Univ. of Freiburg, (1991).

[4] V. Stoks, private communications.

[5] R.A. Arndt, R.L. Workman, and M.M. Pavan, (submitted).

[6] D.V. Bugg, A. Hasan, and R.L. Shypit, Nucl. Phys. A477 (1988) 546, and private communication.

[7] N. Hiroshige, W. Watari, and Y. Yonezawa, Prog. Theor. Phys. 72 (1984) 1146.

[8] R.A. Arndt, Z. Li, L.D. Roper, and R.L. Workman, Phys. Rev. Lett. 65, 157 (1990).

[9] M.L. Goldberger, H. Miyazawa, and R. Oehme, Phys. Rev. 99 (1955) 986.

[10] A. Donnachie and P.V. Landshoff, Phys. Lett. B296 (1992) 227. The older parametrization of Höhler was also used.

[11] R. Koch, private communications.

[12] A comparison with the solution of Ref.[11] reveals that the $C^+$ DR was more challenging in the Karlsruhe analysis as well.

Few-Body Systems, Suppl. 7, 68—79 (1994)

# PERIODIC ORBITS AND RECURRENCES:
## AN INTRODUCTION AND REVIEW

J. B. Delos
Physics Department
College of William and Mary
Williamsburg, VA 23187-8795

The study of periodic classical orbits of quantum systems is a branch of the field called "quantum chaos," which is the study of quantum-mechanical systems whose classical counterparts exhibit chaotic motion. Many examples have now been examined: a one-electron atom in magnetic field[1], an atom in an oscillating electric field[2], any molecule in a high vibrational state[3], a "quantum billiard," such as an electron in a stadium-shaped microjunction[4], and numerous model systems, including the Henon-Heiles oscillator[5], quantum maps[6], or geodesic motion of a particle on a surface of constant negative curvature[7]. Always the central issues are: How do we use information about classical orbits to generate information about quantum wave-functions? Can we interpret observations on a quantum system using classical or semiclassical mechanics? How does classically chaotic behavior manifest itself in quantum properties of a system?

We know that in a chaotic classical system, we can make detailed predictions only for short times, and we can only make statistical predictions for long times. For example, we can make reasonably reliable predictions of the weather a few days in advance, and we can easily predict the long-time average climate (rainy Winters in Amsterdam, beautiful Spring in Williamsburg).

The detailed short-time order in classical mechanics translates into detailed, orderly large-scale structure in quantum mechanics. The statistical order hiding in the long-time chaos of classical mechanics allows us to make statistical predictions about fluctuations of quantum properties. In this paper we will explain the meaning of these two statements, and give some examples.

In this article we will mainly consider systems with two degrees of freedom governed by a Hamiltonian H(p,q), and we will constantly jump back and forth between classical concepts (trajectories, phase space, periodic orbits) and quantum concepts (eigenfunctions, energy spectrum, density of states, etc.). Large-scale reviews of this subject have been given by Gutzwiller, Ozorio de Almeida, Haake, Reichl, and Nakamura[8].

## I.  Short Time Order and Large-Scale Structure

*"what makes these periodic solutions so valuable, is that they offer, in a manner of speaking, the only opening through which we might try to penetrate into the fortress which has the reputation of being impregnable."*

Poincaré 1892

In a chaotic classical system, the only significant structures are the whole volume of phase-space and the periodic orbits.[*] It follows that a theory which connects quantum behavior with classical behavior must focus on these two classical quantities.

The volume of classical phase-space is related to the quantum density-of-states. Each quantum state "occupies" a volume in classical phase-space equal to $(2\pi\hbar)^d$, where d is the number of degrees-of-freedom of the system. It follows that the number of quantum states having energy less than E, N(E), is proportional to the volume of phase space in which $H(p,q) < E$, $\Gamma(E)$:

$$N(E) = \Gamma(E)/(2\pi\hbar)^d = \int_{H(p,q)<E} dp\,dq/(2\pi\hbar)^d \tag{1a}$$

and the average density of states n(E) is the derivative of this quantity

$$n(E) = dN(E)/dE = [d\Gamma(E)/dE]/(2\pi\hbar)^d \tag{1b}$$

---

[*]"significant" means:  invariant under allowed transformations of viewpoint (canonical transformations), invariant under time-displacement, and humanly comprehensible.

The phase-space volume $\Gamma(E)$ is normally a smooth, slowly-varying function of energy, and eqs. (1) produce a smooth "background" density which might be called "the classical density of quantum states."

Superimposed on this smooth background, classical periodic orbits produce fluctuations -- large-scale-structure -- in the quantum density of states. Related fluctuations occur in the absorption spectrum, and indeed in all other quantum properties when they are measured as a function of the energy of the system. Periodic orbits also produce "scars" in quantum eigenfunctions.

A general framework makes use of the quantum Green-function $G_E(q,q')$ -- the wave that arrives at $q$ from a fixed-energy source at $q'$. A semiclassical approximation to the Green function is easy to calculate and easy to interpret. First we compute trajectories that go out from $q'$ in any direction but with fixed energy. The fireworks display in Fig. 1 shows paths of particles going out from a point and falling in a constant force field. Now we select a relevant point $q$ and identify the trajectories that go from $q'$ to $q$. In Fig. 1 there are two such trajectories (labelled by an index $k$). For each we compute the classical action

$$S(q,q') = \int p(t) \cdot dq/dt \, dt$$

on the path from $q'$ to $q$, the classical particle density $\rho(q,q')$ associated with steady flow of particles along the orbits, and the Maslov index $\mu_k$ of each identified orbit, which (in this case) counts the number of times the trajectory has touched a caustic (a boundary of allowed classical motion). The semiclassical approximation to the Green function is then given by

$$G_E(q,q') = \sum_k |\rho_k(q,q');E)|^{1/2} \exp i[S_k(q,q';E)/\hbar - \mu_k \pi/2] \qquad (2)$$

The sum is over the classical orbits that go from $q'$ to $q$.

If we hold $q$ and $q'$ fixed, and examine $G_E(q,q')$ as a function of energy, we find that it contains oscillations associated with interference among the paths that go from $q'$ to $q$: each term in Eq. (2) oscillates with a wavelength (on the energy axis) equal to

$$\lambda_E = 2\pi\hbar / \frac{\partial S_k(q,q';E)}{\partial E} = 2\pi\hbar / T_k(q,q';E) \qquad (3)$$

where $T_k(q,q';E)$ is the time required for a classical particle to travel from $q'$ to $q$ on the kth path at energy E. These oscillations manifest themselves in measurable properties of the system.

For example the density of states is related to the trace of the Green function

$$n(E) = \text{constants} \times \int G_E(q,q)dq \qquad (4)$$

Stationary phase evaluation of this integral converts the sum over all orbits from $q'$ to $q$ into a sum over periodic orbits only.

Atomic absorption spectra can be calculated from a formula

$$Df(E) = \text{constants} \times <d\phi_i \,|\, G_E \,|\, d\phi_i> \qquad (5)$$

where $\phi_i$ is the initial state of the atom, d is the relevant component of the dipole operator and Df(E) is the "absorption spectrum" expressed as an oscillator-strength-density. This formula is especially useful when the energy is close to the ionization threshold; in that case the initial state is very localized compared to the size of the classical orbits, and at the lowest level of approximation we can say that the matrix element in Eq. (5) is related to $G_E(0,0)$; oscillations arise from every classical orbit that goes out from and later returns to the atom.

Finally, the conductance of a microjunction is related to $G_E(q_1, q_2)$ where $q_1$ and $q_2$ are entrance an exit points to the junction, so oscillations arise from interference between paths that go from entrance to exit.

Let us now review some recent experiments.

1. Atoms in Magnetic Fields

As early as 1968, Garton and Tomkins had observed a large-scale structure in the absorption spectrum of an atom in a magnetic field. In the vicinity of the oscillation threshold, they saw regular oscillations in the absorption as a function of energy. Later, the Welge group measured this absorption with better resolution, and they saw that it consists of multiple oscillations[9]. In Fig. 2, we show on the left the absorption spectrum as a function of energy, and below it, its Fourier transform. In the absorption spectrum, we see seemingly-irregular and uninterpretable fluctuations. However, the Fourier transform consists of clear, well-separated peaks. This

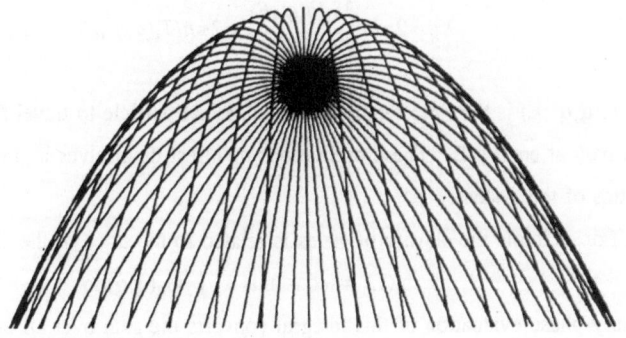

Fig. 1   Particles of fixed energy move out from a source and fall in a
constant force field.

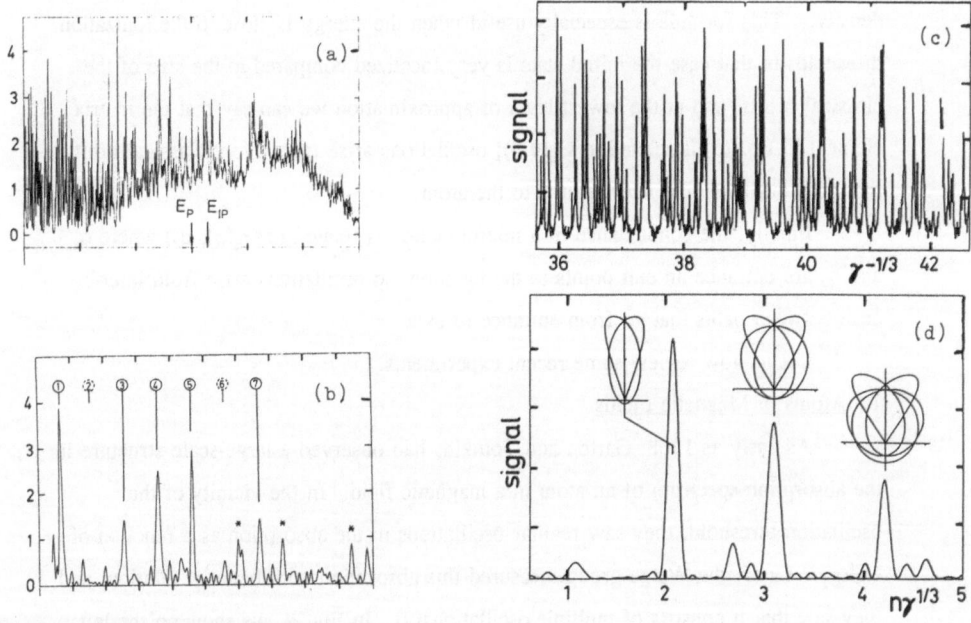

Fig. 2

Left:   Absorption spectrum of Hydrogen in a magnetic field of 6T and its
        Fourier transform.

Right:  Scaled-variable absorption spectrum of Hydrogen.   The peaks in the
        Fourier transform are better resolved.

shows that the absorption spectrum is in fact a superposition of sinusoidal oscillations of the form $\sum_k C_k \sin[(T_k E + \Delta_k)/\hbar]$ . Calculations show that the times $T_k$ are return-times of closed orbits of the electron that go out from and later return to the vicinity of the nucleus. Measurements, therefore, constitute observations of recurrences, with $T_k$ the recurrence time, and $C_k$ the recurrence amplitude, and $|C_k|^2$ the recurrence strength. We now call the Fourier-transform of the absorption spectrum the "Recurrence Spectrum."

More recently, comparable recurrences became visible in real time, using picosecond pulsed lasers. However, a different form of spectroscopy has provided an unprecedented level of detail and an unexpected new view of the structure of recurrences.

For a classical electron which feels the Coulomb force of the nucleus and the magnetic field, there is a scaling law, which asserts that the structure of the classical orbits depends not upon the energy and magnetic field separately, but only upon the scaled energy $\tilde{E} = EB^{-2/3}$. At fixed scaled energy any given returning orbit has a fixed shape, and it changes only its size if we vary B at fixed $\tilde{E}$. Furthermore, the size changes in such a way that the classical action (which represents the phase difference between the outgoing wave and the returning wave) varies as $B^{-1/3}$ at fixed $\tilde{E}$. Therefore if we measure absorption vs. $B^{-1/3}$, varying the laser energy and magnetic field simultaneously so that $\tilde{E}$ is fixed, the absorption spectrum will show perfectly regular sinusoidal oscillations, and the recurrence spectrum will be cleaner and better-resolved. This measurement was also carried out by the Bielefeld group, and is shown in Fig. 2c-d: we call it the recurrence spectrum at fixed scaled-energy.

In Fig. 3, recurrence spectra at many different scaled energies are drawn in a single picture. Again each peak is a recurrence corresponding to one or a few returning orbits. The picture, in effect, tells the classical action and the recurrence strength of various returning orbits as a function of scaled energy.

The most striking phenomenon that is visible in this picture is the proliferation of recurrences with increasing scaled energy. We see individual peaks at low energy develop into extensive mountain ranges at high energy, and we see also isolated peaks appearing in the midst of a flat plain.

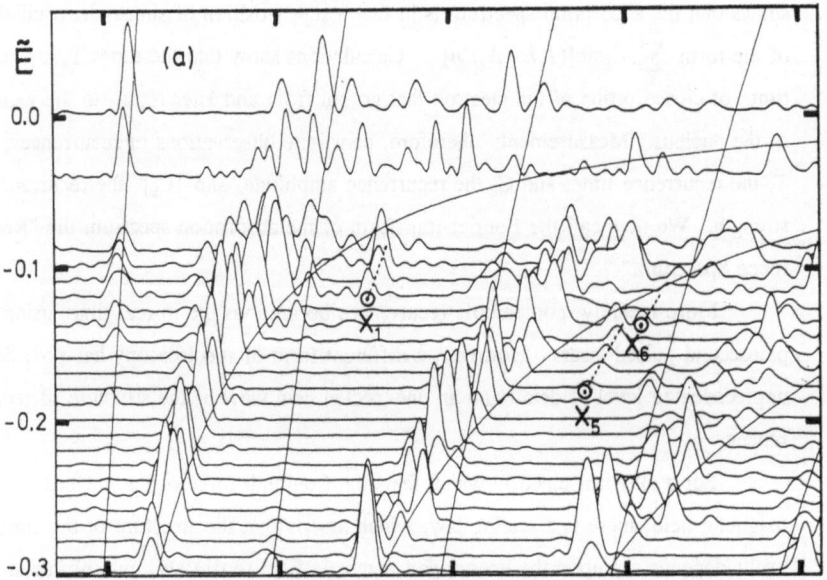

Fig. 3    Proliferation of Recurrences with increasing scaled energy.

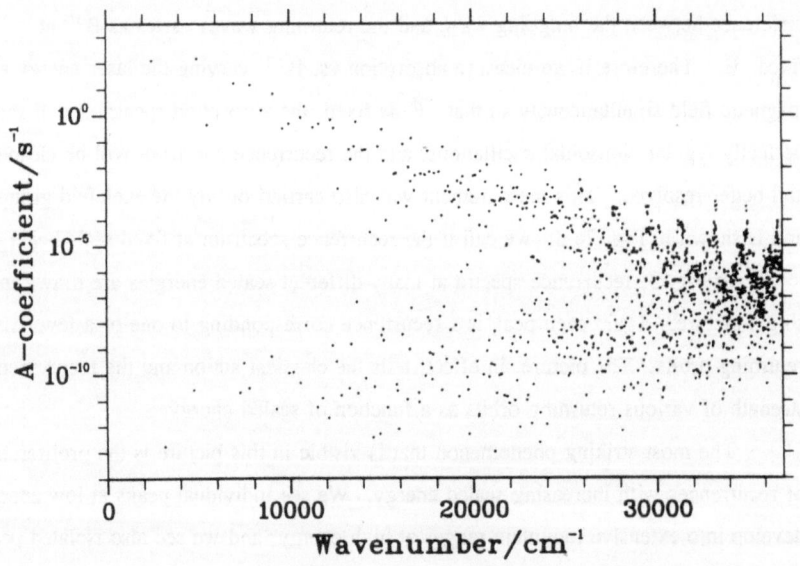

Fig. 4    Einstein A-coefficients for transition from a selected lower state of $H_3^+$
to various vibrational states near the photodissociation threshold.

This proliferation of recurrences is associated with bifurcations of classical periodic orbits as order changes to chaos. As the energy is increased, new classical periodic orbits split out of a given orbit, or, in some cases, a new pair of orbits may be created "out of nowhere." A general theory of bifurcations of periodic orbits in a Hamiltonian system was developed by Meyer in 1970. He showed that typically, bifurcations fall into just five characteristic patterns, which we call saddle-node, period-doubling, touch-and-go (two types), and island-chain bifurcations. The recurrence spectra shown in Fig. 3 have been analyzed in detail using this general framework. Most of the typical types of bifurcation and some non-generic bifurcations have been identified in these measurements[10].

These experiments lead us to a different way of thinking about absorption spectra: instead of trying to understand the structure of individual quantum states, we try to understand directly the large-scale fluctuations in the spectrum, which are connected with the short-time order that always exists even in the most chaotic systems.

## 2. The Vibrational Spectrum of $H_3^+$

LeSueur et al.[11] have calculated the absorption spectrum of $H_3^+$ between a selected set of low-lying levels and a large set of states close to the dissociation threshold (Fig. 4). We see that generally the Einstein A-coefficients decrease with increasing energy of the excited state, but there are fluctuations--if we were to average these coefficients over all states in a small band of energies, a periodic oscillatory pattern would appear. Noting the logarithmic scale in Fig. 4, we realize that these oscillations have a very large amplitude compared to the background of weakly coupled states.

The wavelength $\lambda_E$ of these oscillations can be converted to a time using Eq. (3). This time is found to be the period of a periodic orbit of the $H_3^+$ molecule. In its lowest states, $H_3^+$ is an equilateral triangle, and its small vibrations have the expected patterns of normal modes: a symmetric stretch (breathing mode) and a doubly-degenerate asymmetric stretch. For large amplitudes of vibration, the asymmetric stretch goes over a potential-energy barrier at the collinear configuration and becomes a large-amplitude bending mode. It is this periodic orbit that produces the fluctuations in the absorption spectrum.

The same orbit makes itself visible as a "scar" in some of the eigenfunctions. A theory of scars was developed by Bogomolnyi and tested by Provost and Baranger[12]. Scars are easily understood as another manifestation of the regular fluctuations associated with periodic orbits of the system. (Contrary to appearances, a "scar" of a classical periodic orbit is not particularly associated with any individual quantum state; like a quantum-mechanical resonance, it spreads itself over a range of energies, and it becomes most clearly visible if we sum the squares of eigenfunctions in a small energy band.)

3. <u>A Stadium-Shaped Microjunction</u>[4]

A tiny two-dimensional stadium-shaped microjunction was cut on an Al-Ga-As chip, and the chip was cooled to T $\approx$ $10^{-2}$K and placed in a magnetic field which was varied from 0-2T. The resistance of this junction was measured as a function of the magnetic field (Fig. 5a). This resistance shows wild fluctuations. The resistance was converted to a conductance, g = 1/R, and the square of the Fourier-transform of the conductance was calculated from the measurements (Fig. 5b). Let's call it the "conductance spectrum."

In the long-wavelength portion of the resulting conductance spectrum, we find distinct, strong peaks. The heights of these peaks are not yet fully reproducible in the experiment, but their locations are ~ 10, 25, and 65 cycles/Tesla. For shorter wavelengths, the conductance spectrum is plotted on a logarithmic scale. We see an overall exponential decrease in the conductance strength with decreasing wavelength.

Theoretical calculations which hope to reproduce and interpret this spectrum of conductance fluctuations are being carried out. The calculations begin from the fact that under the circumstances of this measurement, electrons bounce through the microjunction ballistically--at zero magnetic field, they move on straight lines until they hit a wall. Fluctuations are associated with interference of waves associated with the trajectories that go from entrance to exit.

At this moment, the calculations are not complete, but we may tentatively identify some of the low-wavelength peaks as arising from interferences among some of the orbits shown in Fig. 6.

To understand the short-wavelength part of the conductance spectrum, a different approach is needed. This would be associated with interferences among long orbits from entrance to exit. There can be a very large number of such orbits: in a closed stadium one orbit goes directly from the entrance point to the exit point, two bounce off the walls once between entrance and exit, about sixty-five orbits bounce four times between entrance and exit. Clearly a statistical approach is needed.

Classical simulations of long orbits in the stadium with two open leads have been carried out. It is found that: (1) the probability that a particle remains in the stadium after time t decreases exponentially with t; the exponent is easily estimated from the geometrical quantities; (2) the probability that the particle's orbit subtends an oriented area A before escape also decreases exponentially with $|A|$:

$$P(A) \sim \exp[-\alpha|A|] \tag{6}$$

By studying the quantum-classical correspondence for particles moving in magnetic fields, we learn that the exponent $\alpha$ in Eq. (6) may be compared to the rate of exponential decline found in the conductance spectrum. In fact the numbers obtained from simulations are quite comparable with those extracted from the experimental measurements.

This case provides a clear example of the general statement that we can only make statistical predictions about the long-time behavior of a chaotic classical system.

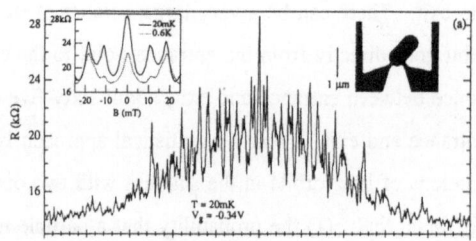

Fig. 5a    The microjunction and its resistance at magnetic fields varying
           from -0.5 to +0.5 T.

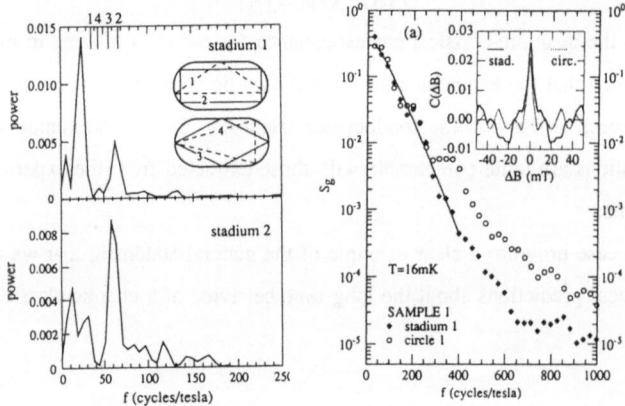

Fig. 5b    Fourier transform of the conductance of the microjunction
           Left:  low-frequency              Right:  higher frequency

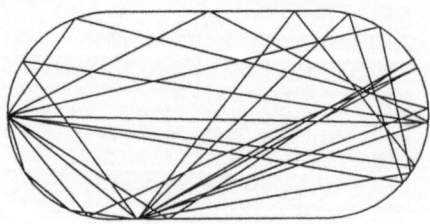

Fig. 6    Some orbits from entrance to exit in the stadium-shaped junction.
          Entrance is on the left and exit at the bottom.

1.    M. L. Du and J. B. Delos, Phys. Rev. A 38, 1896 and 1912 (1988); D. Wintgen and H. Friedrich, Phys. Rev. A35, 1464 (1987 or 36, 131 (1987).

2.    J. Bayfield and P. Koch, Phys. Rev. Lett 33, 258 (1974); R. V. Jensen, Phys. Rev. A30, 386 (1984); N. B. Delone, V. P. Krainov and D. L. Shepelyanskii, Sov. Phys. Usp. 26, 551 (1983).

3.    D. W. Noid, M. L. Koszykowski and R. Marcus, Ann. Rev. Phys. Chem. 32, 267 (1981).

4.    a) R. V. Jensen, Chaos 1, 101 (1991).

      b) C. M. Marcus, A. J. Rimberg, R. M. Westervelt, P. F. Hopkins and A. C. Gossard, Phys. Rev. Lett 69, 506 (1992).

5.    N. Pomphrey, J. Phys. B7, 1909 (1974); R. T. Swimm and J. B. Delos, J. Chem. Phys. 71, 1706 (1979).

6.    M. V. Berry, N. L. Balazs, M. Tabor and A. Voros, Ann. Phys. (N.Y.) 26, 122 (1979).

7.    N. L. Balazs and A. Voros, Phys. Reports 143, 109 (1986); R. Aurich and F. Steiner, DESY preprint 91-044 (1991).

8.    M. Gutzwiller, Chaos in Classical Quantum Mechanics, Springer (1990); A. M. Ozorio de Almeida, Hamiltonian Systems: Chaos and Quantization, Cambridge (1988); F. Haake, Quantum Signatures of Chaos, Springer, Berlin (1991); L. Reichl, The Transition to Chaos in Conservative Classical Systems: Quantum Manifestations, Springer (1992); K. Nakamura, Quantum Chaos: a new paradigm of nonlinear dynamics, Cambridge (1993).

9.    J. Main, G. Wiebusch, A. Holle and K. H. Welge, Phys. Rev. Lett. 57, 2789 (1986); A. Holle, J. Main, G. Wiebusch, H. Rottke and K. H. Welge, Phys. Rev. Lett. 61, 161 (1988).

10.   K. R. Meyer, Tras. Am. Math. Soc. 149, 95 (1970); J. M. Mao and J. B. Delos, Phys. Rev. A45, 1746 (1992); J. Main, G. Wiebusch, K. Welge, J. Shaw and J. B. Delos, Phys. Rev. A (accepted).

11.   C. R. LeSueur, J. R. Henderson and J. Tennyson, Chem. Phys. Lett. 206, 429 (1993).

12.   E. Bogomolnyi, Physica D31, 169 (1988); D. Provost and M. Baranger, Phys. Rev. Lett. 71, 662 (1993).

Few-Body Systems, Suppl. 7, 80—91 (1994)

Few-
Body
Systems
© Springer-Verlag 1994
Printed in Austria

# NUCLEAR ASPECTS OF FEW-BARYON SYSTEMS

B. F. Gibson

*Theory Division, Los Alamos National Laboratory*
*Los Alamos, NM 87545, USA*

### Abstract

Recent progress in understanding the bound state properties of the trinucleons and the alpha particle in terms of a hadron picture of the nucleus is reviewed. The role of three-body forces and meson exchange currents is examined. Novel aspects of few-body hypernuclei as well as unresolved issues in this $S \neq 0$ sector are summarized.

I thank the organizers for their invitation to speak at this convocation of European few-body physicists. The topic of few-baryon bound states covers a wide range of intriguing problems, many more than I will be able to touch upon today. Therefore, for a detailed summary of the trinucleon physical observables and model properties, the audience is referred to a review in the proceedings of the Adelaide International Few-Body Conference.[1] Also, a summary of the work by Sasakawa and collaborators will appear shortly in Few-Body Systems.[2] The trinucleon discussion is based upon calculations by a number of people: Friar, Gibson, Payne, ..., Tjon, ..., Glöckle, Witala, ..., Sauer, Stadler, ..., Sasakawa, Ishikawa, ..., Kim, Pickelsimer, Brandenburg, ..., deSwart & Stoks, .... My brief remarks on $^4$He are based upon the work of Tjon, Carlson , and Glöckle & Kamada, .... My comments in this area will be concentrated on recent results and our growing understanding of the physics. The primary emphasis in the talk will be the strangeness ($S = -1\, \& \, -2$) sector, where hyperons can act as tagged probes of the nucleus. The focus will be on the physics of hypernuclei, where a number of novel features are to be found and a number of unresolved issues remain.

## I. The Strangeness-Zero Sector

Before summarizing the status of our understanding of the low energy physical observables for the conventional ($S = 0$) $A = 3\, \& \, 4$ bound states, a brief glimpse of the history of calculational developments is in order. Tjon [3] gave us the early calculations of $^3$H/$^3$He properties using a realistic potential. The analysis of those impulse approximation charge form factors is still valid today: triton underbinding implies radii that are too large, and the diffraction minima occur at too high a momentum transfer in nonrelativistic Hamiltonian model calculations. Afnan [4] produced the initial $^3S_1$-$^3D_1$, $^1S_0$ (5-channel) calculation for the exisiting realistic force models, using the Adhikari-Sloan expansion; the Ernst-Shakin-Thaler expansion has since been used [5] for contemporary potential model calculations. Recently, the W-matrix expansion of Sandhas, Haberzettl, ... [6] has been shown to give reliable results for a low-rank, small-basis solution and holds promise for speedy nd→nnp breakup calculations. Finally the coordinate-space approach of Kok and Schellingerhout [7] appears to be a computationally efficient means of attacking the 3-body and 4-body problem.

Traditional Model

The traditional approach to nuclear physics can be defined in terms of the constraints placed upon the complexity of the model assumptions:

- Nuclei consist only of nucleons – other degrees of freedom are suppressed.

- Nucleons move slowly within the nucleus – nonrelativistic dynamics prevails.

- Nucleons interact primarily via pairwise (two-body) forces.

This is an enormous simplification of the physics, but it accounts amazingly well for much of the experimental data. Nonetheless, our calculational ability has achieved the precision required to see differences between traditional model predictions and experiment, and much research during the past decade has been focussed upon extensions: meson exchange currents (MEC), three-body forces (3BF), $NN$-$N\Delta$ coupling, relativistic dynamics, quark-gluon substructure, .... I will concentrate on results from the traditional model for binding energies (and implicitly other low energy observables) and address only briefly the MEC, 3BF, and relativity aspects of extensions.

Before turning to specific systems, let us take a quick look at the nonrelativistic assumption. The $A = 3 \& 4$ bound states have charge radii of the size of 1.5-1.8 fm, the same order as the pion compton wave length ($1/m_\pi$). The uncertainty principle permits one to estimate $\Delta p$ from the size of the system: $\langle p \rangle \approx 120$ MeV/c. Also, from the triton kinetic energy of $\approx 40$ MeV, one can estimate $\langle p^2 \rangle^{1/2} \approx 160$ MeV/c. Thus, the average momentum of the system is of the order of $m_\pi$, and

$$\left(\frac{v}{c}\right)^2 \approx \left(\frac{\langle p \rangle/M}{c}\right)^2 \approx \left(\frac{m_\pi}{M}\right)^2 \approx 2\%.$$

*Caveat*: This is an average. It should reflect accurately relativistic corrections to the low energy observables, which are averages over the nuclear volume. However, processes involving high momentum wave function components can suffer much larger relativistic effects.

Status of the Trinucleons

It has now been reasonably well established that the low energy observables "scale" with the trinucleon binding energy.[8, 9] A summary of results for charge radii, wave function probabilities, magnetic moments, Coulomb energy, asymptotic normalization constants, and Nd scattering lengths can be found in Ref. [1]. Because of this scaling property, the discussion will be limited to results for the triton binding energy.

Benchmark results exist from a number of groups [8, 9, 10, 11, 12, 13, 14, 15, 16] for a variety of realistic potentials [17, 18, 19, 20, 21]. In this context realistic implies

- strong spin-isospin dependence ($V_{nn} \neq V_{np}$),

- strong tensor force (OPEP is essential, providing up to 3/4 of the potential energy in $^3$H and $^4$He),

- strong short range repulsion (the probability of $NN$ overlap at such separations should be small),

in addition to a reasonable fit to the $NN$ scattering data. Somewhat surprisingly, the explicitly momentum-dependent models such as Paris and Nij78 are more repulsive in the negative-parity partial waves in a triton binding energy calculation than are the RSC and AV14 models.[22]

The latest development of consequence is a set of local potential models by the Nijmegen group with a $\chi^2/f \approx 1$ for the scattering observables.[23] These charge dependent potential models (with $V_{nn} \neq V_{np}^s \neq V_{np}^t$) have now been used to estimate the triton binding energy to be $B(^3H) = 7.62 \pm .01$ MeV.[24] Thus, a local potential model of the $NN$ interaction that fits the $NN$ observables as well as any phase shift analysis leads to underbinding of the triton by about 0.85 MeV.

Previously such underbinding of the triton by local potential models had led investigators to raise the question of the role of 3BF.[25] Various groups demonstrated that adding a two-pion-exchange ($\pi\pi$) three-body force [26,27] to the Hamiltonian led to overbinding of the triton when the $\pi NN$ vertex cutoff of $5.8m_\pi$, as recommended by the authors of Ref. [26], was used.[28,29,30,31] For example, for the combined RSC/TM model one obtains $B(^3H) \simeq 8.9$ MeV for $j \leq 2$ (or 9.1 MeV for $j \leq 4$). Recently Coon and Peña completed the TM 3BF by deriving the $\pi\rho$ and $\rho\rho$ terms.[32] Now, Stadler et al.[31] have calculated the $^3$H binding energy for the full TM 3BF. Just as the $\rho$ cancels part of the $\pi$ contribution to the $NN$ force, the $\pi\rho$ term cancels part of the $\pi\pi$ term in the $NNN$ force. The result for the RSC/TM model is that 8.9 MeV $\rightarrow$ 8.4 MeV. (Similarly, for the Paris/TM combination one finds 9.1 $\rightarrow$ 8.5 and for the Nij78/TM model one finds 9.3 $\rightarrow$ 8.7.) *Apparently, a nonrelativistic Hamiltonian composed of a local $NN$ potential plus a comparable 3BF can yield approximatley the correct value for* $B(^3H)$.

Where do the Bonn potentials fit in this picture? Let me remind everyone that the Bonn q-space potentials (as approximations to the full Bonn energy-dependent interaction) are not local. That is, they do not depend solely upon the difference $|\vec{p}-\vec{p}'|$. Nonlocal interactions have an added degree of flexibility, which may well be partially mimicked in local Hamiltonians by a 3BF. Recall the theorem by Polyzou and Glöckle [33] that relates the three-body potential and on-shell equivalent two-body interactions. Paraphrasing: If there exist two-body Hamiltonians $H_{12}$ $(= T_1 + T_2 + V_{12})$ and $\bar{H}_{12}$ which have the same binding energies and scattering matrices, are complete, and are related by a unitary transform, then there is a three-body interaction $W$ which relates the three-body Hamiltonians $(T_1 + T_2 + T_3 + V_{12} + V_{13} + V_{23})$ given by $H$ and $\bar{H}' = \bar{H} + W$ such that they have the same binding energies and scattering matrices. In other words, it is quite conceivable that if a local potential plus 3BF describes the physics of the trinucleons, then there is also a nonlocal potential plus different 3BF (perhaps identically zero) that will also describe the physics. The Bonn potentials may already incorporate certain 3BF effects in their nonlocal nature. Using the TM 3BF with the Bonn $NN$ potential is likely double counting at some level, as the result for the Bonn-Q/TM calculation [31] indicates. However, it is also worth noting that the difference between the charge-dependent Bonn-B potential triton, $B(^3H) = 7.92$ MeV, [16] and the nonlocal Nijmegen charge-dependent potential (Nijm I) result, $B(^3H) = 7.72$ MeV [24] is only some 200 keV. Are the model differences so great? The Graz group is examining the electromagnetic response of the $NN$ system in an effort to pin down the proper off-shell behavior.[34]

For the experimentalists, let me reiterate that three-body forces is a "game" for theorists. On the other hand, we now have understood much about the "rules of the game."

### The Alpha Particle

Tjon first established that the $A = 3$ & 4 nucleon system binding energies are strongly correlated.[35] This is in part due to the fact that the $^3S_1$-$^3D_1$ interaction is so dominant. That is, $\langle V^t_{np} \rangle \gg \langle V^s_{np} \rangle \Rightarrow$ the physics of all the s-shell nuclei must be similar.

The GFMC results of Carlson [36] confirm that underbinding for $^3$H ($\simeq 7.7$ MeV for the AV14 model) and for $^4$He ($\simeq 24.9$ MeV) follows the Tjon line. Adding a phenomenological 3BF to fit the triton implies a corresponding fit for the $\alpha$ particle.

Integral equation solutions by Glöckle and Kamada for realistic force models have just appeared.[37] In addition to binding energies for the AV14 model (7.68 MeV and 24.62 MeV), they quote results for several other potentials for $j \leq 3$ $NN$ partial waves. One should note Kamada's talk [38] later in the week as well as Fonseca's review [39] of the $A = 4$ system.

### Meson Exchange Currents

The deviation of the $^3$H/$^3$He magnetic moments from the Schmidt limits provided an early indication of the need to include subnucleon degrees of freedom when nucleons are embedded within the nuclear medium. In one language these effects are due to MEC. Isovector corrections [40] are estimated to be of the order of 15% and are reasonably model independent.[41] The isoscalar MEC contributions to the magnetic moments are expected to be small, of the order of relativisitc corrections. Indeed, the isoscalar magnetic moment combination for $^3$H and $^3$He appears to be reasonably described by impulse approximation.[41]

What is known in the case of MEC corrections to the charge operator? It is tempting to invoke them to obtain a fit to the charge form factor data. An *ad hoc* addition to the impulse approximation form factor of a term which vanishes at $q^2 = 0$ and is negative in the region of the diffraction minimum and beyond will certainly improve the fit to the data. Such a procedure is far from unique. In the case of the deuteron, an isoscalar object, we have seen [42,43] that the difference between relativistic results and the nonrelativistic impulse approximation is small for a significant range of momentum transfer. This is in contrast with the large effect that one sees when relativistic current corrections are included as a perturbation. In his *Les Houches* lectures, Tjon dissected the relativistic calculation for OPE.[44] He demonstrated that including the relativistic corrections to the current operator alone can be extremely misleading. The effect is large, but incorrect. Specifically, the MEC correction is cancelled by a corresponding dynamical correction to the $^2$H wave function – the nucleon can be offshell. When modeling isoscalar (relativistic) corrections to the $^3$H/$^3$He form factors, the practitioner should first check the model results for the $^2$H form factor.

## II. Hypernuclei

A hypernucleus is a baryon system comprised of conventional (nonstrange) nucleons and one or more strange hyperons ($Y = \Lambda$, $\Sigma$, or $\Xi$). The presence of the strangeness degree of freedom (flavor) adds a new dimension to our evolving picture of nuclear physics. Because the $S = -1$ hyperon masses differ markedly from those of the neutron and proton, $SU(3)$ symmetry is clearly broken. How it is broken is a question of fundamental importance to our understanding of the baryon-baryon interaction. The search for the $S \neq 0$ dibaryon states directly bears upon this. Certainly the investigation of strangeness in few-baryon systems will play a significant role in our understanding the strong force.

The physics of hypernuclei has proven to be both novel and puzzling, stretching our intuition and analysis capability beyond that developed during the more than half century that we have explored conventional nuclear physics. The hypernuclear sector of hadronic physics is not just a simple extension of zero-strangeness phenomena. The discoveries have been new; the physics is different. We have observed such novel aspects as:

- obvious charge symmetry breaking in the $^4_\Lambda$He-$^4_\Lambda$H isodoublet,[45, 46]

- anomalously small binding of $^5_\Lambda$He,[45, 47, 48, 49, 50]

- surprisingly narrow structure in the $^4$He(K$^-$,$\pi^\pm$) reactions, a possible $^4_\Sigma$He hypernucleus.[51, 52, 53]

These phenomena stand out clearly without the need for sophisticated theoretical analysis of the data. Furthermore, the following aspects are also interesting:

- The "H" dibaryon ($S = -2$ $uuddss$ object) is a unique prediction of the QCD Bag model.[54]

- Theoretical calculations imply a significant role for 3BF effects in the hypertriton $^3_\Lambda$H.[55, 56]

- A spin inversion between the $A = 4$ ground states and excited states would explain unobserved $\gamma$ rays in heavier hypernuclei.[57, 58, 59]

- Confirmation of a claim for the observation of $^6_{\Lambda\Lambda}$He would argue strongly against the existence of a deeply bound "H".[60]

$\Lambda$ hypernuclear physics has been established as an interesting subject, one which deserves concentrated effort to comprehend. The prospect of exploring $\Sigma$ hypernuclei and the more exotic $S = -2$ systems makes the strange physics sector that much more attractive.

The available data on the few-body $\Lambda$ hypernuclei come primarily from emulsion experiments [45] – binding energies and weak decay properties. We shall discuss only the binding energies today, for we have seen in the $S = 0$ sector that binding energy determines the low energy observables. In the study of hypernuclei, it is customary to quote the $\Lambda$-separation energies

$$B_\Lambda(_\Lambda A) = B(_\Lambda A) - B(A\text{-}1).$$

The accepted values for the s-shell systems are quoted in Table 1 along with the measured $\gamma$-ray deexcitation energies for the two species with particle-stable excited states. The $A = 6$ entry is $B_{\Lambda\Lambda} = B(^6_{\Lambda\Lambda}$He$) - B(^4$He).[60] I would point out that Don Davis of University College London is a great source of information about the emulsion data.[61, 62]

**Table 1.** Hypernuclear $\Lambda$-separation energies and excitation energies in MeV

| hypernucleus | $B_\Lambda$ | $E_\gamma$ |
|---|---|---|
| $^3_\Lambda$H | 0.13±.05 | |
| $^4_\Lambda$H | 2.04±.04 | 1.04±.04 |
| $^4_\Lambda$He | 2.39±.03 | 1.15±.04 |
| $^5_\Lambda$He | 3.10±.02 | |
| $^6_{\Lambda\Lambda}$He | 10.6 | |

### The $YN$ Interaction

Several novel features of the $YN$ interaction come into play in hypernuclear physics. The $\Lambda$ ($T = 0$) and $N$ ($T = 1/2$) cannot exchange a $\pi$ ($T = 1$), so that there is no dominant OPE tensor force in $\Lambda N$ scattering. Shorter range properties of the baryon-baryon interaction play a more important role than in $NN$ scattering. The longest range components are due to the exchange of two pions or one kaon. The shorter range K-exchange potential does admit a tensor-force component, but it is largely cancelled by that from K*-exchange. Thus, tensor-force effects in the $\Lambda N$ interaction are expected to be smaller than those in the $NN$ interaction.[63, 64, 65, 66, 67, 68]

On the other hand, $\Lambda N$-$\Sigma N$ coupling effects are expected to be much more important in the hypernuclear sector than are $NN$-$N\Delta$ coupling effects in the nonstrange sector. The $m_\Sigma$-$m_\Lambda$ mass difference is only some 75 MeV, and the width of the $\Sigma$ is small compared to that of the $\Delta$. Formally eliminating the $\Sigma$ channel from the problem leads to an energy dependent $\Lambda N$ effective interaction and to $\Lambda NN$ three-body forces. Both of these effects have been the subject of recent research in the nonstrange sector.[14]

The experimental data for $\Lambda N$ and $\Sigma N$ scattering consist of some 600 events in the low energy range (momenta of less than 300 MeV/c) [69, 70, 71] and another 250 events in the momentum range 300-1500 MeV/c [72]. A phase shift analysis is not a practical possibility. The low energy data fail to adequately define even the relative sizes of the dominant s-wave spin-triplet and spin-singlet scattering lengths.[69] The limited cross section measurements have been grouped with the abundant $NN$ data in a combined analysis to constrain the $YN$ OBE potential models. The Nijmegen group has developed several OBE models by simlutaneously fitting the $NN$ and $YN$ data and employing $SU(3)$ constraints among the coupling constants. In addition, the Jülich group has now constructed $YN$ potential models with a different character. In particular, the assumption about the F/D ratio (scalar singlet and scalar octet coupling) differs, which affects the relative strengths in the $\Lambda N$ and $\Sigma N$ channels.

The scattering lengths and effective ranges for five of these models are collected in Table 2. The scatter among the parameters gives an indication of how poorly the potentials are constrained by the data. The scattering lengths are clearly negative, so that there is no bound hyperdeuteron. However, one can not determine whether the interaction in the spin-singlet state or the spin-triplet state is the stronger. A cusp is expected in the region of the threshold for $\Sigma$ production, although neither the position nor the magnitude of the peak in this energy region is determined by the data; the cross sections in this energy region exhibit uncertainties of a factor of 2-4. The strong $\Lambda N$-$\Sigma N$ coupling combined with the sizeable $\Sigma^-$-$\Sigma^+$ mass difference and the charge dependence of the $\Sigma N$ OBE force implies a measureable charge symmetry breaking in the $YN$ interaction, which is clearly seen in the $A = 4$ isodoublet.

**Table 2.** The scattering lengths and effective ranges in fm
for the $YN$ potential models listed

| Model | Ref. | $a^s$ | $r_o^s$ | $a^t$ | $r_o^t$ |
|-------|------|-------|---------|-------|---------|
| Nijmegen D | [63] | -1.90 | 3.72 | -1.96 | 3.24 |
| Nijmegen F | [64] | -2.29 | 3.17 | -1.88 | 3.36 |
| Nijmegen SC | [65] | -2.78 | 2.88 | -1.41 | 3.11 |
| Jülich A | [68] | -1.56 | 1.43 | -1.59 | 3.16 |
| Jülich Ã | [68] | -2.04 | 0.64 | -1.33 | 3.91 |

## $\Lambda$ Hypernuclei

The question remains open as to whether we can successfully calculate the properties of the few-body $\Lambda$ hypernuclei utilizing realistic $YN$ and $NN$ interactions. Because neither the $\Lambda N$ nor $\Sigma N$ forces possess sufficient strength to support a bound state, the ground state of the $\Lambda NN$ system (the hypertriton) plays the role of the deuteron. The spin and parity of $^3_\Lambda$H were determined to be $J^\pi = \frac{1}{2}^+$ from an analysis of the angular distribution of pions emitted during its weak mesonic decay. The small $\Lambda$-separation energy (130±50 keV) leads one to expect a loosely bound system − essentially a $\Lambda$ clinging tenuously to a deuteron. As such, one expects the hypertriton to be sensitive to the long range properties of the $YN$ interaction. Because $\Lambda N$-$\Sigma N$ coupling plays a large role in that interaction, one is also led to speculate whether $^3_\Lambda$H is bound only because of the $\Lambda NN$ 3BF that results when the $\Sigma$ channel is eliminated from the formalism.

Because of the weak binding and lack of definitive $YN$ scattering data, many theorists have studied $^3_\Lambda$H using separable potentials fitted to the approximate $\Lambda N$ low energy parameters. (For a partial list, see Ref. [55].) The essential physics can be summarized as follows: 1) The spin-averaged $\Lambda N$ interaction is $V_{\Lambda N} = \frac{1}{4}V^t_{\Lambda N} + \frac{3}{4}V^s_{\Lambda N}$; that is, the spin-singlet interaction dominates. 2) The dispersive energy dependence that results from embedding the $\Lambda N$-$\Sigma N$ coupled-channel potential in a three-body system is repulsive, as is that from the $NN$-$N\Delta$ interaction in the triton. 3) The true 3BF due to the coupling of $\Sigma NN$ states to $\Lambda NN$ states is attractive. 4) The $YN$ tensor force interacts with the $NN$ tensor force in a complex manner, which depends upon the relative sign of the $^3S_1$-$^3D_1$ mixing parameters. This analysis has been confirmed by the recent calculation of Miyagawa and Glöckle [56] using the Jülich Ã model in combination with various realistic $NN$ potentials. Surprisingly, the model $^3_\Lambda$H was found to be unbound. This is surprising because the low energy scattering parameters for Ã in Table 2 would imply that the $\Lambda NN$ system should bind; thus, one must conclude that the short range properties of the realistic potential play an unexpected role in this weakly bound system. The binding of the hypertriton is clearly a balancing act. A 4% enhancement in the strength of $V^s_{\Lambda N}$ would yield a bound state [56], but such a change in the spin-singlet interaction would destroy the model Ã fit to the $YN$ data.

The $^4_\Lambda$He-$^4_\Lambda$H isodoublet provides an even stronger test of our modeling of the $YN$ interaction. The quality of the calculations for the (ground and excited) states of the $A = 4$ system should approach that demonstrated for the $\alpha$ particle. Even greater precision should be obtained for the charge-symmetry-breaking (CSB) difference

$$\Delta B_\Lambda = B_\Lambda(^4_\Lambda He) - B_\Lambda(^4_\Lambda H) .$$

The nominal $\Delta B_\Lambda \simeq 350$ keV is much larger than the $\simeq 100$ keV CSB effect seen in the $^3$He-$^3$H binding energy difference after correcting for the $pp$ Coulomb energy in $^3$He. A key question is whether the CSB can be understood in terms of the free $\Lambda N$ interaction. The question is not trivial. Consider a simple separable-potential model fitted to the $\Lambda p$ and $\Lambda n$ scattering parameters of Nijmegen D listed in Table 3. A folding model approximation for the $\Lambda$-nucleus potential

$$V_{\Lambda A}(k) = \int d^3 p \, \rho_A(\vec{p}) V_{\Lambda N}(p,k) ,$$

which provides effective two-body dynamics in a Schrödinger equation solution, yields a value of $\Delta B_\Lambda = 0.21 \rightarrow 0.24$ MeV when a core compression of $0 \rightarrow 5\%$ is used. In contrast, an exact four-body calculation for the identical potentials yields a value of $\Delta B_\Lambda = 0.43$ MeV − almost twice the folding model result.[46] Few-body dynamics does matter.

The existence of particle-stable excited states for the $A = 4$ system provides a further strong constraint upon our models of the $YN$ interaction. Fitting both the $0^+$ and $1^+$ states within the same model is not a trivial exercise. The spin structure is more complex than one might naively expect; the $^4_\Lambda$H states are not simply $[\Lambda \otimes ^3H]^{[J]}$ configurations. Furthermore, $\Lambda N$-$\Sigma N$ coupling plays an important role because of the composite nature of the trinucleon core states. A simple example follows from the Stepien-Rudza

**Table 3.** The charge-dependent scattering lengths and effective ranges in fm for Nijmegen D

| channel | $a^s$ | $r^s_o$ | $a^t$ | $r^t_o$ |
|---------|-------|---------|-------|---------|
| $\Lambda p$ | -1.77 | 3.78 | -2.06 | 3.18 |
| $\Lambda n$ | -2.03 | 3.66 | -1.84 | 3.32 |

and Wycech [73] separable potential approximation to Nijmegen D. The spin-averaged potential combination that determines the $0^+$ state is $\frac{1}{2}V^s_{\Lambda N} + \frac{1}{2}V^t_{\Lambda N}$, whereas for the $1^+$ state it is $\frac{1}{6}V^s_{\Lambda N} + \frac{5}{6}V^t_{\Lambda N}$. If explicit $\Lambda N$-$\Sigma N$ coupling is neglected, and the $\Lambda N$ effective interactions are constructed from the scattering lengths and effective ranges, then one finds in an exact four-body calculation that

$$B(0^+) \simeq 10.7 \text{ MeV} \qquad \text{and} \qquad B(1^+) \simeq 11.7 \text{ MeV};$$

the ground state and excited states are inverted, the $1^+$ being more bound. However, if one takes into account the composite nature of the $A = 3$ core states and suppresses the $T = 3/2$ trinucleon excited states that couple to the $\Sigma$ but lie 10s of MeV higher in energy, then one finds

$$B(0^+) \simeq 9.6 \text{ MeV} \qquad \text{and} \qquad B(1^+) = 8.2 \text{ MeV},$$

which places the states in proper order.[59] This simplified schematic-model calculation emphasizes the importance of $\Lambda N$-$\Sigma N$ coupling. That has been born out in variational calculations using Nijmegen SC for the $YN$ interaction. For the $0^+$ state, where the singlet and triplet interactions are of equal importance, one finds $B_\Lambda(0^+) \simeq 1.5$ MeV, in qualitative agreement with the data. However, the $1^+$ state is unbound.[74] The spin-triplet interaction which dominates the $1^+$ system derives much of its attraction in Nijmegen SC from the $\Lambda \leftrightarrow \Sigma$ transition term $v_{XN}$ in the coupled channel potential

$$\begin{bmatrix} v_{\Lambda N} & v_{XN} \\ v_{XN} & v_{\Sigma N} \end{bmatrix}$$

which is less effective "in medium" than in free space, just as is tensor coupling in the $NN$ interaction.

The anomalously small binding of $^5_\Lambda$He remains an enigma. In the baryon picture, the four nucleons and $\Lambda$ can coexist in $1s$ states. Simple model calculations based upon $\Lambda N$ potentials parameterized to account for the low energy scattering data, as well as the binding energies of the $A = 3\,\&\,4$ systems, overbind $^5_\Lambda$He by 2-3 MeV.[48,49,50] Explanations of the model overbinding have been proposed in terms of tensor forces that bind $^3$H and $^4$He less than do central forces,[47] and in terms of $\Lambda N$-$\Sigma N$ coupling that is weakened when the nucleon is part of a composite core.[75] The variational calculations by Carlson again fail to bind, because of the strength in the $\Lambda \leftrightarrow \Sigma$ transition potential. The spin averaged interaction is $\frac{3}{4}V^t_{\Lambda N} + \frac{1}{4}V^s_{\Lambda N}$; the spin-triplet potential dominates as it does for the $A = 4$ $1^+$ states. It should also be noted that the reduction needed in the binding has been modeled as a repulsive 3BF,[76] and as a manifestation of quark effects.[77]

A total of three $\Lambda\Lambda$ hypernuclei events have been reported.[78,60,79] The $^6_{\Lambda\Lambda}$He event [60] is the most controversial,[80] although the exact interpretation of the most recent event has been called into question.[81] However, the value for $B_{\Lambda\Lambda}(^6_{\Lambda\Lambda}$He$)$ is consistent with the trend from the other two events and not unreasonable. The 10.6 MeV implies a value for $\langle V_{\Lambda\Lambda} \rangle = B_{\Lambda\Lambda} - 2B_\Lambda(^5_\Lambda$He$) \simeq 4$ MeV, which is a matrix element comparable to that from the $\Lambda N$ potential. That is, the overall strength of the $\Lambda\Lambda$ interaction appears comparable to that of the $\Lambda N$ interaction.[49] Because one would expect a $\Lambda\Lambda$ hypernucleus to decay quickly into an "H" if the "H" mass were much smaller than $2m_\Lambda$, the observation of $\Lambda\Lambda$ hypernuclei argues against the existence of a deeply bound "H". If the existence of $^6_{\Lambda\Lambda}$He is confirmed, then one can surmise that

$$m_H \geq 2m_\Lambda - 10 \text{ MeV}.$$

If it is then bound, the "H" would be more of a nuclear bound state of two $\Lambda$s rather than a deeply bound *uuddss* exotic.

## $\Sigma$ Hypernuclei

While the existence of $\Lambda$ hypernuclei is well established from the observation of many bound states, such is not the case for $\Sigma$ hypernuclei. It was surprising when Hayano [51] reported that the $\pi^-$ spectrum from the $^4$He(stopped $K^-, \pi^\pm$) reactions exhibited narrow structure below the threshold for $\Sigma$ emission. An analysis by Dover and Gal [82] had noted that, if a bound state were to exist, then the $(T = \frac{1}{2}, S = 0)$ $\Sigma NNN$ configuration should lie lower in energy than the $(T = \frac{3}{2}, S = 0)$ configuration, although the latter state was expected to be narrower. The $(K^-, \pi^-)$ reaction can lead to both $T = \frac{1}{2}$ and $T = \frac{3}{2}$ $\Sigma NNN$ states, whereas the $(K^-, \pi^+)$ reactions leads only to $T = \frac{3}{2}$. Thus, because no structure was observed in the $^4$He$(K^-, \pi^+)$ reaction and the $(K^-, \pi^-)$ spin-flip reaction is small, the structure was interpreted as a $^4_\Sigma$He state having quantum numbers $(T = \frac{1}{2}, J^\pi = 0^+)$. New results, which confirm the earlier spectra with stopped kaons, have recently been reported from inflight experiments at the Brookhaven AGS.[83] Furthermore, the data are consistent with the recently reanalyzed bubble chamber data from the exclusive $K^-$ $^4$He $\rightarrow \pi^- \Lambda pd$ measurements,[84] which appear to show a cusp-like enhancement near the $\Sigma^+$ production threshold. The peak of the $\pi^-$ spectrum appears to be centered at $B_{\Sigma^+} = 4 \pm 1$ MeV. (The width is about 10 MeV.) The $\Sigma$ in the inferred hypernucleus would be more bound than is the $\Lambda$ in $^4_\Lambda$He, implying that the $\Sigma^+$ interaction with $^3$H must be more attractive than the corresponding $\Lambda$ interaction with $^3$H or $^3$He. Data were taken at the same time on $^3$He,[85] but no claim for structure has been made.

Following the work of Dover and Gal but prior to the experiment, Harada *et al.* predicted the existence of an $A = 4$ $\Sigma NNN$ bound state with $B_{\Sigma^+} \simeq 4.5$ MeV and a width of about 8 MeV. They predicted no other "bound state" for $A = 2$-5. The only true continuum calculation was published for $\Lambda d$ scattering in the region of the $\Sigma NN$ threshold.[86] Surprisingly, it was found that structure in the elastic scattering cross section would appear *below* the $\Sigma NN$ threshold whether the resonance pole corresponding to a $\Sigma NN$ eigenstate lies above or below the $\Sigma NN$ threshold. That is, although structure in the cross section below threshold does imply the existence of a resonance, it does not guarantee that the pole lies below threshold (that one has a "bound state").

## Acknowledgement

The work of the author was performed under the auspices of the U. S. Department of Energy. He gratefully acknowledges a Research Award for Senior Scientists by the Alexander von Humboldt Stiftung which made possible his stay with the I. K. P. Theorie group at the Forschungszentrum Jülich, where this manuscript was prepared.

# References

[1] Gibson, B. F.: Nucl. Phys. **A543**, 1c (1992).

[2] Sasakawa, T., Ishikawa, S.: Few-Body Sys. (to appear)

[3] Tjon, J. A., et al.: Phys. Rev. Lett. **25**, 540 (1970).

[4] Afnan, I. R., Read, J. M.: Phys. Rev. C **8**, 1294 (1973); **12**, 293 (1975).

[5] Koike, Y., et al., Phys. Rev. C **35**, 396 (1987); Haidenbauer, J., Koike, Y.: Phys. Rev. C **34**, 1187 (1986).

[6] Bartnik, E. A., et al.: Phys. Rev. C **36**, 1678 (1987);

[7] Schellingerhout, N. W.: see contribution to this volume.

[8] Friar, J. L., et al.: Phys. Lett. B **161**, 241 (1985).

[9] Ishikawa, S., Sasakawa, T.: Few-Body Sys. **1**, 3 (1986).

[10] Payne, G. L., et al.: Phys. Rev. C **22**, 823 (1980); Chen, C. R., et al.: Phys. Rev. C **31**, 2266 (1985); Friar, J. L., et al.: Phys. Rev. C **37**,2869 (1988).

[11] Hajduk, C., Sauer, P. U.: Nucl. Phys. **A369**, 361 (1981).

[12] Bömelburg, A.: Phys. Rev. C **28**, 403 (1983).

[13] Ishikawa, S., et al.: Phys. Rev. Lett. **53**, 1877 (1984).

[14] Sauer, P. U.: Prog. in Part. Nucl. Phys. **16**, 35 (1986).

[15] Brandenburg, R. A., et al.: Phys. Rev. C **37**, 1245 (1988).

[16] Witala, H., et al.: Phys. Rev. C **43**, 1619 (1991).

[17] Reid, R. V., Jr.: Ann. Phys. (NY) **50**, 411 (1968); Day, B.: Phys. Rev. C **24**, 1203 (1984) – labeled RSC.

[18] Nagels, M. M., et al.: Phys. Rev. D **17**, 768 (1978) – labeled Nij78.

[19] Lacombe, M., et al.: Phys. Rev. C **21**, 861 (1980) – labeled Paris.

[20] Wiringa, R. B., et al.: Phys. Rev. C **29**, 1207 (1984) – labeled AV14

[21] Machleidt, R., et al.: Phys. Rep. **149**, 1 (1987) – labeled Bonn-R and Bonn-Q; Machleidt, R.: Adv. Nucl. Physics, **19**, 189 (1989) – labeled Bonn-B.

[22] Payne, G. L., Gibson, B. F.: Few-Body Sys. **14**, 117 (1993).

[23] Stoks, V. G. J., Klomp, R. A. M., deSwart, J. J.: private communication.

[24] Friar, J. L., et al.: Phys. Lett. B **311**, 4 (1993).

[25] Gibson, B. F., McKellar, B. H. J.: Few-Body Sys. **3**, 143 (1987).

[26] Coon, S. A., et al.: Nucl. Phys. **A317**, 242 (1979); Coon, S. A., Glöckle, W.: Phys. Rev. C **23**, 1990 (1981) – labeled TM.

[27] Coelho, H. T., et al.: Phys. Rev. C **28**, 1812 (1983); **31**, 646 (1985) – labeled BR.

[28] Chen, C. R., et al.: Phys. Rev. C **33**, 1740 (1986).

[29] Ishikawa, S., Sasakawa, T.: Few-Body Sys. **1**, 143 (1986).

[30] Bömelburg, A.: Phys. Rev. C **34**, 14 (1986).

[31] Stadler, A., Adam, J., Henning, H., Sauer, P. U.: "Triton calculations with $\pi$- and $\rho$-exchange three-nucleon forces" (submitted to Phys. Rev. C).

[32] Coon, S. A., Peña, M. T.: Phys. Rev. C (to appear).

[33] Polyzou, W. N., Glöckle, W.: Few-Body Sys. 9, 97 (1990).

[34] Pauschenwein, J.: Few-Body Sys. Suppl. 6, 195 (1993).

[35] Tjon, J. A.: Phys. Lett. B 56, 217 (1975).

[36] Carlson, J. A.: private communication.

[37] Glöckle, W., Kamada, H.: Phys. Rev. Lett. 71, 971 (1993).

[38] Kamada, H.: see contribution to this volume.

[39] Fonseca, A. C.: see contribution to this volume.

[40] Riska, D. O., Brown, G. E.: Phys. Lett. B 38, 193 (1972).

[41] Friar, J. L. et al.: Phys. Rev. C 37, 2852 (1988).

[42] Zuilhof, M. J., Tjon, J. A.: Phys. Lett. B 84, 31 (1979); Phys. Rev. C 22, 2369 (1980).

[43] Arnold, R. C., et al.: Phys. Rev. C 21, 1426 (1980).

[44] Tjon, J. A.: in: Hadronic Physics with Multi-GeV Electrons, p. 89. New York: Nova Science Publ. (1985); Nucl. Phys. A463, 157c (1987).

[45] Juric, M. et al: Nucl. Phys. B52, 1 (1973).

[46] Gibson, B. F., Lehman, D. R.: Nucl. Phys. A329, 308 (1979).

[47] Bodmer, A.: Phys. Rev. 141, 1387 (1986).

[48] Herndon, R. C., Tang, Y. C.: Phys. Rev. 153, 1091 (1967); 159, 853 (1967); 165, 1093 (1968); Dalitz, R. H. et al.: Nucl. Phys. B47, 109 (1972).

[49] Gibson, B. F. et al.: Phys. Rev. C 6, 741 (1972).

[50] Gal, A.: Adv. Nucl. Physics 8, 1 (1975).

[51] Hayano, R. S. et al: Nuovo Cimento 102A, 437 (1989); Phys. Lett. B 231, 355 (1989).

[52] Harada, T. et al.: Nucl. Phys. A507, 715 (1990).

[53] Hayano, R. S.: Nucl. Phys. A527, 477 (1991).

[54] Jaffe, R. L.: Phys. Rev. Lett. 38, 195 (1977).

[55] Afnan, I.R., Gibson, B. F.: Phys. Rev. C 40, R7 (1989); 41, 2787 (1990).

[56] Miyagawa, K., Glöckle, W.: Phys. Rev. C (to appear).

[57] Bamberger, A. et al.: Nucl. Phys. B60, 1 (1973).

[58] Bejidian, M. et al: Phys. Lett. B 83, 252 (1979).

[59] Gibson, B. F., Lehman, D. R.: Phys. Rev. C **37** 679 (1988).

[60] Prowse, A.: Phys. Rev. Lett. **17**, 782 (1966).

[61] Davis, D. H.: in: Proceedings of the LAMPF Workshop on $(\pi, K)$ Physics, A. I. P. Conf. Proc. **224**, p. 38. New York: Amer. Inst. Phys. (1991).

[62] Davis, D. H., Sacton, J.: in: High Energy Physics, Vol. II, p. 365. New York: Academic Press (1967).

[63] Nagels, M. M. *et al.*: Phys. Rev. D **15**, 2547 (1977).

[64] Nagels, M. M. *et al.*: Phys. Rev. D **20**, 1633 (1979).

[65] Maessen, P. M. M. *et al.*: Phys. Rev. C **40**, 2226 (1989).

[66] Holzenkamp, B. *et al.*: Nucl. Phys. **A500**, 485 (1989).

[67] Holinde, K.: Nuc. Phys. **A547**, 255c (1992).

[68] Reuber, A. G. *et al.*: Czech. J. Phys. **42**, 1115 (1992).

[69] Alexander, G. *et al.*: Phys. Rev. Lett. **13**, 484 (1964); Phys. Rev. **173**, 1452 (1968).

[70] Englemann, R. *et al.*: Phys. Lett. **21**, 587 (1966).

[71] Sechi-Zorn, B. *et al.*: Phys. Rev. **175**, 1735 (1968).

[72] Kadyk, J. A. *et al.*: Nucl. Phys. **27**, 13 (1971).

[73] Stepien-Rudza, W., Wycech, S.: Nucl. Phys. A **362**, 349 (1981).

[74] Carlson, J. A.: in: Proceedings of the LAMPF Workshop on $(\pi, K)$ Physics, A.I.P. Conf. Proc. **224**, p. 198. New York: Amer. Inst. Phys. (1991).

[75] Gibson, B. F. *et al.*: in: Few Particle Problems in Nuclear Interactions, p. 188. Amsterdam: North Holland (1972).

[76] Bodmer, A. R., and Usmani, Q. N.: Nucl. Phys. **A477** 621 (1988); Phys. Rev. C **31**, 1400 (1985); Bodmer, A. R. *et al.*: Phys. Rev. C **29**, 684 (1884).

[77] Hungerford, E. V., Biedenharn, L. C.: Phys. Lett. B **42**, 232 (1984).

[78] Danysz, M. *et al.*: Phys. Rev. Lett. **11**, 29 (1963); Nucl. Phys. **49**, 121 (1963).

[79] Aoki, S. *et al.*: Prog. Theo. Phys. **85**, 1287 (1991).

[80] Dalitz, R. H. *et al.*: Proc. Roy. Soc. London **A426**, 1 (1989).

[81] Dover, C. B. *et al.*: Phys. Rev. C **44**, 1905 (1991).

[82] Dover, C. B., Gal, A.: Phys. Lett. B **110** 443 (1982).

[83] Hayano, R. S.: Nucl. Phys. **A547**, 151c (1992).

[84] Dalitz, R. H. *et al.*: Phys. Lett. B **236**, 76 (1990).

[85] Hungerford, E. V.: Nucl. Phys. **A547**, 157c (1992).

[86] Afnan, I. R., Gibson, B. F.: Phys. Rev. C **47**, 1000 (1993).

Few-Body Systems, Suppl. 7, 92—103 (1994)

Few-
Body
Systems

# ELECTROMAGNETIC FORM FACTORS
# OF TWO-NUCLEON AND THREE-NUCLEON BOUND STATES

Peter U. Sauer and Hartmut Henning

Institute of Theoretical Physics, University of Hannover

Appelstraße 2, D-30167 Hannover, Germany

**Abstract:** The known e.m. deuteron and trinucleon form factors are theoretically described. The description is noncovariant with nucleonic degrees of freedom; single $\Delta$-isobar excitation is allowed in the trinucleon bound states. The importance of two-nucleon exchange contributions to the charge and current operators for a successful account of the data as well as the ambiguities in their theoretical derivation are discussed; the two-nucleon charge operator is of relativistic order. Problems remain in simultaneously describing the existing deuteron and trinucleon data.

## 1. Experimental Situation

Elastic electron scattering from nuclei measures electromagnetic (e.m.) form factors. Unpolarized scattering suffices to determine the form factors of spin-0 and spin-$\frac{1}{2}$ nuclei. Polarization experiments are needed to separate the form factors of different multipolarity in case of nuclei with higher spin. The deuteron, $^3$He and $^3$H form factors are experimentally known up to momentum transfers $Q$ corresponding to the nucleonic rest mass $m_N$; only selected observables, i.e., the deuteron structure functions $A(Q^2)$ and $B(Q^2)$ and the $^3$He charge form factor $F_C^{^3He}(Q^2)$, are measured up to even higher momentum transfers.

With respect to the deuteron, the measurements [1] of the tensor polarization $T_{20}(Q^2)$ allow, together with the data [2] on the structure functions $A(Q^2)$ and $B(Q^2)$, the separation of the deuteron charge monopole and quadrupole form factors. The minimum of the charge monopole form factor is determined to be at $4.39(16)fm$. At momentum transfers $-Q^2 \leq 1fm^{-2}$ the contribution of the quadrupole and the magnetic form factors to the structure function $A(Q^2)$ is less than 1%; thus, the deuteron r.m.s. charge radius can be obtained from $A(Q^2)$ directly with confidence; however, the value of $2.116(6)fm$, usually quoted differently as model-dependent matter radius, comes from the old data of Ref. [3]; they appear in their momentum transfer dependence inconsistent with newer data [2], which favor a larger radius. A larger value is also obtained from the measurement [4] of the isotope shift in the 1S-2S transition of atomic hydrogen and deuterium which sees the nuclear e.m. properties at zero momentum transfer with startling precision, in contrast to the modest accuracy of nuclear physics experiments. However, the theoretical analyses of the isotope shift measurement and of electron scattering at small momentum transfers have to take the polarization of the rather soft deuteron into account, before an accurate charge radius can be extracted.

Rosenbluth separation has usually been employed in order to determine the longitudinal and transverse structure functions individually in electron scattering without polarization. In contrast, the analysis of the trinucleon data in Ref. [5] does not rely on Rosenbluth separation any longer for the extraction of the charge and magnetic form factors $F_C^A(Q^2)$ and $F_M^A(Q^2)$ with $A = {}^3He, {}^3H$. Instead, Ref. [5] fits a quite general, but physically sound ansatz for the form factors to all existing experimental cross sections; the ansatz is taken to be a sum of Gaussians for the charge and magnetization distributions in configuration space and corresponds to our picturesque understanding of nucleonic distributions in nuclei; the ansatz allows data at various momentum transfers to cooperate in determining the $Q^2$-dependence of the form factors. The fit of such a form factor ansatz makes effective use of kinematically uncorrelated data; it has the same advantage over the Rosenbluth separation, as the energy-dependent phase shift analysis of two-nucleon scattering [6] has compared with the energy-independent one. As a consequence, the experimental data on the trinucleon form factors are not given any more by data points with error bars at definite momentum transfers, but by a functional form with error bands. The trinucleon e.m. radii result [7] from that analysis; they still have rather large errors, since all existing data are taken into account, i.e., $r_C({}^3He) = 1.959(30)fm$, $r_C({}^3H) = 1.755(86)fm$, $r_M({}^3He) = 1.965(153)fm$ and $r_M({}^3H) = 1.840(181)fm$. Assuming perfect isospin symmetry, the trinucleon form factors can easily be split [8] into isoscalar (IS) and isovector (IV) parts, i.e.,

$$F_C^{IS,IV}(Q^2) = \frac{1}{2}\left[Z({}^3He)F_C^{{}^3He}(Q^2) \pm Z({}^3H)F_C^{{}^3H}(Q^2)\right], \qquad (2a)$$

$$F_M^{IS,IV}(Q^2) = \frac{1}{2}\left[\mu({}^3He)F_M^{{}^3He}(Q^2) \pm \mu({}^3H)F_M^{{}^3H}(Q^2)\right], \qquad (2b)$$

$Z(A)$ and $\mu(A)$ being the trinucleon charges and magnetic moments, respectively. A possible isospin impurity due to the Coulomb interaction between the two protons in

[3]He affects [9] the [3]He form factors only within its existing experimental errors; thus, the isospin separation of Eqs. (2) is not endangered. The isoscalar trinucleon form factors can cleanly be compared with the deuteron form factors: Both are determined by the same charge and current operators. The isovector trinucleon form factors have a correspondence in the structure functions of deuteron electrodisintegration; furthermore, $\mu$-capture by [3]He as well as the weak process [3]He$(e^-, {}^3\text{H})\nu_e$ measure the weak trinucleon form factors; however, their experimental determination requires the knowledge of the weak charge-changing vector current which is given by the isovector e.m. trinucleon form factors.

The experimental deuteron and trinucleon form factors as well as other observables are displayed in the figures of Sects. 2 and 3 together with their theoretical description.

## 2. Framework of Theoretical Description

A description of the deuteron and trinucleon form factors in terms of effective hadronic degrees of freedom is promising, at least for momentum transfers up to $-Q^2 \approx m_N^2$. The microscopic quark-gluon degrees of freedom have been employed in QCD motivated quark models; they are usually applied to the deuteron only [10]; their application to the three-nucleon system [11] lacks consistency in the simultaneous description of e.m. and strong processes; in general, the application of those quark models cannot match yet the sophistication which calculations with hadronic degrees of freedom achieve.

Charge $\rho(\mathbf{Q})$ and spatial current $\mathbf{j}(\mathbf{Q})$ which form the Lorentz four-vector $j^\mu(\mathbf{Q})$ determine the form factors; their calculation for momentum transfers up to $-Q^2 \approx m_N^2$ requires wave function components with nucleonic momenta which are not small compared with the nucleonic rest mass $m_N$. The nonrelativistic nucleonic bound-state problems can be solved with high precision and with the addition of the $\Delta$-isobar degree of freedom [12], of explicit pionic degrees of freedom [13] and of irreducible three-nucleon forces [14] for the three-nucleon system; however, combined with a single-nucleon charge operator of nonrelativistic order, such a description fails for the trinucleon charge form factors, irrespectively of the computed binding energy as Fig. 1 demonstrates. Thus, the nonrelativistic description of the deuteron and of the trinucleon e.m. properties has to be amended for relativistic corrections. This conclusion is firm, though a strictly nonrelativistic description of the trinucleon magnetic properties which includes exchange currents and the $\Delta$-isobar degree of freedom is quite successful according to Fig. 2.

A fully relativistic description of the two- and three-nucleon bound states were advisable, at least as a necessary check on a nonrelativistic treatment with the incorporation of relativistic corrections. Indeed, such a description of the two-nucleon system has been developed by Tjon and collaborators [21] in the frame work of the Bethe-Salpeter equation with ladder approximation for meson exchange; alternative relativistic descriptions are discussed in Ref. [22]; further results for deuteron form factors are given in Refs. [23] and [24]. However, no relativistic calculation of corresponding quality exists yet for the three-nucleon bound states and their form factors. At present, only nonrelativistic calculations with relativistic corrections can provide a simultaneous description of two- and three-nucleon bound states with comparable quality.

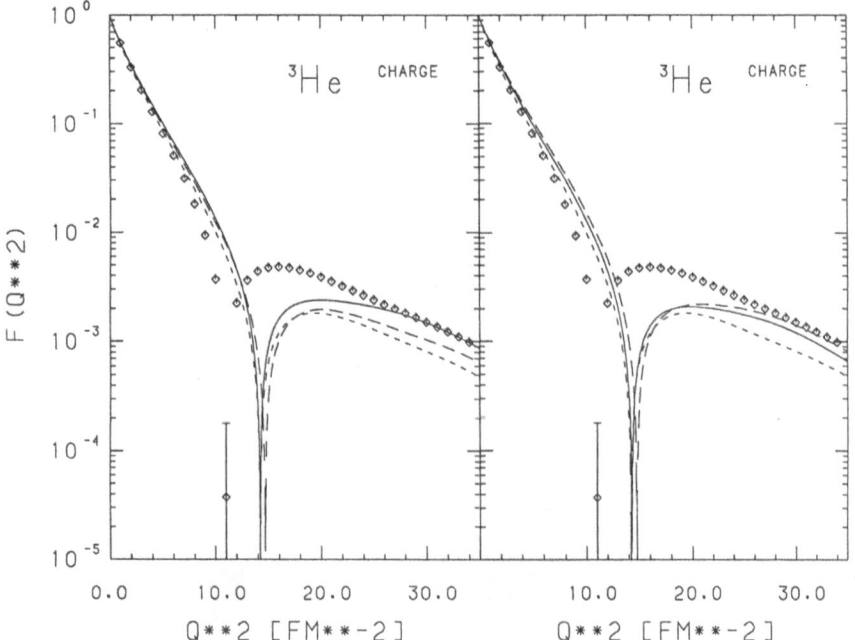

Fig. 1: $^3$He charge form factor $F_C^{^3He}(Q^2)$ in nonrelativistic impulse approximation as function of $\mathbf{Q}^2$. It is obtained in the Breit frame. The result for the two-nucleon Paris potential [15] (triton energy $E_T = -7.38\,MeV$) is given by the short-dashed curves in both plots. In the calculation of the trinucleon wave function three-nucleon forces of various sorts are included: In the left side of the plot, the $\Delta$-isobar degree of freedom is added and is the essential mechanism for the three-nucleon force; the long-dashed curve is derived from the force model A2 of Ref. [12] ($E_T = -7.72\,MeV$), the solid curve from the force model A2 with an additional irreducible three-nucleon force of the Tucson-Melbourne type in which the $\Delta$-isobar contribution contained explicitly in A2 is subtracted [16] ($E_T = -8.05\,MeV$). In the right side of the plot, irreducible three-nucleon forces of the Tucson-Melbourne type are added; the long-dashed curve is derived from the Paris potential and the pion-exchange three-nucleon force [14] ($E_T = -9.06\,MeV$), the solid curve from the Paris potential and the three-nucleon force based on the combined pion and rho exchange [17] ($E_T = -8.49\,MeV$). The charge operator is the single-nucleon one in nonrelativistic order using the Pauli and Dirac nucleon form factors of Ref. [18]; for the results of the left side the $\Delta$-isobar degree of freedom is also kept in the single-baryon charge as in Ref. [12]. The experimental data are taken from Ref. [5].

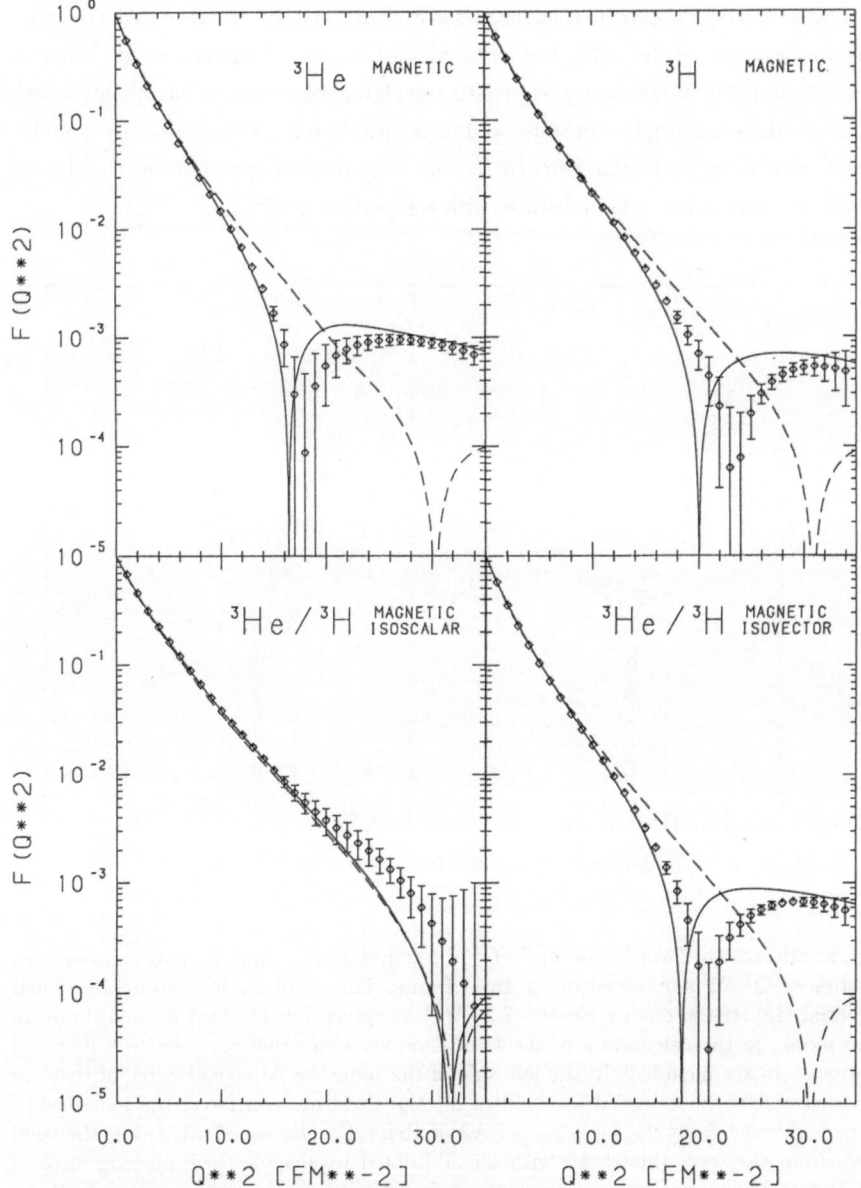

Fig. 2: Trinucleon magnetic form factors $F_M^{^3He}(Q^2)$ and $F_M^{^3H}(Q^2)$ in nonrelativistic description as function of $\mathbf{Q}^2$. They are obtained in the Breit frame. The purely nucleonic calculation is based on the Paris potential [15]; it includes the standard two-nucleon exchange currents derived from pion and rho exchange without any $\Delta$-isobar contribution according to Fig. 3 further down; its results are displayed by the dashed curves. The solid curves contain the additional effects of the $\Delta$-isobar. The theoretical predictions are taken from Ref. [19], the experimental data from Refs. [5] and [8]. The agreement is better for the corresponding calculation of Ref. [20] whose spatial current operator is more consistently tuned to the underlying hadronic interaction.

There are three major techniques [25,26] for introducing relativistic corrections into nonrelativistic calculations, i.e., the quasi-potential method, the equation of motion method and the extended S-matrix method. All eliminate the explicit antibaryon and mesonic degrees of freedom and define currents up to second order in the meson-baryon coupling constants $g$. The calculations presented in this talk are based on the third method, the extended S-matrix method [26]. That method is sketched for the nucleon as only baryonic degree of freedom, it is graphically displayed in Fig. 3. The method defines a two-nucleon potential $V$ by one-boson exchange and a single-nucleon current; then the two-nucleon exchange current results from the requirement that a noncovariant calculation with relativistic kinetic energies $K$ and global propagators yields the same S-matrix for the processes of second order in the coupling constants as relativistic field theory does. The one- and two-nucleon currents satisfy the continuity equation. The two-nucleon current contains important remainders of the relativistic propagation of nucleons; however, the defined potential does not contain them; it would only do so in higher than second order of $g$ and would then receive an energy-dependent two-nucleon part and a three-nucleon contribution; the quantitative consequences of that neglect are still a point of controversy [27]. And even the employed one-boson exchange definition of the potential is an ambiguous procedure: The defining Feynman amplitude is to be computed between on-mass-shell states and contains an energy conserving $\delta$-function; the noncovariant description with global propagators requires the potential off-energy-shell; the ambiguity of its off-shell extension and the simultaneous freedom of choosing any mixture between pseudoscalar and pseudovector pion-nucleon coupling is reflected by the parameters $(\tilde{\mu}\nu)$; as a consequence, the two-nucleon exchange current depends on the same ambiguity parameters. Thus, an infinity of technically different, but physically equivalent force and current models results. Choosing the two-nucleon operator $M(\tilde{\mu}\nu, \tilde{\mu}_0\nu_0)$ according to

$$V(\tilde{\mu}\nu) - V(\tilde{\mu}_0\nu_0) = i\,[K, M(\tilde{\mu}\nu, \tilde{\mu}_0\nu_0)]\,, \tag{2.1}$$

the two-nucleon Hamiltonians $H(\tilde{\mu}\nu) = K + V(\tilde{\mu}\nu)$ and $H(\tilde{\mu}_0\nu_0)$ are unitarily related, their binding energy and phase shifts are the same up to the second order in $g$, i.e.,

$$H(\tilde{\mu}\nu) = U^+(\tilde{\mu}\nu, \tilde{\mu}_0\nu_0)H(\tilde{\mu}_0\nu_0)U(\tilde{\mu}\nu, \tilde{\mu}_0\nu_0) + O(g^4), \tag{2.2a}$$

$$U(\tilde{\mu}\nu, \tilde{\mu}_0\nu_0) = exp[iM(\tilde{\mu}\nu, \tilde{\mu}_0\nu_0)], \tag{2.2b}$$

and the current matrix elements are unitarily related by

$$\langle\psi_f(\tilde{\mu}\nu)|j^\mu(\mathbf{Q}, \tilde{\mu}\nu)|\psi_i(\tilde{\mu}\nu)\rangle = \langle\psi_f(\tilde{\mu}_0\nu_0)|j^\mu(\mathbf{Q}, \tilde{\mu}_0\nu_0)|\psi_i(\tilde{\mu}_0\nu_0)\rangle + O(g^4). \tag{2.3}$$

In Eq. (2.3) $|\psi_i(\tilde{\mu}\nu)\rangle$ and $|\psi_f(\tilde{\mu}\nu)\rangle$ are eigenstates of the two-nucleon Hamiltonian $H(\tilde{\mu}\nu)$. With respect to nucleonic momenta, the charge and current operators are

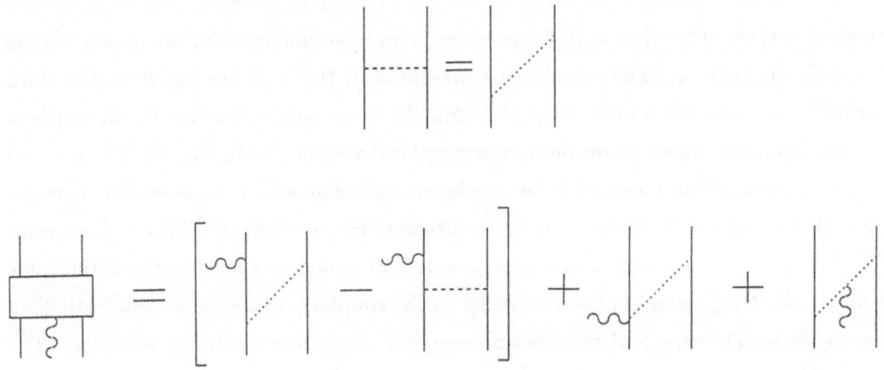

Fig. 3: Definition of the two-nucleon one-boson exchange potential and of the two-nucleon exchange current in the extended S-matrix method up to second order in the meson-baryon coupling constant. The potential is denoted by a horizontal dashed line, the exchange current by a horizontal box with wavy photon line. The dotted lines denote field-theoretically active mesons in retarded propagation. The exchange current receives contributions from the Born (in square brackets), contact and meson-in-flight processes; its Born contribution contains the difference between the relativistic single-nucleon and the noncovariant global propagators; the potential does not carry any effect of relativistic single-nucleon propagation. Pseudovector coupling is used for pion exchange. The spatial exchange current gets contributions from the contact and meson-in-flight processes in nonrelativistic order $(1/m_N)$. The exchange charge gets a Born contribution in leading relativistic order $(1/m_N)^2$; it is not due to pairs, but due to the positive-frequency part of relativistic single-nucleon propagation.

usually expanded to leading relativistic order in $(1/m_N)$. In a pion-exchange model the violation of unitary equivalence by that limited expansion in coupling constants and $(1/m_N)$ is found [26] to set in strongly already at momentum transfers $-Q^2 \approx 20\,fm^{-2}$. Nevertheless, we shall use those charge and current operators up to even higher momentum transfers; we are aware that sizeable relativistic corrections may be left out by the employed expansion. In that expanded form unitary equivalence is also observed for the charge and current operators of the three distinct approaches used for defining relativistic corrections. That observation [25,26] is nontrivial, since the three techniques use different procedures for the definition of the potential and of the charge and current operators. The observation proves that the three approaches have an equivalent physics basis.

## 3. Numerical Results

We attempt a consistent noncovariant description of the deuteron and trinucleon e.m. form factors with leading relativistic corrections in the charge operator. We shall indicate to what extent our results are also characteristic for the calculations of other groups. We have complete results for the Reid soft-core (RSC) [28], the Paris [15] and the Bonn OBEPQ, A, B and C [29] potentials. The Bonn OBEPQ potential and its consistent current account for the known deuteron observables satisfactorily, but it fails for the trinucleon properties; in contrast, the Paris potential describes the trinucleon properties rather well, but then misses the deuteron. Selected theoretical predictions of the Bonn OBEPQ potential for the deuteron are displayed in Fig. 4; those of the Paris potential for the trinucleon charge form factors in Fig. 5. The conclusion is: The theoretical apparatus employed by us is able to obtain the elastic deuteron and trinucleon e.m. properties separately correct, but not consistently together.

Fig. 6 proves this claim for the Paris potential in case of the deuteron charge monopole and the trinucleon isoscalar charge form factors: Whereas the nonrelativistic impulse approximation fails for the trinucleon charge properties, it accounts pretty well for the deuteron charge monopole form factor; in contrast, the exchange charge of leading relativistic order improves the theoretical description of the trinucleon charge properties significantly, but then destroys the agreement achieved for the deuteron in nonrelativistic impulse approximation. Ref. [31] finds the same result.

Figs. 6 and 7 prove another unexpected point: The Paris and Bonn B potentials, whose two-nucleon phase shifts are close, can be considered almost unitarily equivalent [32] in the sense of Eqs. (2.2) and (2.3). Fig. 7 demonstrates that the deuteron wave function of both potentials can be transformed into each other by a unitary transformation of the form (2.1) and (2.2b); it is of order $g^2$ and $(1/m_N)^2$; the fact that the unitary equivalence is not perfect at small relative distances is therefore understandable. Both potentials exhibit entirely different deuteron and trinucleon predictions in nonrelativistic impulse approximation; however, as Fig. 6 shows, the employed charge corrections of leading relativistic order make the predictions of both potentials in the more complete calculation quite the same.

Using the different two-nucleon potentials of this section, remarkable linear relations [33] are found between corresponding characteristic properties of the deuteron charge monopole and the isoscalar trinucleon charge form factors, i.e., for the minimum and maximum positions and for the height of the secondary maximum. One of these relations is displayed in Fig. 8; the relations hold in nonrelativistic impulse approximation and with the inclusion of one- and two-nucleon charge corrections of relativistic order $(1/m_N)^2$; they are conceptually interesting; we do not have yet a proper theoretical understanding of them. The displayed linear relation for the diffraction minimum

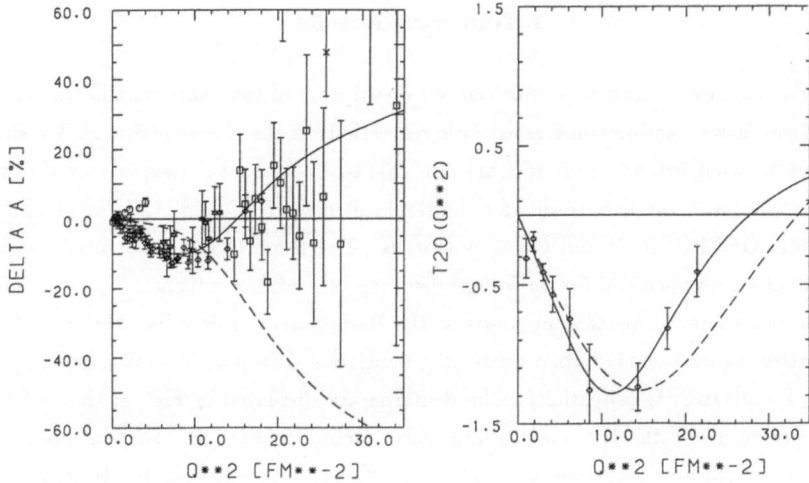

Fig. 4: Deuteron structure function $A(Q^2)$ and tensor polarization $T_{20}(Q^2)$ as function of $\mathbf{Q}^2$. They are obtained in the Breit frame. The structure function is displayed in the form $(A(Q^2) - A_{NRIA}(Q^2, Paris))/A_{NRIA}(Q^2, Paris)$ which shows structural details. The theoretical predictions are derived from the Bonn OBEPQ potential; the dashed curves show the nonrelativistic impulse approximation, the solid curves include the exchange charge defined in Fig. 3 with the ambiguity parameters $(\tilde{\mu}\,\nu) = (-1, 1/2)$; the Born contribution of the pion is quite important; the corresponding rho contribution is considered of higher order and left out; as in Ref. [2] the neutron charge form factor is used to tune the theoretical prediction. The experimental data are taken from Refs. [1] and [2]

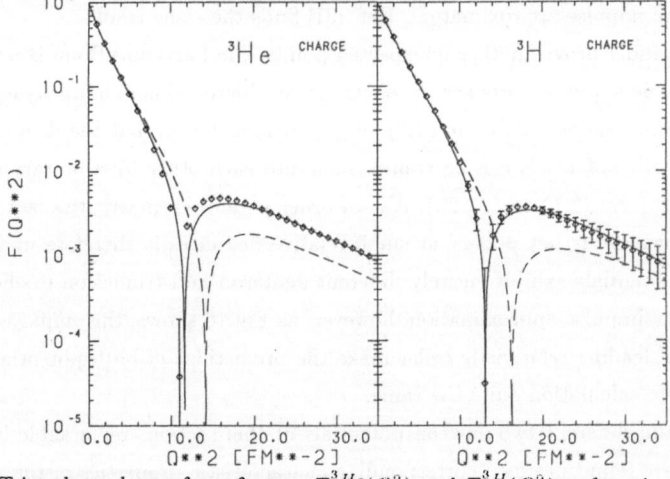

Fig. 5: Trinucleon charge form factors $F_C^{3He}(Q^2)$ and $F_C^{3H}(Q^2)$ as function of $\mathbf{Q}^2$. They are obtained in the Breit frame. The theoretical predictions are derived from the Paris potential; the dashed curves show the nonrelativistic impulse approximation, the solid curves include the exchange charge defined in Fig. 3; the Born contribution of the pion is used with full strength in contrast to the suggested [26] ambiguity parameters $(\tilde{\mu}\,\nu) = (0, 1/2)$; the corresponding rho contribution is kept. The experimental data are taken from Ref. [5]. The present calculation improves the older results of Ref. [19] by including more mesons. The corresponding calculation of Ref. [30] achieves a similar satisfactory agreement with the data.

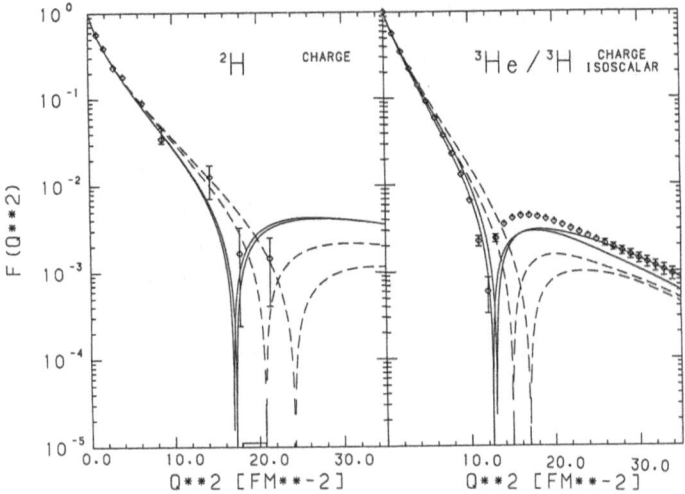

Fig. 6: Deuteron charge monopole and isoscalar trinucleon charge form factors $F_C^d(Q^2)$ and $F_C^{I,S}(Q^2)$ as function of $\mathbf{Q}^2$. They are obtained in the Breit frame. The theoretical predictions are derived from the Paris $(0, 1/2)$ and Bonn B $(-1, 1/2)$ potentials; the ambiguity parameters $(\tilde{\mu}\,\nu)$ used for the present consistent calculation are indicated. The dashed curves show the nonrelativistic impulse approximation, the solid curves include the exchange charge defined in Fig. 3; the Born contribution of the pion is quite important; the corresponding rho contribution is considered of higher order and left out; the predictions with the diffraction minimum at lower (higher) momentum transfers correspond to the Paris (Bonn B) potential. The experimental data are taken from Refs. [1] and [8].

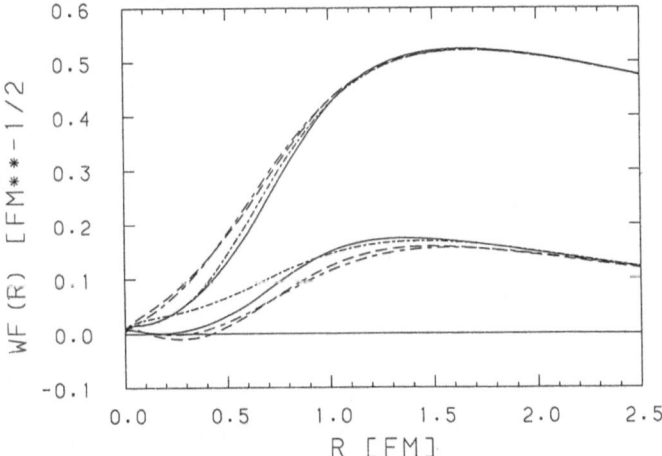

Fig. 7: Deuteron S and D wave functions as function of relative distance between the nucleons. Wave functions derived from the respective potentials and those unitarily transformed into each other are shown. The upper set of curves represent the S-wave, the lower set the D-wave, with the Paris, Paris transformed, Bonn B and Bonn B transformed wave functions given by the solid, long-dashed, point-long-dashed and point-short-dashed curves. Approximate unitary equivalence is established, since the wave functions for the Paris and Bonn B transformed as well as for Bonn B and Paris transformed lie almost on top of each other.

passes the experimental data in their present error bars by a substantial margin; that fact is another illustration for the earlier observation that the existing range of noncovariant potentials together with the present understanding of meson-exchange currents is unable to simultaneously account for the e.m. properties of the two- and three-nucleon systems. That observation relies on the experimental deuteron charge form factor as derived from the measured tensor polarization [1]. That experiment is an admirable technical achievement, but additional data with decreased error bars in the already measured range of momentum transfers and additional data at larger momentum transfers would be highly welcome for an improved determination of the minimum position in the deuteron charge monopole form factor and for a first determination of the position and height of its secondary maximum. Those experiments are under way [34].

**Acknowledgement:** The results of this talk are derived in a collaboration with J. Adam Jr. and A. Stadler. The calculations were performed at Regionales Rechenzentrum für Niedersachsen (RRZN) and Rechenzentrum der Universität Kiel.

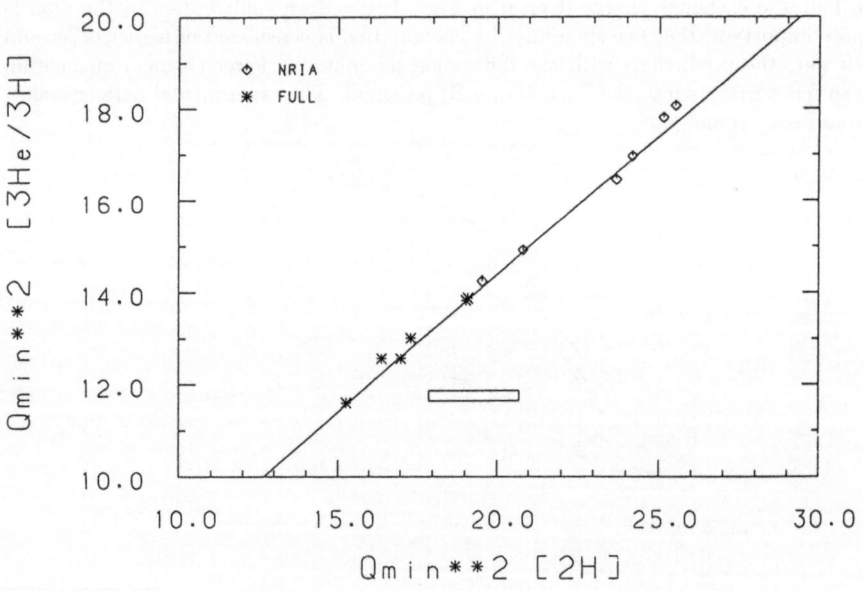

Fig. 8: Linear relations between the minimum position $\mathbf{Q}^2_{min}$ of the deuteron charge monopole and the isoscalar trinucleon charge form factors. The results refer to the RSC, Bonn C, Paris, Bonn B, Bonn OBEPQ and Bonn A potentials from lower left to upper right; the full calculation contains the exchange charge consistently as described in more detail for Fig. 6. The linear relation still holds when three-nucleon forces as described for Fig. 1 are included. The experimental data are indicated by a square box.

# References

[1] The, I., et al.: Phys. Rev. Letters **67**, 173 (1991), and References there

[2] Platchkov, S.K., et al.: Nucl. Phys. **A510**, 740 (1990), and References there

[3] Simon, G.G., et al.: Nucl. Phys. **A364**, 285 (1981)

[4] Schmidt-Kaler, F., et al.: Phys. Rev. Letters **70**, 2261 (1993)

[5] Platchkov, S.K., et al.: Saclay-Basel Preprint 1993

[6] Rijken, Th.A.: Invited Talk at this Conference

[7] Sick, I.: Private Communication

[8] Amroun, A., et al.: Phys. Rev. Letters **69**, 253 (1992)

[9] Friar, J.L., Gibson, B.F., Tomusiak, E.L., Payne, G.L.: Phys. Rev. **C 24**, 665 (1981)

[10] Buchmann, A., Yamauchi, Y., Faessler, A.: Nucl. Phys. **A496**, 621 (1989)

[11] Kisslinger, L.S.: Lecture Notes in Physics **260**, 432 (1986)

[12] Hajduk, Ch., Sauer, P.U., Strueve, W.: Nucl. Phys. **A405**, 581 (1983)

[13] Peña, M.T., Sauer, P.U., Stadler, A., Kortemeyer, G.: Phys. Rev. **C**, in Print

[14] Stadler, A., Glöckle, W., Sauer, P.U.: Phys. Rev. **C44**, 2319 (1991)

[15] Lacombe, M., et al.: Phys. Rev. **C21**, 861 (1980)

[16] Stadler, A., Sauer, P.U.: Phys. Rev. **C46**, 64 (1992)
    Stadler, A.: Private Communication

[17] Stadler, A., Adam Jr., J., Henning, H., Sauer, P.U.: In Preparation

[18] Simon, G.G., et al.: Nucl. Phys. **A333**, 381 (1980)
    Galster, S. et al.: Nucl. Phys. **B32**, 221 (1971)

[19] Henning, H., Sauer, P.U., Theis, W.: Nucl. Phys. **A537** (1992) 367

[20] Schiavilla, R., Pandharipande, V.R., Riska, D.O.: Phys. Rev **C40** (1989) 2294;

[21] Tjon, J.A.: Few Body Systems Suppl. **5**, 5 (1992), and References there

[22] Wallace, S.J.: Invited Talk at this Conference

[23] Arnold, R.G., Carlson, C.E., Gross, F.: Phys. Rev. **C21**, 1426 (1980)

[24] Chung, P.L., Coester, F., Keister, P.D., Polizou, W.N.: Phys. Rev. **C37**, 2000 (1988)

[25] Friar, J.L.: Phys. Rev. **C 22**, 796 (1980)

[26] Adam Jr., J., Göller, H., Arenhövel, H.: Phys. Rev. **C 48**, 370 (1993)

[27] Tjon, J.A., Zuilhof, M.J.: Phys. Lett. **84B**, 31 (1979)

[28] Reid, R.V.: Ann. Phys. (N.Y.) **50**, 411 (1968)

[29] Machleidt, R.: Adv. Nucl. Phys. **19** (1989) 189, and References there

[30] Schiavilla, R., Pandharipande, V.R., Riska, D.O.: Phys. Rev. **C41**, 309 (1990)

[31] Schiavilla, R., Riska, D.O.: Phys. Rev. **C43**, 437 (1991)

[32] Adam Jr., J.: Private Communication

[33] Henning, H., Adam Jr., J., Sauer, P.U., Stadler, A.: Hannover Preprint 1993

[34] Jones, C.: Invited Talk at this Conference

Few-Body Systems, Suppl. 7, 104—111 (1994)

# ELECTRON SCATTERING WITH POLARIZED $^3$He TARGETS

R. D. McKeown

W. K. Kellogg Radiation Laboratory

California Institute of Technology

Pasadena, CA 91125, USA

## ABSTRACT

The recent development of polarized $^3$He targets for use in electron scattering experiments has opened up a wide range of possibilities for new experiments. A variety of fundamental issues in both nucleon structure and the nuclear structure of $^3$He will be addressed in these experiments. Some recent preliminary measurements as well as several new proposals will be discussed.

Inclusive electron scattering has been a tremendously useful tool for the study of hadronic and nuclear structure. Although there is substantial kinematic flexibility, one can only access two experimental quantities by using the technique of Rosenbluth separation: the longitudinal response function and the transverse response function. More detailed information on the electromagnetic response requires either detection of final state hadrons in coincidence with the scattered electron or utilization of polarization degrees of freedom (or both).

The formalism for electron scattering with spin degrees of freedom has been developed to a point where the physics potential of this technique is now clearly evident [1]. In addition, the study of correlations of final state hadrons emitted in coincidence with the scattered electron from a polarized target offers a powerful technique for studying the multipole structure of spin dependent response functions [2].

$^3$He is a particularly interesting nucleus for polarization studies. The spin-dependent properties are often dominated by the neutron because the $^3$He wave function is predominantly a spatially symmetric $S$-state and antisymmetrization of the wave function requires that the protons be in a spin singlet state. If the $^3$He wave function were entirely $S$-state, the spin of the nucleus would be carried solely by the unpaired neutron. In this case measurement of the spin-dependent quantities in inclusive scattering of polarized electrons from polarized $^3$He would directly yield information on the neutron electromagnetic structure. There are small admixtures of other states in the $^3$He wave function which introduce a dependence on the proton electromagnetic form factors, but realistic calculations for the three-body system can give a reliable estimate of these contributions. In addition, as shown below, further coincidence measurements with polarized $^3$He can be used to test and constrain these small and interesting pieces of the three nucleon wave function.

This discussion will focus on recent and future work in spin-dependent quasielastic electron scattering from $^3$He. There is also a great deal of interest and activity in $N^*$ electroproduction and deep inelastic scattering using polarized $^3$He targets. Spin-dependent deep inelastic scattering from $^3$He is the subject of another contribution to this conference [3].

## Polarized $^3$He Target Development

There are two techniques that have been recently developed for laser optically pumped polarized $^3$He targets. One has been developed at Caltech, with further work now taking place at MIT and Mainz as well as Caltech. The other was developed at Princeton and later Harvard by T. Chupp and his co-workers.

The technique used by our group at Caltech [4] involves optical pumping of the $2^3S$ metastable state of the He atom. These states are created with a weak discharge at a concentration of about $10^{-6}$ and are polarized by pumping to the $2^3P$ states with 1083 nm polarized laser light. The $^3$He nuclei become polarized through the hyperfine interaction. The metastables then collide with ground state atoms, exchange atomic states, but leave the polarized nucleus in an atomic ground state with no electronic angular momentum. In this way, the nuclear polarization is transferred to the much larger population of atoms in the ground state. Substantial polarizations $(50 - 80\%)$ are achieved using high power lasers that have recently been developed.

The group at Mainz [5] has developed a higher pressure external polarized $^3$He target by mechanically compressing the gas. They have achieved over 40% polarization at 1 atm using a Toepler pump compression scheme.

Chupp and his co-workers [6] have developed a technique that involves optical pumping of Rb vapor in the $^3$He gas. The polarized Rb atoms transfer angular momentum to the $^3$He nuclei during collisions. A nitrogen buffer gas is necessary to reduce radiation trapping. Higher densities of $> 1$ atm are achievable with this technique. This group also employs a two-cell system to decouple the optical pumping region from the electron beam.

## Quasielastic Scattering

In this section we present the formalism for quasielastic scattering of longitudinally polarized electrons from polarized $^3$He.

### Theory of Inclusive Scattering

The general form for inclusive scattering of longitudinally polarized electrons from a polarized spin 1/2 target is given by [1]

$$\frac{d\sigma}{d\Omega d\omega} = \Sigma \pm \Delta(\theta^*, \phi^*), \tag{1}$$

where $\omega$ is the energy transfer and the angles $\theta^*$ and $\phi^*$ define the target spin direction relative to the momentum transfer direction. The plus(minus) sign corresponds to positive (negative) helicity incident electrons. The spin-independent cross section $\Sigma$ is given by the usual Rosenbluth formula

$$\Sigma = 4\pi\sigma_{Mott}[v_L R_L(q,\omega) + v_T R_T(q,\omega)], \tag{2}$$

in which $q$ is the momentum transfer, $v_L$ and $v_T$ are kinematic factors [1], $R_L$ is the longitudinal response function, and $R_T$ is the transverse response function. These response functions contain the nuclear electromagnetic structure information.

The spin-dependent cross section $\Delta$ contains two new response functions:

$$\Delta = -4\pi\sigma_{Mott}[\cos\theta^* v_{T'} R_{T'}(q,\omega) + 2\sin\theta^* \cos\phi^* v_{TL'} R_{TL'}(q,\omega)], \tag{3}$$

where $v_{T'}$ and $v_{TL'}$ are kinematic factors independent of nuclear structure [1]. The $R_{TL'}$ response function is of particular interest since it results from interference of longitudinal and transverse amplitudes. Thus it can be especially helpful in determining longitudinal amplitudes when the transverse amplitudes are dominant. This is the case in quasielastic scattering in $^3$He where $R_{TL'}$ is sensitive to the value of the neutron electric form factor. On the other hand, the response function $R_{T'}$ in quasielastic scattering from $^3$He is primarily determined by the neutron magnetic form factor.

Experimentally one measures the asymmetry in the cross section under reversal of electron helicity:

$$A \equiv \frac{\Delta}{\Sigma}. \tag{4}$$

Preliminary theoretical studies of spin-dependent quasielastic $\vec{e}^- - {}^3\vec{\text{He}}$ scattering in the plane-wave impulse approximation (PWIA) by Blankleider and Woloshyn [7] explored the sensitivity to nuclear structure as well as neutron form factors. In their study, the excitation of the residual $2N$ system was treated in the closure approximation. Friar *et al.* [8] used a spin density-matrix approach (integrated over the nucleon momenta and energies) to examine the effect of various ground state wave functions.

More sophisticated calculations have recently been performed by Ciofi degli Atti and co-workers [9] as well as by Schulze and Sauer [10]. These groups both employ a full spin-dependent spectral function to describe the ground state and compute the spin-dependent quasielastic scattering in PWIA. In addition, they prefer a new extraction scheme for computing the response functions; this new method yields different results for the $R_{TL'}$ interference response function than the one used by Blankleider and Woloshyn.

All of these calculations show that the spin dependent cross section for $\vec{q}$ parallel to the target spin is dominated by the neutron contribution proportional to $(G_M^n)^2$. The major effect of the $^3$He nucleus is a dilution of the asymmetry due to the contribution of the protons to the unpolarized cross section and a smearing of the quasielastic strength due to the momentum distribution of the nucleons in the nucleus. In addition, small components of the $^3$He wave function in which the protons are polarized contribute to the asymmetry.

The spin dependent cross section for $\vec{q}$ perpendicular to the target spin is sensitive to the neutron electric form factor. The early calculations of Blankleider and Woloshyn [7] indicated that the neutron contribution was quite substantial which would imply that information on $G_E^n$ could be reliably extracted from measurements of this quantity. However, the more recent calculations [9,10] of the response function $R_{TL'}$ obtain a result with a much smaller neutron contribution, which would imply a reduced sensitivity to the neutron electric form factor.

One should also consider the possibility that processes not included in these calculations can diminish the expected sensitivity to the neutron form-factors. The inclusive measurements should not be as sensitive to final state interactions (FSI) as exclusive measurements because there is an experimental integration over final states. One can expect that the theoretical treatment of the three-body continuum in nuclear physics will advance to the point where these issues can be quantitatively addressed during the next few years. Indeed, first calculations of the (unpolarized) quasielastic response in the $A = 3$ system by solving the Fadeev equations in the continuum have recently been published [11]. The possible influence of meson exchange currents remains to be investigated theoretically.

### Inclusive Measurements

In 1990, we performed the first measurements of spin dependent electron scattering from a polarized $^3$He target at Bates [12,13]. We simultaneously measured the longitudinal asymmetry $A_{T'}$ (proportional to $R_{T'}$) at $Q^2 = 0.2$ $(GeV/c)^2$ with one magnetic spectrometer and the transverse asymmetry $A_{TL'}$ (proportional to $R_{TL'}$) at $Q^2 = 0.16$ $(GeV/c)^2$ with another. Similar measurements were subsequently performed using a Rb spin-exchange target by a Harvard-MIT collaboration [14].

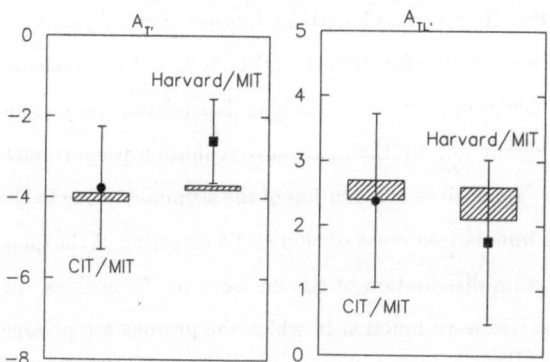

Fig. 1 Experimental results and theoretical calculations for: (a) the longitudinal asymmetry $A_{T'}$, and (b) the transverse asymmetry $A_{TL'}$. The calculations are from Ref. 9 and the experiments are from the Caltech-MIT group [12,13] and the Harvard-MIT group [14].

The results are shown in Fig. 1 along with theoretical predictions. Good agreement of the longitudinal asymmetry $A_{T'}$ with theory is evidence that the reaction mechanism is well understood. The transverse asymmetry $A_{TL'}$ is sensitive to the neutron electric form factor but the statistical precision of the measurements is insufficient to draw any firm conclusions about this quantity.

New high-precision measurements using the target developed at Caltech were performed recently at Bates. In this new experiment, a Wein spin-rotator in the polarized injector allowed the delivery of single-pass longitudinally polarized beam to the polarized $^3$He target. The beam energy was 370 MeV, and typically $25\mu$A at 37% polarization was available. Higher beam intensity coupled with more reliable operation of the single-pass beam enabled the attainment of much higher statistical precision than in the earlier runs. The target polarization was higher (typically 38%) and more stable in the new experiment due to the use of new LNA laser technology. The MEPS spectrometer was used to study the longitudinal asymmetry $A_{T'}$ at $Q^2 \simeq 0.20$ (GeV/c)$^2$, and preliminary results indicate that the measured asymmetry is in excellent agreement with recent calculations. The experimental uncertainty for $A_{T'}$ is approximately 15%. The OHIPS spectrometer was used in the new experiment to study the asymmetry $A_{TL'}$ where the spin is perpendicular to $\vec{q}$ at momentum transfer of $Q^2 \simeq 0.16$ (GeV/c)$^2$. In addition to measurements at the quasielastic peak, data were obtained for the elastic asymmetry and the threshold inelastic region $\omega \sim 50$ MeV. The elastic results are quite precise ($\sim 15\%$ uncertainty) and in very good agreement with the asymmetry calculated from the well-known $^3$He elastic form factors [15]. The data in the threshold inelastic region are sensitive to the theoretical technique for calculation of the response function $R_{TL'}$ in PWIA [9,10]. Further inclusive measurements of quasielastic asymmetries are proposed for both Bates [16] and CEBAF [17].

Exclusive Reactions

The detection of final state hadrons in coincidence with the scattered electron from a polarized target offers a powerful technique for studying the multipole structure of these response functions [2]. Such experiments that use the spin degrees of freedom and correlations of final state particles in coincidence with the scattered electron will fully exploit the polarized, high-intensity, CW beams that will be available in the near future at CEBAF, Bates, and Mainz.

Measurements of the exclusive quasielastic reactions $(e, e'p)$ and $(e, e'n)$ will allow detailed testing of the nuclear effects (both final and initial states) mentioned above. Studying the $(e, e'n)$ reaction with $\vec{q}$ parallel to the target spin $\vec{S}$ will test the effect of final state interactions of the outgoing nucleon in the spin-dependent response (the neutron magnetic form factor and the initial neutron wave function are relatively well known *a priori* in this case). Laget has performed calculations [18] that indicate final state interactions are a small correction for $Q^2 > 0.2$ GeV/c$^2$. Comparison of $(e, e'n)$ asymmetries with measurements of the inclusive $(e, e')$ asymmetries and the neutron energy dependence should help reveal the contributions of final state interactions.

The small amplitudes with the protons in spin $S = 1$ states referred to above can be studied in a rather direct fashion using the quasielastic $(\vec{e}, e'p)$ reaction on a polarized $^3$He target. (In the plane-wave impulse approximation the quasielastic asymmetry would vanish if the protons are in spin $S = 0$ states only.) Therefore, one can quantitatively study the $S'$ and $D$ state effects on the spin dependent response in $(e, e'p)$ (note that the proton form factors are well known). In fact, preliminary calculations by Laget [18] show that the $(e, e'p)$ asymmetries are quite sensitive to these wave function components and that final state effects and meson exchange corrections are quite small.

Armed with the results of this type of analysis, one can then confidently proceed to apply nuclear corrections to the inclusive $(e, e')$ and exclusive $(e, e'n)$ data in order to reliably extract the neutron form factors with good precision. The calculations by Laget [18,19] indicate that this technique for studying the neutron form factors appears very promising. The measurements require CW polarized beam as well as a large acceptance detector system. Experiments have been proposed for both the CEBAF Large Acceptance Spectrometer (CLAS) [17] and the Bates Large Acceptance Spectrometer Toroid (BLAST) [16].

An experiment at Mainz using the polarized $^3$He target developed there is just beginning. Preliminary asymmetry measurements have been performed and the results are very promising [20]. The goal is to study the electric form factor of the neutron in the range $0.2 \leq Q^2 \leq 0.6$ (GeV/c)$^2$ by measuring the quasielastic $(e, e'n)$ asymmetry with $\vec{q} \perp \vec{S}$. The scattered electrons are detected in a lead glass calorimeter with a gas Cerenkov detector for particle identification. The neutrons emitted in coincidence near the $\vec{q}$ direction are detected in an array of plastic scintillator detectors. Production runs are expected to take place during the next year.

<u>Conclusion</u>

In this brief survey, I have attempted to show how the present and future applications of polarized $^3$He targets in electronuclear experiments will yield exciting information on a variety of physics topics. Clearly, we are just beginning and I expect that as the experiments discussed here are completed and as the techniques improve there will be a next generation of new topics to explore with these powerful tools.

This work is supported in part by the National Science Foundation, Grant No. PHY91-15574.

## References

1. Donnelly, T.W., Raskin, A.S., Annals of Physics **169**, 247 (1986).

2. Raskin, A.S., Donnelly, T.W., Annals of Physics **191**, 78 (1989).

3. Meziani, Z.E., these proceedings.

4. Milner, R.G., McKeown, R.D., Woodward, C.E., Nucl. Instrum. Methods **A257**, 286 (1987); and Nucl. Instrum. Methods **A274**, 56 (1989).

5. Otten, E., Heil, W., private communication.

6. Chupp, T.E., *et al.*, AIP Conference Proceedings **187**, 1320 (1989).

7. Blankleider, B., Woloshyn, R.M., Phys. Rev. **C29**, 538 (1984).

8. Friar, J.L. *et al.*, Phys. Rev. **C43**, 2310 (1990).

9. Ciofi degli Atti, C., Pace, E., Salme, G., Phys. Rev. **C46**, R1591 (1992); proceedings of the workshop on "Perspectives in Nuclear Physics at Intermediate Energies," to be published (World Scientific, Singapore).

10. Schulze, R., Sauer, P., Phys. Rev. **C48**, 38 (1993).

11. Van Meijgaard, E., Tjon, J.A., Phys. Lett. **B228**, 307 (1989).

12. Woodward, C.E. *et al.*, Phys. Rev. Lett. **65**, 698 (1990).

13. Jones, C.E. *et al.*, Phys. Rev. **C47**, 110 (1993).

14. Thompson, A.K. *et al.*, Phys. Rev. Lett. **68**, 2910 (1992).

15. Beck, D. *et al.*, Phys. Rev. Lett. **59**, 1537 (1987).

16. Bates proposal 88-25, Chupp, T., Milner, R., spokesmen; Bates proposal 89-12, Milner, R.G., van den Brand, J., spokesmen.

17. CEBAF proposal 91-020, McKeown, R., spokesman.

18. Laget, J.M., Phys. Lett. **B273**, 367 (1991).

19. Laget, J.M., private communication.

20. Klein, F., private communication; Walcher, T., these proceedings.

Few-Body Systems, Suppl. 7, 112—119 (1994)

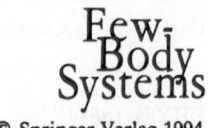

# ELECTRON SCATTERING FROM POLARIZED DEUTERIUM AT VEPP-3

C. E. Jones, K. P. Coulter,[a] R. Gilman,[b] R. J. Holt,
E. R. Kinney,[c] R. S. Kowalczyk, D. H. Potterveld, L. Young
*Argonne National Laboratory, Argonne, Illinois USA*

V. V. Frolov, S. I. Mishnev, D. M. Nikolenko, S. G. Popov,
I. A. Rachek, D. K. Toporkov, E. P. Tsentalovich, B. B. Wojtsekhowski,[d]
*Budker Institute for Nuclear Physics, Novosibirsk, Russia*

C. W. de Jager, G. Retzlaff,[e] J. Theunissen, and H. de Vries
*NIKHEF, Amsterdam, The Netherlands*

V. V. Nelyubin, V. V. Vikhrov
*St. Petersburg Institute for Nuclear Physics, Gatchina, Russia*

V. N. Stibunov, A. V. Osipov
*Tomsk Polytechnic Institute, Tomsk, Russia*

## Abstract

The experiment to measure the tensor analyzing power $T_{20}$ of the deuteron using a tensor-polarized internal target at the VEPP-3 electron storage ring in Novosibirsk is currently in its second phase. Tensor-polarized deuterium atoms from an atomic beam source feed an active storage cell in the electron ring, serving as the target for a beam of 2 GeV unpolarized electrons. Analysis of the first Phase 2 data in the kinematic range $Q^2 = 0.36 - 0.50$ $(GeV/c)^2$ is reported here. Plans for the third phase of the experiment, which will use a high-density, laser-driven polarized deuterium target, are also discussed. The measurement of $T_{20}$ from the second and third phases of the experiment will cover the kinematic region $0.36 \leq q \leq 0.90$ $(GeV/c)^2$ where $T_{20}$ is particularly sensitive to isoscalar meson exchange currents.

## Overview

The electromagnetic structure of the deuteron is characterized by three form factors which can be measured in elastic electron-deuteron scattering; $F_C$, the charge monopole form factor, $F_M$, the magnetic form factor, and $F_Q$, the charge quadrupole form factor. Although the deuteron is a much-studied system, both in terms of experimental and theoretical studies, the models predict a wide range of values for the electromagnetic form factors even at modest momentum transfer. For example, see the papers presented in these proceedings alone [1]. Experimental data on polarization observables constrain models of the deuteron wave function, the effect of non-nucleonic degrees of freedom and the size of relativistic corrections. As data at higher momentum becomes available, it is also possible to test calculations based upon quark degrees of freedom and ultimately to test QCD.

Unpolarized scattering measurements are unable to fully separate the three form factors because the unpolarized cross section,

$$\left(\frac{d\sigma}{d\Omega}\right)_0 = \sigma_{Mott}\left(A(Q^2) + \tan^2(\theta/2)B(Q^2)\right), \tag{1}$$

depends upon two structure functions,

$$A(Q^2) = F_C^2 + (8\tau^2/9)F_Q^2 + (2\tau/3)F_M^2 \tag{2}$$

and

$$B(Q^2) = 4\tau(1+\tau)F_M^2/3, \tag{3}$$

where

$$\tau = Q^2/4m_d^2. \tag{4}$$

It is not possible to independently determine both $F_C$ and $F_Q$ without the introduction of polarization degrees of freedom into the scattering process. However, knowledge of both the unpolarized cross section and the tensor analyzing power $T_{20}$ in elastic scattering of unpolarized electrons from tensor-polarized deuterium is sufficient to separate all three form factors. The expression for $T_{20}$ in terms of the form factors is

$$T_{20} = -\sqrt{2}[x(x+2) + y/2]/[2(x^2+y)+1], \tag{5}$$

where

$$x = \frac{2}{3}\tau\left(\frac{F_Q}{F_C}\right) \tag{6}$$

and

$$y = \frac{\tau}{3}(1 + 2(1+\tau)\tan^2(\theta/2))\left(\frac{F_M}{F_C}\right)^2. \tag{7}$$

In principle, a measurement of the tensor polarization of the outgoing deuterons in elastic scattering using unpolarized beam and target provides the same information about the structure of the deuteron, and this technique was employed in a recent measurement at Bates [2].

Above $Q^2 \sim 0.40$ (GeV/c)$^2$ the tensor analyzing power $T_{20}$ in elastic $e-d$ scattering is very sensitive to the deuteron model used in the calculations because, in the impulse approximation, $F_C$ is predicted to have a zero in this kinematic region and the position of the first minimum is quite sensitive to the introduction of isoscalar meson exchange currents [3]. Fig. 1 shows a plot of $T_{20}$ calculated using a number of different theoretical models to show the range of the predictions. The dot-dash line is the calculation of Schiavilla and Riska using the Argonne V14 potential which includes meson exchange currents explicitly [3], the dotted line is the coupled channel calculation (model D′) of Blunden et al. [4], the short dashed line is the relativistic calculation of Chung et al. using light-front dynamics to calculate the deuteron wave function from a nonrelativistic potential [5], and the solid line is the relativistic calculation of Hummel and Tjon using a covariant wave function calculated from the one-boson exchange model [6]. The long dashed line is a recent calculation based

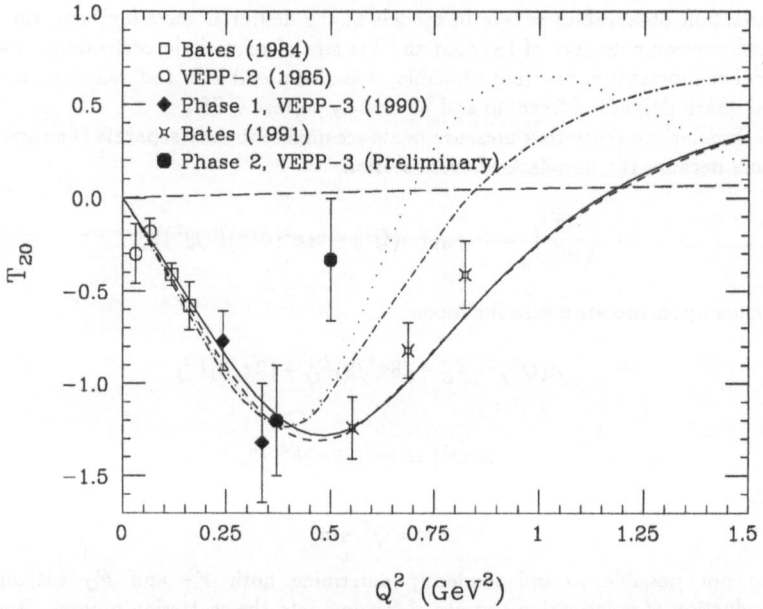

Fig. 1. Models of the tensor analyzing power in elastic $e - d$ scattering $T_{20}$, calculated by Schiavilla and Riska (dotdash) [3], Blunden et al. (dot) [4], Chung et al. (short dash) [5], Hummel and Tjon (solid) [6], and Brodsky and Hiller (long dash) [7]. Also shown are data from previous measurements [2,8,9,10] and the preliminary analysis of the data from the ongoing internal target experiment at VEPP-3.

on perturbative QCD by Brodsky and Hiller [7]. Also shown are data from previous measurements [2,8,9,10].

Because the nucleon-nucleon model serves as input for "exact" calculations of the electromagnetic structure of the three-body system, one should be able to describe both the two and three-body systems starting from the same model of the deuteron. Fig. 2 shows the experimental data on the deuteron charge form factor and the three-body isoscalar charge form factor along with a consistent calculation for $A = 2$ by Schiavilla and Riska [3] and for $A = 3$ by Schiavilla, Pandharipande and Riska [11] performed both in the impulse approximation and with meson exchange currents. The inclusion of meson exchange currents shifts the minimum in $F_C$ to lower momentum for both the two- and three-body system. A measurement of $T_{20}$ in the range $3.8 \leq q \leq 4.6$ fm$^{-1}$ at the MIT-Bates electron accelerator using an unpolarized target and a recoil polarimeter [2] determined that the node in $F_C$ lies at $q = 4.39 \pm 0.16$ fm$^{-1}$, a value that is consistent with impulse approximation calculations which include no meson exchange currents. However, the experimental data from the three-body system [12], shown as a light band to indicate the size of the experimental uncertainty, clearly indicate the necessity of including meson exchange currents. The shift to lower momentum of the minimum in $F_C$ for the deuteron is a general feature of models that include meson exchange currents [13], and presently there is no model of the deuteron that both agrees with the deuteron experimental

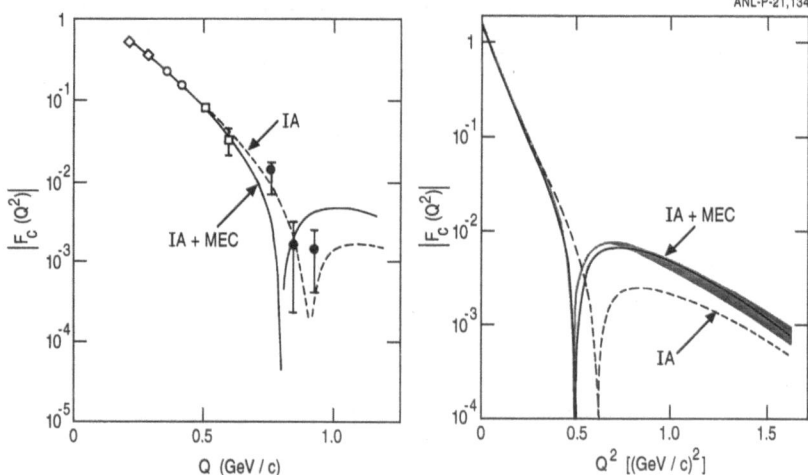

Fig. 2. The isoscalar charge form factor for the deuteron calculated by Schiavilla and Riska (left) [3] and for the three-body system calculated by Schiavilla, Pandharipande and Riska (right) [11] in the impulse approximation (dashed line) and including meson exchange currents (solid line). The experimentally-determined values of the $A = 3$ isoscalar charge form factor are shown as a band to indicate the size of the experimental uncertainty [12].

data and, when used as input for three-body calculations, successfully describes the existing data on the three-body isoscalar form factors. It is noted that relativistic corrections to the impulse approximation for the $A = 3$ isoscalar charge form factor shift the minimum to higher momentum [14], worsening the agreement between theory and experiment and supporting the claim that meson exchange currents are important in the description of nuclear dynamics.

### The VEPP-3 Experiment

The experiment to measure $T_{20}$ at the VEPP-3 electron storage ring at the Budker Institute for Nuclear Physics in Novosibirsk, Russia, has been underway since 1988, when the internal target technique was pioneered [15]. Results from Phase 1 have been reported [10], and the second phase of the experiment, which started in 1990, is ongoing. Installation of Phase 3 is scheduled to begin in summer 1994. The experiment uses an internal tensor-polarized deuterium target and a beam of 2 GeV electrons.

The major difference between the different phases of the experiment is the type of target used. In the first phase an atomic beam source (ABS) coupled to a passive storage cell was used, with a resulting useful target thickness of $3 \times 10^{11}$ cm$^{-2}$. Currently the same ABS is coupled to an active storage cell which opens during beam filling and closes once the electrons are stored, providing a useful target thickness of $2 \times 10^{12}$ cm$^{-2}$. The third phase of the experiment will use a high density, laser-driven polarized deuterium source coupled to a passive storage cell. Flow rates of up

116

to $9 \times 10^{17}$ s$^{-1}$ have been demonstrated with the laser-driven source [16], so it is likely that the target thickness in Phase 3 will be limited by the vacuum requirements of the storage ring. Using the laser-driven source, a target thickness of $4 \times 10^{14}$ cm$^{-2}$ is expected.

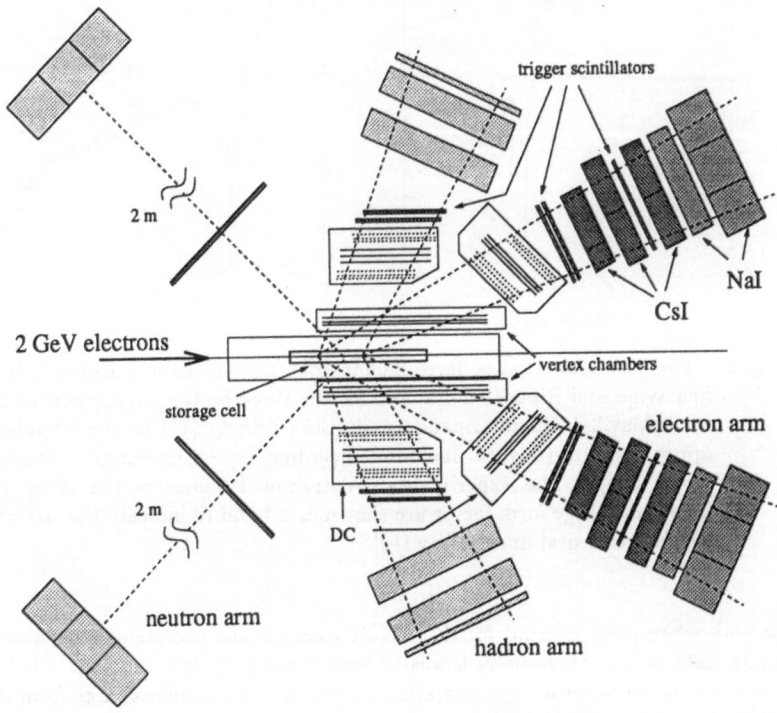

Fig. 3. Detector layout for Phase 2 and Phase 3 of the VEPP-3 experiment.

Fig. 3 shows the detector layout for the experiment. The detector system has two electron arms and two hadron arms arranged symmetrically around the electron beamline. The arms are nearly identical and provide redundancy to minimize systematic errors. The electron arm covers the angular range $20° - 30°$ and the hadron arm covers the range $60° - 70°$. Immediately outside the scattering chamber there are vertex chambers that record events for both the hadron and electron arms. The electron arm consists of drift chambers, thin scintillators for the event trigger, and a calorimeter that uses CsI and NaI blocks. The hadron arm has drift chambers, trigger scintillators, and thick scintillators for hadron detection. The system also contains two neutron arms, consisting of thick scintillators, in the backward direction $(132° - 138°)$ which are used in inelastic triggers. Not shown in Fig. 3 are small forward angle detectors installed at the end of the second data run which serve as a polarization monitor by measuring small angle $e - d$ elastic scattering. Initially there was a high background in the detector because of beam halo scattering from the cell walls, however collimators installed in the beamline upstream of the interaction region effectively suppressed the background in the detectors, reducing the rate on the trigger scintillators of the electron and hadron arms approximately an order of magnitude.

The detector system used in Phase 1 was not optimized for an internal target, having been designed for an earlier experiment that used a gas jet target, so the effective target thickness seen by the detectors was only approximately a fifth of the total target. In the current experiment the detector system has been altered to view most of the central region of the target, so that the effective target thickness is half the total thickness. In the first run of Phase 2 the effective target thickness was $(2.0 \pm 0.5) \times 10^{12}$ cm$^{-2}$. Because the reliability of the active storage cell has proven problematic, a passive cell will be used during the next phase.

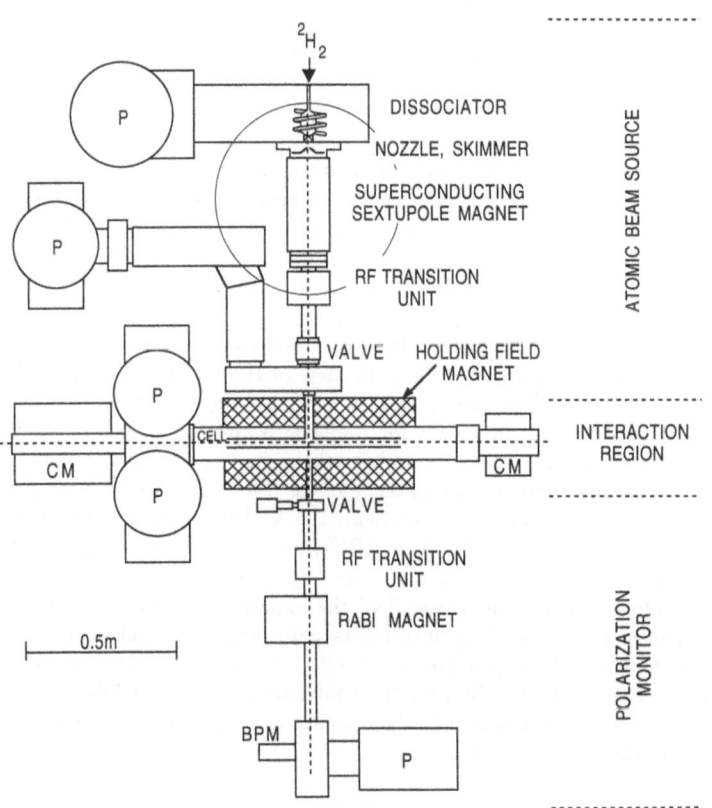

Fig. 4.  Schematic of the target system, consisting of an atomic beam source coupled to a storage cell and a Rabi polarimeter. The vacuum pumps are denoted by 'P' and the compensating magnets by 'CM.'

The layout of the target region is shown in Fig. 4. The polarized deuterium from an atomic beam source flows into a storage cell. Holding field magnets that provide a 900 Gauss holding field in the target region are placed outside the scattering chamber. Opposite the entrance to the storage cell is a small hole which allows atoms to pass into a Rabi polarimeter [17], so that the polarization of the beam from the ABS can be continuously monitored. The beam from the ABS has a tensor polarization of $p_{zz} = \pm 0.98 \pm 0.05$. Both the ABS and the polarimeter can be valved off from the beam line for maintenance.

## Results and Conclusions

In the current phase of the experiment two data runs have been taken and a third is planned. The results discussed here are from the preliminary analysis of the data from the first run only. The results are shown as the solid circles in Fig. 1. Because the extraction of $T_{20}$ from the experimental asymmetry requires knowledge of the average $p_{zz}$ of the atoms in the target cell, while the polarimeter measures the $p_{zz}$ mainly of the atoms coming directly from the ABS, the average $p_{zz}$ is determined by normalizing the data point at $Q^2 = 0.37$ (GeV/c)$^2$ to the theoretical value calculated in the impulse approximation from the Paris potential. The normalization procedure yields $p_{zz} = 0.67 \pm 0.19$. The events constituting the two data points shown in Fig. 1 are simultaneously within the momentum acceptance of the detector system. The error bars on the higher momentum point include the systematic uncertainty from the normalization of the lower momentum datum. The value of $p_{zz}$ extracted from the first data run by normalization is in agreement with a direct polarization measurement using the small angle detectors installed at the end of the second data run, which gave a measured average of $p_{zz} = 0.58 \pm 0.17$.

With the data from the first run only, the statistical accuracy of this new datum at the higher momentum transfer is not sufficient to indicate the need for isoscalar meson exchange currents, although the point lies above the results of the previous measurement in this kinematic region. In view of the fact that the deuteron charge form factor extracted from the currently available data is incompatible with the calculations and experimental data for the $A = 3$ system, it is important both for experimentalists to verify the data and for theorists to improve the nuclear models. Because the polarized internal target technique is different from the recoil polarimeter technique used in the Bates measurement, the VEPP-3 measurement will serve to experimentally verify the results from MIT-Bates using an experimental technique subject to different systematic errors. The improved statistics from the rest of the data from Phase 2 and from Phase 3 of the experiment, where a factor of $5 - 10$ improvement in the figure of merit of the target is expected, will provide a stringent constraint theory in the region from $0.4 \leq Q^2 \leq 0.9$ (GeV/c)$^2$ where isoscalar meson exchange currents significantly alter the analyzing power. Hopefully, a combination of new data and theory can rectify the existing discrepancy between descriptions of the two- and three-body systems.

## Acknowledgements

This work is supported in part by the U.S. Department of Energy, Nuclear Physics Division, under contract W-31-109-ENG-38, and by the Netherlands' Organization for Scientific Research, under contract 713-119.

## References

[a] Current address: University of Michigan, Ann Arbor, Michigan, USA
[b] Current address: Rutgers University, Piscataway, New Jersey, USA
[c] Current address: University of Colorado, Boulder, Colorado, USA
[d] Current address: Rensselaer Polytechnic Institute, Troy, New York, USA

<sup>e</sup> Current address: Saskatchewan Accelerator Laboratory, Saskatoon, Canada

[1] See, for example, the contribution of P. U. Sauer and of S. J. Wallace.

[2] I. The et al., Phys. Rev. Lett. **67**, 173 (1991).

[3] R. Schiavilla and D. O. Riska, Phys. Rev. C **43**, 437 (1991).

[4] P. G. Blunden et al., Phys. Rev. C **40**, 1541 (1989).

[5] P. L. Chung et al., Phys. Rev. C **37**, 2000 (1988).

[6] E. Hummel and J. A. Tjon, Phys. Rev. Lett. **63**, 1788 (1989).

[7] Stanley J. Brodsky and John R. Hiller, Phys. Rev. D **46**, 2141 (1992).

[8] M. E. Schulze et al., Phys. Rev. Lett. **52**, 597 (1984).

[9] V. F. Dmitriev et al., Phys. Lett. 157B, 143 (1985).

[10] R. Gilman et al., Phys. Rev. Lett. **65**, 1733 (1990).

[11] R. Schiavilla, V. R. Pandharipande, and D. O. Riska, Phys. Rev. C **41**, 309 (1990).

[12] A. Amroun et al., Phys. Rev. Lett. **69**, 253 (1992).

[13] H. Henning, J. Adams, Jr., P. U. Sauer, and A. Stadler, submitted to Phys. Rev. C.

[14] George Rupp and J. A. Tjon, Phys. Rev. C **45**, 2133 (1992).

[15] R. Gilman et al., Nucl. Instrum. Meth. **A327**, 277 (1993).

[16] M. Poelker et al., ANL Preprint Phy-7140-ME-93.

[17] A. V. Evstigneev, S. G. Popov, and D. K. Toporkov, Nucl. Instrum. Meth. **A238**, 12 (1985).

Few-Body Systems, Suppl. 7, 120—127 (1994)

Few-
Body
Systems
© Springer-Verlag 1994
Printed in Austria

# Results of (e,e′x) experiments on ⁴He

## H. P. Blok

Department of Physics and Astronomy, Vrije Universiteit,
de Boelelaan 1081, 1081 HV Amsterdam, The Netherlands
and
NIKHEF-K, P.O. Box 41882, 1009 DB Amsterdam, The Netherlands

### Abstract

Results of (e,e′x) (x=p,n,d,t,³He) experiments on ⁴He are discussed. The description of the final continuum state seems to be a crucial ingredient. Charge exchange plays an important role in the 1N and 3N knockout reactions. A microscopic description of the (e,e′d) reaction largely overstimates the measured cross section, presumably because the description of the final state is too simple.

## 1 Introduction

The nucleus ⁴He constitutes an excellent nuclear laboratory for testing various ingredients of both nuclear structure and reactions. On the one hand ⁴He is small enough so that realistic calculations can be done, at least for the ground state, and hopefully in the future also for continuum states. On the other hand it is a strongly bound system, in which the nucleon-nucleon correlations, as visible for instance in the high-momentum components in the nucleon momentum density distribution, are already almost as large as for a heavy nucleus like ²⁰⁸Pb.

The best way to probe various aspects of ⁴He is to use electron induced reactions since the electromagnetic interaction is known and relatively weak. The latter entails that initial-state interactions, which for instance obscure the interpretation of pion-absorption reactions, are negligible. The use of virtual photons allows one to study in detail such properties as the single-nucleon density of ⁴He, (nucleon-nucleon) correlations, the role of two-body currents in the interaction, etc.

From the analyses presented it will become clear that ⁴He has also a drawback: approximations like an optical-model description for the distortions of the outgoing-nucleon waves, which is a reasonably good approximation for heavy nuclei where there are many open channels, are poor for ⁴He. Especially the charge-exchange channel plays an important role, as will be demonstrated.

## 2    Proton and neutron knockout from $^4$He

If one describes the $^4$He(e,e'N) (N=p,n) process as quasi-elastic scattering and uses plane waves for the electrons and the outgoing proton (PWIA), the cross section can be written as [1]

$$\frac{d^6\sigma}{de'dp} = K\sigma_{ep}(q)S(\mathbf{p}_m, E_m) \tag{1}$$

where K is a kinematical (phase-space) factor, $\sigma_{ep}$ is the (relativistic) off-shell electron-proton cross section and $S(\mathbf{p}_m, E_m)$ is the spectral function. For a transition to a discrete state the spectral function reduces to the momentum distribution $\rho(\mathbf{p}_m)$, which depends on the $\langle A|A-1\rangle$ overlap (bound-state wave function). The PW formula given above is no longer valid when distortions of the outgoing proton or two-body currents are present. In the spirit of eq. (1) one then defines a reduced cross section as the measured cross section divided by $K\sigma_{ep}$. One can also just integrate the cross section over a discrete peak, which then gives a five-fold cross section. Further details about the formalism can be found in refs. [1, 2].

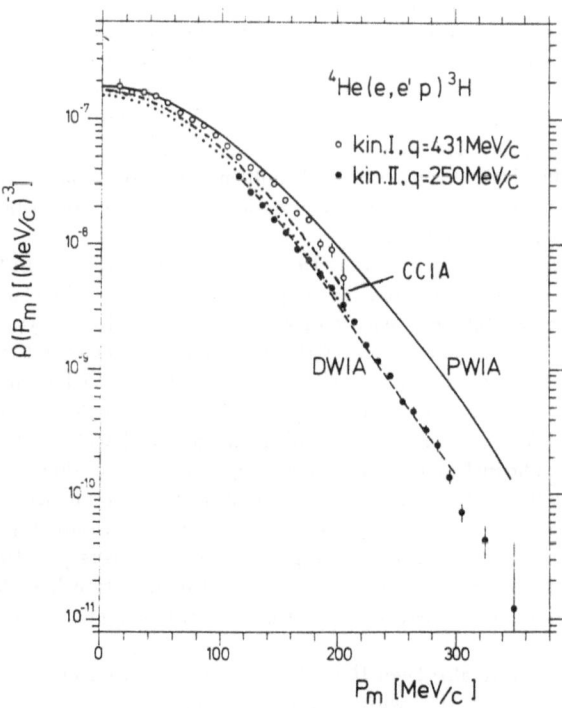

Figure 1: Data for the $^4$He(e,e'p)$^3$H reaction in two kinematics.

In order to illustrate what one has learned and what problems one encounters in the interpretation of $^4$He(e,e'N) data, we focus on (e,e'p) data taken in two kinematics, which have been the subject of several discussions [3] - [8]. The data were taken at an electron energy of appr. 450 MeV with the two-spectrometer set-up of NIKHEF and a cryogenic gas target. The missing-energy resolution was about 0.4 MeV. Parts of the missing momentum distribution were measured at a p-$^3$He center-of-mass energy of 75

MeV for two values of the virtual-photon polarization: $\epsilon = 0.48$ (kin. I) and $\epsilon = 0.80$ (kin. II).

The data are reproduced in fig. 1. The dotted and dashed curves are the results of a DWIA analysis that includes a $\langle{}^3H|{}^4He\rangle$ overlap calculated by Schiavilla with the variational Monte Carlo method [9] and a p-${}^3$H optical potential due to van Oers [10]. The DWIA calculations clearly can not describe the larger cross sections measured in the more transverse kinematics. It is known that in such kinematics $(e,e'n)(n,p)$ contributions can play a non-negligible role [11]. These can easily be evaluated within the Lane formalism. This leads to the coupled equations:

$$(E - T - V)\chi_p = \mu_p f + V_{pn}\chi_n \tag{2}$$
$$(E - T - V)\chi_n = -\mu_n f + V_{pn}\chi_p \tag{3}$$

where the $\chi$'s are the optical-model wave functions and the $f$'s the source terms for an $(e,e'p)$ reaction, which contain the overlap function. The extra minus sign in front of $\mu_n$ is due to antisymmetrization. It is easily seen that in first order (where $\chi_p$ and $\chi_n$ are just proportional to each other) the influence of the $(e,e'n)$ channel leads to an additional distorting potential in the proton channel of $-\frac{\mu_n}{\mu_p}V_{pn}$. Since $V_{pn}$ is repulsive and absorptive, the net distortions get smaller and the calculated cross sections move towards the PW results. This means an increase in this case, as can be seen from the dash-dotted curve for kin. I in fig. 1. (The increase in kin. II is only a few percent.) Part of the discrepancy between the two kinematics is thus explained. Laget has performed calculations [12], in which he also uses Schiavilla's overlap function, treats the final-state interaction (FSI) as single rescattering thus including charge-exchange (CE), and includes meson-exchange currents (MEC). The latter have a noticeable influence only for $p_m$ larger than 200 MeV/c, although even there the effect is small. The data are rather well described in kin. I, and somewhat underestimated in kin. II.

It is now interesting to look at the ${}^4$He(e,e'n) reaction, since there the influence of charge exchange is expected to be much more prominent, as the $(e,e'p)$ channel is relatively strong. Unfortunately, taking $(e,e'n)$ data with low-duty factor ($\leq 1\%$) accelerators is difficult. Data have been obtained at NIKHEF for $p_m$ values of 176 and 210 MeV/c by detecting the recoiling ${}^3$He nucleus with a special detector for low-energy particles in the focal plane of the QDQ spectrometer [13]. The data, which have large error bars, are compared to DWIA (no CE), CCIA (CE included) and Laget's calculation (also CE included) in fig. 2. The inclusion of charge exchange makes a large difference and improves the description. It should be mentioned that the optical-model potential used in the DW and CC calculations has unrealistic diffuseness parameters, presumably mimicking in a pure *elastic* description the influence of charge exchange. In this respect these calculations could be improved. Some exploratory attempts in this direction have been made [14]. Further it is questionable if a first order treatment of FSI as applied by Laget is sufficient (see e.g. [15]).

The (e,e'p) data have also been the subject of several other analyses. Schiavilla [6] performed a DWIA calculation with charge exchange added in a Lane formalism and including MEC's. He mentions that the inclusion of CE hardly affects the calculated cross sections in kin. I and even decreases them in kin. II. In view of the formalism presented above this is presumably due to the fact that he just *added* the charge-exchange potential to the optical potential. The opening of the extra channel $(e,e'p)(p,n)$ then removes proton flux, while the $(e,e'n)(n,p)$ route brings some back. In longitudinal kinematics (II) the latter is small, so the net effect is a reduction, while in kin. I they approximately balance each other. Furthermore Schiavilla has investigated the effect of non-orthogonality of the bound and scattering states of the proton by using correlated orthogonal scattering states (OCS). The effects are small, especially at low values of $p_m$.

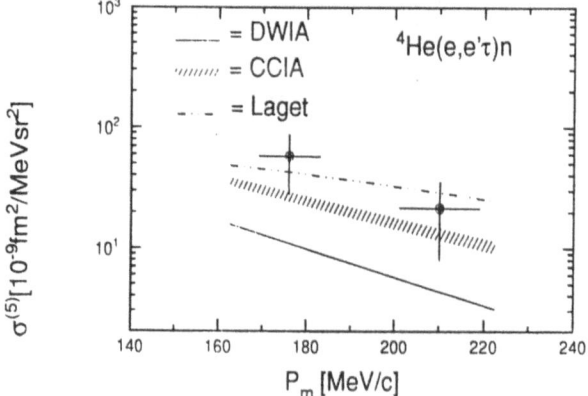

Figure 2: Data for the $^4$He(e,e'n)$^3$He reaction taken by detecting the residual $^3$He nucleus.

Warmann and Langanke [7] performed Resonating Group calculations with the modified Minnesota force, thus having a consistent description of the bound ($^4$He) and scattering ($^3$H+p, $^3$He+n) states. They get perfect agreement with the data, but it is not clear how their $\langle ^3H|^4He\rangle$ overlap compares to the one from Schiavilla and certainly their calculation neglects loss of flux into channels like p+d+n. Furthermore they do not include MEC's.

Finally, Buballa *et al.* [8] have performed an RPA type (e,e'p) calculation, which thus includes charge exchange and yields a rather good description with the data. Here there are also some questions about the overlap function and FSI effects beyond RPA.

Many more data have been obtained in other kinematics both at NIKHEF [5] and Saclay [16, 17]. The widest analysis has been done using Laget's calculations. On average these give a reasonable description, but deviations of up to 50% with the measured cross sections occur. As previously mentioned the first order treatment of rescattering is the main point to improve. A good and necessary requirement seems to be that p+$^3$H and n+$^3$He scattering data are described correctly. This consistency requirement is not fulfilled by several calculations. The unsatisfying conclusion at this moment must be that uncertainties in the description of the FSI, i.e., the scattering state, preclude conclusions about the bound-state properties, i.e., the overlap, for which rather advanced calculations exist.

## 3   The $^4$He(e,e'd) reaction

Correlations between nucleons can be probed with the (e,e'd) reaction. In a cluster-type description, in which the deuteron exists as an entity already in the initial nucleus (quasi-elastic deuteron knockout), the cross-section formula is similar to eq. (1), with the proton replaced by a deuteron. In a microscopic (i.e. based on separate protons and neutrons) formalism [18] the cross section is derived from the T-matrix element (in PWIA):

$$T = \langle \phi_{e'}\phi_d\Psi_{A-2}|V_{ep}|\phi_e\Psi_A\rangle \tag{4}$$

where the $\phi$'s are plane waves. (For simplicity we have assumed only a charge interaction.) By expanding the $\langle A - 2|A\rangle$ overlap in terms depending on the center-of-mass **R** and

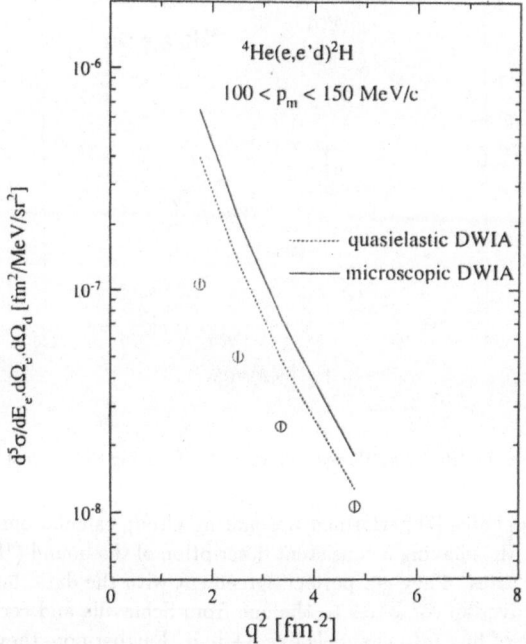

Figure 3: Measured cross sections as a function of $q$ for the $^4$He(e,e'd)$^2$H reaction.

relative coordinate **r** of the p-n pair this can be written as the product of two factors

$$T = \int e^{-i\mathbf{p}_d.\mathbf{R}} e^{i\mathbf{q}.\mathbf{R}} \Phi_{NL}(\mathbf{R}) d\mathbf{R} \int \varphi_d^*(\mathbf{r}) F_p(q) e^{i\mathbf{q}.\mathbf{r}/2} \varphi_{pn}(\mathbf{r}) d\mathbf{r}, \tag{5}$$

where $F_p(q)$ is the proton form factor. The first factor is the spectral function for the c.o.m. motion of the p-n pair in the initial nucleus (in DWBA the plane wave for the outgoing deuteron in this formula is replaced by a distorted wave). The second one is a sort of $pn - d$ transition matrix element, which, if the initial p-n pair is already a deuteron, leads to the $\sigma_{ed}$ factor in the cross section formula for quasi-elastic knockout.

It is interesting to note that a similar description of the (e,e'p) reaction leads to an equivalent formula except that $\varphi_d(\mathbf{r})$ in eq. (5) is replaced by the relative wave function (in PWIA a plane wave $e^{i\mathbf{p}.\mathbf{r}}$) of the final p-n pair. It is thus seen that both (e,e'd) and (e,e'pn) probe the same relative p-n motion ($\varphi_{pn}(\mathbf{r})$) in the initial nucleus, though in different ways. The (e,e'd) reaction has the special characteristic that it acts as an isospin filter. Thus the transition to the $^2$H final state in the $^4$He(e,e'd) reaction proceeds only through the T=0 part of the interaction of the electron with the nucleons, while the transition to the p-n continuum just above threshold is dominantly a $\Delta S = 0$, $\Delta T = 1$ interaction.

The dependence of the cross section for the $^4$He(e,e'd)$^2$H transition on the value of the momentum transfer $q$, which reflects $\varphi_{pn}$, has been investigated [19] at a fixed missing momentum of the deuteron of 75 MeV/c. The data are shown in fig. 3. DWIA calculations with a folding type d-d optical potential fitted to measured elastic scattering data were performed, where the $\langle A-2|A\rangle$ overlap was taken from ATMS variational calculations [20]. In the first (quasi-elastic type) calculation a (distorted) spectral function was calculated, which was multiplied by $K\sigma_{ed}$. In the second calculation the full microscopic T-matrix

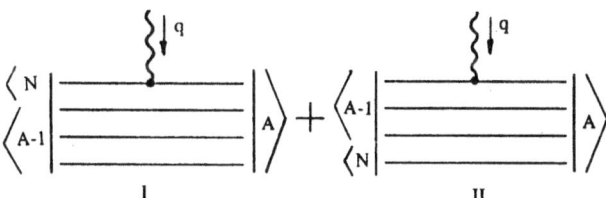

Figure 4: The "direct" and "recoil" diagrams contributing to an (e,e'p) reaction.

element, including (anti)symmetrization and D-states for ⁴He and the deuterons, was evaluated. The calculated cross sections severely overestimate the measured ones. The origin of this large discrepancy is not yet clear. There is an effect from non-orthogonality between the bound and scattering states, but it is not large [21]. Since the data were taken in largely longitudinal kinematics, no large effects of MEC's are expected. Contributions from sequential knockout (e,e'p)(p,d) can not be ruled out, but it is not clear then why these are not visible in the ⁶Li(e,e'd) reaction, which is perfectly well described [22]. Moreover RGM calculations by Warmann and Langanke [23], which include this channel, show some effect, but by far not large enough to explain the discrepancy. A point could be that the FSI distortions were calculated in a factorised form, which may not be a good approximation in the non-parallel kinematics in which the data were taken. All these points will be studied for the ³He(e,e'd) reaction, for which exact calculations can be performed [24].

## 4    The ⁴He(e,e't) and (e,e'³He) reactions

A study of the ⁴He(e,e't) and also of the (e,e'³He) reaction is of importance first of all to investigate the mechanism of multi-nucleon cluster knockout, but also because in certain kinematics of the ⁴He(e,e'p)³H reaction triton knockout may give nonnegligible contributions. This is illustrated in fig. 4. The cross section is determined by the coherent sum of the two amplitudes. When data are taken under such conditions that $p_m$ is large and $p_{p'}$ is relatively small, contributions from the process where the virtual photon has interacted with the triton, the detected proton being a spectator, may be appreciable. The second amplitude is often called the antisymmetrization term, since it arises when one uses a fully antisymmetrized formalism. In fact one gets this term already without antisymmetrization, since the virtual photon may interact with any of the nucleons in the initial nucleus. Antisymmetrization then gives precisely the same terms again. This can easily be seen from the following antisymmetrized formula for the T-matrix element:

$$T = \langle A[\mathrm{p}(1)\ B(2,...,Z,Z+1,..,A)]|\sum_{i=1}^{Z} V_{ei}|A(1,...,Z,Z+1,...,A)\rangle, \qquad (6)$$

where $A$ means antisymmetrization and the states A and B are assumed to be internally antisymmetrized already. Without antisymmetrization the sum over the protons gives two different terms, a direct term for i=1, where the interaction took place with the detected proton, and Z-1 identical "body" terms for i=2...Z, where the interaction took place with a proton in the residual nucleus B. Since one only has to antisymmetrize proton 1 with

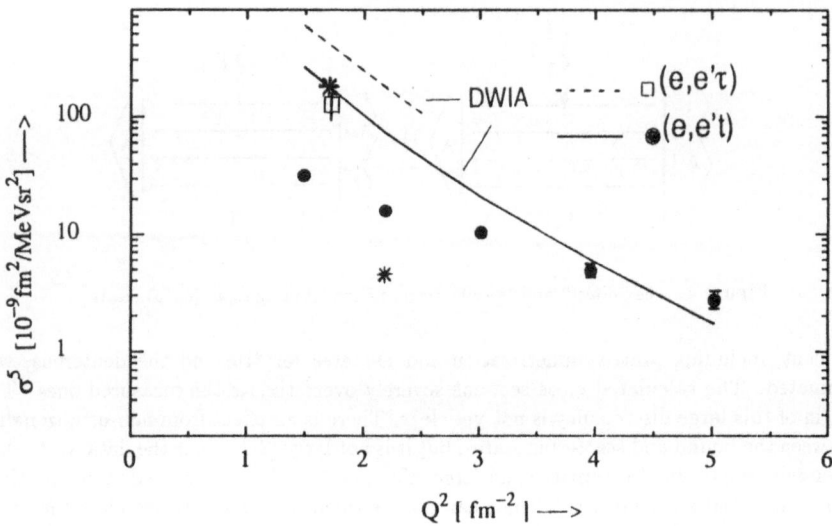

Figure 5: Measured cross sections as a function of $q$ for the ${}^4\text{He}(e,e't)$ and ${}^4\text{He}(e,e'{}^3\text{He})$ reactions.

the other Z-1 ones, the antisymmetrization operator is

$$A = \frac{1}{\sqrt{Z}}\left(1 - \sum_{i=2}^{Z} P_{1i}\right),\tag{7}$$

where $P_{kl}$ permutates nucleons k and l. As now the numbering is irrelevant, the sum of the interactions can be replaced by Z times the interaction with proton 1. One then finds

$$T = \sqrt{Z}\,[\langle p(1)\,B(2,...,Z,Z+1,...,A)|V_{e1}|A(1,...,Z,Z+1,...,A)\rangle$$
$$-(Z-1)\langle p(2)\,B(1,3,..,Z,Z+1,...,A)|V_{e1}|A(1,...,Z,Z+1,...,A)\rangle].\tag{8}$$

The $\sqrt{Z}$ factor is cancelled by the difference in normalization constants of the Slater determinants for the nuclei A and B.

The ${}^4\text{He}(e,e't)$ and ${}^4\text{He}(e,e'{}^3\text{He})$ reactions in "direct" 3N knockout kinematics have been studied at a $p_m$ value of 180 MeV/c (i.e. the residual nucleon has a momentum of 180 MeV/c) and c.o.m. energies of the final system of 40 MeV and 25 MeV, respectively (see [3, 14] for further details). The data are shown in fig. 5. In a simple quasi-elastic picture one expects the $(e,e'{}^3\text{He})$ cross section to be a factor of about 3 larger than the one for $(e,e't)$ at $Q^2 = 2.2$ fm${}^{-2}$ (a factor of 4 due to the difference in charges times 0.75 due to the difference in form factors of the triton and ${}^3\text{He}$ at this value of $Q^2$). The experimental ratio is much larger. The magnitude of the separate data is also largely overestimated by quasi-elastic DWIA calculations, in which a formula similar to eq. (1) is used.

One may expect that charge-exchange plays a role also in these reactions. Using similar first order arguments in the Lane model as in section 2 one finds that in this case both cross sections will be decreased, but the one for $(e,e't)$ more than for $(e,e'{}^3\text{He})$. A first calculation using the same optical potential, bound-state wave function and charge-exchange potential as for the ${}^4\text{He}(e,e'p)$ reaction indeed yields some decrease in the cross section of the $(e,e'{}^3\text{He})$ channel and a much larger decrease, which depends sensitively on the details of the calculations, in the $(e,e't)$ channel, see the asterixes in fig. 5. However, one should also evaluate processes like $(e,e'p)(p,t)$, $(e,e'p)(p,^3He)$ and MEC's, plus the fact that the triton and $^3He$ in $^4He$ are different from the free ones, before one can make definite statements about the description of these knockout reactions. This means, on the other hand, that the interpretation of $^4He(e,e'p)^3H$ measurements at high values of $p_m$, where "direct" triton knockout starts to contribute, is still uncertain.

*Acknowledgements.* I would like to thank J.F.J. van den Brand, M. Daman, their supervisor E. Jans, and R. Ent, part of whose thesis data have been used here.

# References

[1] S. Frullani and J. Mougey, Adv. Nucl. Phys. **14**, 1 (1984)

[2] S. Boffi, C. Giusti and F.D. Pacati, Phys. Rep. **226**, 1 (1993)

[3] J.F.J. van den Brand, Thesis, University of Amsterdam, 1988

[4] J.F.J. van den Brand et al., Phys. Rev. Lett. **60**, 2006 (1988)

[5] J.F.J. van den Brand et al., Nucl. Phys. **A534**, 637 (1991)

[6] R. Schiavilla, Phys. Rev. Lett. **65**, 835 (1990)

[7] T. Warmann and K. Langanke, Phys. Lett. **B273**, 193 (1991)

[8] M. Buballa et al., Phys. Rev. **C44**, 810 (1991)

[9] R. Schiavilla, V. R. Pandharipande and R. B. Wiringa, Nucl. Phys. **A449**, 219 (1986)

[10] W.T.H. van Oers et al., Phys. Rev. **C25**, 390 (1982)

[11] G. van der Steenhoven et al., Phys. Lett. **B191**, 227 (1987)

[12] J.F.J. van den Brand et al., Phys. Rev. Lett. **66**, 409 (1991)

[13] J.J.M. Steijger et al., Nucl. Instr. Meth. **A295**, 123 (1990)

[14] M. Daman, Thesis, University of Amsterdam, 1991

[15] E. van Meijgaard and J.A. Tjon, Phys. Rev. **C42**, 96 (1990)

[16] A. Magnon et al., Phys. Lett. **B222**, 352 (1989)

[17] J.M. Legoff et al., Proc. 4th Workshop on Perspectives in Nuclear Physics at Intermediate Energies, Trieste, 1989, p. 376
see also: E. Jans, Nucl. Phys. **A508**, 433c (1990)

[18] R. Ent, Thesis, Vrije Universiteit, Amsterdam, 1990 and R. Ent et al., to be published

[19] R. Ent et al., Phys. Rev. Lett. **67**, 18 (1991)

[20] H. Morita , Y. Akaishi and H. Tanaka, Prog. Theor. Phys. **79**, 863 (1987)

[21] W. Leidemann et al., Phys. Lett. **B279**, 212 (1992)

[22] R. Ent et al., Phys. Rev. Lett. **57**, 2367 (1986)

[23] T. Warmann and K. Langanke, Z. f. Physik **A342**, 489 (1990)

[24] W. Gloeckle et al., to be published in Proc. Workshop on Electron-nucleus scattering, Elba, July 1993

Few-Body Systems, Suppl. 7, 128—135 (1994)

# SEPARATION OF ELECTROMAGNETIC RESPONSE FUNCTIONS OF FEW-BODY NUCLEI IN (e,e'p) REACTIONS

C. Marchand

Service de Physique Nucléaire (DAPNIA/SPhN), Centre d'Etude de Saclay
91191 GIF-SUR-YVETTE CEDEX, FRANCE

Abstract We report on latest results in electro-induced reactions on few-body nuclei, especially recent experimental determinations of the cross sections for the reactions $^2$H(e,e'p)n and $^3$He(e,e'p)$^2$H for which three out of four electromagnetic structure functions have been separated. These new precise data allow a thorough study of the reaction mechanism of the (e,e'p) reaction at momentum transfers up to one GeV. The data are compared to theoretical descriptions of the (e,e'p) cross sections which include final state interactions, meson exchange currents and relativistic effects by consistently solving the scattering problem. We will discuss the relevance of FSI at low q and the importance of a relativistic treatment of the current operator at high q and the longitudinal-transverse interference structure function.

## 1. Introduction

The (e,e'p) reaction has been extensively studied in the last decades to extract information on the nuclear structure, in particular nucleon energy and momentum distributions. The advantage over hadron probes is its weak coupling to the nucleus, avoiding thus initial state interaction corrections. Nevertheless, final state interactions of the ejected proton have to be known to extract reliable information like spectroscopic factors or momentum distributions from the measured cross sections, not to speak about meson exchange currents.

The usual way to deal with FSI in medium to heavy nuclei is to treat the distortion of the proton wave by mean of an optical model. The validity of this approach seems justified by the good description it gives of the experimental shapes of momentum distributions in $^{40}$Ca, $^{90}$Zr or $^{208}$Pb [1], but it fails to reproduce the data on $^4$He [2]. A detailed study of the q and $T_p$ dependence of the ${}^4$He(e,e'p)${}^3$H cross section [3] shows that the deduced momentum distribution is at least as sensitive to the value of q as $T_p$, and is well reproduced by a recent calculation [4] where the proton rescattering is consistently treated to first order in the T-matrix expansion of the scattering amplitudes.

A real breakthrough in the understanding of the (e,e'p) reaction has been made since the advent of high precision experiments where the different helicity components of the cross section have been separated. In particular, the above cited calculation [4] fails to separately reproduce the longitudinal and transverse components by about 20% [5]. This behavior has also been observed in ${}^6$Li and ${}^{40}$Ca, where differences up to 35% are reported between the experimental data and the present best models of the (e,e'p) reaction. These observations are at variance with the results reported in [1] about the longitudinal/transverse ratio, which are correctly reproduced when taking into account final state interactions and coulomb distortion effects. Our opinion is that no theoretical description is at hand which would simultaneously reproduce the separated structure functions in all kinematical conditions. New precise results for the reactions ${}^2$H(e,e'p)n and ${}^3$He(e,e'p)${}^2$H are an incontestable benchmark to test the quantitative importance of FSI, meson exchange currents and relativistic effects. These can be handled on a much firmer ground for light nuclei.

## 2. Quasi-elastic (e,e'p) reaction

A detailed description of the (e,e'p) formalism can be found in [6] and we will focus here on the features of interest for the discussion.

### 2.1 General form of the cross section

In the Born approximation, where the electron interaction is assumed to occur via the exchange of *one* virtual photon, the (e,e'p) cross section can be decomposed into the polarization states of this virtual photon:

$$\frac{d^6\sigma}{d\Omega_{e'}\,de'\,d\Omega_{p'}\,dp'} = p'^2\sigma_{Mott}\left(W_L + \epsilon^{-1}W_T + W_{TT}\cos 2\Phi + \sqrt{1+\epsilon^{-1}}W_{LT}\cos\Phi\right) \quad (1)$$

where the subscripts L and T refer to the longitudinal (in the direction of the momentum transfer $\vec{q}$) and the transverse (orthogonal to $\vec{q}$) polarization states of the photon, $\epsilon$ is its longitudinal polarization and $\Phi$ the angle between the scattering plane (e,e') and the reaction plane (q,p'). The structure functions $W_i$ (i = L,T,TT,LT) depend not only upon q and $\omega$ like it is the case for the proton, but also on the ejected proton momentum p' and it's angle with q, $\gamma$. In comparison with elastic scattering on the free proton, where only the $W_L$ and $W_T$ contribute, two interference structure functions appear in the case of (e,e'p) on a nucleus. These are proportional to the sine of $\gamma$.

The latter vanish in so called parallel kinematics for which q//p' ($\gamma = 0$). In that case, the experimental determination of $W_L$ and $W_T$ is done through the Rosenbluth separation from two measurements at $q,\omega,p'$ fixed, but at two different values of $\epsilon$, i.e. different scattering angles $\Theta_{e'}$.

The complete determination of the four structure functions would require measurements at angles $\Phi$ different from 0 or $\pi$, what has only marginally been attempted so far by lack of experimental equipment enabling out of plane measurements. The longitudinal-transverse interference structure function $W_{LT}$ can however been determined from two measurements at all kinematical parameters fixed, except $\Phi = 0$ and $\pi$, that is the proton is detected on both sides of the virtual photon (also called left-right separation). It should be noted that the structure functions $W_L$ and $W_T$ are obtained at $\gamma = 0$, whereas by definition, $W_{LT}$ can only be measured when $\gamma \neq 0$.

## 2.2 Description of the Plane Wave Impulse Approximation

The most economical description of the A(e,e'p)B reaction mechanism is the Plane Wave Impulse Approximation, for which the energy $\omega$ and momentum q is transferred to a single proton with binding energy $E_m = \omega - T_{p'} - T_B$ and momentum $-p_m$, which then leaves the nucleus without further interaction with the residual system B. In that case, the (e,e'p) cross section factorizes as:

$$\frac{d^6\sigma}{d\Omega_{e'}de'd\Omega_{p'}dp'} = p'^2\widetilde{\sigma_{ep}}S\left(p_m, E_m\right) \qquad (2)$$

where $\widetilde{\sigma_{ep}}$ is the cross section for electron scattering on a "free" moving and off-shell proton and S the spectral function, i.e. the probability to find the proton in the nucleus with initial momentum $p_m$ and binding energy $E_m$. Calculations of the off-shell proton cross section $\widetilde{\sigma_{ep}}$ suffer all from the fact that the IA current is not gauge invariant, but in quasi-elastic kinematics and for small values of $p_m$, differences between different prescriptions are of the order of a few % and the usual choice is to take de Forest's CC1 [7]. In the PWIA framework, the factorization (2) holds also for the individual structure functions $W_i$ defined by equation (1): $W_i = W_i^p \times S_i$, and the $S_i$'s obey the following relation:

$$S_L\left(\equiv \frac{W_L}{\widetilde{W_L^p}}\right) = S_T\left(\equiv \frac{W_T}{\widetilde{W_T^p}}\right) = S_{LT}\left(\equiv \frac{W_{LT}}{\widetilde{W_{LT}^p}}\right) = S_{TT}\left(\equiv \frac{W_{TT}}{\widetilde{W_{TT}^p}}\right) = S\left(p_m, E_m\right) \quad (3)$$

The spectral function should be the same independently of which structure function it is deduced from, and of course should not depend upon q and $\omega$, or in other worlds, the q dependence of the $W_i$'s should be that of the "free" proton. Any departure from the above relation is thus a direct measure of the non validity of PWIA.

## 2.3 Treatment of non PWIA effects on few-body nuclei

The limitations of the PWIA to describe in full detail the (e,e'p) reaction are well known:

- the interaction of the outgoing proton with the residual nucleus can not be neglected and one has to take into account Final State Interactions;

- the current operator constructed from individual off-shell nucleon currents is not conserved and one needs to include at least two-body Meson Exchange Currents to restore gauge invariance.

Contrary to nuclei with more than four constituents for which the FSI have to be treated in an optical potential fitted on p-nucleus scattering, the final state $|\Psi_f\rangle$ can be computed exactly for few-body nuclei by solving the time dependent non relativistic Schrödinger equation with the condition that $|\Psi_f\rangle \rightarrow |p'\rangle|(A-1)^*\rangle, r, t \rightarrow \infty$ (Lipmann-Schwinger equations for $^2$H or Faddeev equations in the continuum for $^3$He). In such an approach, the initial and final states are non relativistic and a non-relativistic reduction of the current operator in expansion of $\frac{1}{M}$ is needed to compute the T-matrix elements. In all these calculations, the terms up to $\frac{1}{M^2}$ are included [8, 9, 10, 11, 12], unless it is especially specified non-relativistic [13]. In this case only the leading term of the expansion is retained, with all the ambiguities about the choice of Dirac or Sachs form factors which are of course not present in a relativistic treatment. Recently, fully relativistic calculations have been performed for the case of $^2$H by solving the covariant Bethe-Salpeter equations [14].

## 3. (e,e'p) structure functions for $^2$H and $^3$He

We will now compare how the "exact" theoretical calculations of the (e,e'p) cross section compare to recent experimental data.

### 3.1 q-dependence of the longitudinal and transverse structure functions

The q-dependence of the longitudinal and transverse structure functions for the reaction $^2$H(e,e'p)n has been measured both at NIKHEF and SACLAY. The NIKHEF data have been taken on a target consisting of a thin film of heavy-water [15]. The validity of equation (3) for deuterium can not be inferred from these data because the value of missing momentum $p_m$ was not kept constant in the measured range of $q^2$ (0.05 to 0.27 $[GeV/c]^2$). Instead, the results are presented in terms of ratios $W_{L,T}^{exp}/W_{L,T}^{PWIA}$, which is not only sensitive to FSI, but also to the correct description of the spectral function. Nevertheless, a reasonable agreement is found between the measured L and T structure functions and the relativistic calculation of [14].

Recent data from SACLAY obtained on a liquid deuterium target [16] allow a much more detailed study of the q-dependence of the structure functions, because $W_L$ and $W_T$ have been measured over a wide range of q (200 to 900 MeV/c) at fixed values of missing momentum $p_m$ (20, -100, +100 MeV/c). In that case, according to equation 3, any departure from a flat q-dependence of $S_{L,T}^{exp}$, or a ratio $S_L/S_T$ different from 1. is a direct measure of non PWIA effects, independently of the precise knowledge of the absolute value of the spectral function.

The experimental spectral functions $S_L^{exp}$ and $S_T^{exp}$ deduced from the measured structure functions $W_{L,T}^{exp}$ for the reaction $^2$H(e,e'p)n are shown in Fig. 1. The effects of FSI appear clearly at low values of q as a sizable reduction of both $S_L^{exp}$ and $S_T^{exp}$. The trend of this reduction is correctly accounted for by the theoretical calculations cited in 2.3 [8, 9, 10, 13, 14] which treat the final state interactions by a consistent description of the scattering states. It should be noted that the comparison of $S^{exp}$ and $S^{theor}$ does not

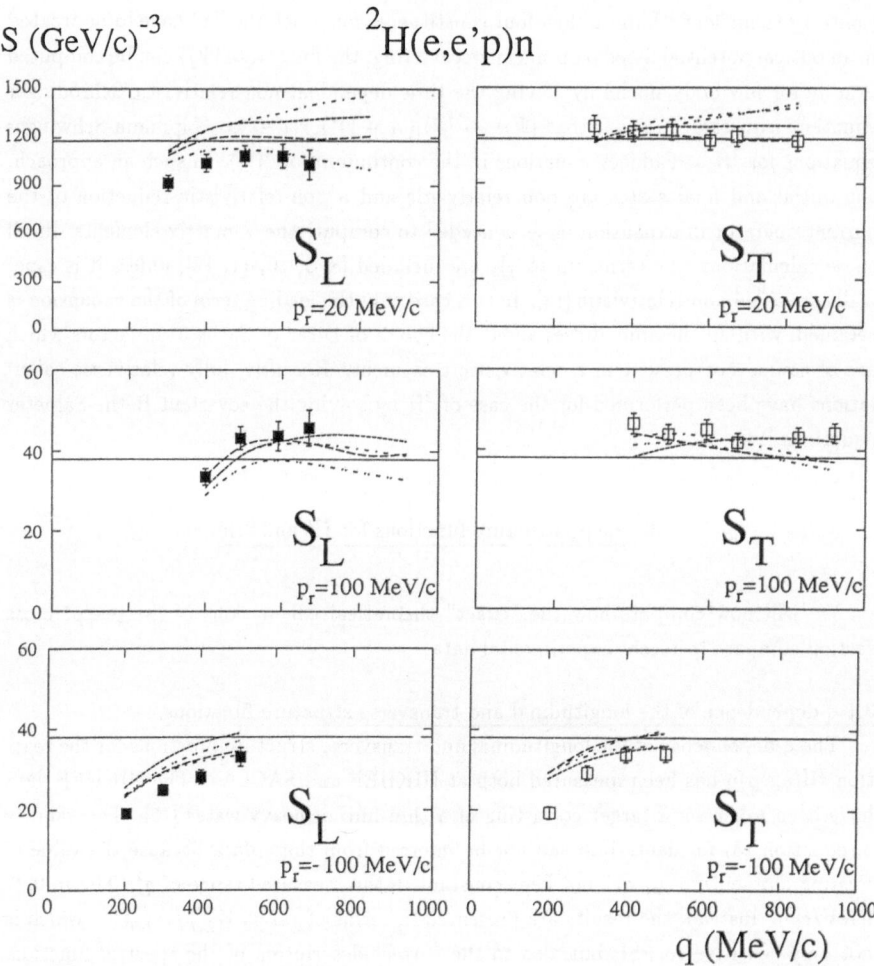

Figure 1: $S_L^{exp}$ and $S_T^{exp}$ for the reaction $^2H(e,e'p)n$ as a function of q. The legend of the curves is the following: the full line is the value of the spectral function computed in PWIA with the $^2H$ wave function obtained from the Paris potential; the dash-dotted curve is the non-relativistic calculation from H. Arenhövel; the long-dashed curve is the calculation from J.M.Laget and the dash-dot-dotted curve the calculation from B. Mosconi and P. Ricci, both including relativistic corrections up to order $\frac{1}{M^2}$; the short-dashed curve is the fully relativistic calculation from E. Hummel and J.A. Tjon .

suffer from off-shell ambiguities in the definition of $\widetilde{W^p}$, because the same $\widetilde{W^p}$ is used to define them.

The influence of meson exchange currents, which require to add two-body currents to the naive PWIA nucleon current, is found to be less than 5% in the quasi-elastic

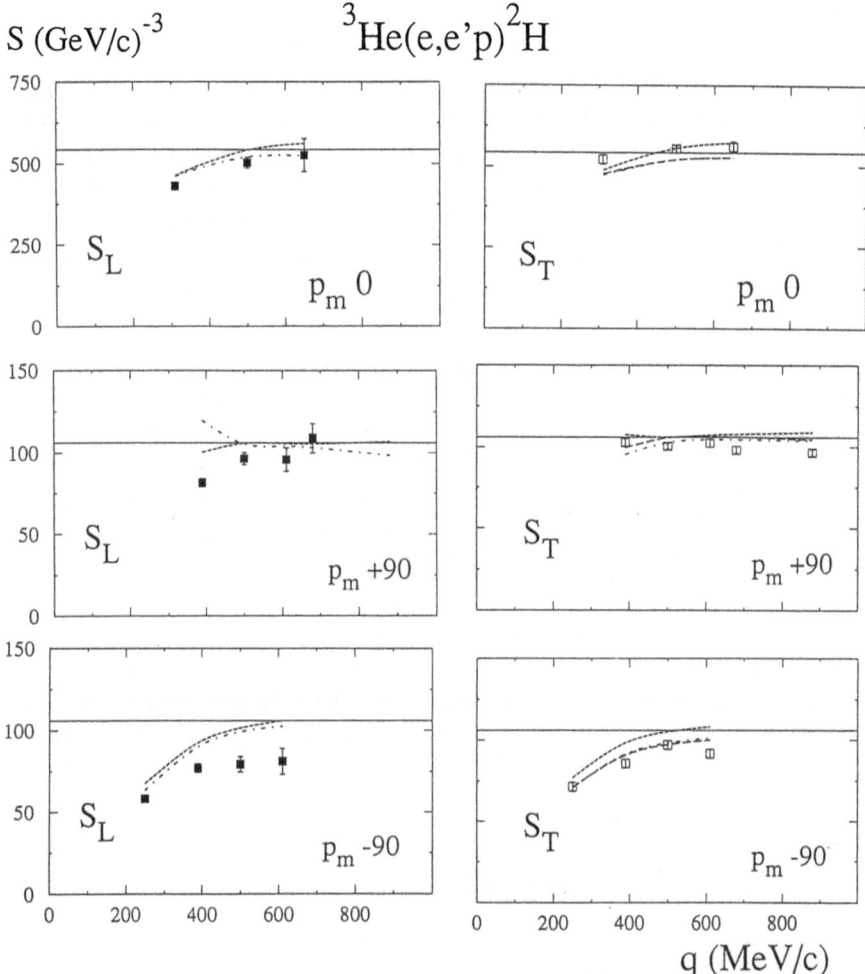

$S \ (GeV/c)^{-3}$     $^{3}He(e,e'p)^{2}H$

Figure 2: $S_{L}^{exp}$ and $S_{T}^{exp}$ for the reaction $^{3}He(e,e'p)^{2}H$ as a function of q. The legend of the curves is the following: the full line is the value of the spectral function computed in PWIA with the $^{3}He$ wave function obtained from the Argonne potential; the short-dashed curve is the calculation of Meijgaard and Tjon, the dot-dashed and long-dashed curves are the calculations of J.M. Laget without and with MEC.

kinematics of this experiment [13, 8]. Above q = 500 MeV/c, the data show no longer a significant q-dependence, and $S_L = S_T$ to within 10 %. That means that PWIA describes well the reaction $^{2}H(e,e'p)n$ at q > 500 MeV/c. The quite large scattering of the theoretical calculations at large values of q reflects the limitation of non-relativistic reductions and calls for relativistic covariant treatments of the (e,e'p) cross section at

134

CEBAF energies.

At SACLAY, the reaction $^3$He(e,e$'$p)$^2$H has been studied in kinematics comparable to those of the $^2$H(e,e$'$p)n reaction. The results are shown on Fig. 2, together with two different predictions from ref. [11, 12]. Both approaches include relativistic corrections up to order $\frac{1}{M^2}$, but differ somewhat in their treatment of final state interactions: J.M. Laget restricts the summation of p-$^2$H rescattering amplitudes to the S,D and P channel, whereas Meijgaard and Tjon completely determine the p-$^2$H final state by solving the Faddeev equations in the continuum. It appears that both predictions are very similar, except for $S_L$ at $p_m = 90$ MeV/c for the lowest value of q = 400 MeV/c, for which the proton kinetic energy is 50 MeV and the truncation of J.M. Laget is questionable.

As to the reaction $^2$H(e,e$'$p)n, the estimated effect of meson exchange currents is small at the kinematics of the experiment, and the reduction of $S_L$ and $S_T$ at low momentum transfer q is entirely due to the final state interactions. Both theoretical calculations reproduce well the measured structure functions on the whole, except in antiparallel kinematics where the experimental values are below. As in the case of deuterium, the effect of FSI dies out at q > 500 MeV/c, and the measured spectral functions are in good agreement with the prediction of ref. [18].

## 3.2 The longitudinal-transverse structure function

It has been stressed by several authors (e.g. [9]) that the longitudinal-transverse interference structure function is more sensitive to relativistic reductions of the current operator, as the relativistic corrections appear quadratically in $W_L$ and $W_T$, whereas

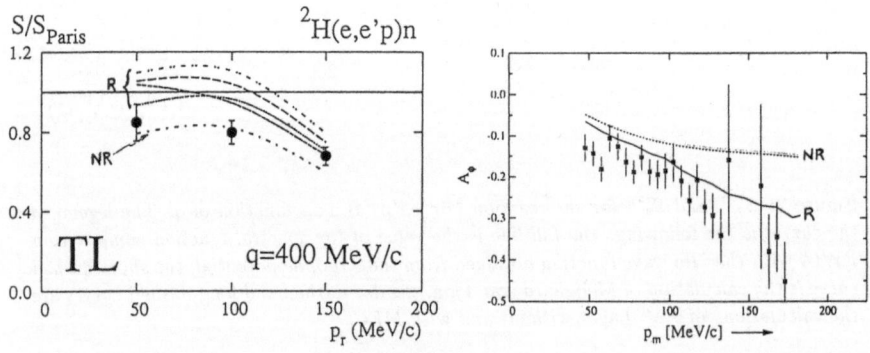

Figure 3: *LT interference structure function for the reaction $^2$H(e,e$'$p)n. The left inset represents SACLAY results at q = 400 MeV/c expressed in terms of the ratio $S_{TL}^{exp}/S_{Paris}$; the signification of the curves is the same as in Fig. 1, the dotted curve is Beck and Arenhövel's calculation including relativistic corrections up to order $\frac{1}{M^2}$. The right inset represents NIKHEF results at q = 460 MeV/c expressed in terms of the asymmetry $A_\Phi$; the full curve is the fully relativistic calculation from E. Hummel and J.A. Tjon, the dotted curve is the non-relativistic calculation of Arenhövel.*

they enter linearly in $W_{LT}$. This $W_{LT}$ structure function has been measured at both SACLAY [16] at q = 400 MeV/c and NIKHEF [19] at q = 460 MeV/c for the reaction $^2$H(e,e'p)n as a function of $p_m$. The results for the reaction $^2$H(e,e'p)n are shown in Fig. 3.

It is surprising that SACLAY results favour the non-relativistic calculation of Ref. [13], whereas NIKHEF results clearly show better agreement with the fully relativistic calculation of Ref. [14]. It should be noted that only statistical error bars are displayed, and that the left-right separation technique used to deduce the longitudinal-transverse structure function is very sensitive to systematic effects. These are of the order of the statistical errors for SACLAY results, what brings the apparent disagreement with relativistic calculations within $2\sigma$. For the NIKHEF result, the systematic error is +- 0.05 on the measured assymetry $A_\Phi = W_{LT}/(aW_L + bW_T + cW_{TT})$, where a,b,c are kinematical constants; in that case, the non-relativistic calculation is clearly not reproducing $A_\Phi$ even taking the systematic error into account.

### References

1. L. Lapikas: Nucl. Phys. **A553**, 297c (1993).
2. J. van den Brand *et al.*: Phys. Rev. Lett. **60**, 2006 (1988).
3. J. van den Brand *et al.*: Phys. Rev. Lett. **66**, 409 (1991).
4. J.M. Laget: Nucl. Phys. **A497**, 391c (1989).
5. A. Magnon *et al.*: Phys. Lett. **B222**, 352 (1989).
   J.E. Ducret *et al.*: Nucl. Phys. **A556**, 373 (1993).
6. S. Frullani and J. Mougey: Adv. Nucl. Phys. **14**, 1 (1984).
7. T. de Forest: Nucl. Phys. **A392**, 232 (1983).
8. J.M. Laget: Phys. Lett. B **199**, 493 (1987).
9. B. Mosconi and P. Ricci: Nucl. Phys. **A517**, 483 (1990).
10. G. Beck and H. Arenhövel: Few Body Syst. **13**, 165 (1992).
11. J.M. Laget: Phys. Lett. B **151**, 325 (1985).
12. E. van Meijgaard and J.A. Tjon: Phys. Rev. **C42**, 74 (1990).
13. H. Arenhövel: Nucl. Phys. **A384**, 287 (1982).
14. E. Hummel and J.A. Tjon: Phys. Rev. Lett. **63**, 1788 (1989); Phys. Rev. **C42**, 423 (1990) and to be published.
15. M. van der Schaar *et al.*: Phys. Rev. Lett. **66**, 2855 1991.
16. J.E. Ducret *et al.*: Nucl. Phys. **A553**, 697c (1993) and to be published.
17. L. Lakehal-Ayat *et al.*: 5$^{th}$ Workshop on Perspectives in Nuclear Physics at Intermediate Energies, Eds S. Boffi *et al.*, Trieste, May 1991, p.218; Nucl. Phys. **A553**, 693c (1993) and to be published.
18. R. Schiavilla, V.R. Pandharipande and R.B. Wiringa: Nucl. Phys. **A449**, 219 (1986).
19. M. van der Schaar *et al.*: Phys. Rev. Lett. **68**, 776 (1992).

Few-Body Systems, Suppl. 7, 136—143 (1994)

# A SIMULTANEOUS MEASUREMENT OF THE (γ,n) AND (γ,p) REACTIONS IN ⁴HE

J. Asai, G. Feldman, R.E.J. Florizone, E.L. Hallin, D.M. Skopik and J.M. Vogt

Saskatchewan Accelerator Laboratory, University of Saskatchewan
Saskatoon, Saskatchewan, Canada S7N 0W0

R.C. Haight and S.M. Sterbenz

Los Alamos National Laboratory, Los Alamos, New Mexico, USA 87545

## Abstract

A significant charge asymmetry in the nuclear force has been attributed to differences in the $^4$He$(\gamma,n)^3$He and $^4$He$(\gamma,p)^3$H cross sections below $E_\gamma \sim 35$ MeV. To investigate this claim, we have performed a simultaneous measurement of these cross sections by measuring the recoiling $^3$He and $^3$H nuclei in the same detector system consisting of windowless $\Delta$E proportional counters backed by silicon E detectors. We find, within our errors, that for $25 < E_\gamma < 35$ MeV the ratio of the cross sections $\sigma(\gamma,p)/\sigma(\gamma,n)$ is consistent with charge symmetry and with the predictions of shell-model calculations.

## Introduction

Experiments designed to measure violations of charge symmetry (CS) in the nuclear force have for the most part found small amounts. There is one notable exception, the analysis of the ratio $R = \sigma(\gamma,p)/\sigma(\gamma,n)$ by Calarco *et al.* [1] which gave a value of $R \sim 1.8$ near the peak of the cross section at $E_\gamma \sim 26$ MeV. Their recommended peak cross section values which give rise to such a large ratio are $\sigma(\gamma,p) \sim 1.8$ mb and $\sigma(\gamma,n) \sim 1.0$ mb. This result has been a contentious issue for the past fifteen years since it requires either a large amount of isospin mixing in the continuum or a direct charge-symmetry breaking of the nuclear force, neither of which is consistent with other experimental results [2,3,4].

Several recent experiments [5,6] have supposedly resolved this problem, reporting new values of the $\sigma(\gamma,p)$ and $\sigma(p,\gamma)$ cross sections that are much closer to the $\sigma(\gamma,n)$ value of 1 mb recommended by the Calarco analysis (hereafter referred to as CBD). In contrast to these results, a recent photon scattering experiment [7] reported a total absorption cross section that is inconsistent with the sum of the CBD value for $\sigma(\gamma,n)$ and the new proton data. These authors point out that all of the experiments which directly address the ratio problem have not reported absolute measurements for both channels and that differential cross sections measured at a laboratory angle of 90° should be converted to total cross sections, taking into account the shift in angle and the corresponding asymmetry in the cross section. Without a direct measurement of the angular distribution, a theoretical asymmetry, including terms through E2 absorption, needs to be folded into the data in order to form total cross sections. When properly done, however, the correction is only a few percent different from assuming pure E1 absorption for 90° laboratory cross sections.

In this paper we present the first truly simultaneous measurement of absolute $(\gamma,p)$ and $(\gamma,n)$ cross sections using real, monoenergetic photons. The experiment was done by detecting the charged recoil particles, $^3$He and $^3$H, that define the proton and neutron channels, in arrays of $\Delta$E-E detectors.

## Experimental Arrangement

The experiment was carried out at the Saskatchewan Accelerator Laboratory (SAL). The accelerator complex houses a 300 MeV linac, energy compressor and a pulse stretcher ring that produces a monoenergetic ($\Delta E/E \sim 10^{-4}$), high duty factor (> 70%) electron beam. The essential features of the new facility are described in references [8] and [9]. A plan view of the accelerator facility and the experimental areas is shown in Fig. 1. The photon tagging system is shown in this figure in its present, permanent location in EA2; for this experiment, however, the tagging system was located in EA3.

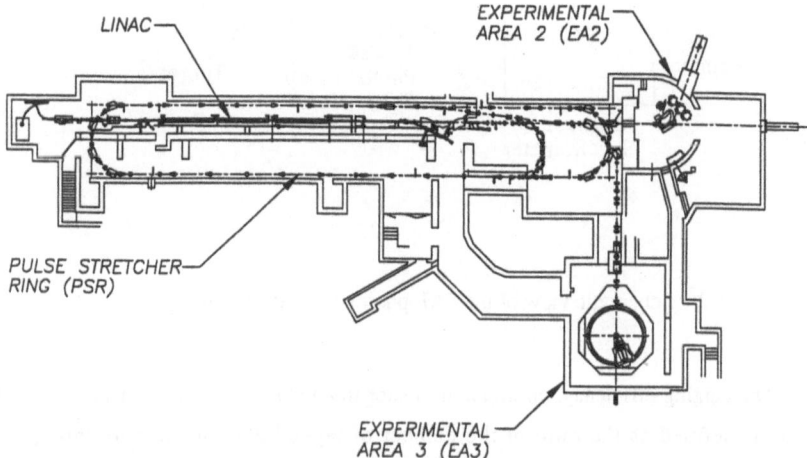

Figure 1: A plan view of the accelerator, pulse stretcher ring and the experimental areas at SAL. EA2 houses the photon tagging spectrometer, EA3 is used for electron scattering experiments.

<u>The photon tagger</u>

A nearly continuous 119 MeV electron beam was incident upon a thin, 0.1% Al radiator, generating a complete spectrum of bremsstrahlung photons. The scattered electrons which produced a tagged photon that initiated a reaction event were measured in a 62-channel plastic scintillator array located in the focal plane of a magnetic spectrometer. The photon energy was determined to within the channel width of ~0.6 MeV, over a corresponding photon (laboratory) energy interval of 25.98-57.96 MeV. Fig. 2 shows the general photon tagging arrangement used in this experiment [10].

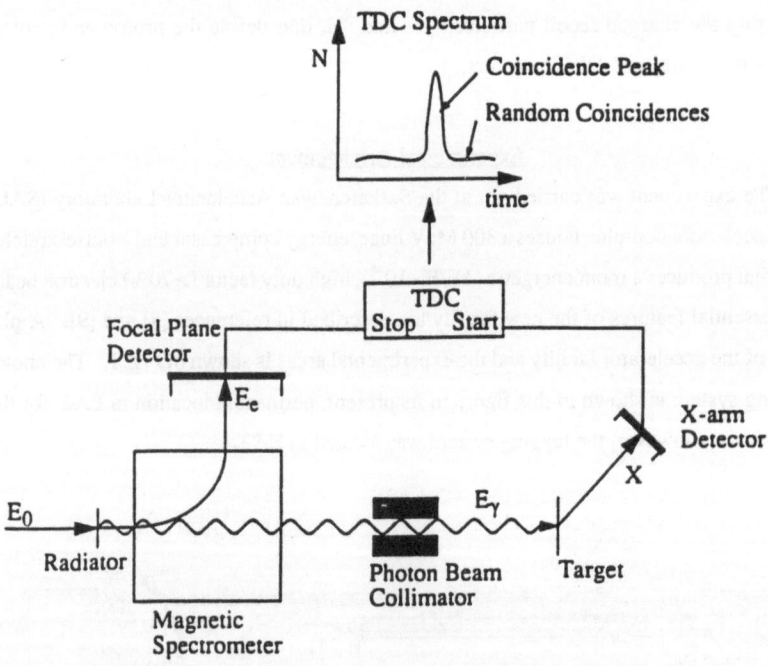

Figure 2: Schematic view of the SAL photon tagging system

The tagging efficiency, an important factor in an absolute measurement such as this one, is defined as the ratio of the number of tagged photons passing through the collimating system to the total number of photons tagged. It was measured periodically by placing a lead glass detector in the photon beam and directly comparing the number of lead glass events to the number of electrons measured in the focal-plane detector array. For the tagging efficiency measurements, the lead glass was assumed to be 100% efficient for photon detection, an assumption that has been verified by EGS simulations. Each tagging efficiency determination consisted of two measurements, one with the radiator in and one with the radiator out. The radiator out runs measured the general room background and served as a check on the overall beam focussing and quality.

<u>Target and detector arrangement</u>

The detectors were four $\Delta$E-E detector arrays located at 90° with respect to the incident photon beam. A layout of the experimental arrangement is given in Fig. 3. The $\Delta$E detectors were windowless proportional chambers immersed directly in the target gas (helium). A judicious choice of quench gas (hydrogen) allowed us to measure the $^3$He and $^3$H recoil particle down to ~ 1.5 MeV. This idea of placing the wire chambers directly in the target gas was the quintessential feature of the experiment; it allowed us to extend the cross section measurement down to low energies which were hitherto unattainable in a direct, simultaneous cross section determination.

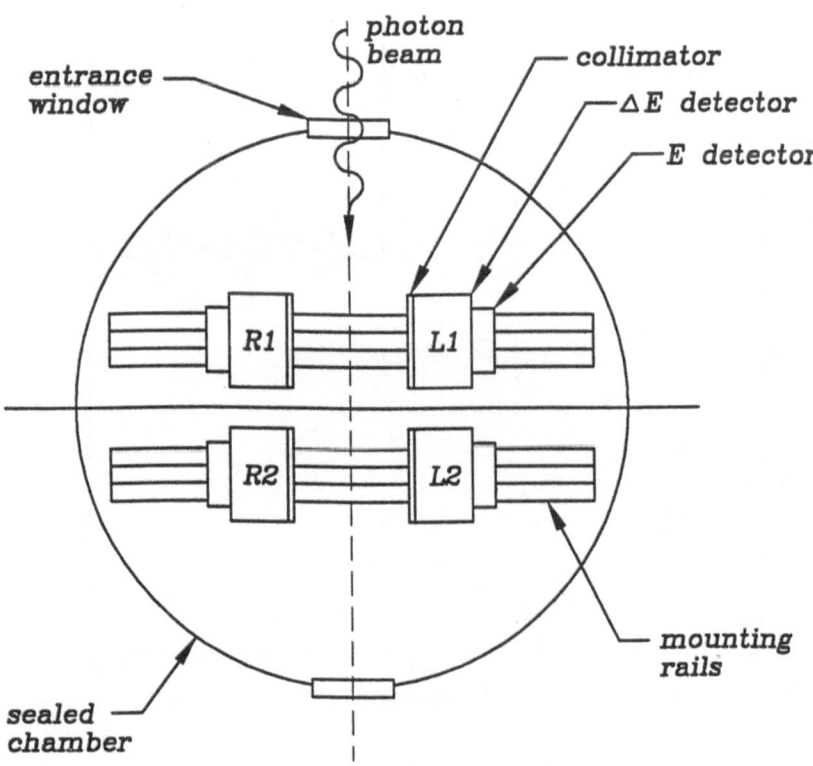

Figure 3: A plan view of the target chamber and the detector systems used in the experiment

## Data Acquisition and Analysis

Data were accumulated with a CAMAC-based acquisition system described in reference [11]. Standard particle identification techniques using energy loss and time-of-flight (TOF) were employed to identify the $^3$He and $^3$H recoils. The E-detectors were calibrated with an $^{241}$Am α source and the TOF of the recoils was measured with respect to the tagged photon beam. The TOF information allowed us to correct the coincidence timing peak to a few ns. Independently measuring the recoil energy permitted us to over-determine the kinematics, since the tagger measures the photon energy separately. This kinematic condition, combined with the timing information, gave an ΔE-E event histogram shown in Fig. 4. The $^3$H and $^3$He recoils are clearly resolved from all discernible background. Final $^3$H and $^3$He yields were obtained by placing background cuts on each side of the timing spectrum.

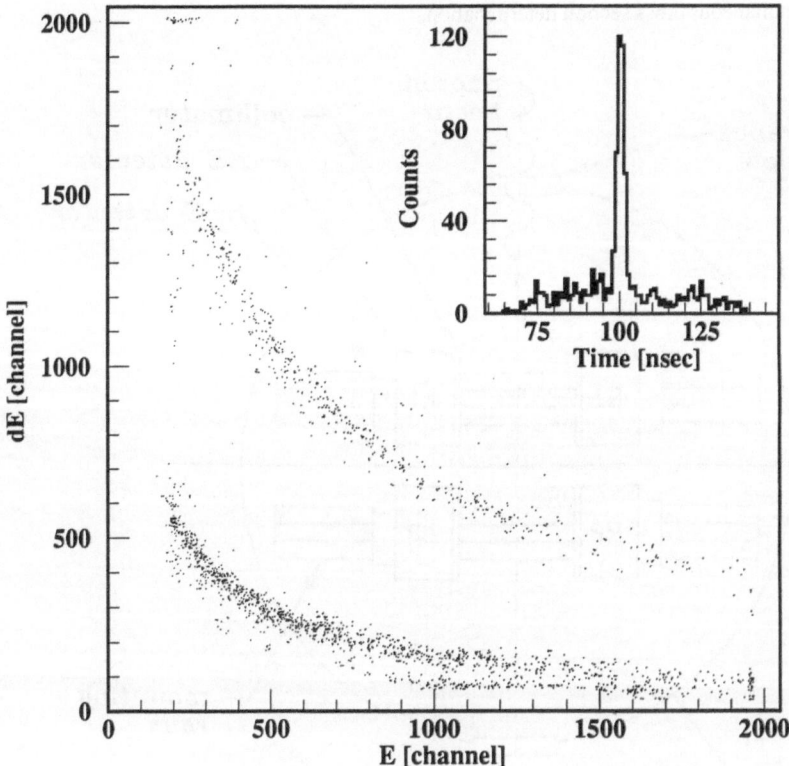

Figure 4:   A ΔE-E spectrum showing the clear separation of the $^3$He and $^3$H bands from each other and the background. A typical timing spectrum showing the good true/random ratio in the experiment is shown in the inset.

At this stage of the analysis standard normalizations, the photon flux and efficiency are straightforwardly done. The acceptance of the detectors (solid angle * effective path length) is not as straightforward. It was calculated by two independent methods which agreed to 0.5%; one used a geometrical analysis, the other a standard Monte Carlo technique.

Total cross sections were found by first transforming the $^3$He and $^3$H 90° differential cross sections to the centre-of-mass system and then applying a correction for the asymmetry in the angular distributions. Most theories give similar results for this asymmetry (it does not depend upon calculating the absolute cross section) and we chose to use the calculation by Wachter et al. [12].

## Results and Discussion

Our values for the ratio of cross sections are shown in Fig. 5. The lowest-energy data points (solid triangles) were taken with a gas pressure of 150 torr. This pressure allowed us to detect recoiling $^3$He nuclei down to an energy of about 1.5 MeV, and permitted us to histogram data with a "centre-of-bin" photon energy (determined independently by the tagger) as low as $E_\gamma = 26.4$ MeV. This low target gas pressure was sufficient to provide enough ionization to clearly separate the $^3$H recoils from the background, and since the $^3$He particles are four times more effective at producing ion pairs in the gas, both recoil mass groups were always clearly identified in the measurement. The cross section ratio data shown as open circles correspond to a gas pressure of 400 torr. Over the entire energy interval that was measured, our cross section ratios are clearly not unity and also do not resemble the recommended CBD results. At the lower energies our results are consistent with the shell-model calculations for $^4$He [12,13] but our error bars also overlap the CBD analysis.

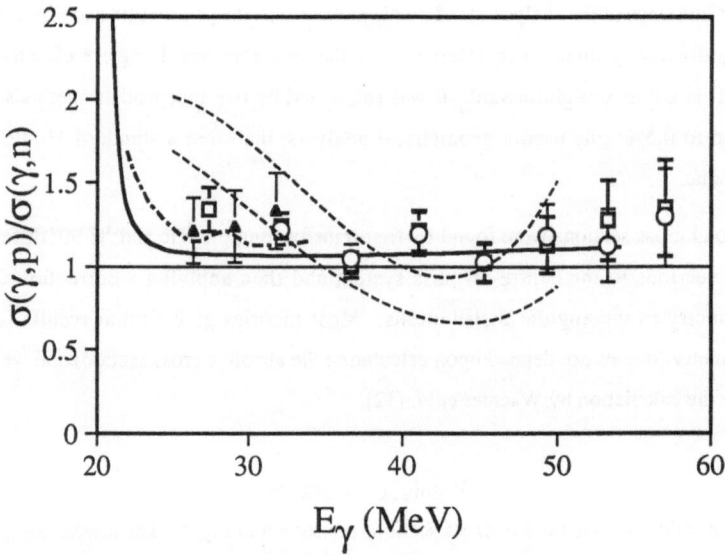

Figure 5: Ratio of the cross sections obtained in this experiment. The recommended CBD values are shown by the dashed lines. The solid squares are the ratio of all of the data that have been combined to form total cross sections. The solid and dash-dot curves are the shell model calculations of Ref. [13] and [14] respectively.

The other feature of our experiment is that we measured absolute cross sections. This allows us to compare directly the sum of our $\sigma(\gamma,p)$ and $\sigma(\gamma,n)$ cross sections to the new total absorption measurement of Illinois [7], which constrains the sum of these cross sections to be $2.86 \pm 0.12$ mb at the peak of the photo-absorption cross section ($E_\gamma \sim 26$ MeV), a result which is consistent with the CBD analysis but not with the latest proton data. Table 1 summarizes these results. Our new simultaneously measured absolute cross sections are consistent with the new Illinois total absorption cross section.

Table 1: Cross sections, ratios and sums near the peak value

| Experiment | $\sigma(\gamma,p)$ | $\sigma(\gamma,n)$ | $\sigma(\gamma,p)+\sigma(\gamma,n)$<br>$2.86\pm0.12$ mb[a] |
|---|---|---|---|
| CBD | 1.80±0.12 | 1.10±0.10 | 2.90±0.16 |
| Ref. [5] & CBD | 1.27±0.08 | 1.10±0.10 | 2.37±0.13 |
| Ref. [6] & CBD[b] | 1.15±0.03 | 1.10 0.10 | 2.25±0.10 |
| This Experiment[c] | 1.50±0.05 | 1.12±0.10 | 2.62±0.11 |

[a]Ref. [7].
[b]The lowest measured photon energy in this experiment was 28.6 MeV.
[c]Low and high pressure data combined in 7 tagger channels (~3 MeV).

## Summary

Within the present errors our new experiment supports charge symmetry in the nuclear force and our sum cross section agrees with the recent total absorption measurement from Illinois. Ours is the only experiment performed to date that determines a truly simultaneous measurement of the ratio R near the peak of the cross sections, and it is the first ever simultaneous set of absolute cross sections to be reported. Our future plans are to measure a complete angular distribution, reduce the error bars in the direct ratio determination and extend the energy range to below the peak value of the cross sections.

## Acknowledgment

This work has been supported by the Natural Sciences and Engineering Research Council of Canada (NSERC) and the Department of Energy of the United States.

## References

[1]  J.R. Calarco, B.L. Berman and T.W. Donnelly, Phys. Rev. C27, 1866 (1983).

[2]  M. Spahn et al., Phys. Rev. Lett. 63, 1574 (1989).

[3]  C. L. Blilie et al., Phys. Rev. Lett. 57, 543 (1986).

[4]  B. A. Raue, Investigation of the Isospin Response of the $^4$He Continuum using the $^4$He(p,p'X) at $T_p$ = 100 MeV, Ph.D. thesis, Indiana University Cyclotron Facility, (1993) unpublished.

[5]  G. Feldman et al., Phys.Rev. C42, R1167 (1990).

[6]  R. Bernabei et al., Phys. Rev. C38, 1990 (1988m).

[7]  D. P. Wells et al., Phys.Rev. C46, 449 (1992)

[8]  L.O. Dallin, High Duty Factor, Monochromatic Extraction from EROS, Ph.D. thesis, Univ. of Sask., (1990) unpublished

[9]  D. M. Skopik, Few-Body systems, Suppl. 6, 460 (1992)

[10]  J. Vogt et al., Nucl. Instr. & Meth. A324, 198 (1993)

[11]  Saskatchewan Accelerator Laboratory, LUCID Data Acquisition and Analysis System User's Guide, Version 1.2 (1991)

[12]  B. Wachter, T. Mertelmeier and H. M. Hoffman, Phys. Rev. C38, 1139 (1988).

[13]  D. Halderson and R. J. Philpott, Phys. Rev. C28, 1000 (1983).

[14]  J.T. Londergan and C. M. Shakin, Phys. Rev. Lett. 28, 1729 (1972).

Few-Body Systems, Suppl. 7, 144—150 (1994)

Few-
Body
Systems
© Springer-Verlag 1994
Printed in Austria

# ELECTRODISINTEGRATION OF THE DEUTERON

B. H. Schoch

*Physikalisches Institut, Universität Bonn, Nussallee 12, D-53115 Bonn, Germany*

## Abstract

The potential of future measurements is demonstrated by using large acceptance hadron detectors in combination with an electron beam. Results of a measurement of the asymmetry $A_\phi$ are presented. These results stress the necessity to include relativistic corrections in calculations based on a nonrelativistic framework.

## Introduction

With the availability of high duty factor electron beams more - kinematically complete - experiments can be performed. Thereby, the study of the deuteron via electron induced reactions is motivated for the following reasons:

(1) Structure of the deuteron

(2) N-N interaction probed "off the mass shell"

(3) Interplay of nuclear and subnuclear degrees of freedom

Due to the well known electromagnetic interaction the results extracted from electron induced reactions can be interpreted in a rather transparent way. The interaction takes place by the exchange of a virtual photon described by a 4-vector potential $A_\mu$ (see Fig. 1), coupling to a 4-vector current

$j_\mu$ which can be decomposed into three pieces: Coulomb ($\rho$), convection ($\frac{e}{m}\vec{p}$) and spin ($\vec{\sigma}$).

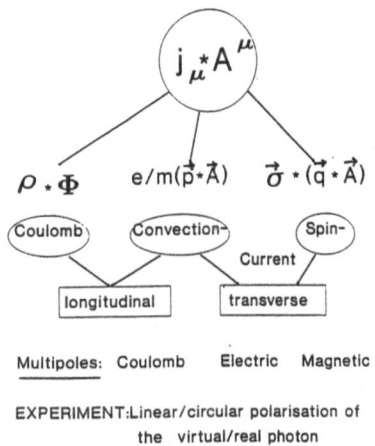

Fig. 1: The different pieces of the hadronic current explored by the electromagnetic probe

Because of the polarization degrees of freedom of the virtual photons three different pieces of the interaction amplitude can be separated experimentally. In general, for an experimental separation of the three pieces - Coulomb, transverse electric, magnetic - the polarization of the virtual photon has to be explored. By performing a coincidence experiment of the type $d(e,e'x)$ the partial linear polarization of the virtual photon in the scattering plane ((e-e')-plane) can be explored by detecting the produced particle x out of plane. This kind of experiment will be described in chapter 2. In order to study spin currents or magnetic transitions in general circular polarized virtual/real photons should be used. With the different pieces of the polarization of the virtual photon at hand different physics questions can be addressed for otherwise the same kinematic conditions like energy and momentum transfer.

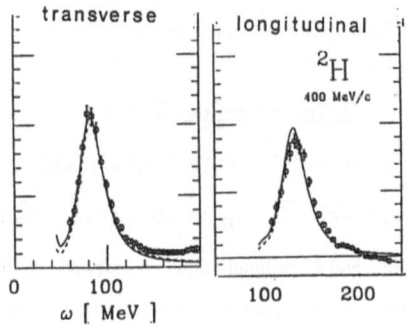

Fig. 2: The longitudinal and transverse absorption cross sections
for virtual photons
Experiment: BATES/MIT [1], Calculation: Arenhövel et al.

Fig. 2 shows the inclusive cross sections for the reaction d(e,e') separated in its longitudinal and transverse pieces. The longitudinal cross section can be described fairly well in a nonrelativistic calculation [1] and using a one body current. The transverse cross section contains additional contributions due to meson exchange and isobar currents. The last aspect gets amplified in the coincidence cross section of the reaction d(e,e'p) covering the excitation region of the (3.3)-resonance [2] (see Fig. 3).

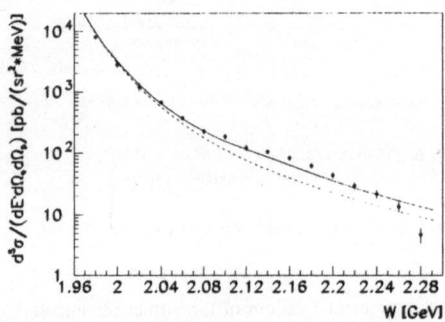

Fig. 3: Results of a d(e,e'p)-measurement
covering the (3.3)-resonance region
Calculation: Arenhövel [3]

Therefore, longitudinal photons are very well suited to address the topics (1) and (2), transverse photons are better to investigate certain aspects of topic (3).

In the past, most calculations concerning reactions on the deuteron have been performed within a nonrelativistic framework. Relativistic corrections are needed, however, in order to describe the results of experiments in special kinematic regions [4, 5]. Very recent results [6] on an observable sensitive to these effects will be be presented in chapter 3.

## 2. The (e,e'p)-Cross Section

The absorption cross section for the virtual exchanged photon $\gamma^*$ can be written [7] as

$$\sigma = \sigma_T + \epsilon\sigma_L + \sqrt{\epsilon(\epsilon+1)} \cdot \sigma_{TL} \cdot \cos\phi + \epsilon\sigma_{TT} \cdot \cos2\phi \qquad (1)$$

Thereby, $T$ stands for transverse and $L$ for longitudinal, $\epsilon$, describes the degree of $\gamma^*$-polarization. The classical solution to separate the longitudinal/transverse pieces of the cross section due to the Rosenbluth-method asks for accurate absolute cross section measurements in different kinematical regions.

Coincidence measurements out of the (e,e')-reaction plane allow an exploration of the linear polarization degrees of freedom of the virtual photon, and therefore, an extensive investigation of e. g. the convection current at the target nucleus.

The detectors for the nucleon should cover a large angular acceptance around the direction of the virtual photon allowing the determination of the φ-dependance of the cross section (see Fig. 4).

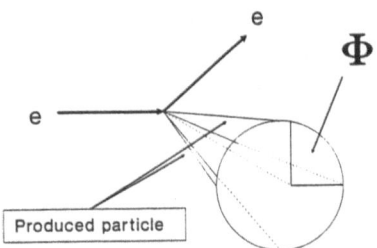

Fig. 4: Out of plane detection of a produced particle
in order to explore the partial linear polarization
of the virtual photon

Such a set-up can be realized with a large acceptance time-of-flight spectrometer for the product particle combined with a magnetic spectrometer for the electron. Such a device is under construction for experiments with the ELSA beam. Large bars of scintillators will be used to detect protons, neutrons and other particles. In a test-set-up (see Fig. 5) for the larger system a measurement was performed in order to measure the $\sigma_{TL}$-piece of the d(e,e'p)-cross section.

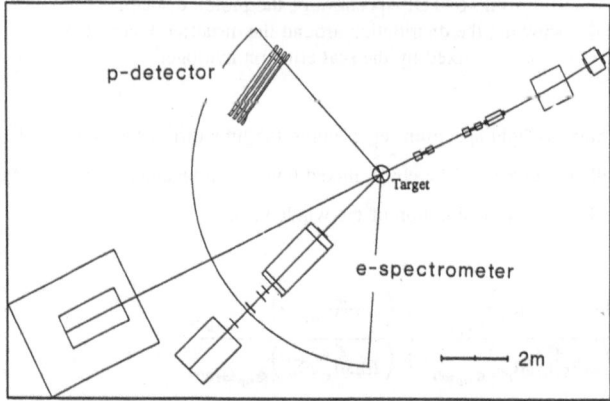

Fig. 5: Set-up for a measurement of the asymmetry $A_\phi$

### 3. A Measurement of the Asymmetry

According to formula (1) in measuring the cross section at the angles $\phi = 0°$, and $\phi = 180°$, the sensitivity on $\sigma_{TL}$ will be emphasized. This quantity, however, should be sensitive to relativistic corrections in an otherwise nonrelativistic description of the process. Thereby, relativity enters [8] as relativistic contributions to the current operators and, furthermore, in the wave functions. For the latter case one may distinguish between the internal dynamics of the rest frame wave function and the boost operation transforming the rest frame wave function into a moving frame.

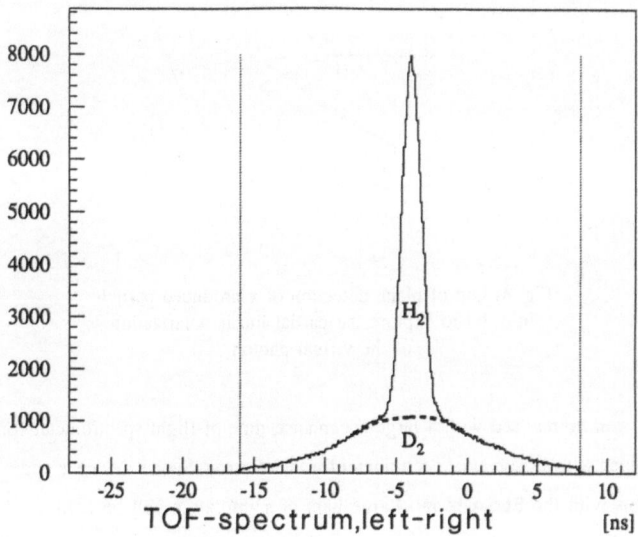

Fig. 6: TOF-spectrum of the proton detector
showing the distribution around the quasifree kinematics
fixed by the scattering on hydrogen

Fig. 6 shows a time-of-flight spectrum representing the time difference between the left and right side of the scintillation detector. Thereby, a mixed liquid hydrogen/deuterium target has been used, the hydrogen peak served as calibration of the whole set-up.

The asymmetry

$$A_\phi = \frac{\left(\frac{d^5\sigma}{dE'_e d\Omega_{e'} d\Omega_{p'}}\right)_{\Phi_{np}=0°} - \left(\frac{d^5\sigma}{dE'_e d\Omega_{e'} d\Omega_{p'}}\right)_{\Phi_{np}=180°}}{\left(\frac{d^5\sigma}{dE'_e d\Omega_{e'} d\Omega_{p'}}\right)_{\Phi_{np}=0°} + \left(\frac{d^5\sigma}{dE'_e d\Omega_{e'} d\Omega_{p'}}\right)_{\Phi_{np}=180°}}$$

extracted from these data is shown in Fig. 7 as a function of the internal (reconstructed) proton momentum. With the data the results of calculations [9] are shown.

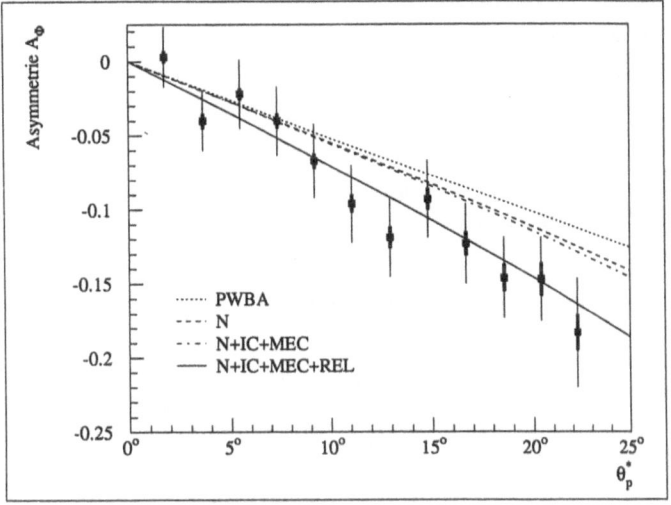

Fig. 7: The asymmetry extracted from the measurement,
calculations from [9]
The total error bars (thin lines) contain the contributions
from systematic and statistical (thick lines) contributions.

The data clearly favour the calculation with the inclusion of relativistic corrections. Furthermore, the accuracy achieved with this measurement gives hope to disentangle all structure functions (1) with the TOF-set-up under construction.

150

## References

1. Von Reden, K., et al.: Bates Linear Accelerator Center, Annual Report, 106, 1985

2. Boden, B., et al.: Nucl. Phys. A 267, 471 (1992)

3. Arenhövel, H.: private communication

4. Hughes, R.J., et al.: Nucl. Phys. A 267, 329 (1976)

5. Van der Schaar, M., et al.: Phys. Rev. Lett. 68, 776 (1992)

6. Frommberger, F.: Phd-thesis, Univ. Bonn (1993)

7. Lyth, D.H., Donnachie, A., and Shaw, G. (eds.): Electromagnetic Interactions with Hadrons I. New York: Plenum Press 1978

8. Wilbois, T., et al.: preprint Univ. Mainz, MKPH-T-93-10 (1993)

9. Wilbois, T., et al.: private communication, 1993

Few-Body Systems, Suppl. 7, 151—159 (1994)

Few-
Body
Systems
© Springer-Verlag 1994
Printed in Austria

# EXCLUSIVE ELECTRON SCATTERING FROM DEUTERIUM AT HIGH $Q^2$

J.F.J. van den Brand

Department of Physics
University of Wisconsin-Madison
Madison, Wisconsin 53706, USA

## Abstract

Cross sections are presented for the reaction $^2$H(e,e'p)n for momentum transfers in the range $1.0 \leq Q^2 \leq 6.8 \, (\text{GeV/c})^2$ and for missing momenta from 0 to 250 MeV/c. The longitudinal-transverse interference structure function has been separated at $Q^2 = 1.2$ $(\text{GeV/c})^2$. The observables are compared to calculations performed in non-relativistic and relativistic frameworks. The best description of the data is found using a fully relativistic treatment.

## 1. Introduction

The (e,e'p) reaction can be used to explore single-particle properties of the nucleus through a measurement of the spectral function, or for a study of the identity of the nucleon embedded in the nuclear medium. Until recently, quasi-elastic (e,e'p) scattering experiments were performed at relatively moderate four-momentum transfer, typically $Q^2 \approx 0.2 \, (\text{GeV/c})^2$. With the advent of a new generation of electron-scattering facilities (CEBAF, Mainz, MIT-Bates, and NIKHEF), capable of delivering higher beam energies, experiments are foreseen at higher momentum transfer. These experiments may be sensitive to new phenomena, for example the QCD prediction of color transparency[1] or manifestations of quark-exchange effects[2]. In order to interpret such new data at high $Q^2$ correctly, the (e,e'p) cross section must be precisely understood in this kinematical regime.

The analysis of the quasi-elastic (e,e'p) experiments has generally been performed in a non-relativistic framework, in order to facilitate the calculation of nuclear wave functions in the initial and final state. For the current operator often a non-relativistic reduction is used, truncated at lowest orders in $p/M$, where $p$ $(M)$ is the proton momentum (mass). At high $Q^2$ this approach is bound to fail since the proton momentum is not small compared to its mass. It has been claimed[3] that already at low $Q^2$ ($\approx$ 0.2 $(\text{GeV/c})^2$) relativistic effects may become important in the longitudinal-transverse part of the cross section, whereas for the separated longitudinal and transverse structure functions[4] the non-relativistic and relativistic calculations gave about the same results. Apart from relativistic ingredients in the (e,e'p) reaction mechanism, new phenomena may occur due to the fact that the wavelength of the photon becomes small ($\lambda^* \approx 0.08$ fm at $Q^2 = 6.8$ (GeV/c)$^2$) with respect to the size of the nucleon ($< r >_{RMS} \approx 0.8$ fm). It is thus not a priori clear, that quasi-elastic scattering still can be described in terms of the impulse approximation using nucleon-meson fields. At the highest momentum transfer probed in this experiment, $Q^2 = 6.8$ (GeV/c)$^2$, the electromagnetic form factors of the proton are known to follow counting-rule behavior[5]. This may indicate that quark-gluon degrees of freedom are important for a correct description of the reaction cross section.

In the present experiment the (e,e'p) cross section is measured for quasi-elastic scattering off the deuteron for $Q^2$ values of 1.0, 1.2, 3.0, 5.0 and 6.8 (GeV/c)$^2$. The longitudinal-transverse interference structure function, $W_{LT}$, is obtained at $Q^2 = 1.2$ (GeV/c)$^2$. Exact calculations for realistic nucleon-nucleon (NN) potentials can be performed for the two-body system. This allows for an important benchmark study of the (e,e'p) reaction mechanism for quasi-elastic scattering off nuclei. The dependence of the cross section on $Q^2$ is used to investigate the validity of the impulse approximation for quasi-elastic scattering while $W_{LT}$ has been extracted in order to determine the importance of relativistic effects in the cross section.

## 2. Theory

It is well-known that in the one-photon exchange approximation one can write the spin-averaged five-fold (e,e'p) coincidence cross section as[6]

$$\frac{d^5\sigma}{d\Omega_e d\Omega_p dE} = K R \sigma_{Mott} \frac{Q^2}{q^2} \{W_L + \frac{1}{\varepsilon}W_T - \sqrt{\frac{\varepsilon+1}{\varepsilon}} cos(\phi)W_{LT} + cos(2\phi)W_{TT}\} \quad (1)$$

in which K represents a kinematical factor, R the recoil factor (i.e. the Jacobian $\partial E_p/\partial E_m$, with $E_p$ ($E_m$) the energy of the knocked-out proton (the missing energy)), $\varepsilon$ equals the virtual-photon polarization, and $\phi$ the angle between the electron scattering plane and the reaction plane, defined by the direction of the momentum transfer and the knocked-out proton. For scattering off a stationary *free* nucleon, expression (1) reduces to the Rosenbluth formula which depends solely on $W_L$ and $W_T$. The interference structure functions $W_{LT}$ and $W_{TT}$ are induced for scattering of a free nucleon when its rest frame is boosted such that it has a non-zero momentum component, $p_\perp$, perpendicular to the virtual-photon momentum. The interaction of the perpendicular nucleon current, $J_\perp$, with the $e_1$ polarization component of the electromagnetic field (i.e. the component of the magnetic field normal to the scattering plane) gives rise to the $\phi$-dependence of the electron scattering cross section. Hence, the measurement of $W_{LT}$ may allow the separation of the contribution of the convection current to the cross section from that of the spin current.

In the case of proton knockout from a complex nucleus a dynamical model is needed to calculate the nuclear structure functions $W_i$. In the plane-wave impulse approximation (PWIA) the virtual photon couples to the nucleon, of which the electromagnetic form factors are chosen equal to the free nucleon form factors. In this case, the cross section reduces to

$$\frac{d^5\sigma}{d\Omega_e d\Omega_p dE} = K R \sigma_{ep} \rho(|\mathbf{p_m}|) \quad (2)$$

where $\sigma_{ep}$ represents the off-shell electron-proton cross section and $\rho(|\mathbf{p_m}|)$ is the momentum distribution of the proton inside the deuterium nucleus. From equation (1) it is clear that one can determine the value of $W_{LT}$ by performing measurements of the cross section at $\phi$ equals 0 and 180°, keeping all other kinematical quantities constant. In order to reduce systematic uncertainties we present the data in terms of the asymmetry $A_\phi$:

$$A_\phi = \frac{\sigma(0) - \sigma(\pi)}{\sigma(0) + \sigma(\pi)} \left( = \frac{\sqrt{(\varepsilon + 1)\varepsilon}W_{LT}}{\varepsilon W_L + W_T + \varepsilon W_{TT}} \right). \tag{3}$$

In PWIA this asymmetry exclusively depends on the nuclear current operator. In order to assess the sensitivity to choices in the off-shell current operator we present the asymmetry calculated with different models in Fig. 1.

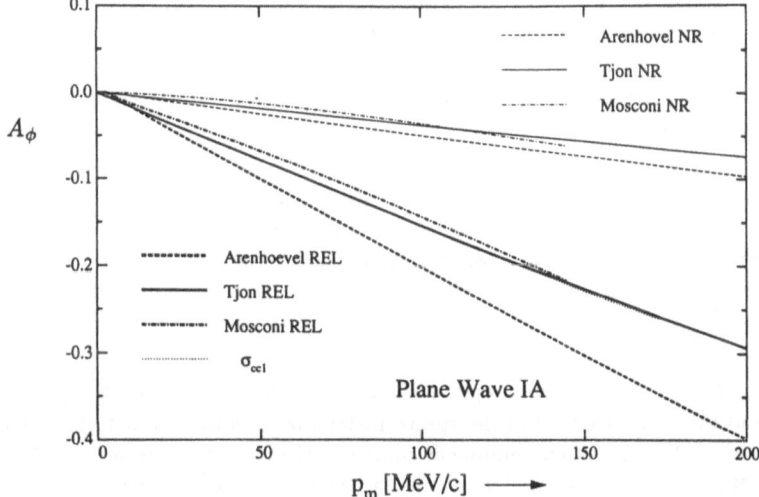

**Figure 1.** *The asymmetry $A_\phi$ for the reaction $^2H(e,e'p)n$ as a function of the missing momentum at $Q^2 = 1.2\,(GeV/c)^2$ for various non-relativistic (upper set of curves labeled NR) and relativistic (lower set of curves labeled REL) models.*

In Tjon's relativistic field theoretical model[7] the Bethe-Salpeter equation is solved for a one-boson exchange potential. In order to reduce the computational complexities the quasi-potential approximation is used, in which the two-nucleon propagator is replaced by an effective propagator[8]. 'Relativistic' and 'non-relativistic' contributions to the observables can be separated by making a non-relativistic reduction of the current operator through a second-order Foldy-Wouthuysen[9] transformation and projecting out the positive-energy states of the proton wave function. The non-relativistic calculations of Mosconi[10] and Arenhövel[11] start from the Schrödinger equation. They use non-relativistic wave functions and the McVoy-van Hove[12] current operator. Their 'relativistic' calculations include the following corrections. Firstly, they calculate the nucleonic part in the current operator by using the Foldy-Wouthuysen transformation extended up to order $(p/M_p)^3$. Secondly, they include contributions from the boost operator that boosts the initial state wave function from the laboratory frame to the center-of-mass frame. Further details on these calculations can be found in Refs. 7,10,11. For comparison, also the asymmetry according to the off-shell prescription for the current operator $\sigma_{cc1}$ of de Forest[13] is shown which is close to the result of Tjon's relativistic calculation.

It is remarkable that the differences are sizeable even in PWIA, reflecting the theoretical uncertainties in the treatment of the off-shell nuclear current operator. All non-relativistic calculations yield an asymmetry that is approximately four times smaller than the relativistic calculations, indicating that an important ingredient is missing in the non-relativistic reduction scheme when applied to the LT-interference strength.

## 3. Experimental Set-Up

The experiment has been performed using the spectrometer setup at endstation A of the Stanford Linear Accelerator Center (SLAC). The NPI gun supplied the electron beam, which was accelerated to energies between 2.0 and 5.1 GeV and had a duty factor of 0.02 %. The scattered electrons and knocked-out protons were detected in two magnetic spectrometers which were able to detect particles with a maximum momentum of 1.6 GeV/c and 8 GeV/c, respectively.

The field of the electron spectrometer was generated by a dipole magnet with a 90° bend angle and shaped pole tips that gave the desired quadrupole and sextupole magnetic field components. The detection equipment consisted of three pairs of drift chambers with wires in the x- and y-direction which were used to determine the particle trajectory, two scintillator hodoscopes that gave the time of the event, a $CO_2$ gas Čerenkov detector, and two layers of lead glass shower counters to identify the particle as an electron.

Two dipole magnets, with a bend angle of 15° each, and three quadrupole magnets generated the magnetic field of the spectrometer that was used to detect the knocked-out protons. The detection equipment consisted of three arrays of fast plastic timing scintillators that were separated over a distance of 3.84 m, ten MWPCs of which five with wires in the x-direction and five in the u,v-direction, a plastic scintillator hodoscope and a Freon Čerenkov detector. The timing scintillators determined the time of the event as well as the velocity of the particle. The resolution in velocity $\beta = v/c$ was better than 0.03 (FWHM) and allowed clean identification of protons for $Q^2 < 5.0$ (GeV/c)$^2$. For $Q^2 \geq 5.0$ (GeV/c)$^2$ particle identification was obtained from the Čerenkov detector.

With this setup, a resolution in coincidence time was achieved of better than 0.8 ns (FWHM). The obtained kinematical uncertainties (FWHM) are 0.1 % for the beam energy, 0.5 mrad for the electron scattering angle, 0.5 mrad for the proton scattering, 9 (28) MeV in missing energy and on average 10 (45) MeV/c in missing momentum at $Q^2 = 1.0$ (6.8) (GeV/c)$^2$. The centroid of the missing-momentum distribution could be determined up to 1.3 MeV/c. The central kinematics used in the experiment are given in Table 1.

The data have been corrected for radiative effects by applying a Monte Carlo[14] code in which the distribution in missing momentum and missing energy is generated using a PWIA description that is based on $\sigma_{cc1}$ of de Forest and the proton momentum distribution in the initial state derived from the Bonn potential. The Monte Carlo code took into account the experimental resolutions and detection volume. Wasson et al.[15] extended the approach of Mo and Tsai[16] in order to describe the radiative effects in the

**Table 1.** *Kinematics for the* $^2$H(e,e'p)n *experiment performed at SLAC.*

| $E_{BEAM}$ [ GeV ] | $E'$ [ GeV ] | $p'$ [ GeV/c ] | $\theta_e$ [ Degr. ] | $\theta_p$ [ Degr. ] | $Q^2$ [ (GeV/c)$^2$ ] |
|---|---|---|---|---|---|
| 2.02 | 1.35 | 1.28 | 38.8 | 35.9 - 50.2 | 1.20 |
| 2.02 | 1.39 | 1.20 | 35.5 | 43.4 - 54.6 | 1.04 |
| 3.19 | 1.47 | 2.45 | 47.7 | 27.7, 30.5 | 3.06 |
| 4.21 | 1.47 | 3.54 | 53.4 | 20.9 | 5.00 |
| 5.12 | 1.47 | 4.49 | 56.6 | 16.7 | 6.77 |

(e,e'p) reaction. The soft photon approximation is used and proton radiation is included. The extraction of $W_{LT}$ from the coincidence cross section is sensitive to the treatment of radiative effects. The procedure has been extensively tested using $^1$H(e,e'p) data. More details on the experimental setup and the data analysis can be found in Ref. 17.

## 4. Results

Next we compare the data for $W_{LT}$ and the five-fold differential cross sections for the reaction $^2$H(e,e'p)n with the results of various models. Fig. 2 shows the experimental asymmetry and the results of the full calculations (including final-state interaction (FSI), meson-exchange currents (MEC) and isobar configuration (IC) effects) of Arenhövel and Tjon and the PWIA calculations of Mosconi.

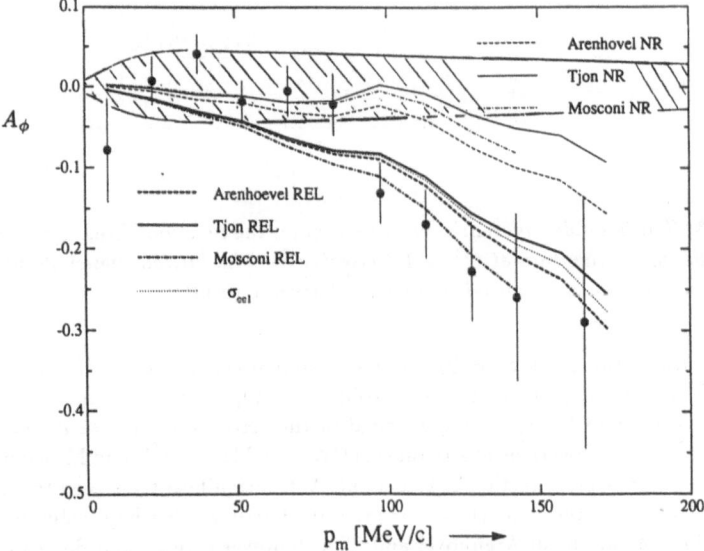

**Figure 2.** *The asymmetry $A_\phi$ for the reaction $^2$H(e,e'p)n as a function of the missing momentum at $Q^2 = 1.2$ (GeV/c)$^2$.*

The shaded band indicates the experimental uncertainty, which is mainly due to imperfect knowledge of the kinematics. The theoretical calculations are folded over the

156

acceptance of the experimental setup. In comparison with Fig. 1 a smaller asymmetry is observed at low missing momentum, since the component of the missing momentum parallel to the momentum transfer is not negligible. The systematic error is large at low missing momentum where it obscures the interpretation of the results. For $|\mathbf{p_m}|$ > 100 MeV/c, the data clearly prefer the relativistic descriptions of the asymmetry over the non-relativistic descriptions, but within the uncertainty no preference can be given between the three relativistic models. Note, that $\sigma_{cc1}$ of de Forest also yields a satisfactory description.

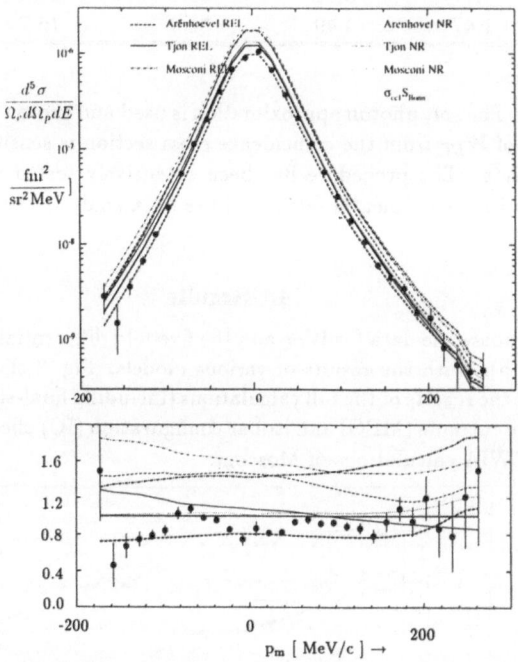

**Figure 3.** *The five-fold differential cross section for the reaction $^2H(e,e'p)n$ as a function of the missing momentum at $Q^2 = 1.2$ $(GeV/c)^2$. The bottom panel shows the data and the calculations normalized to Tjon's relativistic result.*

The cross section is shown in Fig. 3. The systematical error of 8.2 %, which is mainly due to the uncertainty in the detection volume (5 %), the proton angle (4 %) and the radiative corrections (3 %), is not included in the error bars. It is seen that both the non-relativistic cross sections of Arenhövel (DWIA + MEC + IC) and Mosconi (PWIA), and the relativistic cross section of Mosconi (PWIA) significantly exceed the data. Also for this observable, the non-relativistic treatment fails at this high value of $Q^2$. The relativistic calculations of Arenhövel and Tjon however give a good description of the cross section. We have investigated the influence of the NN potential on the calculated $W_{LT}$ using Arenhövels treatment including FSI, MEC and IC contributions. The results for $A_\phi$ using the Paris, Bonn, Nijmegen, and Argonne V14 NN potential differ by less than 0.02 and therefore do not influence our conclusions. In addition, we studied the influence of the reaction mechanism in both Tjons and Arenhövels non-relativistic and relativistic model. The results in PWIA, distorted wave impulse approximation (DWIA)

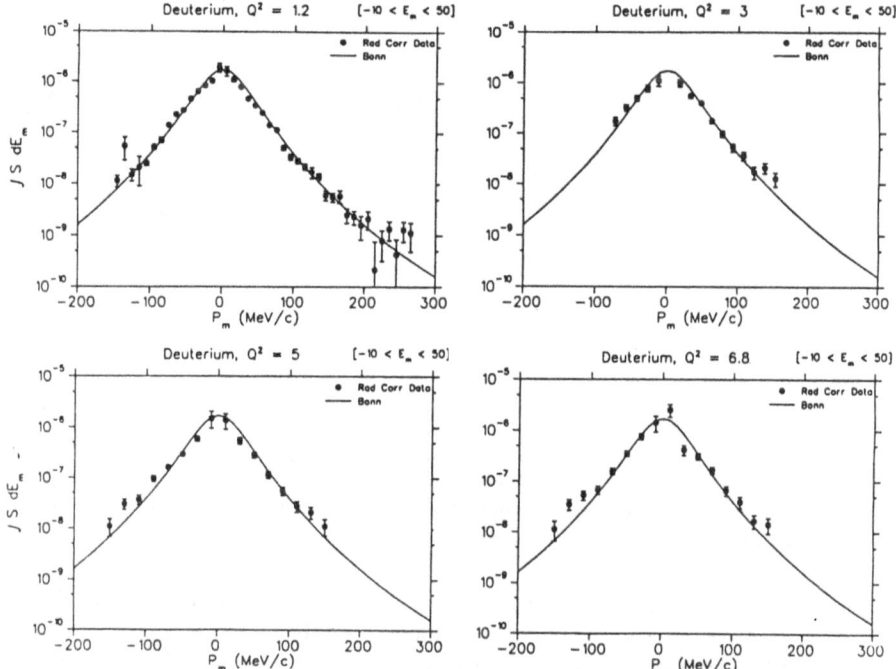

**Figure 4.** *Preliminary reduced cross section for the reaction $^2H(e,e'p)n$ as a function of the missing momentum for various values of the four-momentum transfer in the range $1.0 \le Q^2 \le 6.8\ (GeV/c)^2$.*

with and without MEC and IC effects affect $A_\phi$ by less than 0.04. Therefore, also the reaction mechanism is sufficiently well behaved to allow a clear interpretation of our data.

Fig. 4 shows the reduced coincidence cross section as a function of missing momentum for momentum-transfer values $Q^2 = 1.0$, 3.1, 5.0, and 6.8 $(GeV/c)^2$. The reduced cross section is obtained from the five-fold differential cross section by dividing out, on an event-by-event basis the kinematical factor K, the Jacobian R, and $\sigma_{cc1}$. Therefore, assuming a correct description of the nuclear current operator the reduced cross section would in PWIA represent the proton momentum distribution of deuterium. The curves show $\rho(|\mathbf{p_m}|)$ based on the Bonn potential. Fair agreement is obtained and the data validate the quasi-free scattering mechanism.

Next, we fit the data to the results of our Monte Carlo code, based on $\sigma_{cc1}$ and the Bonn spectral function, with as free parameter the normalization. The results are shown in Fig. 5. In order to estimate the contribution of rescattering effects in the final state, the figure also shows the result of a Glauber calculation[18]. Within the 8.2 % systematic error, the data are in good agreement with the Glauber result (0.92) and no anomalous $Q^2$ dependence is observed.

In summary, the present $^2H(e,e'p)n$ experiment extends the range of $Q^2$ by more than a factor of 30. It is found that the LT-interference structure function is sensitive to the description of the nuclear current operator, and especially to relativistic ingredients of the theory. The relativistic models of Refs. 7, 10, and 11 give a satisfactory description

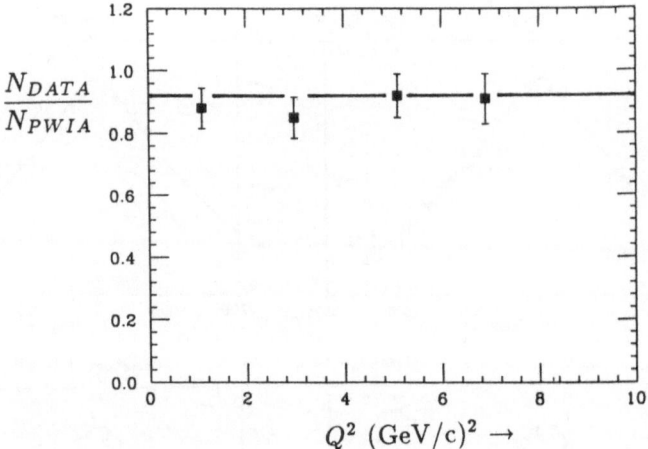

**Figure 5.** *Integrated five-fold differential cross section normalized to the PWIA cross section as a function of four-momentum transfer. The error bars include the systematical error. The line is the result of a Glauber calculation*[18]

of the asymmetry, whereas the non-relativistic descriptions fail to do so. Both the $Q^2$ dependence of the cross section and the asymmetry can be described at the level of 10 % in the PWIA approximation, using the $\sigma_{cc1}$ off-shell prescription of de Forest and the Bonn spectral function for deuterium.

More conclusive results on relativistic effects in the (e,e'p) reaction can be obtained by experiments that combine high luminosity with precise determination of the kinematical quantities. In the near future, such experiments will become feasible with the high-resolution spectrometers in Hall A at CEBAF.

The author acknowledges interesting discussions with Henk Jan Bulten. The research is supported by the National Science Foundation under Contract No. PHY-9019983.

## References

[1] A.H. Mueller, Proc. of the Moriond Conference, 1982; S.J. Brodsky, XIII Int. Symp. on Multiparticle Dynamics, 1982.

[2] L.Ya. Glozman *et al.*, Phys. Lett. **B252**, 23 (1990).

[3] M. van der Schaar *et al.*, Phys. Rev. Lett. **68**, 776 (1992).

[4] M. van der Schaar *et al.*, Phys. Rev. Lett. **66**, 2855 (1991).

[5] S.J. Brodsky and G.L. Farrar, Phys. Rev. Lett. **31**, 1153 (1973).

[6] S. Frullani and J. Mougey, Adv. Nucl. Phys. **14**, 1 (1984).

[7] J.A. Tjon, Few-Body Systems **Suppl. 5**, 17 (1992); E. Hummel and J.A. Tjon, Phys. Rev. **C42**, 423 (1990); E. Hummel, Ph. D. thesis, Rijks Universiteit Utrecht (1991), unpublished.

[8] R. Blankenbeckler and R. Sugar, Phys. Rev. **142**, 1051 (1966); A.A. Logunov and A.N. Tavkhelidze, Nuovo Cim. **29**, 380 (1963).

[9] L.L. Foldy and S.A. Wouthuysen, Phys. Rev. **78**, 29 (1950).

[10] B. Mosconi and P. Ricci, Nucl Phys **A517**, 483 (1990).

[11] H. Göller and H. Arenhövel, Few-Body Systems **13**, 117 (1992).

[12] K.W. McVoy and L. van Hove, Phys. Rev **125**, 1034 (1962).

[13] T. de Forest, Jr., Nucl. Phys. **A392**, 232 (1983).

[14] N. Makins *et al.*, private communications.

[15] D. Wasson *et al.*, to be published;

[16] L.W. Mo and Y-S. Tsai, Rev. Mod. Phys. **41**, 205 (1969).

[17] N. Makins, T. O'Neill *et al.*, private communications.

[18] N.N. Nikolaev, private communications.

Few-Body Systems, Suppl. 7, 160—163 (1994)

Few-
Body
Systems

# PROTON–DEUTERON BREAK UP INCLUDING COULOMB EFFECTS

E. O. Alt and M. Rauh[†]

Institut für Physik, Universität Mainz, D-55099 Mainz, Germany

**Abstract**: The first results of the calculation of proton–deuteron break-up cross sections
are presented and compared with experimental data.

The theoretical investigation of the proton–deuteron (pd) break-up reaction represents one of the major fields of application of three–body theories. But despite of great
efforts the results obtained up to now have not yet lived up to the expectations. Either
the theoretical situation is well under control, as in the deuteron break-up by neutrons,
but the experimental data are of a quality which does not suffice to draw interesting
conclusions. Or when, as in the proton–induced reaction, a multitude of excellent experimental data exist then the employed theoretical analyses are inadequate in that in
most cases the influence of the Coulomb repulsion between the two protons has not
at all or only partly been taken into account (however, see [1] and references therein).
Therefore, the comparison of theoretical nd break-up calculations with experimental pd
results conventionally put into practice can only be considered as an attempt to obtain
a qualitative to semiquantitative understanding of this reaction. This is obvious unless,
of course, it is known from somewhere that, under the kinematic conditions considered,
Coulomb effects play a negligible role only; then even quantitative conclusions could be
inferred from such an analysis. This concerns, e.g., possible signatures of three–nucleon
forces, the inadequacy of on-shell aspects of the two–nucleon force in channels which
are difficult to study in nucleon–nucleon scattering, or limitations of the validity of the
charge–independence hypothesis.

In order to obtain a quantitative estimate of the importance or unimportance of
Coulomb effects in the pd break-up reaction we have solved the equations proposed
by ASZ [2] which allow one to take into account the long–ranged Coulomb potential
in a mathematically well–defined and consistent manner. The basic idea consists in
screening the Coulomb potentials (R denotes the screening radius), renormalising the
resulting break-up amplitudes and performing the zero–screening limit numerically.

Let us briefly sketch some structural aspects of the formalism used. We calculate
the (screened) amplitude $T_{01}^{(R)}$, describing the break up of a bound state of the proton
2 and the neutron numbered particle 3, by the impinging proton 1 with center–of–mass

momentum $\vec{q}_1$, from the screened two-fragment amplitudes $T_{\gamma 1}^{(R)}$ via quadrature,

$$T_{01}^{(R)}(\vec{p}_3', \vec{q}_3', \vec{q}_1; E + i0) = V_{01}^{(R)'}(\vec{p}_3', \vec{q}_3', \vec{q}_1; E + i0) +$$
$$\sum_\gamma \int d^3 q_\gamma'' \, V_{0\gamma}^{(R)}(\vec{p}_3', \vec{q}_3', \vec{q}_\gamma''; E + i0) \, G_{0\gamma}^{(R)}(\vec{q}_\gamma'') \, T_{\gamma 1}^{(R)}(\vec{q}_\gamma'', \vec{q}_1; E + i0). \quad (1)$$

Here, $V_{01}^{(R)}$ is the break–up potential and $G_{0\gamma}^{(R)}$ the propagator of the correlated pair $\gamma$. For explicit expressions we refer to [2]. Furthermore, $\vec{p}_3', \vec{q}_3'$ are the Jacobi momenta characterising the final state, with the two protons forming subsystem 3. The two–fragment amplitudes $T_{\beta 1}^{(R)}$ satisfy the set of coupled AGS integral equations [3] which in operator notation read

$$T_{\beta 1}^{(R)}(E + i0) = V_{\beta 1}^{(R)}(E + i0) + \sum_{\gamma=1}^{3} V_{\beta \gamma}^{(R)}(E + i0) \, G_{0\gamma}^{(R)}(E + i0) \, T_{\gamma 1}^{(R)}(E + i0), \quad (2)$$

with $V_{\beta\gamma}^{(R)}(E + i0)$ denoting the effective arrangement potentials. As discussed in [2] the elastic part $V_{\alpha\alpha}^{(R)}(\vec{q}_\alpha', \vec{q}_\alpha; E + i0)$ contains, for $\alpha \neq 3$, as its longest–ranged part the so–called center–of–mass Coulomb potential which for exponential screening looks as $v_\alpha^R(\vec{q}_\alpha', \vec{q}_\alpha) = e_1 e_2 / 2\pi^2[(\vec{q}_\alpha' - \vec{q}_\alpha)^2 + R^{-2}]$. (Its occurrence is the reason for the failure of standard scattering theory in the present problem.) Let $t_\alpha^R$ be the amplitude obtainable by solving a two–body Lippmann–Schwinger equation with $v_\alpha^R$. It is a genuine two–body amplitude describing the Coulomb scattering of the proton $\alpha$ ( $= 1$ or 2) off the total charge of the deuteron concentrated in its center of mass. Then from the solution of eq. (2) one calculates the Coulomb-modified short-range amplitude

$$T_{\beta 1}^{SR}(\vec{q}_\beta', \vec{q}_1; E + i0) = T_{\beta 1}^{(R)}(\vec{q}_\beta', \vec{q}_1; E + i0) - \delta_{\beta 1} t_1^R(\vec{q}_1', \vec{q}_1), \quad (3)$$

for which the applicability of the renormalization procedure has been proven [2]. Insertion of the decomposition (3) into (1) leads to the following representation for the break–up amplitudes

$$T_{01}^{(R)}(\vec{p}_3', \vec{q}_3', \vec{q}_1; E + i0) = B_{01}^R(\vec{p}_3', \vec{q}_3', \vec{q}_1; E + i0) + T_{01}^{SR}(\vec{p}_3', \vec{q}_3', \vec{q}_1; E + i0). \quad (4)$$

Denoting by $|\vec{q}_{1,R}^{(+)}\rangle$ the scattering state due to $v_1^R$, the driving term $B_{01}^R(\vec{p}_3', \vec{q}_3', \vec{q}_1; E+i0)$, which represents the pure Coulomb break–up, is given on the energy shell as

$$B_{01}^R(\vec{p}_3', \vec{q}_3', \vec{q}_1; E + i0) = \int d^3 q_1'' \, V_{01}^{(R)}(\vec{p}_3', \vec{q}_3', \vec{q}_1''; E + i0) \langle \vec{q}_1'' | \vec{q}_{1,R}^{(+)} \rangle. \quad (5)$$

The Coulomb-modified short-range break–up part $T_{01}^{SR}(\vec{p}_3', \vec{q}_3', \vec{q}_1; E + i0)$ follows from the Coulomb-modified short-range arrangement amplitudes by quadrature

$$T_{01}^{SR}(\vec{p}_3', \vec{q}_3', \vec{q}_1; E + i0) = \sum_\gamma \int d^3 q_\gamma'' \, V_{0\gamma}^{(R)}(\vec{p}_3', \vec{q}_3', \vec{q}_\gamma''; E + i0) \, G_{0\gamma}^{(R)}(\vec{q}_\gamma'') \times$$
$$T_{\gamma 1}^{SR}(\vec{q}_\gamma'', \vec{q}_1; E + i0). \quad (6)$$

Then, as shown in [2], after renormalization by suitable, explicitly known renormalization factors $Z^{(R)}(p_3')$ and $Z_1^{(R)}(q_1')$, the zero–screening limit exists in both

$$Z^{(R)^{-\frac{1}{2}}}(p_3') B_{01}^R(\vec{p}_3', \vec{q}_3', \vec{q}_1; E + i0) Z_1^{(R)^{-\frac{1}{2}}}(q_1') \xrightarrow{R \to \infty} B_{01}^C(\vec{p}_3', \vec{q}_3', \vec{q}_1; E + i0) =$$
$$\int d^3 q_1'' \, \psi_{C,\vec{p}_3'}^{(-)*}(\vec{k}'[\vec{q}_3', \vec{q}_1'']) \chi_1(\vec{k}[\vec{q}_3, \vec{q}_1'']) \psi_{C,\vec{q}_1}^{(+)}(\vec{q}_1''), \quad (7)$$

and

$$Z^{(R)^{-\frac{1}{2}}}(p_3') T_{01}^{SR}(\vec{p}_3', \vec{q}_3', \vec{q}_1; E + i0) Z_1^{(R)^{-\frac{1}{2}}}(q_1) \xrightarrow{R \to \infty} T_{01}^{SC}(\vec{p}_3', \vec{q}_3', \vec{q}_1; E + i0). \quad (8)$$

In (7) the momenta $\vec{k}$ and $\vec{k}'$ have to be expressed as the standard linear combinations of the indicated momenta, and the form factor $\chi_1$ is related to the incoming deuteron bound state wave function in the usual way. $\psi_{C,\vec{q}_1}^{(+)}(\vec{q}_1'')$ denotes the unscreened center-of-mass momentum space Coulomb wave function depending on the indicated incoming momentum $\vec{q}_1$. Similarly for the wave function $\psi_{C,\vec{p}_3'}^{(-)*}(\vec{k})$ for the two outgoing protons. In this way we obtain the full break–up amplitude for *unscreened* Coulomb potentials as

$$\mathcal{B}_{01}^C(\vec{p}_3', \vec{q}_3', \vec{q}_1; E + i0) + \mathcal{T}_{01}^{SC}(\vec{p}_3', \vec{q}_3', \vec{q}_1; E + i0) = \mathcal{T}_{01}(\vec{p}_3', \vec{q}_3', \vec{q}_1; E + i0). \qquad (9)$$

The practical procedure is now as follows. We solve the appropriately antisymmetrized equation (2) and the two–body Lippmann–Schwinger equation for $t_\alpha^R$ to get the Coulomb-modified short-range arrangement amplitude (3), introduce this result in (6), and repeat the whole procedure for increasing values of R until the renormalised Coulomb-modified short-range break–up amplitude, the l.h.s. of (8), becomes independent of the screening radius R, in this way numerically performing the limit in (8). To this we add (7) which has been evaluated in the coordinate space representation, to finally end up via relation (9) with the desired physical proton–deuteron break–up amplitude. We remark that, while (2) was solved with the approximation which consists in replacing the pp Coulomb T–matrix occurring in $\mathcal{V}_{\beta\gamma}^{(R)}$ and $\mathcal{G}_{0\gamma}^{(R)}$ by its Born term, the break–up potentials $\mathcal{V}_{01}^{(R)}$ have been calculated exactly.

We used separable nucleon–nucleon potentials of rank one [4]. In this way we have obtained the differential cross sections $d^5\sigma/d\Omega_1^L d\Omega_2^L dE_S$, where $\Omega_1^L$ and $\Omega_2^L$ are the laboratory angles of the measured protons and $E_S$ is the arc length along the kinematically allowed curve, for various interesting kinematic configurations. As an example we present in Fig. 1a cross sections for a np final state interaction (FSI), and in Fig. 1b for a pp quasifree scattering (QFS) configuration, as function of the $E_S$, at 16 MeV proton laboratory energy. The experimental data are from Klein et al [5]. For comparison the cross section for the neutron–induced reaction, obtained by switching off the Coulomb interaction in the above equations, are included. The following conclusions can be drawn.

(i) In the various peak regions the Coulomb corrections are definitively nonnegligible and bring the theoretical calculations into rather good agreement with the experimental data. This is of semiquantitative significance since our nd calculations, despite of the simplicity of the ansatz for the nuclear interaction employed, are rather close to those obtained with more sophisticated NN–forces [6,7]. Outside of the peak regions the Coulomb interaction appears to be of minor importance.

(ii) In both configurations the main effects stem from the Coulomb corrections in the Coulomb-modified short-range arrangement amplitudes $\mathcal{T}_{\beta 1}^{SR}$, and thus are not accounted for in Migdal–Watson type models. However, in other kinematic configurations also the Coulomb contribution from the break–up potentials can become important.

(iii) In [5], a nd QFS calculation had to be multiplied by a factor of 0.776 to get agreement with the observed pd data. Our *calculated* pd to nd cross section ratio is 0.79 at the maximum; but in contrast to the above phenomenological procedure, the calculated ratio tends towards 1 for small and large arc lengths.

(iv) As is apparent the experimental data can to reasonable accuracy be explained by the inclusion of Coulomb effects. Hence not much room is left for "unconventional" effects.

† Supported by the Deutsche Forschungsgemeinschaft, Project no. Al 250/1-3

[1] Kröger, H., Nachabe, A. M., Slobodrian, R.J.: Phys. Rev. C **33**, 1208 (1986)
[2] Alt, E.O., Sandhas, W. Ziegelmann, H.: Phys. Rev. C **17**, 1981 (1978)

[3] Alt, E.O., Grassberger, P., Sandhas, W.: Nucl. Phys. **B2** , 167 (1967)

[4] Alt, E.O., Sandhas, W., Ziegelmann, H.: Nucl. Phys. **A445** , 429 (1985)

[5] Klein, H., Eichner, H., Helten, H.J., Kretzer, H., Prescher, K., Stehle, H., Wohlfarth, W.W.: Nucl. Phys. **A 199**, 169 (1973)

[6] Frank, T.N., Haberzettl, H., Januschke, T., Kerwath, U., Sandhas, W.: Phys. Rev. **C 38**, 1112 (1988)

[7] Witala, H., Cornelius, T., Glöckle, W.: Phys. Rev. C **39**, 384 (1989)

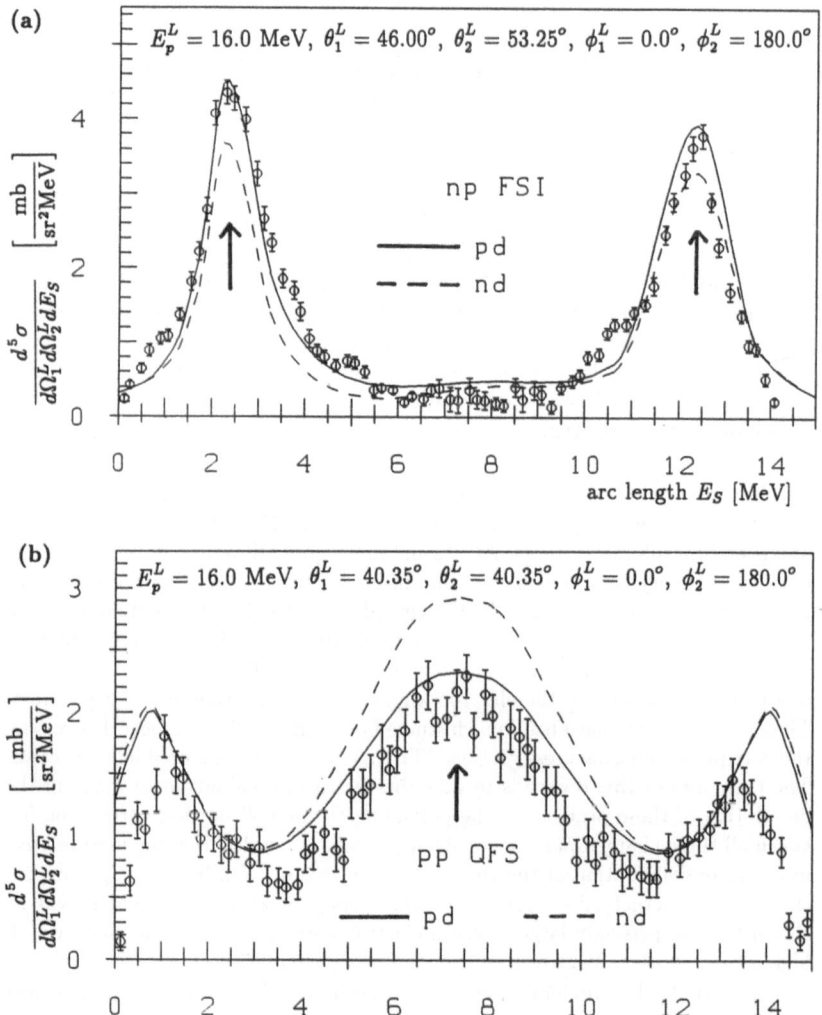

Fig. 1: Differential cross sections for the reaction d(p,pp)n at 16 MeV laboratory kinetic energy, as function of the arc length: (a) for the np FSI configuration, (b) for a pp QFS configuration. Full line: pd results including Coulomb effects, dashed line: nd results. Data are from Klein et al [5]. Arrows indicate the FSI, resp. QFS condition.

Few-Body Systems, Suppl. 7, 164—176 (1994)

# STRUCTURE AND FRAGMENTATION OF FEW-BODY COULOMB SYSTEMS

J.S.Briggs

Fakultät für Physik, Hermann-Herder-Str. 3, 79104 Freiburg i. Br.

## ABSTRACT

The problem of the structure and the fragmentation of few-body Coulomb systems remains a difficult outstanding problem of atomic and molecular physics. Recent progress in the resolution of this problem, both theoretical and experimental, is described. Particular attention is given to the fundamental three-body Coulomb problems of the photo-ionisation of helium and the electron impact ionisation of atomic hydrogen.

## 1 INTRODUCTION

In recent years, significant progress has been made in our understanding of the structure of multiply-differential cross-sections (MDCS) for the fragmentation of few-body Coulomb systems. As is usual in a rapidly developing subject, experiment and theory have advanced 'hand-in-hand', the one providing the impetus for the other. Here attention will be focussed on particular atomic fragmentation processes leading to two free electrons and a doubly or singly charged residual ion in the final state. The processes of main interest are a) the double photoionisation of a neutral atom and b) the electron-impact single ionisation of a neutral atom. These two processes have been much studied experimentally and recently even detailed triply-differential cross-sections (TDCS) have been measured in some cases. The main aim of this review is to describe the theoretical advances made in the description of these processes. The pervading theme will be that correlation between all three charged particles is always significant in deciding the final relative momentum distribution of the three-body continuum. Whilst one might expect this feature energetically close to the three-body break-up threshold it will be shown that surprisingly large electron-electron correlation effects persist even in high-energy (with respect to the ionisation energy) collisions.

Essentially the problem separates into two parts. The first is the preparation of the collision complex leading to two electrons emanating from the positive ion core. For double photoionisation this is relatively simple; the initial state is multiplied by the sum of the electron-dipole operators corresponding to the sudden (first-order perturbation theory) absorption of a photon. For electron-impact ionisation it is more complicated since the incident electron interacts both with the target electron and the positive ion core and the preparation of the collision complex cannot be viewed as a perturbative process. The second part of the problem is the description of the correlated three-body motion describing two electrons breaking off from the collision complex. It is here that the long-range correlation due to Coulomb forces come into play.

The problem of two-electron ionisation at threshold has mostly been treated separately from the higher-energy region, although quite what constitutes 'the threshold region' has never been defined clearly. Such threshold treatments go under the general heading of 'Wannier theory', following Wannier's 1953 analysis[1] of the problem, using classical mechanics. Subsequent semi-classical treatments of the problem[2,3,4] provide the energy dependence of the threshold cross-section and make predictions as to the form of the angular distribution[5,6]. Such treatments are generally quite involved and require several approximations whose validity is difficult to assess. Here it will be shown that a much simpler approach to the electron correlation problem provides a good description of most features of the relative momentum distribution of the two continuum electrons, even down to total energies less than 1 eV above threshold.

In section 2 the general parametrisation of the TDCS is discussed and the importance of bipolar harmonics in the expansion of the continuum 3-body wavefunction is emphasised. In section 3 some previous analyses of the 3-body wavefunction in the Wannier configuration are reviewed. The process of double photoionisation is discussed in detail in section 4. The more difficult e-2e ionisation process is the subject of section 5 and section 6 contains some remarks on the relationship of the three-body continuum states to the doubly-excited resonant states existing just below the three-body break-up threshold. Atomic units will be used throughout.

## 2 THE MULTIPLY-DIFFERENTIAL CROSS-SECTION

Klar and Fehr[7] have given a general parametrisation of MDCS for both double photoionisation and electron impact ionisation. The simpler case of photoionisation is considered first. The idea is that if the initial state of the neutral atom is randomly orientated and if the orientation of the residual atom is not detected, then the MDCS is a scalar quantity. Then it depends only on certain scalar invariants formed from the unit vectors $\hat{k}_a$, $\hat{k}_b$ (defining the emission directions of electrons 'a' and 'b') and the unit polarisation vector $\hat{\varepsilon}$ of the photon. In terms of Racah tensors $\underline{C}_l$, the invariants are

$$I_{l_a l_b L} = [\underline{C}_{l_a}(\hat{k}_a) \wedge \underline{C}_{l_b}(\hat{k}_b)]_L \cdot \underline{C}_L(\hat{\varepsilon}) \tag{1}$$

$$= \sum_M \mathcal{Y}^{l_a l_b}_{L,M}(\hat{k}_a, \hat{k}_b) C_{LM}(\hat{\varepsilon}) \tag{2}$$

where $\mathcal{Y}^{l_a l_b}_{LM}$ is a bipolar spherical harmonic

$$\mathcal{Y}^{l_a l_b}_{L,M}(\hat{k}_a, \hat{k}_b) = \sum_{m_a m_b} < l_a m_a l_b m_b \mid LM > C_{l_a m_a}(\hat{k}_a) C_{l_b m_b}(\hat{k}_b) \tag{3}$$

and the $C_{lm}$ are proportional to spherical harmonics. The double photoionisation cross-section for fixed energies of the two electrons is the TDCS

$$\frac{d\sigma}{d\hat{\mathbf{k}}_a d\hat{\mathbf{k}}_b dE_a} = 4\pi^2 \alpha \frac{k_a k_b}{\omega} \sum_{M_f} \frac{1}{2J_i + 1} \sum_{M_i} |< \psi_f^-(\hat{\mathbf{k}}_a, \hat{\mathbf{k}}_b) \,|\, \hat{\boldsymbol{\varepsilon}} \cdot \sum_i \nabla_i \,|\, \phi_i >|^2 \qquad (4)$$

where $\alpha$ is the fine-structure constant, $\omega$ the photon frequency, $J_i$ the angular momentum of the initial state and $\psi_f^-(\hat{\mathbf{k}}_a, \hat{\mathbf{k}}_b)$ the exact scattering state of a pair of electrons moving in the doubly-charged ion core with quantum numbers $J_f$, $M_f$. In the particularly simple case of double ionisation of helium the TDCS reduces to

$$\frac{d\sigma}{d\hat{\mathbf{k}}_a d\hat{\mathbf{k}}_b dE_a} = 4\pi^2 \alpha \frac{k_a k_b}{\omega} |< \psi_f^-(\hat{\mathbf{k}}_a, \hat{\mathbf{k}}_b) \,|\, \hat{\boldsymbol{\varepsilon}} \cdot (\nabla_a + \nabla_b) \,|\, \phi_i >|^2 \qquad (5)$$

where $\psi_f^-$ is an exact scattering state describing the motion of two free electrons in the field of the bare nucleus of charge two. Since the cross-section (4) is bi-linear in $\hat{\boldsymbol{\varepsilon}}$, only the L values 0 and 2 are allowed, so that in terms of the scalar invariants (2), with $L = 0$ or 2, the TDCS may be written

$$\frac{d\sigma}{d\hat{\mathbf{k}}_a d\hat{\mathbf{k}}_b dE_a} = \sum_l A_l P_l(\hat{\mathbf{k}}_a \cdot \hat{\mathbf{k}}_b) + \sum_{l_a l_b} B_{l_a l_b} \sum_M \mathcal{Y}_{2M}^{l_a l_b}(\hat{\mathbf{k}}_a, \hat{\mathbf{k}}_b) C_{2M}(\hat{\boldsymbol{\varepsilon}}) \qquad (6)$$

If one restricts to linear polarisation and chooses the z-axis along $\hat{\boldsymbol{\varepsilon}}$ this parametrisation assumes the simpler form

$$\frac{d\sigma}{d\hat{\mathbf{k}}_a d\hat{\mathbf{k}}_b dE_a} = \sum_l A_l P_l(\hat{\mathbf{k}}_a \cdot \hat{\mathbf{k}}_b) + \sum_{l_a l_b} B_{l_a l_b} \mathcal{Y}_{20}^{l_a l_b}(\hat{\mathbf{k}}_a, \hat{\mathbf{k}}_b) \qquad (7)$$

where the coefficients $A_l$ and $B_{l_a l_b}$ are functions of the energies $E_a$ and $E_b$. In ref. 7 a similar parametrisation is given in the case of circularly-polarised light. It is also shown how, by integrating over the angles $\hat{\mathbf{k}}_b$, the doubly-differential cross-section DDCS may be written

$$\frac{d\sigma}{d\hat{\mathbf{k}}_a dE_a} = \frac{1}{4\pi} \frac{d\sigma}{dE_a} [1 + \beta_{20} P_2(\hat{\mathbf{k}}_a \cdot \hat{\boldsymbol{\varepsilon}})] \qquad (8)$$

where $\beta_{20} \equiv B_{20}/A_0$ and $d\sigma/dE_a$ is the singly-differential cross-section (SDCS), equal to $16\pi^2 A_0$. Finally, in the case of ionisation by a beam of electrons with initial momentum $\mathbf{k}_i$, the TDCS for the e-2e process can be parametrised as

$$\begin{aligned} \frac{d\sigma}{d\hat{\mathbf{k}}_a d\hat{\mathbf{k}}_b dE_a} &= \sum_{l_a l_b L} B_{l_a l_b L} I_{l_a l_b L} \\ &= \sum_{l_a l_b L M} B_{l_a l_b L} \mathcal{Y}_{LM}^{l_a l_b}(\hat{\mathbf{k}}_a, \hat{\mathbf{k}}_b) C_{LM}(\hat{\mathbf{k}}_i) \end{aligned} \qquad (9)$$

or, with $\hat{\mathbf{k}}_i$ as the z-direction, in the simpler form

$$\frac{d\sigma}{d\hat{\mathbf{k}}_a d\hat{\mathbf{k}}_b dE_a} = \sum_{l_a l_b L} B_{l_a l_b L} \mathcal{Y}_{L0}^{l_a l_b}(\hat{\mathbf{k}}_a, \hat{\mathbf{k}}_b) \tag{10}$$

and again the coefficients $B_{l_a l_b L}$ are functions of the energies $E_a$, $E_b$ and $E_i$.

For later use it is also useful to note that the final-state two-electron wavefunction itself can be expanded in bipolar harmonics

$$< \mathbf{r}_a, \mathbf{r}_b \mid \psi_f^-(\mathbf{k}_a, \mathbf{k}_b) >= \sum_{l_a, l_b, L, M} \mathcal{Y}_{L,M}^{l_a, l_b\,*}(\hat{\mathbf{k}}_a, \hat{\mathbf{k}}_b) \Big\{ \mathcal{Y}_{L,M}^{l_a, l_b}(\hat{\mathbf{r}}_a, \hat{\mathbf{r}}_b) R_{L,M}^{l_a, l_b, k_a, k_b}(r_a, r_b) \tag{11}$$

$$\pm \mathcal{Y}_{L,M}^{l_a, l_b}(\hat{\mathbf{r}}_b, \hat{\mathbf{r}}_a) R_{L,M}^{l_a, l_b, k_a, k_b}(r_b, r_a) \Big\}$$

where the $+$ sign denotes the singlet and the $-$ sign the triplet state. Such an expansion in terms of states of well-defined two-electron angular momentum L is the generalisation of the usual expansion of a one-electron wavefunction of given momentum $\mathbf{k}$ in terms of partial waves.

## 3 ANALYSIS OF SYMMETRIES IN THE WANNIER CONFIGURATION

Several authors[8,9,10,11,12] have examined the properties of the two-electron wavefunction, as functions of $r_a, r_b$ in the configuration $\mathbf{r}_a \approx -\mathbf{r}_b$ considered by Wannier to be important in the region of the double-ionisation threshold. Note that strictly speaking the spatial wavefunction itself is not the object of interest (since its coordinates are integrated over in the transition matrix element). Rather it is the symmetries of particular $\mathcal{Y}_{LM}^{l_a l_b}(\hat{\mathbf{k}}_a, \hat{\mathbf{k}}_b)$ which decide the detected angular distribution, However from (11) one sees that the spatial and momentum bipolar harmonics appear symmetrically so that the properties in the region $\hat{\mathbf{k}}_a \approx -\hat{\mathbf{k}}_b$ are the same as those in the classical Wannier configuration $\hat{\mathbf{r}}_a \approx -\hat{\mathbf{r}}_b$. Alternatively one assumes that the directions $\hat{\mathbf{k}}_a, \hat{\mathbf{k}}_b$ are identical with the directions $\hat{\mathbf{r}}_a, \hat{\mathbf{r}}_b$ *asymptotically* and this is the region deciding the form of the cross-section.

The Wannier configuration wavefunction has been examined in a variety of coordinate systems. Here a form will be used that, at least in my opinion, most directly brings out the essential symmetries and has been used extensively to describe resonant states[13]. The key is to recognise that $\mathbf{r}_a = -\mathbf{r}_b$ implies

$$\mathbf{r} = \frac{1}{2}(\mathbf{r}_a + \mathbf{r}_b) = 0 \tag{12}$$

i. e. the centre-of-mass (and charge) of the two electrons is at the origin. The second orthogonal Jacobi coordinate is then naturally

$$\mathbf{R} = \mathbf{r}_a - \mathbf{r}_b \tag{13}$$

the interelectronic distance. These Jacobi coordinates are then expressed in terms of body-fixed coordinates $R$, $\eta = (r_a + r_b)/R$, $\gamma = (r_a - r_b)/R$ and Euler angles $\psi, \theta, \phi$, the latter angle $\phi$ corresponding to the projection m of the total angular momentum L on the body-fixed axis $\hat{\mathbf{R}}$. The projection on a space-fixed axis is M. Expressing a wavefunction of given L, S and total parity $\pi$ in these coordinates one can show[12] (see also refs. 8–11)

a) Only states with $(-1)^l = \pi(-1)^S = 1$ can be finite at the Wannier saddle $\hat{\mathbf{r}}_a = -\hat{\mathbf{r}}_b$.

b) Additionally only states with *body-fixed* $m = 0$ are finite at $\hat{\mathbf{r}}_a = -\hat{\mathbf{r}}_b$. All others have a zero at this point. Note that for this rule to hold the space and body-fixed frames must coincide. As a corollary of a) and b) one can show that only states with $\pi(-1)^L = 1$ can be finite at $\hat{\mathbf{r}}_a = -\hat{\mathbf{r}}_b$. One also has the result

c) Only $^3S^e$ and $^1P^e$ states have a node at $r_a = r_b$.

As an example of the application of these rules one can consider the pure $^1P^o$ wavefunction arising from double photoionisation of the $^1S^e$ ground state of helium. From a) this state is zero along $\hat{\mathbf{r}}_a = -\hat{\mathbf{r}}_b$, or relative angle $\theta_{ab} = 180°$. The $^1P^o$ states have $| M | = 0$ or 1 and therefore $| m | = 0$ or 1 also. As the double-ionisation threshold is approached (infinitely slow electrons) one can expect the space-fixed and body-fixed frames to coincide (no rotation of the interelectronic axis) so that according to rule b) only $^1P^o$ states with $| M | = | m | = 1$ will be populated. This leads immediately to the result that if only one electron is detected its angular distribution will be of $| m | = 1$ i. e. $\sin^2 \theta$ form. Hence in the DDCS of equation (8) one has $\beta_{20} = -1$. This result was arrived at by Greene[11] from a rather involved argument based on Herrick's[15] $SO_4$ classification of doubly-excited states. It has been shown[13] that Herrick's T quantum number is identical with the body-fixed m quantum number.

## 4 DOUBLE PHOTOIONISATION

Attention will be focussed here on the 'pure' three-body Coulomb problem represented by the double photoionisation of helium ($H^-$ would also satisfy this condition). Measurements exist of the DDCS (8)[16,17] and very recently of the TDCS (5)[18]. The measurements of the DDCS near to threshold will be described first.

The main parameter of interest in the DDCS is the asymmetry parameter $\beta_{20}$ of equation (8). In this section the notation will be simplified and this parameter called $\beta$. Greene[11] was the first to suggest that at threshold this parameter should take the value $-1$, as explained in section 3. Experiment[16] appeared not to agree with this prediction, rather giving values $\beta \approx -0.5$. A possible explanation was provided on the basis of calculations[19] of the photoionisation cross-section using an explicit 3-body correlated (3C) wavefunction for the final state. Since the use of this wavefunction will figure prominently in the results to be presented on angular distributions, both in photoionisation and in impact ionisation, its form will be given in detail.

The correlated 3-body wavefunction has the symmetric form of a product of three two-body on-shell Coulomb wavefunctions, one for each pair of interacting particles i. e.

$$
\begin{aligned}
\Psi_{3C} &= (2\pi)^{-\frac{3}{2}} \exp(-\pi\alpha_a/2)\Gamma(1 - i\alpha_a)e^{i\mathbf{k}_a\cdot\mathbf{r}_a}{}_1F_1(i\alpha_a; 1; -i[k_a r_a + \mathbf{k}_a \cdot \mathbf{r}_a]) \\
&\times (2\pi)^{-\frac{3}{2}} \exp(-\pi\alpha_b/2)\Gamma(1 - i\alpha_b)e^{i\mathbf{k}_b\cdot\mathbf{r}_b}{}_1F_1(i\alpha_b; 1; -i[k_b r_b + \mathbf{k}_b \cdot \mathbf{r}_b]) \\
&\times \exp(-\pi\alpha_{ba}/2)\Gamma(1 - i\alpha_{ba}){}_1F_1(i\alpha_{ba}; 1; -i[k_{ba}r_{ba} + \mathbf{k}_{ba} \cdot \mathbf{r}_{ba}])
\end{aligned} \tag{14}
$$

with

$$
\alpha_a = \frac{-Z}{k_a} \qquad \alpha_b = \frac{-Z}{k_b} \qquad \alpha_{ba} = \frac{1}{2k_{ba}} \quad \text{and} \quad k_{ba} = \frac{|\mathbf{k}_b - \mathbf{k}_a|}{2} \tag{15}
$$

This wavefunction has been used in expression (5) for the TDCS in helium and, after integration over the emission angles of electron 'b', in (8) to calculate the energy dependence of the $\beta$ parameter. The initial-state wavefunction was a relatively simple Hylleraas $^1S^e$ correlated wavefuntion. The results for the TDCS

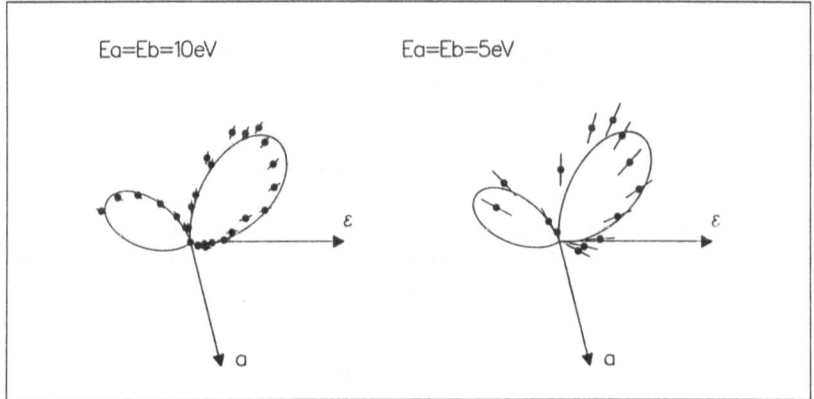

**Fig. 1:** The angular distribution of electron 'b' following double photoionisation of helium. Electron 'a' is emitted with equal energy in a direction fixed with respect to the polarisation vector $\hat{\mathbf{e}}$. Data from ref. 18. The continuous curve is theory from ref. 20.

for equal final energies are shown in figure 1 where excellent agreement in both length and velocity forms is obtained with the experimental results[18]. In fact, it has been shown[20] that even though the total energy above threshold is as low as 10 eV, the angular distribution is essentially explained by a form

$$
\frac{d\sigma}{d\hat{\mathbf{k}}_a d\hat{\mathbf{k}}_b dE_a} \sim (\cos\theta_a + \cos\theta_b)^2 C(\theta_{ab}) \tag{16}
$$

where $C(\theta_{ab}) = |N(k_{ab})|^2$ is the Coulomb density of states (CDS) factor for the repulsive electron-electron interaction. It is this fundamental form that explains the oriented double-lobe structure of the TDCS shown in figure 1. The $\beta$ values of the DDCS calculated with wavefunction (14) do indeed approach the limit $-1$ at threshold[19], but only if threshold is approached in such a way that the ratio $R = E_a/E_b$ is fixed. In addition the approach to $\beta = -1$ is very slow, being most quickly approached in the $R \approx 1$ equal energy case. Calculated $\beta$ values agree with experiment over a range of total energies above threshold from less than 0.1 eV to of the order of 80 eV as shown in figure 2. The major features of the shape of $\beta$ in this region arise from the effects of the electron-electron interaction in the final state. Proulx and Shakeshaft[22] have used a product of only two electron-nucleus wavefunctions, but with momentum-dependent effective charges designed to mimic this repulsion effect, and obtained almost the same quality of agreement with experiment.

In the case of absorption of one photon from a $^1S^e$ state, the TDCS can be written in the form[22]

$$\frac{d\sigma}{d\hat{\mathbf{k}}_a d\hat{\mathbf{k}}_b dE_a} \sim \mid g(k_a, k_b, \cos\theta_{ab})\cos\theta_a + g(k_b, k_a, \cos\theta_{ab})\cos\theta_b \mid^2 \qquad (17)$$

which clearly reduces to (16) for $k_a = k_b$. On the basis of Wannier theory, Selles et al.[10] have argued that g is a symmetric function of $k_a$ and $k_b$ for all energy-sharing ratios in the threshold region. On this basis they derive a cross-section form identical to (16) but with the factor $C(\theta_{ab})$ of Gaussian form. Proposed forms of this Gaussian[5,6] do not give good agreement with experiment however[18].

The particular case of double photoionisation with left or right circularly polarised light has been examined in detail by Berakdar et al.[23]. They point out the interesting result that the TDCS exhibits circular dichroism in that the momentum distribution is sensitive to the circular polarisation.


## 5 e-2e CROSS-SECTIONS

The major features of the TDCS for electron-impact ionisation at higher energies and for very asymmetric energy sharing have been understood for some time. For fixed scattering angle of the projectile electron, a slow ionised electron angular distribution shows a double-peak structure; the binary peak at forward angles and the recoil peak at backward angles. It is also well-appreciated that the first Born cross-section fails to reproduce the fine details of this distribution, even at impact energies $\approx 500$ eV on hydrogen. Higher-order processes must be taken into account in order to predict accurately the shape and position of the binary and recoil peaks.

In addition to the use of the 3C correlated wavefunction (14) for the final state, various types of distorted-wave theory have been developed which more- or less-successfully describe the high-energy TDCS[24,25].

Here only two new features of higher-energy e-2e processes that have emerged recently will be discussed. The first is the recognition of the importance of explicit

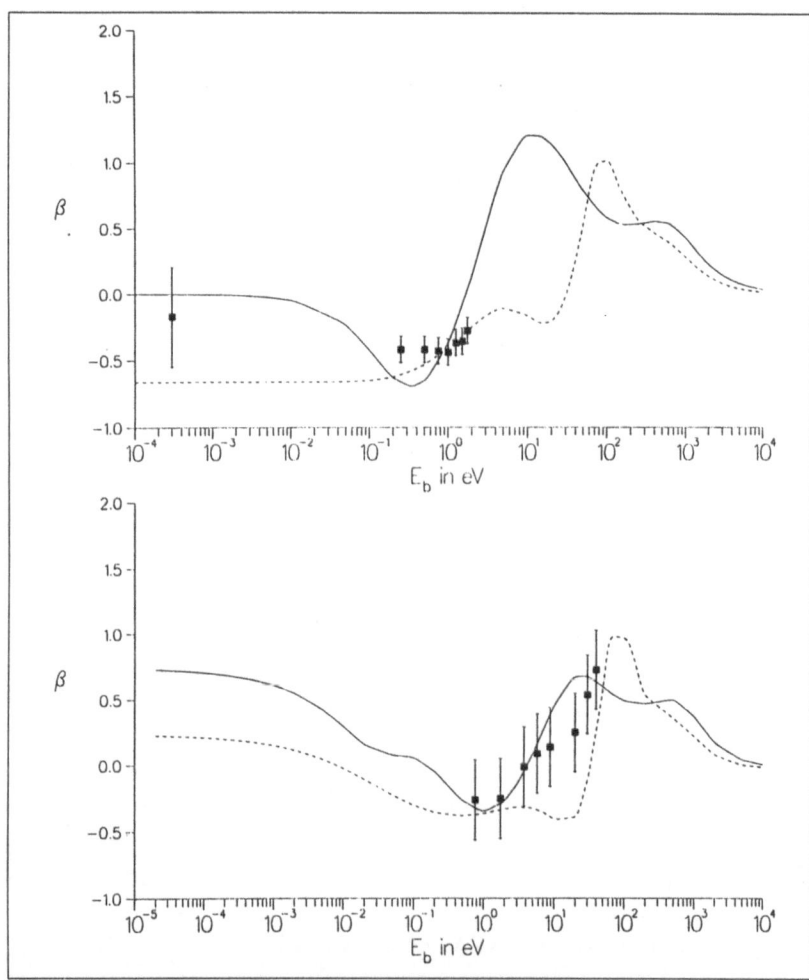

**Fig. 2:** The asymmetry parameter of equation 8. Upper curve: data from ref. 16 with fixed $E_a = 0.25$ eV. Lower curve: data from ref. 17 with fixed $E_a = 1.24$ eV. Theory is from ref. 21; continuous curve, velocity form; dashed curve, orthogonalised form.

double-binary[26] collisions giving rise to peaks in TDCS. A secondary peak, arising only in second Born calculations, was identified ten years ago in large-angle co-planar symmetric (equal energy, emission angles equal and opposite in sign) e-2e processes[27]. Later the appearance of this peak close to an angle of 135° was

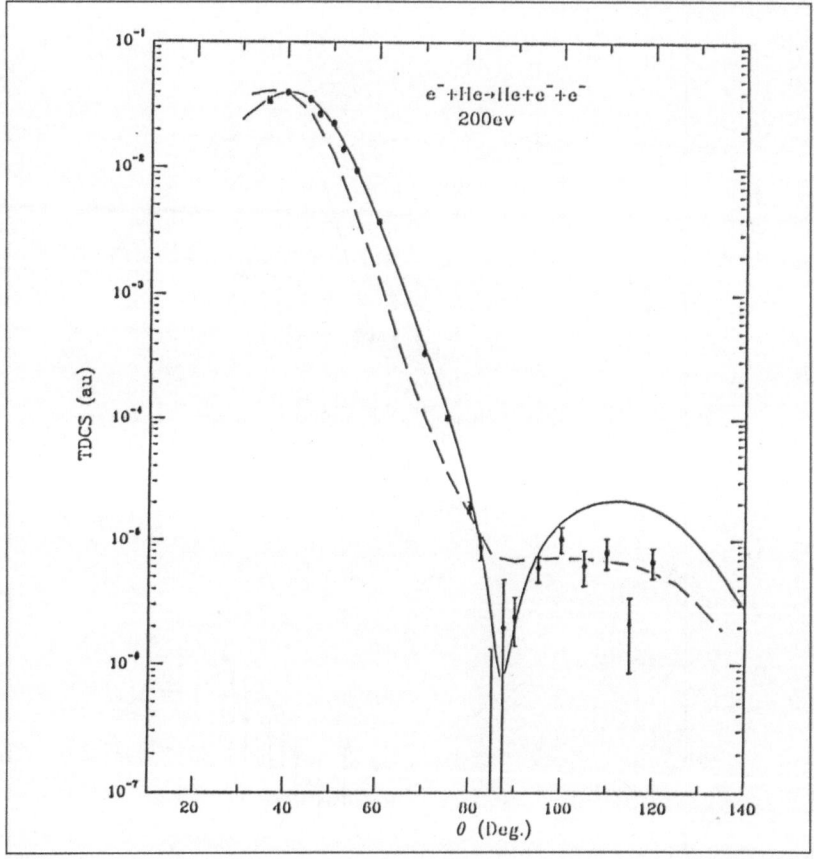

**Fig. 3:**   TDCS for the electron impact ionisation of helium in coplanar symmetric geometry at incident energy 200 eV (from ref. 32). Theory:
continuous curve, DWBA from ref. 32; dashed curve, second Born theory from ref. 31.

explained[26] as arising from a double-binary collision. The fast incident electron
scatters first elastically through 180° off the nucleus. As it emerges, it makes a
second binary collision with the target electron so that the electrons with equal
energy appear at 90° to each other and therefore at 135° to the beam direction.
The symmetric co-planar e-2e measurement is shown in figure 3.

The major peak is around 45° and is due essentially to a single electron-
electron binary collision. The single and double binary processes interfere and
produce a remarkably sharp dip. Again distorted-wave theories are successful in

explaining the shape[28,32].

The double-collision process has been shown to be important in two other situations. The first is in asymmetric geometry but where the angular distribution of the fast projectile (exchange can be neglected) electron is measured for fixed momentum of the slow electron. Then it has been shown[29] that peaks due to double-binary collisions should be visible in this angular distribution. The second situation is where the initial beam direction and the final electron momenta are not co-planar. For example, in 'equatorial' geometry the two electrons are detected at 90° to the beam direction. Clearly in this case a close collision with the nucleus is necessary to steer the fast incident electron through a large angle. A subsequent collision with the target electron leads to ionisation. Peaks due to such double-binary events have been observed[30] and calculated[31,32].

One feature is not well-explained by distorted-wave or second Born theories[32]. This is the sharp fall in the cross-section for small emission angles i. e. when two electrons of the same energy are emitted close to the forward direction. Then it is clear that it is the electron-electron repulsion that is causing the dip, the analogue of the well-known cusp electrons seen in impact ionisation by positively-charged particles. This effect has been studied in detail by Pan *et al.*[33] where it was shown that the 3C wavefunction, since it includes electron correlation explicitly, describes the dip in the cross-section well.

More challenging for the theorist has been the explanation of new data[34-36] in the intermediate to low-energy regime. It is fair to say that the intermediate energy data (impact energy $\approx$ 50 eV) for asymmetric energy sharing is not well explained by theory. Surprisingly perhaps, in certain geometries near threshold (a few eV above) TDCS are quite well-described by both the 3C and distorted-wave theories[34,28,37]. This is important since the geometries are those in which the two electrons have equal energies and emerge at fixed angle to each other. For relative angles around 180° this is precisely the configuration that should be important near threshold according to Wannier's theory.

The TDCS for a hydrogen target with electrons emerging with 1 eV each at fixed relative angle of 180° and variable angle with respect to the incident beam direction, is shown in figure 4 in comparison with the calculated TDCS from the 3C wavefunction. The cross-section is smooth, with a minimum at 90° and maxima in the forward and backward directions. This tendency of the *relative* momentum to stay aligned along the beam direction can be explained simply by a plane-wave as final state[34]. The same tendency is also seen when the relative angle is fixed at 150°. The TDCS exhibits a maximum when the relative momentum is along the beam direction and a minimum when at 90°.

The shape of the TDCS for fixed angle of 180° in hydrogen indicates predominantly P-wave in the final state. The TDCS for a helium target in the same geometry shows the same overall tendency of maximum cross-section in the forward and backward directions. However, precisely at 90° there is a subsidiary maximum indicating an admixture of even-L waves. Distorted-wave calculations [28,37] allow-

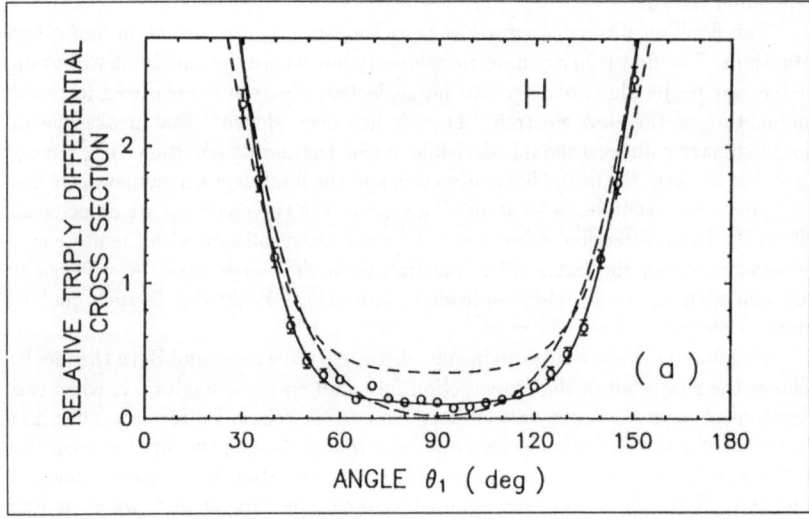

**Fig. 4:** Relative TDCS for $E_a = E_b = 1$ eV and relative angle fixed at 180°. The angle $\theta_1$ is that of the interelectronic axis relative to the beam direction (from ref. 28). Data and theory (long dashed curve) from ref. 34. Theory: continuous curve, ref. 28; dashed curve ref. 37.

ing for scattering of the two electrons in the $He^+$ core reproduce this maximum.

## 6 THE THRESHOLD REGION

The traditional approach to the threshold region is to employ the semi-classical version of Wannier's classical theory[2,4]. Here it is assumed that the dominant configuration at threshold is one in which the electrons appear predominantly at 180° to each other. Furthermore, for a given total energy, all combinations of individual electron energies are assumed to be equally probable. The Wannier theory provides threshold laws[38], angular distributions[3,4,6] and has also been used to calculate TDCS[39]. The theory is quite different from the approaches described in section 5, which however give accurate results for the shape of the TDCS near threshold. There appears to be an urgent need for an analysis indicating precisely the conditions upon which Wannier theory is valid and whether these conditions are experimentally realisable. Friedman *et al.*[40] have thrown doubt upon experimental results which appear to establish the Wannier threshold law. Hence the situation close to threshold is still unclear. Although the theory using the 3C wavefunction predicts angular distributions very well, it appears not to give absolute cross-section values accurately, or to describe the relative probability of energy sharing correctly.

One aspect of the threshold region has been clarified recently however and that is the connection of the energy region just above threshold with the infinity of doubly-excited resonant states residing just below threshold. In the two-body Coulomb problem the bound state Rydberg spectrum extrapolates smoothly through the continuum and quantum defect theory makes the formal smooth connection between states just above and just below the two-particle break-up threshold. Although the doubly-excited states in helium or $H^-$ are unbound they can have very long lifetimes and it appears to have been assumed generally that the character of these states also connects smoothly with the Wannier mode $r_a \approx -r_b$, assumed to be the dominant mode leading to three-body breakup. Recently, very accurate calculations have become available for rather high-lying states of helium[41]. An analysis of the structure of these states, together with a knowledge of the nature of the underlying classical mechanics has shown, however, that the symmetric doubly-excited states are essentially *orthogonal* to the Wannier mode. Whilst it is true that there is a large probability for the electrons to be located near the Wannier ridge i. e. $r = -r_b$, the predominant character of the motion is not a symmetric stretch as in the Wannier mode, but an asymmetric stretch[42,43]. The consequence that in this mode the electrons avoid being simultaneously in the presence of the nucleus (a highly unstable configuration) explains the relative stability of the doubly-excited states.

## REFERENCES

1. G.H.Wannier, Phys.Rev. 90, 817, (1953)
2. R.Peterkop, J.Phys.B 4, 513 (1971)
3. A.R.P.Rau, Phys.Rev.A 4, 207 (1971)
4. J.M.Feagin, J.Phys.B 17, 2433 (1984)
5. A.R.P.Rau, J.Phys.B 9, L283 (1976)
6. A.Huetz, P.Selles, D.Waymel and J.Mazeau, J.Phys.B 24, 1917 (1991)
7. H.Klar and M.Fehr, Z.Phys.D 23, 295 (1992)
8. C.H.Greene and A.R.P.Rau, Phys.Rev.Letts. 48, 533 (1982)
9. A.D.Stauffer, Physics Letters 91A, 114 (1982)
10. P.Selles, J.Mazeau and A.Huetz, J.Phys.B 20, 5183 (1987)
11. C.H.Greene, J.Phys.B 20, L357 (1987)
12. J.M.Rost and J.S.Briggs, J.Phys.B 24, L393 (1991)
13. J.M.Feagin and J.S.Briggs, Phys.Rev.A 37, 4599 (1988)
14. J.M.Rost and J.S.Briggs, J.Phys.B 24, 4293 (1991)
15. D.R.Herrick, Adv.Chem.Phys. 52, 1 (1983)
16. R.I.Hall, L.Avaldi, G.Dawber, M.Zubek, K.Ellis and G.C.King, J.Phys.B 24, 115 (1991)
17. R.Wehlitz, F.Heiser, O.Hemmers, B.Langer, A.Menzel and U.Becker, Phys.Rev.Letts. 67, 3764 (1991)
18. O.Schwarzkopf, B.Krässig, J.Elmiger and V.Schmidt,

Phys.Rev.Letters 70, 3008 (1993)

19. F.Maulbetsch and J.S.Briggs, Phys.Rev.Letts. 68, 2004 (1992)
20. F.Maulbetsch and J.S.Briggs, submitted to J.Phys.B
21. F.Maulbetsch and J.S.Briggs, J.Phys.B 26, 1679 (1993)
22. D.Proulx and R.Shakeshaft, Phys.Rev.A, to be published
23. J.Berakdar, H.Klar,A.Huetz and P.Selles, J.Phys.B 26, 1463 (1993)
24. I.E.McCarthy Zeit.für Phys. 23, 287 (1992)
25. X.Zhang, C.T.Whelan and H.R.J.Walters, Zeit. für Phys. 23, 301 (1992)
26. J.S.Briggs, J.Phys.B 19, 703 (1986)
27. Byron et al. , J.Phys.B 16, L769 (1983)
28. C.Pan and A.F.Starace, Phys.Rev.Letts. 67, 185 (1992)
    —, Phys.Rev.A 45, 4588 (1992)
29. M.Brauner and J.S.Briggs, J.Phys.B, to be published
30. A.J.Murray and F.H.Read, Phys.Rev.Letts. 68, 2912 (1992)
    —, J.Phys.B 25, 3021 (1992)
31. F.Mota Furtado and P.F.O'Mahony J.PhysB 22, 3925 (1989)
32. X.Zhang, C.T.Whelan and H.R.J.Walters, J.Phys.B 23, L509 (1990)
33. G.Pan, P.Hvelplund, H.Knudsen,Y.Yamazaki,M.Brauner and J.S.Briggs,
    Phys.Rev.A 47, 1531 (1993)
34. M.Brauner, J.S.Briggs, H.Klar, J.T.Broad, T.Rösel, K.Jung and H.Ehrhardt,
    J.Phys.B 24, 657 (1991)
35. M.Cherid, F.Gelebart, A.Pochat, R.J.Tweed, X.Zhang, C.T.Whelan
    and H.R.J.Walters, Zeit.für Phys.D 23, 347 (1992)
36. T.Rösel, P.Schlemmer, J.Röder, L.Frost, K.Jung and H.Ehrhardt,
    Zeit.für Phys.D 23, 359 (1992)
37. S.Jones, D.H.Madison and M.K.Srivastava, J.Phys.B 25, 1899 (1992)
    T.Rösel, J.Röder, L.Frost, K.Jung, H.Ehrhardt, S.Jones and D.H.Madison,
    Phys.Rev.A 46, 2539 (1992)
38. H.Klar, Zeit.für Phys.A 307, 75 (1982)
39. D.R.J.Carruthers and D.S.F.Crothers, Zeit.für Phys.D 23, 365 (1992)
40. J.R.Friedman, X.Q.Guo, M.S.Lubell and M.R.Frankel, Phys.Rev.Letts. (1991)
41. J.M.Rost, R.Gersbacher, K.Richter, J.S.Briggs and D.Wintgen,
    J.Phys.B 24, 2455 (1991)
42. G.S.Ezra, K.Richter, G.Tanner and D.Wintgen, J.Phys.B 24, L431 (1991)
43. D.Wintgen, K.Richter and G.Tanner, CHAOS 2, 19 (1992)

Few-Body Systems, Suppl. 7, 177—192 (1994)

© Springer-Verlag 1994
Printed in Austria

# RECENT PROGRESS ON THE FOUR-NUCLEON FRONTIER

*A.C. Fonseca*

Centro de Física Nuclear da Universidade de Lisboa

Av. Prof. Gama Pinto 2, 1699 Lisboa Codex, PORTUGAL

## ABSTRACT

The most recent developments involving numerically "exact" solutions of the Yakubovsky equations in coordinate and momentum space are reviewed and compared with alternative methods both exact and approximate. Future perspectives are discussed together with a summary of the physics we have learned.

## INTRODUCTION

Prior to 1989 one finds excellent reviews on the four-body problem involving both theory and applications. This started at the Karlsruhe conference[1] with a report on the achievements of different few-body methods in calculating the $^4$He binding energy and corresponding wave functions, and was followed in 1985/86 by three very complete reviews on the integral equation approach to the calculation of four-nucleon bound states and scattering observables[2-4].

More recently one finds reviews on the ATMS approach[5], Green's Function Monte Carlo (GFMC) and Variational Monte Carlo (VMC) calculations[6,7], Coupled Reaction Channel (CRC) calculations[8], and Integrodifferential Equation approach[9] (IDEA) as well as text books[10,11] on the theory of few-body equations and their application to the solution of a number of physics problems.

As pointed out by Friar[12] "the field of few-nucleon physics has been driven by the desire to obtain accurate solutions of the Schrödinger equation for physically interesting problems". To some this desire is viewed as a form of intelectual "weight lifting", while to others is an essential step in our understanding of the physics that binds nucleons together, or leads to the emission of a pion as few nucleons scatter each other.

Until J. Carlson[13] showed that GFMC can be safely and accurately used for a system of fermions, variational methods based on a sophisticated ansatz or on the expansion of the wave function in a complete set of functions that spans the space, where the sole hope one had to calculate an upper-bound for the binding energy of $^4$He with realistic interactions between nucleons. Although GFMC still imposes some limitations on the nature of the potential one may use, we know that it gives 0.5 to 1 MeV more binding than the corresponding variational calculations. On the other hand work based on the expansion of the wave function in a complete set of basis functions has always been plagued with convergence related uncertainties to the exception of CRC[8]which, for the three-nucleon system, has been shown to be as accurate as GFMC or Faddeev calculations. Therefore, until very recently, GFMC seems to be the method of choice for the alpha particle, since numerically converged solutions of the Yakubovsky equations only existed for one term separable two-nucleon interactions acting in channels $^1S_0$ and $^3S_1$ - $^3D_1$[14] or for Mafliet-Tjon I/III[15]. A gigantic step has been recently undertaken by Kamada and Glöckle[16-18] following an equally important achievement by Schellingerhout, Schut and Kok[19] who pioneered precise solutions of four-body equations in three dimensions. In this review talk we will cover most of their work and stress the major highlights. For comparison we also quote some of the best VMC and GFMC results including a very recent relativistic calculation. Finally we make a very short outline of the status of low energy scattering calculations involving four nucleons.

## CONFIGURATION-SPACE YAKUBOVSKY EQUATIONS

The configuration-space Yakubovsky equations were first implemented by Merkuriev, Yakovlev and Gignoux[20,21] who also studied the corresponding boundary conditions. Since most people are familiar with the momentum space version of these equations we will use the notation of ref.[4] to set them up; this same reference should be consulted for further details. Starting with the Schrödinger equation

$$(H_0 + \sum_i v_i)\Psi = E\Psi \quad , \tag{1}$$

one first writes

$$(E - H_0)\Psi = \sum_i v_i\Psi \quad , \tag{2}$$

or

$$\Psi = G_0 \sum_i v_i \Psi \quad , \tag{3}$$

where $G_0$ is the four free particle Green's function and $v_i$ is the potential between particles in pair i which runs from one to six. Equation (3) suggests the Faddeev decomposition

$$\Psi = \sum_i \Psi_i \quad , \tag{4}$$

where

$$\Psi_i = G_0 v_i \Psi \quad . \tag{5}$$

Substituting (4) in (5) one gets

$$(1 - G_0 v_i)\Psi_i = G_0 v_i \sum_{j \neq i} \Psi_j \quad , \tag{6}$$

which, once inverted, leads to

$$\Psi_i = G_0 t_i \sum_{j \neq i} \Psi_j \quad , \tag{7}$$

where $t_i$ is the two-body t-matrix for pair i. For three-particles Eq. (7) is equivalent to the Faddeev equations. For four-particles one needs to decompose the $\Psi_i$'s even further in order to bring about a connected four-particle Kernel. This was first achieved by Yakubovsky[22] who defined

$$\Psi_i^\rho = G_0 t_i \sum_{\substack{j \neq i \\ j \subset \rho}} \Psi_j \quad , \tag{8}$$

where $\rho$ is a two cluster partition that contains i. If for example i is pair (12), Eq. (7) becomes

$$\Psi_{12} = G_0 t_{12} (\Psi_{13} + \Psi_{23} + \Psi_{14} + \Psi_{24} + \Psi_{34}) , \tag{9}$$

which with the help of (8) , one may write (9) as

$$\Psi_{12} = \Psi_{12}^{(123)4} + \Psi_{12}^{(124)3} + \Psi_{12}^{(12)(34)} \quad , \tag{10}$$

where (123)4, (124)3 and (12)(34) are the two-cluster partitions that contain pair 12. Therefore one may write

$$\Psi_i = \sum_{\rho \supset i} \Psi_i^\rho \quad . \tag{11}$$

If we now substitute (11) in (8) and use the operator identity

$$G_0 t_i = G_i v_i , \tag{12}$$

we get

$$\Psi_i^\rho = G_i v_i \sum_{\substack{j \neq i \\ j \subset \rho}} \sum_{\sigma \supset j} \Psi_j^\sigma \quad . \tag{13}$$

The differential version of this equation emerges naturally by multiplying by $G_i^{-1}$ leading to

$$(H_0 + v_i - E)\ \Psi_i^\rho = -\ v_i \sum_{\substack{j \neq i \\ j \subset \rho}} \sum_{\sigma \supset j} \Psi_j^\sigma\ . \tag{14}$$

As known, this is a set of eighteen coupled equations that, for four identical particles, reduces down to two coupled equations in three vector variables. In Eq.(14) the indices $\rho$,i uniquely specify a set of three Jacobi variables[2] that are most suited to represent $\Psi_i^\rho$. Therefore the left side of the equation in best represented in the Jacobi variables specified by $\rho$,i while on the right side one has to take matrix elements of $v_i$ between sets of Jacobi variables specified by $\rho$,i and $\sigma$,j. This is best handled by using permutation operatores that transform one set of variables into another. Schellingerhout, Schut and Kok (SSK) proceed to solve these equations numerically using triple partial wave expansion (tripolar harmonics that involve the coupling of three spherical harmonics, one for each Jacobi coordinate), s-wave approximation for all underlying orbital angular momentum and the expansion of wave function components as a product of cubic Hermite splines in each of the three remaining continuous variables times an unknown coefficient that depends on all indices.

Although in ref.[19] they do not follow exactly the same procedure, we outline their steps in a simplified manner. First one writes.

$$< xyz\ |\Psi> = \sum_\alpha (xyz)^{-1}\ \phi_\alpha\ (xyz)\ \xi_\alpha (\hat{x}\hat{y}\hat{z})\ , \tag{15}$$

$$\xi_\alpha (\hat{x}\hat{y}\hat{z}) = \Big[[Y_{\ell_x}(\hat{x}) \otimes Y_{\ell_y}(\hat{y})] \otimes Y_{\ell_z}(\hat{z})\ \Big]_{LM}, \tag{16}$$

where $\ell_x$, $\ell_y$ and $\ell_z$ are the orbiral angular momentum associated with the Jacobi variables x, y and z respectively, which are set to zero in the s-wave approximation. Since centrifugal terms are zero, the Laplacian in $H_0$ becomes $\Delta = \partial^2/\partial x^2 + \partial^2/\partial y^2 + \partial^2/\partial z^2$. Finally

$$\phi_\alpha (xyz) = \sum_{ijk} S_i(x)\ S_j(y)\ S_k(z)\ a_{\alpha\ ijk}\ , \tag{17}$$

where the S's are Hermite splines. This final step reduces the differential equation to a generalized eigenvalue problem which is subsequently solved with the help of a "tensor trick"[23] made possible by the choice of spline representation, and the Lanczos algoritm. For further details on numerical methods one should follow the review at this conference by Schellingerhout himself.

The equations are solved for two versions of the Mafliet-Tjon (MT) V potential which is spin independent, MT-I/III and Afnan and Tang (S3) potentials which are spin dependent. In the frame work of the s-wave approximation they use, the N-N interaction is taken in channels $^1S_0$ and $^3S_1$ alone, and all remaining orbital angular momentum are set to be zero. This means that the $^4$He bound state wave function they calculate has only spatial symmetric and mixed symmetric components with total spin zero and total angular momentum zero. The results are shown in Table I for both the trinucleon and four-nucleon systems and are compared with other calculations such as: s-wave integrodifferential equations (SIDE); Faddeev equations in momentum space (FMS) or coordinate space (FCS); momentum space Yakubovsky equations (YMS) that also make use of the s-wave approximation.

At the three-nucleon level the agreement is excellent among all modern calculations that take advantage of new numerical techniques. Since SIDE is equivalent to the corresponding Faddeev equation where all orbital angular momentum have been set to zero, the results have to be the same. Nevertheless, at the four-nucleon level, only the two most recent solutions of the Yakubovsky equations[16, 19] coincide within about 20 KeV. Although the work in ref [16] is discussed below, we already quote part of their results here. As mentioned before all calculations involve the s-wave approximation, but SIDE only includes two-body correlations. Therefore SIDE leads to less binding than YMS or YCS.

**Table I** - Three and four-nucleons binding energies for different test potentials between nucleons

|  | Method | Reference | MT-V$^a$ | MT-V$^b$ | S3$^{av}$ | MT-I/III[.] |
|---|---|---|---|---|---|---|
| $^3$H | FCS | [19] | 7.540 | 8.043 | 6.409 | 8.536 |
|  | FCS | [24] | 7.540 |  |  |  |
|  | FMS | [16] |  |  | 6.41 | 8.54 |
|  | FMS | [25] |  |  |  | 8.56 |
|  | SIDE | [26] | 7.54 | 8.04 | 6.41 | 8.54 |
| $^4$He | YCS | [19] | 28.781 | 30.063 | 25.675 | 30.312 |
|  | YMS | [16] |  | 30.07 | 25.69 | 30.29 |
|  | YMS | [25] |  |  |  | 29.6 |
|  | YMS | [15] |  |  |  | 30.36 |
|  | YCS | [20] | 29.1 |  | 25.5 |  |
|  | SIDE | [26] | 28.47 | 29.74 | 25.38 | 29.74 |

## MOMENTUM-SPACE YAKUBOVSKY EQUATIONS

Although Yakubovsky was the first to propose the decomposition of the wave function in eighteen components as shown in Eq.(8), it was Alt, Grassberger and Sandhas[27] (AGS) who developed the very useful matrix notation in terms of operators that have the appropriate right-left off-shell properties that allow for a reduction in the dimensionality of the equations each time the underlying subamplitudes are expressed in a separable form. Nevertheless, if one aims at solving the equations in three continuous variables, the AGS formulation has no great advantage and one may use any other operator structure.

If one uses (12) to write (13) as

$$\Psi_i^\rho = G_0 \, t_i \sum_{\substack{j\neq i \\ j\subset\rho}} \sum_{\sigma\supset j} \Psi_j^\sigma \quad , \tag{18}$$

and proceed to write the momentum space Yakubovsky equations by bringing to the left side of (18) the terms where $\sigma = \rho$ and leaving on right side all other where $\sigma \neq \rho$ we get

$$\Psi_i^\rho - G_0\, t_i \sum_{\substack{j\neq i \\ j\subset\rho}} \Psi_j^\rho = G_0\, t_i \sum_{\substack{j\neq i \\ j\subset\rho}} \sum_{\sigma\supset j} \tilde{\delta}_{\sigma\rho}\, \Psi_j^\rho, \qquad (19)$$

which, in AGS three-body matrix notation, reads

$$[1 - G_0\, V^\rho]\, \Psi^\rho = G_0\, V^\rho\, \mathbb{R}, \qquad (20)$$

where

$$[V^\rho]_{ij} = t_i\, \tilde{\delta}_{ij}, \qquad (21)$$

for $i,j \subset \rho$ and zero otherwise. This matrix is therefore confined to all $i,j \subset \rho$.

Defining the operator $\mathbf{K}^\rho$ as

$$(1 + G_0\, \mathbf{K}^\rho)\,(1 - G_0\, \mathbf{V}^\rho) = 1, \qquad (22)$$

such that

$$K_{ij}^\rho = t_i\, \tilde{\delta}_{ij} + \sum_{k\neq j} K_{ik}^\rho\, G_0\, t_k, \qquad (23)$$

one may multiply (20) by $1 + G_0\, \mathbf{K}^\rho$ to obtain

$$\Psi_i^\rho = \sum_j G_0\, K_{ij}^\rho \sum_{\sigma\supset j} \tilde{\delta}_{\sigma\rho}\, \Psi_j^\sigma, \qquad (24)$$

which is the original Yakubovky equation expressed in terms of breakup operators for subsystem $\rho$. As shown in (23), $K_{ij}^\rho$ is the sum of all diagrams that end with an interaction in pair i and start with an interaction in any pair k other than j. As shown in ref.[4] the operator K relates to the AGS operator U

$$K_{ij}^\rho = t_i\, G_0\, U_{ij}^\rho \qquad (25)$$

where U satisfies the well known AGS equation

$$U_{ij}^\rho = G_0^{-1}\, \tilde{\delta}_{ij} + \sum_k \tilde{\delta}_{ik}\, t_k\, G_0\, U_{kj}^\rho, \qquad (26)$$

If $\rho$ is a 3+1 two cluster partition, the U's lead to all relevant three-body amplitudes for "particle + bound pair" going to "particle + bound pair" while the K's to all amplitudes for "particle + bound pair" going to "particle + broken pair", that is all breakup amplitudes.

If one substitutes (25) in (24) one gets the AGS equation

$$\Psi_i^\rho = \sum_j G_0\, t_i\, G_0\, U_{ij}^\rho \sum_{\sigma\supset j} \tilde{\delta}_{\sigma\rho}\, \Psi_j^\sigma, \qquad (27)$$

which allows for a reduction in the dimensionality of the equations each time $t_i$ or $U_{ij}^\rho$ is expressed as a finite rank operator.

The gigantic step Kamada and Glöckle (KG) were able to undertake involves the solution of Eq.(24) for a number of realistic and test potentials in three-continuous variables and including, in a few cases, all partial waves needed for convergence. Since their aim is the solution of the equations in three continuous variables, they start with Eq.(24) and proceed to make its momentum space representation, followed by triple partial wave expansion and spline interpolation. From the technical point of view the work is similar to the solution of the momentum space three-nucleon problem with additional complication resulting from one extra dimension, more channels, and two subamplitudes ($K^{3+1}$ and $K^{2+2}$) in two-variables, instead of one subamplitude in one variable. For this reason KG make extensive use of cubic spline interpolation in addition to Pade to calculate $K^{3+1}$ and $K^{2+2}$ by summing the Born series. As expected KG encounter similar problems as SSK, including Jacobi variable transformation, which they handle by the use of appropriate permutation operators. The major difference between the two works is the number of channels considered by KG leading, for the first time, to converged results for a given potential. This was achieved for MT-V where in Table II we show the rate of convergence as the maximum orbital angular momentum associated with a given Jacobi variable is increased from zero to eight. The value of -30.07 corresponds to the s-

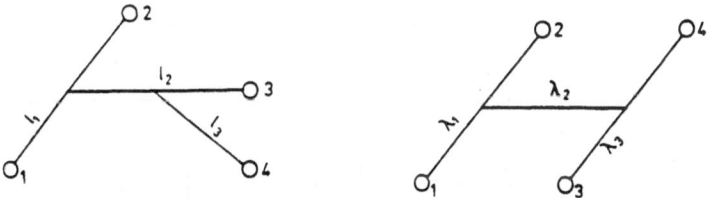

**Fig.1**    Jacobi variables and their respective orbital angular momentum.

**Table II** - Binding energy versus number of channels for four-boson interacting via MT-V potential. The angular momentum $\ell_1$, $\ell_2$, $\ell_3$ and $\lambda_1$, $\lambda_2$, $\lambda_3$ correspond to the Jacobi variables shown in Fig.1. In column one we show the number of 3+1 and 2+2 channels separately

| # channels | $\ell_1^{max}$ | $\ell_2^{max}$ | $\ell_3^{max}$ | $\lambda_1^{max}$ | $\lambda_2^{max}$ | $\lambda_3^{max}$ | E(MeV) |
|---|---|---|---|---|---|---|---|
| 1+1 | 0 | 0 | 0 | 0 | 0 | 0 | -30.07 |
| 3+2 | 0 | 2 | 2 | 0 | 2 | 2 | -30.07 |
| 5+3 | 0 | 4 | 4 | 0 | 4 | 4 | -30.07 |
| 2+2 | 2 | 2 | 0 | 2 | 0 | 2 | -31.11 |
| 3+6 | 4 | 4 | 0 | 4 | 0 | 4 | -31.20 |
| 4+4 | 6 | 6 | 0 | 6 | 0 | 6 | -31.21 |
| 5+5 | 8 | 8 | 0 | 8 | 0 | 8 | -31.21 |
| 19+13 | 6 | 6 | 2 | 6 | 2 | 6 | -31.30 |
| 39+24 | 6 | 6 | 4 | 6 | 4 | 6 | -31.35 |
| 58+34 | 6 | 6 | 6 | 6 | 6 | 6 | -31.36 |

wave approximation of SSK and is already shown in Table I. The converged results shown last in Table II may be compared with -31.357 MeV (CRC ref.[8]); -31.3 ± 0.2 MeV (GFMC ref.[28]); -31.19 ± 0.05 MeV (VMC ref.[29]); -31.36 (ATMS ref.[5]); -30.98 (IDEA ref.[26]). Except for IDEA that only includes two-body correlactions, the agreement one gets with CRC, GFMC and ATMS seems to indicate that for MT-V one has really calculated the four-boson binding with an accuracy of about 10 KeV. As shown in Table II the number of channels grows rapidly with the maximum value for the angular momentum, which in turn increases the size of the calculation by several orders of magnitude (few minutes to several hours). When they add spin and isospin quantum numbers to be able to treat the real $^4$He with realistic interactions, the number of channels grows even more rapidly. Although KG have developed a perfectly valid procedure to count channels in terms of maximum orbital momentum $\ell_3^{max}$ and $\lambda_2^{max}$ allowed for fixed $j_1^{max}$ ($\vec{j_1} = \vec{\ell_1} + \vec{s_1}$), we suggest a different procedure that uses instead the quantum numbers of the 3+1 subsystem $j_2^{max}$ as the limiting number, together with $j_1^{max}$, and allow $\ell_3$ and $\lambda_2$ to take all possible values that are consistent with $J = 0^+$ for $^4$He. This has the advantage of saturating the contribution of (2+2) and (3+3) subsystem amplitudes as we go up in $j_1^{max}$ and $j_2^{max}$. This is shown in Table III for $j_1^{max} = 1^+$, 1, 2 and $j_2^{max} = \frac{1^+}{2}, \frac{3^+}{2}, \frac{3}{2}, \frac{5}{2}$. The number of 3+1 channels is extremely easy to calculate since, for given $j_2^{\pi}$, it involves the corresponding number of three-nucleon channels times the multiplicity of that subsystem quantum number in the four nucleon problem. Given that $^4$He has $J = 0^+$, the $\frac{1^+}{2}$ trinucleon subamplitude may only couple with the fourth nucleon in $\ell_3 = 0$,

**Table III -**  Four-nucleon channels for increasing $j_1^{max}$ and $j_2^{max}$ where $j_2^{max} = \frac{3^+}{2}$ means $\frac{1^+}{2}$ and $\frac{3^+}{2}$.

| | $j_1^{max} = 1^+$ | $j_1^{max} = 1$ | $j_1^{max} = 2$ |
|---|---|---|---|
| $j_2^{max} = \frac{1^+}{2}$ | | | |
| (3+1) | 5 | 10 | 18 |
| (2+2 | 9 | 34 | 130 |
| **Total** | 4 | 44 | 148 |
| $j_2^{max} = \frac{3^+}{2}$ | | | |
| (3+1) | 12 | 24 | 48 |
| (2+2) | 9 | 34 | 130 |
| **Total** | 21 | 58 | 178 |
| $j_2^{max} = \frac{3}{2}$ | | | |
| (3+1) | 24 | 48 | 96 |
| (2+2) | 9 | 34 | 130 |
| **Total** | 33 | 82 | 226 |
| $j_2^{max} = \frac{5}{2}$ | | | |
| (3+1) | 38 | 76 | 156 |
| (2+1) | 9 | 34 | 130 |
| **Total** | 47 | 110 | 286 |

while $\frac{1^-}{2}$ in $\ell_3 = 1$, $\frac{3^+}{2}$ in $\ell_3 = 2$, and $\frac{3^-}{2}$ in $\ell_3 = 1$. Since $j_1^{max} = 1^+$ involves 5 channels for $\frac{1^+}{2}$, 5 for $\frac{1^-}{2}$, 7 for $\frac{3^+}{2}$, and 7 for $\frac{3^-}{2}$, the total number of 3+1 channel for $j_1^{max} = 1^+$ and $j_2^{max} = \frac{3}{2}$ is

24. The number of (2+2) channels may also be calculated "à la" KG but given the strengh of the N-N tensor-force and the results in ref.[14], it is best to include all $\lambda_2$ that are needed to couple any two pairs of nucleons to $J = 0^+$.

In their first manuscript [16] KG did not follow this prescription and limit themselves to $\ell_3^{max} = \lambda_2^{max} = 0$ which, for $j_1^{max} = 1^+$, reduces the number of (2+2) channels to 5 instead of 9. This may explain why they got less binding than in ref.[14], when they compared results for separable Yamaguchi potentials with 5.5% and 7% deuteron d-state probability; Fonseca got -29.10MeV and -26.56 MeV respectively, while KG obtained -28.87 MeV and -26.32 MeV. Although the discrepancy may also be attributed in part to differences in the numerical methods, the bulk may come from the neglect by KG of the $\lambda_2 = 2$ term in the coupling of two $^3S_1$ - $^3D_1$ pairs to $J = 0^+$.

In their subsequent work [17] for realistic interactions, KG solved the full Yakubovsky equations with 9 channels in (2+2) and increasing three-nucleon subsystem quantum numbers: $\frac{1^+}{2}$ ; $\frac{1^\pm}{2}$ $\frac{3^-}{2}$ ; $\frac{1^\pm}{2}$ $\frac{3^\pm}{2}$ $\frac{5^+}{2}$. Since they only include the positive parity two-nucleon channels with $j_1^{max} \leq 1^+$, that is $^1S_0$ and $^3S_1$ - $^3D_1$, 9 channels in the (2+2) subamplitude completely exhaust the contribution of the 2+2 partitions. Therefore increasing the number of (3+1) subamplitudes clearly shows the convergence of the calculation in terms of 3+1 partitions. This is shown in Table IV for nine realistic potentials. The reason why they have (15+9) and (27+9) channels instead of (17+9) and (31+9) is due to the neglect of $\ell_2 > 2$ components in the (3+1)

**Table IV -** $^3H$ and $^4He$ (no Coulomb) binding energies in MeV for different realistic interactions and increasing number of channels

| Potential | $P_D\%$ | $B_3$5ch | $B_4$(5+5)ch | $B_4$ (15+9)ch | $B_4$(27+9)ch |
|-----------|---------|----------|--------------|----------------|---------------|
| Reid | 6.47 | 7.02 | 21.79 | 22.20 | 22.19 |
| Paris | 5.77 | 7.30 | 23.20 | 23.63 | 23.60 |
| Nijmegen | 5.39 | 7.49 | 24.16 | 24.55 | 24.53 |
| SSC(A) | 4.43 | 7.56 | 24.90 | 25.24 | 25.23 |
| AV14 | 6.08 | 7.45 | 23.36 | 23.77 | 23.75 |
| Bonn B | 4.99 | 8.17 | 26.88 | 27.32 | 27.30 |
| Bonn A | 4.38 | 8.38 | 28.11 | 28.52 | 28.51 |

subamplitudes whose contribution in the three-nucleon problem in known to be small at very low energies.

The important conclusion from Table IV is that, irrespective of the potential used, the step from (5+5) to (15+9) channels increases the binding by about 400 KeV. It is nevertheless unfortunate that KG did not calculate an intermediate step corresponding to (5+9) or (15+5) channels in order to isolate the contribution of $\lambda_2 = 2$ in the coupling of two triplet pairs, and the importance of the negative parity three-nucleon subamplitudes $\frac{1-}{2}$ and $\frac{3-}{2}$. In previous work the contribution of negative parity states to $^4$He binding was found negligible[30], while $\lambda_2 = 2$ gave a contribution of about 0.5%[14] extra binding. A similar investigation with realistic interactions may provide, not only a savings in computing time, but also further insight into the physics of bound $^4$He. An important additional check is the calculation of S-, S'-, P- and D-state percentage in the $^4$He wave function, as well as the contribution of given (3+1) or (2+2) subamplitudes to each (L,S) component.

The results of Table IV were plotted in Fig.2 to show the correlation between $^3$H and $^4$He binding (Tjon line[25]). As in ref.[14], KG find that results from potentials fit to pp and np data lie on different lines. They subsequently study Charge Indepence Breaking (CIB) with the help of a simple effective two-nucleon t-matrix

$$t_{eff} = \frac{2}{3} t_{nn} + \frac{1}{3} t_{np}$$

in channel $^1S_0$. Using Bonn A and Bonn B potentials, they observe a 600 KeV shift in $^4$He binding due to CIB forces, which is a factor of three bigger than in $^3$H.

Finally comparing the results of this calculation with the best VMC or GFMC calculations that take on the full potential, not just a few N-N partial waves, one naturally finds a small difference that soon may disappear when KG surprise us with a converged

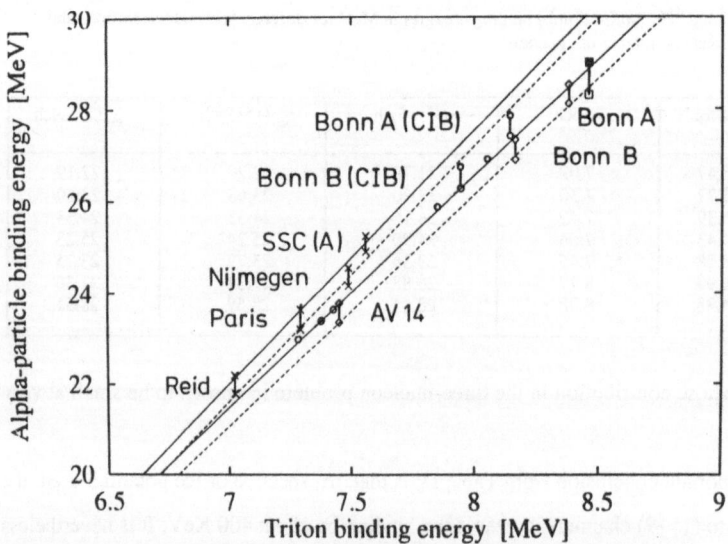

**Fig.2**   4N versus 3N binding for different potentials and number of channels.

Yakubovsky calculation in three continuous variables. Until then we can only quote some of the available results in the literature which, for the same potential, keep changing with improved numerics and faster computers. Some of the results are shown in Table V for VMC, GFMC and YMS (by KG). Given that GFMC calculations include Coulomb, one gets about $24.89 \pm 0.2$ MeV for $^4$He with AV 14 and no Coulomb.

**Table V -** $^4$He Binding in MeV for different NN potentials and calculation methods. To compare with YMS with VMC and GFMC one should add about 0.7 MeV to VMC and GFMC results.

|  | RSCV8 | AV8 | AV14 |
|---|---|---|---|
| VMC | 23.1 ref.[6]<br>23.6 ± 0.1 ref.[7] | 24.6 ref.[6] | 23.2 ref.[6]<br>23.54 ref.[7] |
| GFMC | 24.6 ref.[6]<br>24.5 ± 0.1 ref.[7] | 25.03 ± 0.04 ref.[31] | 24.17 ± 0.2 ref.[31] |
| YMS(No Coulomb) |  |  | 23.75 ref.[17] |

Therefore, for AV14, the contribution of the N-N partial waves beyond $^1S_0$, $^3S_1$ - $^3D_1$ should be about 1.1 MeV which compares with 230 KeV[32] in the trinucleon for the same potential. It should be interesting to investigate how the contribution from higher partial waves changes, both $^3$H and $^4$He binding and how it depends on the N-N potential. Given the findings of KG, concerning the correlation between $^3$H and $^4$He binding, the only hope to observe a "Tjon band" instead of "Tjon line" may come from higher partial waves and three-body force effects. At this conference KG reported calculations with partial waves up to $j_1^{max} \leqslant 3$ (190 channels) leading to 24.62 MeV binding for $^4$He with AV 14 potential. Comparing with GFMC in Table V one finds a small discrepancy whose origine is important to investigate.

Although the study of the three-nucleon force in $^4$He has long been undertaken by variational experts using ATMS, VMC and GFMC, KG[18] have recently added the Tucson-Melbourne (TM) $2\pi$-exchange three-nucleon force with a typical cut-off value of $\lambda_\pi = 5.828 \mu$ ($\mu = 139.6$ MeV and corresponds to the pion mass) to some of the two-body interactions considered before. To simplify the calculation they only take $j_1^{max} = 1^+$ ($^1S_0$, $^3S_1$ - $^3D_1$) and keep $\ell_3^{max} = \lambda_2^{max} = 0$. This corresponds to (5+5) channels where the $\frac{1^-}{2}$ (3+1) subamplitude is fully included, but the $\lambda_2 = 2$ terms in the (2+2) subamplitude are neglected. In order to include in the Yakubovsky equations the three-nucleon force

$$W = W_{123} + W_{124} + W_{134} + W_{234},$$

where each term splits naturally into 3 parts, they are able to arrange the 12 terms into 6 groups of two, such that each pair behaves under permutation exactly like a given 2N-force. This way the number of Yakubovsky components remains the same and the structure of the equations is only slightly modified by the presence of W.

The results are shown in Table VI and display for the first time the effect of the 3NF in $^4$He for a broad range of 2N potentials. This clearly shows that the 3NF has a stronger impact

**Table VI-** 3N and 4N binding energies with and without 3WF's.

| Potential | 3N | | 4N | |
|---|---|---|---|---|
| | without | with 3NF | without 3NF | with 3NF |
| RSC | 7.02 | 7.58 | 21.79 | 24.83 |
| Paris | 7.30 | 8.12 | 23.20 | 28.10 |
| Njimegen | 7.49 | 8.52 | 24.16 | 30.15 |
| AV14 | 7.45 | 8.31 | 23.36 | 28.11 |
| Bonn B | 8.17 | 9.80 | 26.88 | 37.17 |

in $^4$He than in $^3$H. While in $^3$H the 3NF increases the binding by 0.56 MeV to 1.63 MeV, in $^4$He the energy increase due to 3NF ranges from 3.04 MeV to 10.29 MeV.

In particular the energy gain in $^4$He is about six fold the corresponding increase in $^3$He. Since in $^4$He there are four 3NF terms, instead of one in $^3$H, one should perhaps naively expect a factor of four instead of six in the binding energy gain due to 3NF. Similar findings have recently been achieved by VMC and CFMC experts as shown in Table VII. Nevertheless

**Table VII** - VMC and GFMC calculations for 3N and 4N system with and without 3N-force. All $^4$He results include the Coulomb energy($\simeq$. 75 MeV).

| | VMC ref.[ ] | | GFMC ref.[7] |
|---|---|---|---|
| | AV$_{14}$ | AV$_{14}$ + UVIII | AV$_{14}$ + UVIII |
| $^3$H | 7.45(1) | 8.21(2) | |
| $^4$He | 23.54(4) | 27.23(6) | 28.3 ± 0.2 |

their results are less dramatic than KG's because the UVIII 3NF is weaker that TM's 3NF and has been tuned to $^3$H, $^4$He and nuclear matter data with the help of VMC and GFMC calculations. Given that, in the absence of 3NF, higher AV$_{14}$ N-N partial waves give 230 KeV aditional binding to $^3$H, which rises to about 1 MeV when a 3NF is added[32], one expects the results in Table VI to increase by a few MeV when all channels are considered. By then one may be confronted with energies that grossly overbind $^4$He, depending of the combination of 2N - and 3N - force used. Nevertheless, the results in Table VI are already sufficiently indicative that a greater theoretical ambiguity exist in the choice of 3NF versus 2NF and points to the need for a consistent theory of 2N - and 3N - force.

## CONTINUUM CALCULATIONS

Unlike the recent progress in the solution of the bound state 4N problem, continuum calculations are still at a relative simplified stage concerning force models. Even for Yamaguchi separable interactions in channels $^1S_0$ and $^3S_1$ - $^3D_1$ there are no numerically converged solutions of the AGS equations for either isospin $T = 1$ or $T = 0$ reactions.

The last published development[34] involves the solution of AGS equations by taking the four-nucleon channels that are associated with the two-nucleon tensor-force in first order perturbation theory. One starts by constructing the four-nucleon Kernel for a one term separable Yamaguchi potential in channels $^1S_0$ and $^3S_1$ - $^3D_1$ corresponding to 4% d-state probability in the deuteron. Next one drops all terms where a triplet pair couples to a third nucleon in $\ell_2 = 2$, $s_2 \neq j_2$, or two negative parity 3+1 states are coupled, and solves exactly the resulting 4N Kernel. Finally they include the neglected terms in first order perturbation, together with left and right off-shell rescattering through the unperturbed four-particle t-matrix. The resulting amplitudes are used to calculate Tensor Analysing Powers (TAP) for dd $\rightarrow$ n$^3$He and dd $\rightarrow$ dd as well as cross sections. A typical result is shown in Fig.3 for $T_{20}$ and $T_{21}$. The agreement is qualitatively good but an exact calculation is needed to allow for concluding remarks.

More recently came to our attention a work by E-Uzu, S. Oryu and M.Tanifuji[35] on the exact calculation of the AGS equations for a similar problem, but where the Yamaguchi potential was substituted by PEST 1. They show TAP's in the 30 - 90 KeV region but

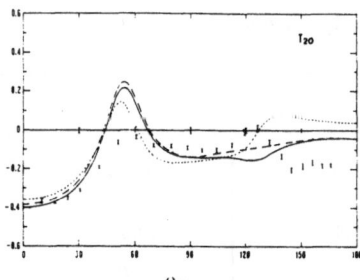
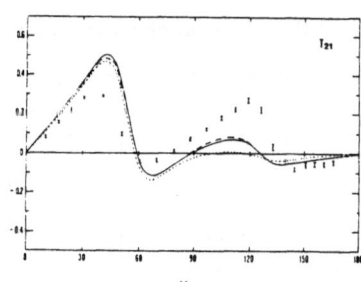

**Fig.3**   $T_{20}$ and $T_{21}$ for dd $\rightarrow$ n$^3$He at 6MeV. The curves correspond to different rescaterring terms.

agreement with data implies a multiplication of calculated TAP's by a factor of three. The absence of Coulomb at these energies may be a serious handicap.

Now that Schellingerhout, Schut, Kok, Kamada and Glöckle have set higher standarts in four-body work, may be old "cronies" in the field and enthusiastic newcomers feel the challenge to reach out for better calculations.

## RELATIVISTIC HAMILTONIAN FOR $^4$He

One of the most recent developments in the calculation of $^4$He bindind is the consideration of a relativistic (R) Hamiltonian[36].

$$H = \sum_i \left[ (m^2 + p_i^2)^{1/2} - m \right] + \sum_{i<j} \vartheta_{ij} (P_{ij}) + \sum_{i<j<k} W_{ijk} (P_{ijk}) ,$$

$$P_{ij} = p_i + p_j ,$$
$$P_{ijk} = p_i + p_j + p_k,$$

which is a simple generalization of a corresponding nonrelativistic (NR) Hamiltonean. The interactions depend upon the total momentum of the interacting particles and are written as

$$\vartheta_{ij}(P_{ij}) = \tilde{\vartheta}_{ij} + \delta \, \vartheta_{ij} (P_{ij}) ,$$

$$W_{ijk} (P_{ijk}) = \tilde{W}_{ijk} + \delta \, W_{ijk} (P_{ijk}) ,$$

where $\tilde{\vartheta}_{ij}$ ($\tilde{W}_{ijk}$) is the interaction in the $P_{ij} = 0$ ($P_{ijk} = 0$) "rest frame" and $\delta \, \vartheta_{ij} (P_{ij}=0) = 0$ as well as $\delta \, W_{ijk} (P_{ijk} = 0) = 0$. The rest frame interaction $\tilde{\vartheta}_{ij}$ and $\tilde{W}_{ijk}$ can be determined by fitting the data and relativistic covariance is used to obtain the momentum dependent parts $\delta \, \vartheta_{ij}$ ($P_{ij}$) and $\delta \, W_{ijk}$ ($P_{ijk}$) from $\tilde{\vartheta}$ and $\tilde{W}$.

Based on AV$_{14}$ nonrelativistic potential, they develop $\tilde{\vartheta}_{ij}$ and make a new fit to the NN scattering data and deuteron properties using the relativistic kinetic energy term. Then they proceed to use a VMC anzatz to minimize E for both $^3$H and $^4$He Hamiltonians which contain relativistic kinetic energy terms, $\tilde{\vartheta}_{ij}$ and $\tilde{W}_{ijk}$ which was taken to be the U VIII potential. Subsequently they calculate the contribution of $\delta \, \vartheta_{ij}(P_{ij})$ in firts order perturbation theory.

**Table VIII-** Relativistic and nonrelativistic calculation for $^3$H and $^4$He.

| | $^3$H | | $^4$He | |
|---|---|---|---|---|
| | $\tilde{\vartheta}$ | $\tilde{\vartheta} + W_{ijk}$ | $\tilde{\vartheta}$ | $\tilde{\vartheta} + W_{ijk}$ |
| $E_v$ | -7.27 | -8.41 | -22.68 | -27.64 |
| $<\delta \vartheta>$ | .31 | .33 | 1.30 | 1.70 |
| $E_0^R$ | -6.96 | -8.08 | -21.38 | -25.97 |
| $E_0^{NR}$ | -7.38 | -8.42 | -23.16 | -27.98 |

The momentum dependence of $\delta \, \vartheta_{ij}$ is calculated in the framework of classical relativistic mechanics and power series expansion in powers of $P^2$. The expression they use is

$$\delta \, \tilde{\vartheta}(\mathbf{P}, r) = - \frac{P^2}{8m^2} \, \tilde{\vartheta}(r) + \frac{1}{8m^2} \, (\mathbf{P} \cdot \mathbf{r}) (\mathbf{P} \cdot \nabla) \tilde{\vartheta}(r) \, .$$

and $\delta W = 0$

The results are shown in Table VIII for the binding energies of $^3$H and $^4$He, leading in both cases to the conclusion that relativity leads to less binding. This is contrary to the outcome of a calculation by Rupp and Tjon[37] where they use Bethe-Salpeter-Faddeev equations, together with finite rank representations of realistic interactions that have been made covariant, to find that relativity contributes to more attraction.

Therefore, relativistic effects may strongly depend on the method used to generalize the NR Hamiltonian for use in relativistic calculations as well as on the importance of nucleon negative energy states in nuclei.

## CONCLUSIONS

At this conference one may be able to recognize that our ability to calculate $^4$He binding is finally approaching the stage $^3$H was ten years ago when the first converged Faddeev calculations emerged for the first time for realistic interactions.

Nevertheless progress will come faster because numerical methods tested in FCS and FMS calculations come into play in YCS and YMS developments. Therefore one can expect to have fully convergent Yakubovsky calculations of $^4$He for a number of realistic two-body interactions and three-nucleon forces within one or two years. Threshold continuum calculations may follow, but converged low energy (below four-particle breakup) cross sections and TAP's for all $T = 0$ and $T = 1$ reactions may only be possible in the next few years with simple potentials or finite rank representations of local realistic interactions.

In $^4$He we definitely need the whole picture which implies the calculations of S-, S'-, P- and D-sate components, assymptotic norms, D-to S- ratio $\eta_\alpha$ , rms radius, Coulomb energy, charge density, electromagnetic form factors and eventually $\Delta$ degrees of freedom. Nevertheless most of the above is no simple task. Therefore the important thing we have to understand, before jumping into hours of CPU time, is what we can learn from $^4$He that $^3$H does not teach us. The key evidence is the observation of a "Tjon band" instead of a "Tjon line" and where does the experimental point lies. The next key evidence is 2N - versus 3N - force. If the pattern observed in Table VI gets not only confirmed, but reenforced by converged calculations, we have to stop and think about a consistent theory of 2N - and 3N force. It is also possible that relativity "à la" Carlson, Pandharipande and Schiavilla may cure the overbinding, but this is uncertain right now.

Another important issue is the use of finite rank representations of realistic forces to be used in continuum calculations. Their accuracy can now be tested in $^4$He against the results of KG. This should reduce CPU time by many hours and make possible the interpretation of most low energy data in T = 0 and T = 1 reactions. With the developments of the past two years YMS and YCS may surpass GFMC as the method of choice for $^4$He.

On the contary IDEA, which three years ago looked very promissing for $^4$He, given its simplicity and quality of results "vis a vis" the small size of the calculations, has not kept momentum being surpassed by Correlated Hyperspherical Harmonics[38] in both $^3$H and $^4$He.

[1]     A.C. Fonseca, Nucl. Physics A416, 421c (1984) and references therein.
[2]     A.C. Fonseca in "Few-Body Methods: Principles and Applications, Ed. by T. Lim, C. Bao, D. Hou and S. Huber (world Scientifique, Singapore (1986).
[3]     H. Fiedeldey, Nucl. Phys. A463, 335 c (1987).
[4]     A.C. Fonseca, Lecture Notes in Physics 273, 161 (1986).
[5]     Y. Akaishi, Few-Body Systems, Suppl. 2, 64 (1987); Lecture Notes in Physics 273, 324 (1986).
[6]     J. Carlson, Nucl. Phys. A508, 141c (1990).
[7]     R.B. Wiringa, Nucl. Phys. A543, 199c (1992).
[8]     M. Kamimura and H. Kameyama, Nucl. Phys. A508, 17c (1990).
[9]     H. Fiedeldey, Few-Body Systems, Suppl. 6, 557 (1992).
[10]    S.K. Adhikari and K.L. Kowalski. "Dynamical Collision Theory and its Applications"(Academic Press, New York 1991).
[11]    V.B. Belyaev, Lectures on the Theory of Few-Body Systems (Springer-Verlag, Berlin 1986).
[12]    J.L. Friar, Few-Body Systems, Suppl. 6, 538 (1992).
[13]    J.Carlson, Phys. Rev. C36, 2026 (1987).
[14]    A.C. Fonseca, Phys. Rev. C40, 1390 (1989).
[15]    S.A. Sofianos, H. Friedeldey, H. Haberzettl and W. Sandhas, Phys. Rev. C26, 228 (1982).
[16]    H. Kamada and W. Glöckle, Nucl Phys. A548, 205 (1992).
[17]    H. Kamada and W.Glöckle, Phys. Lett. B 292 , 1 (1992).
[18]    H. Kamada and W. Glöckle, Accepted for Nucl. Phys. A
[19]    N. W. Schellingerhout, J.J. Schut and L.P. Kok, Phys. Rev. C46, 1192 (1992).
[20]    S.P. Merkuriev, S.L. Yakovlev and C. Gignoux, Nucl. Phys. A31, 125 (1984).
[21]    S.P. Merkuriev  and S.L. Yakovlev, Yad Fiz. 39, 1580 (1984) [Sov. J. Nucl. Phys. 39, 1002 (1984)].
[22]    O.A. Yakubovsky, Yad, Fiz. 5, 1312, (1967) [Sov. J. Nucl. Phys. 5, 937 (1967)].
[23]    N.W. Schellingerhout, L.P. Kok and G.D. Bosveld, Phys. Rev. A40, 5568 (1989).
[24]    J.L. Friar, B.F. Gibson and G.L. Payne, Phys. Rev. C24, 2279 (1981).
[25]    J.A. Tjon, Phys. Lett. 56B , 217 (1975).
[26]    W. Oehm, S. Sofianos, H. Fiedeldey and M. Fabre de la Ripelle, Phys. Rev. C42, 2322 (1991).
[27]    P. Grassberger and W. Sandhas, Nucl. Phys. B2, 181 (1967); E.O. Alt, P.Grassberger and W. Sandhas, JINR Report. n°. E4-6688 (1972).
[28]    J. Lomnitz-Adler, V.R. Pandharipande and R.A. Smith, Nucl. Phys. A361, 399 (1981)
[29]    J.Carlson and V.R. Pandharipande, Nucl. Phys. A371, 301 (1981).
[30]    A.C.Fonseca, Few-Body Systems 1, 69 (1986).
[31]    J. Carlson, private communication.
[32]    B.F. Gibson, Nucl. Phys. A543, 1c (1992).
[33]    R.B. Wiringa, Phys. Rev. C43,1585 (1991).
[34]    A.C. Fonseca, Phys. Rev. Lett. 63, 2036 (1989).
[35]    E. Uzu, S. Oryu and M. Tanifuji, submitted for publication
[36]    J.Carlson, V.R. Pandharipande and R. Schiavilla, Phys. Rev. C47, 484 (1993).
[37]    G. Rupp and J.A. Tjon, Phys. Rev. C45, 2133 (1992).
[38]    A. Kievsky, M. Viviani and S. Rosati, Nucl. Phys. A551, 241 (1993); Il Nuovo Cimento A105, 1473 (1992)

Few-Body Systems, Suppl. 7, 193—200 (1994)

Few-
Body
Systems

# PHOTO-INDUCED REACTIONS ON ⁴He

S.A. Sofianos[a], H. Fiedeldey[a], G. Ellerkmann[b], and W. Sandhas[b]

[a]Physics Department, University of South Africa, P.O.Box 392, Pretoria 0001, South Africa

[b]Physikalisches Institut der Universität Bonn, Germany

### Abstract

The photodisintegration of the $\alpha$-particle and more specifically the ⁴He($\gamma$,p)³H and the ⁴He($\gamma$,n)³He reactions, have been at the center of theoretical as well as experimental studies since the early fifties. Various aspects of the problem will be discussed. A brief historical overview of the experimental and theoretical controversies surrounding the photoneutron and photoproton cross-sections in the low energy region will be presented. Results obtained by means of the integral (momentum space) and integrodifferential (configuration space) equations will be given.

## 1 Introduction

Among the five possible photodisintegration modes of the $\alpha$-particle, the ⁴He($\gamma$,p)³H and the ⁴He($\gamma$,n)³He reactions and their inverse processes, are of particular interest. The first experimental evidence of the radiative capture of protons by tritons to form ⁴He was obtained in the early fifties [1]. Immediately afterwards an attempt was made by Flowers and Mandl [2] to investigate these reactions in a first order perturbation theory approximation. In their work the existence of a sharp Giant Dipole Resonance (GDR) at low incident photon energies ($E_\gamma \sim 30$ MeV) was predicted. Since then, a great wealth of experimental information has been accumulated (see, for example [3,4,5,6] and refs therein) and special attention was given to the photoproton and photoneutron total cross sections and their ratio $R_\gamma = \sigma(\gamma, p)/\sigma(\gamma, n)$. Despite the plethora of experimental data compiled by various groups, significant discrepancies and uncertainties, especially for the ratio $R_\gamma$, were recorded. This ratio is expected to be $\sim 1$. According to early compilations, however, it was found to be in the range of 1.6 to 1.9 in the peak region. Furthermore the results exhibited two trends, namely, a pronounced GDR for the photoproton cross-section and a relatively flat structure for the corresponding photoneutron cross section. These trends were discussed in detail by Calarco et al [7] in their critical review of the data available up to the early 80's. While a $\sim 10\%$ difference between the $(\gamma, n)$ and the $(\gamma, p)$ could be attributed to Coulombic effects, a larger ratio would mean that there exist a large charge asymmetric component in the nuclear force.

On the theoretical front the situation was equally confusing with most calculations, mainly of the Shell Model (SM)- type [8,9,10], giving the 'correct' $R_\gamma \sim 1$. At the same time, however, these calculation were persistent in predicting a GDR behaviour at low incident photon energies for both the $(\gamma, n)$ and $(\gamma, p)$ reactions.

An alternative method for treating this problem has been provided by Casel and Sandhas [11,12,13] on the basis of the exact four-nucleon AGS formalism [14]. The first application of the method in the early eighties, predicted a flat behaviour of the $(\gamma, n)$ cross section over an extended region - which was a surprise at that time. This result was obtained by employing a separable interaction and it has been later confirmed [15] by employing the Malfliet-Tjon (MT I+III) potential [16], where an even more pronounced flattening was predicted in the controversial energy region.

It is clear that new and reliable experimental data should be compiled using modern techniques and that reliable theoretical calculations should be performed, which rely to the large extent possible on an exact microscopic theory while employing the free nucleon-nucleon force and not an effective force with adjustable parameters.

New experimental results indeed have now appeared [17,18] showing no sign of a sharp GDR behaviour. At the same time, new theoretical results obtained by Wachter et. al. [19] by means of Resonating Group Method (RGM), were in favour of the older generation of data i.e exhibiting a pronounced GDR behaviour. Similar results were obtained by the Kharkov group [20] via field theoretical methods. A more recent RGM calculation by Unkelbach and Hofmann has meanwhile appeared [21] in which the results of [19] have been adjusted to the latest experimental data. We shall return to these results and discuss certain ambiguities of this method later.

Finally we must mention that yet another experimental measurement has recently appeared where a simultaneous measurement of the $(\gamma, p)$ and $(\gamma, n)$ cross sections was made [22]. These data were analysed by using dispersion relations and some doubts were raised on whether one can decide with present day data if the ratio $R_\gamma = \sigma(\gamma, p)/\sigma(\gamma, n)$ is around one or not.

In what follows we will present two theoretical methods used by our group to handle these photoprocesses. In the first method, we employ the AGS integral equations to obtain the transition amplitude. This method is, of course, exact and can therefore provide the required reliable results. The second method is implemented in configuration space and use is made of the recently developed Integrodifferential Equation Approach (IDEA) to obtain the three- and four-nucleon bound state wave functions [23,24], while the outgoing 3+1 scattering states were obtained via an optical model accurately fitted to the scattering data. The latter approach has the advantage of being capable of extension to larger systems unlike the first one, which is restricted to systems of up to four particles. However, comparing both approaches it is possible to make a careful test of the accuracy of the IDEA plus optical potential approximation to larger systems.

In the next section we shall describe our approach to the problem. In section 3 we shall present our results with some discussions. Our conclusions will be summarised in section 4.

## 2 Formalism

Let us first recall some basic relations of the problem. The differential cross-section for the photodisintegration of $^4$He by an unpolarised incident photon beam is given by

$$\frac{d\sigma}{d\Omega} = \frac{\mu q}{2\pi\hbar^2 k_\gamma} \frac{1}{2} \sum_{\lambda=1}^{2} |M^\lambda(\mathbf{q})|^2, \tag{1}$$

where the factor 2 stems from the fact that only neutrons are counted experimentally. Here $\mu$ denotes the reduced mass of the outgoing fragments, $k_\gamma$ and $\lambda$ the momentum and the polarization of the incident photon, and $M^\lambda(\mathbf{q})$ the photodisintegration amplitude

$$M^\lambda(\mathbf{q}) = {}^{(-)}< \mathbf{q}; \Psi_3 | \mathcal{H}_{em} | \Psi_4 > . \tag{2}$$

$|\mathbf{q}; \Psi_3 >^{(-)}$ denotes the full scattering state associated with the channel state $|\mathbf{q} > |\Psi_3 >$ in which the nucleon is moving with momentum $\mathbf{q}$ relative to the three-nucleon bound state $|\Psi_3 >$; $\mathcal{H}_{em}$ is the relevant electromagnetic interaction, and $|\Psi_3 >$ and $|\Psi_4 >$ are the three- and four-nucleon bound state wave functions, respectively.

It is generally accepted that at low energies, the process under consideration takes place primarily via an electric dipole transition, which contribute up to $\sim 97\%$ to the total cross-section [25]. Thus, bearing in mind Siegert's theorem , we are led to the following choice of the electromagnetic interaction

$$\mathcal{H}^\lambda_{em} = \frac{e(E_f - E_i)}{2\imath\hbar c} \sum_{j=1}^{4} \hat{\varepsilon}_\lambda \cdot \mathbf{r} \left[ \frac{1}{3}(\tau_z^1 + \tau_z^2 + \tau_z^3) - \tau_z^4 \right], \tag{3}$$

where $E_i$ and $E_f$ are the bound state energies of the initial and final bound systems respectively i.e $(E_f - E_i) = E_\gamma$, $E_\gamma$ being the incident photon energy. The $\hat{\varepsilon}_\lambda$ is the polarization direction of the photon and $\mathbf{r}$ is the coordinate canonically conjugate to the relative momentum $\mathbf{q}$.

As mentioned in the introduction, in order to evaluate the disintegration amplitude $M^\lambda(\mathbf{q})$, two methods were employed namely the integral equation approach (AGS) and the Integrodifferential Equation Approach together with an Optical Potential (IDEA+OP). In what follows we shall describe the two methods briefly.

## 2.1 Integral Equation Method

The crucial point in this formalism is to express the scattering states $|\mathbf{q}_\sigma; \Psi_3 >^{(-)}$ in terms of the (free) channel states $|\mathbf{q}_\sigma > |\Psi_3 >$ describing the situation where the particle $\imath$ is moving relative to the $(jkl)-$ bound system i.e the $\sigma = (i, jkl)$ fragmentation. This can be achieved by using the Möller operator $\Omega_\sigma^{(-)}$ [26]

$$^{(-)}< \mathbf{q}_\sigma; \Psi_3 | = < \mathbf{q}_\sigma | < \Psi_3 | \Omega_\sigma^{(-)\dagger} = < \mathbf{q}_\sigma | < \Psi_3 | \left[ 1 + \sum_\rho K^{\sigma\rho} \Omega_\rho^{(-)\dagger} \right], \tag{4}$$

where $K^{\sigma\rho}$ is simply the AGS kernel for the $(3 + 1)$ -partition of the four-nucleon system describing the rearrangement of the cluster $\rho$ going over to cluster $\sigma$.

The AGS formalism is exact. However, in view of the complexity of the problem and the herculean computational effort needed, we used the W-matrix separable representation of the two-body t-matrix [28] and the Energy Dependent Pole Approximation (EDPA)[29,30] for the 3+1 and the 2+2 transition amplitudes both of which have been shown to be quite accurate in numerous applications. In this way the AGS scattering equations for rearrangment processes are reduced to an effective two-body Lippmann-Schwinger (LS) -type equation

$$T^{\sigma\rho} = V^{\sigma\rho} + \sum_\tau V^{\sigma\tau} \mathcal{G}_0^\tau T^{\tau\rho}. \tag{5}$$

Note that the kernel $K^{\sigma\tau} = V^{\sigma\tau} \mathcal{G}_0^\tau$ incorporates the (2)-, the (3+1)-, and the (2+2)- subamplitudes in its structure.

In analogy to this reduction and by using eq.(4), we obtain for the photodisintegration amplitude $M^{\lambda,\sigma}$ and for a specific fragmentation $\sigma$ , the following expression

$$M^{\lambda,\sigma} = B^{\lambda,\sigma} + \sum_\tau V^{\sigma\tau} \mathcal{G}_0^\tau M^{\lambda,\tau}, \tag{6}$$

where the Born term $\mathcal{B}^{\lambda,\sigma}$ is given by

$$\mathcal{B}^{\lambda,\sigma}(\mathbf{q}) = <\mathbf{q}_\sigma| < \Psi_3|\mathcal{H}_{em}^\lambda|\Psi_4 > . \tag{7}$$

The kernels of equation (5) and (6) are identical and thus we may use the same techniques to solve the latter as the one employed for the four-nucleon bound state and rearrangement scattering [30,31]. The schematic structural form of equation (6), after symmetrization, is as follows

$$\mathcal{M}^{3+1} = \bar{\mathcal{B}}^{3+1} + \int V^{3+1,3+1}\mathcal{G}_0^{3+1}\mathcal{M}^{3+1} + \int V^{3+1,2+2}\mathcal{G}_0^{2+2}\mathcal{M}^{2+2} \tag{8}$$

$$\mathcal{M}^{2+2} = \bar{\mathcal{B}}^{2+2} + \int V^{2+2,3+1}\mathcal{G}_0^{3+1}\mathcal{M}^{3+1}. \tag{9}$$

To present the dynamical features of these coupled equations in a more transparent way, we substitude the second into the first to obtain

$$\mathcal{M}^{3+1} = \mathcal{B}^{3+1} + \int C^{3+1,3+1}\mathcal{G}_0^{3+1}\mathcal{M}^{3+1}, \tag{10}$$

where the Born term is now given by

$$\mathcal{B}^{3+1} = \bar{\mathcal{B}}^{3+1} + \int V^{3+1,2+2}\mathcal{G}_0^{2+2}\bar{\mathcal{B}}^{2+2} \tag{11}$$

and the effective interaction for the $(3 + 1)$−transition by

$$C^{3+1,3+1} = V^{3+1,3+1} + \int V^{3+1,2+2}\mathcal{G}_0^{2+2}V^{2+2,3+1}. \tag{12}$$

It is seen that both, the Born term and the effective interaction include also effects from the $(2+2)$− channel.
We must stress once more that the method is in principle exact which, apart from the numerical approximations involved, leaves no room for adjustable parameters. However, to obtain the full solution for energies above the $2+1+1$ break-up threshold is a formidable problem. Furthermore it has the disadvantage that at present Coulomb forces cannot be included and thus only the $(\gamma, n)$ cross sections can be computed. Therefore one has to search for alternative methods which, while preserving the microscopic description of the reaction, incorporate simplifications which reduce the problem to manageable proportions.

## 2.2 Integrodifferential Equation Approach

A particularly simplifying approach to the above exact method, is to approximate the photodisintegration amplitude $M^\lambda(\mathbf{q})$ by

$$M^\lambda(\mathbf{q}) = {}^{(-)}< \mathbf{q}| < \Psi_3|\mathcal{H}_{em}|\Psi_4 >, \tag{13}$$

i.e. in this model approach we take the Final State Interaction (FSI) into account via the approximation ${}^{(-)} < \mathbf{q}; \Psi_3| \simeq {}^{(-)}< \mathbf{q}| < \Psi_3|$. Thus we may use the IDEA[23,24] to obtain the bound state wave-functions for the three-and four-body systems and an optical model to construct the scattering states ${}^{(-)} < \mathbf{q}|$ corresponding to a nucleon scattered off a bound three-body system.
In the IDEA method the A-nucleon wave function $\Psi(\mathbf{x})$ is written as a sum of subamplitudes $\Psi(\mathbf{x}) = \sum_{i<j\leq A} \Psi(\mathbf{r}_{ij}, r)$ obeying the Faddeev-type equation

$$\left[T + \bar{V}_0\right] \Psi(\mathbf{r}_{ij}, r) = -[V(\mathbf{r}_{ij}) - V_0(r)] \sum_{i<j\leq A} \Psi(\mathbf{r}_{ij}, r), \tag{14}$$

where r is the hyperradius of the system, $r^2 = \frac{2}{A} \sum r_{ij}^2$, $\bar{V}_0(r) = \frac{A(A-1)}{2} V_0(r)$, $V_0(r)$ being the hypercentral potential representing the first term of the Potential Harmonic expansion of the interaction which takes into account, in a good approximation, the correlations stemming from the coupling of the spectator's particle relative orbitals with that of the $ij-$ pair. The above equation can be solved by projecting it onto the $r_{ij}$- space. In the case of spin dependent forces, eq (14) reduces to a coupled,two-variable, integrodifferential equation

$$\left[\frac{\hbar^2}{m}\nabla_0^2 - \bar{V}_0(r) + E\right] P_0^\alpha(z,r) = \left[\frac{V^{1+} + V^{3+}}{2} - V_0(r)\right] \Pi_0^\alpha(z,r) + \left[\frac{V^{1+} - V^{3+}}{2}\right] \Pi_0^\beta(z,r)$$

(15)

where

$$\Pi_0^i(z,r) = P_0^i(z,r) + \int_{-1}^{+1} f_0(z,z') P^i(z',r) dz',$$

(16)

with $\alpha, \beta, i = S, S'$ the singlet and triplet even states and $z = (2r_{ij}^2/r^2 - 1)$. More details concerning these equations can be found in refs [23,24]. It should be mentioned that this method produces better results for $^4He$ and larger nuclei than the extended Shell Model, even in its most sophisticated form at a minute fraction of the computational effort [32]. Furthermore, results obtained with the MT I+III forces, are in close agreement with those obtained via the Faddeev-Yacubovsky equations [33].

The optical potential used to obtain the scattering wave function $\chi_q^{(-)}(\mathbf{r}) = {}^{(-)}< q|\mathbf{r} >$ is the one employed by Podmore and Sherif [34] for the four nucleon scattering processes but with depth adjusted to reproduce the $L = 1$, $S = 0$, and $T = 1$ phase shifts relevant to the E1-transition.

## 3 Results and Discusions

The choice of the nucleon-nucleon interaction is of utmost importance since the three- and four-nucleon bound state equations must not only provide the correct binding energy, but also good wave functions with the correct shape and range. Unfortunately the realistic forces give binding energies which are up to $\sim$1.5 MeV lower than the correct three-nucleon binding energy and up to $\sim$ 8 MeV for the four-nucleon system. To overcome this difficulty we opted for the semirealistic MT I+III potential which is known to have in its structure the most important features of the nucleon-nucleon interaction (such as a repulsion at short distances) and provides reasonable scattering and bound state results for the three- and four-nucleon systems.

In Fig. 1 we present the results obtained by using the integral equation formalism. Note that due to the complicated cut structure of the kernel, the calculations, which take into account the FSI exactly, could only be performed below the three-fragment break-up threshold corresponding to an incident photon energy $E_\gamma = 26.3$ MeV. The solid line corresponds to the full solution i.e with FSI effects included, whereas the broken line corresponds to the Born approximation, (eq. 11). The agreement with the experimental $(\gamma, n)$ data is astonishing while the nonexistence of a GDR is once more demonstrated. The difference between the total cross section with FSI and the Born one, $\Delta\sigma = \sigma_{FSI} - \sigma_B$, explicitely shows that in the region around 24 MeV incident photon energies, the FSI plays an important role. Thereafter there is a rapid drop of its importance, and simple extrapolation indicates that its role beyond the $\sim$ 50 MeV should be insignificant. This can be seen in Fig. 2 where the Born results are plotted. The broken line corresponds to the integral equations results whereas the broken-dotted line to the IDEA. The experimental results are well reproduced by both methods which are also in excellent agreement with each other. We must mention here that apart from the $(\gamma, p)$ results of Bernabei et al [17], ($\square$), and Jones [36], ($\nabla$), the $(\gamma, p)$, ($+$), and $(\gamma, n)$, ($\times$), results of Gorbunov [37], belong to the older generation of data.

In Fig. 3 we present our configuration space (IDEA+OP) results for the MT I+III potential.The experimental results are the recent $(\gamma, p)$ data of Feldman et al [18], ($\bullet$), and of Bernabei et al [17], ($\square$). The solid line was obtained by adjusting the optical potential to Tombrello's p-wave phase shifts as described above. The broken line corresponds to the results obtained by switching off the Coulomb interaction. This clearly indicates that the differences between the $(\gamma, p)$ and $(\gamma, n)$ results as shown in Fig. 1 and Fig. 3, can not be attributed to pure Coulomb effects but rather to other nuclear observables involved in the $(3 + 1)$- fragmentation.

In Fig. 4 we show the results obtained by adjusting the optical potential to reproduce the RGM phase shifts of Reichstein et al [38], (solid line), and the Generator Coordinate Method (GCM) of Furutani et al [39],(broken-dotted line). It is seen that the results based on the RGM phase shifts exhibit a GDR behaviour and in this respect they are in substantial agreement with the RGM results obtained by Wachter et al [19] which, however, have recently been adjusted by Unkelbach and Hofmann [21] to the new experimental data of Feldman et al [18]. In these full RGM calculations [19,21], there is an enormous discrepancy between the calculated binding energies and the experimental values, which has been exploited to fit first the old giant dipole resonance behaviour and now the new relatively flat shape of the total cross-section. In other words, this approach resembles a fitting procedure which is rather undesirable in few-body photoprocesses.

# 4 Conclusions

We have shown that both our schemes gave results which are in agreement with the most recent experimental data and among themselves. Since the much simpler approximation IDEA+OP scheme is much more easily extended to larger systems, this is a most encouraging result.

The use of the exact integral equation method revealed that the $(\gamma, n)$ total cross-section does not exhibit any GDR behaviour. Even if one allows a 10% variation in the results due to omitted input information such as higher order transitions, tensor forces etc, it is unlikely to change the low energy nonresonant behaviour of the results predicted by this microscopic approach.

Our configuration space calculations produce results which are in good agreement with the $(\gamma, p)$ data. These results exhibit a sensitivity to the input phase shifts which shows that the purely nuclear and photonuclear observables are described in a consistent manner. This sensitivity is more pronounced in photonuclear processes than in purely nuclear reactions.

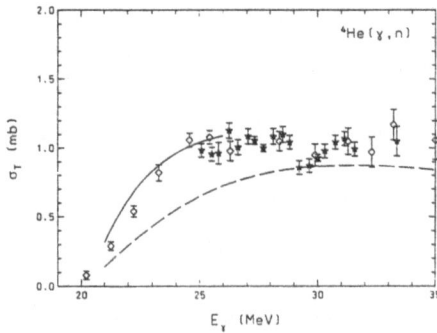

**Fig 1.** The $(\gamma, n)$ integral equation results. Full solution with FSI, (———); Born approximation, (- - -). The experimental data are from refs [4], ($\Diamond$), and [5], ($\star$).

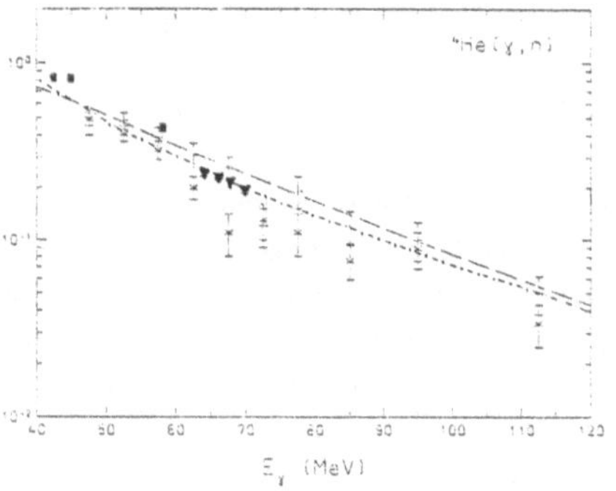

**Fig 2.** The Born approximation results for the integral equation method,(- - -) and the IDEA+OP scheme,(-.-.-). The $(\gamma, p)$ experimental data are from refs [17], ($\blacksquare$), [36], ($\blacktriangledown$), and ref [37], (+); the $(\gamma, n)$ data are also from ref [37], ($\times$).

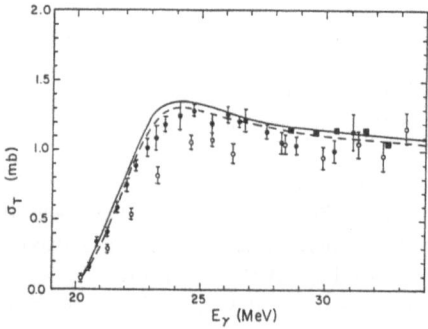

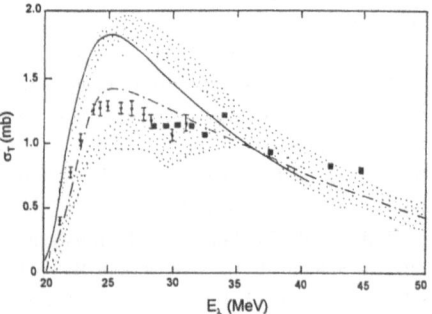

**Fig 3.** Our main $(\gamma, p)$ IDEA+OP results based on the experimental p-wave phase-shifts of Tombrello [35]. The experimental data are from refs [18], ($\bullet$), and [17], ($\blacksquare$).

**Fig 4.** As in Fig 3 but based on the RGM phase-shifts of ref [38], (———), and the GCM phase-shifts of ref [39], (-.-.-). The shaded areas are the ones recommended by Calarco et al [7], for the $(\gamma, p)$ (upper) and $(\gamma, n)$ (lower) data.

# References

[1] A.V.Argon et al, Phys. Rev. **78**, 691(1950).

[2] B. H. Flowers and F. Mandl, Proc. R. Soc. London ,Ser **A206**, 131(1950).

[3] W.E. Meyerhof, M. Suffert, and W. Feldman ,Nucl. Phys. **A148**, 211(1970).

[4] B.M. Berman, D.D.Faul, P. Meyer , and D.L. Olson, Phys. Rev.**C22**, 2273(1980).

[5] L. Ward et al, Phys. Rev. **C24**, 317(1981).

[6] J.R. Calarco et al, Phys. Rev. **C28**, 483(1983).

[7] J.R. Calarco, B.M. Berman, and T.W. Donnelly , Phys. Rev. **C27**, 1866(1983).

[8] J.T.Londergan and C.M. Shakin, Phys. Rev. Lett. **28**, 1729(1972).

[9] A.H.Chung, R.G. Johnson, and T.W. Donnelly, Nucl. Phys. **A235**, 1(1972).

[10] D. Halderson and R.J. Philpott, Nucl. Phys. **A359**, 365(1981); Phys. Rev. **C28**, 1000(1983).

[11] A. Casel and W. Sandhas, in *Proc. Ninth Int. Conference on Few-Body Problem*, Vol I, ed. M.G. Moravsik, Eugene, Oregon, 1980, p33.

[12] W.Böttger,A Casel,and W Sandhas, Phys. Lett. **B92**,11(1980).

[13] A. Casel and W. Sandhas, Czech. J. of Phys., **B36**,300(1986).

[14] P. Grassberger and W. Sandhas, Nucl. Phys. **B2**, 181(1967); E.O. Alt, P. Grassberger, and W. Sandhas, Phys. Rev. **C1**, 85(1970).

[15] S.A. Sofianos, H. Fiedeldey, H Haberzettl, and W. Sandhas, in *Proc. of the Int. Workshop on Microscopic Methods in Few-body Systems*, Kallinin, USSR, 1988.

[16] R. A. Malfliet and J.A. Tjon, Nucl. Phys. **A127**, 161(1969)

[17] R. Bernabei et al, Phys. Rev. **C38**, 1990(1988).

[18] G. Feldman et al, Phys. Rev. **C42**, R1167(1990).

[19] B. Wachter, T. Mertelmeier, and H.M. Hofmann, Phys. Rev. **C38**, 1139(1988).

[20] A.A.Zayats et al, KFTI 91-4, Kharkov Institute of Technology.

[21] M.Unkelbach and H.M. Hofmann, Nucl. Phys. **A549**, 550(1992).

[22] D.P. Wells et al Phys. Rev. **C46**, 449(1992).

[23] M. Fabre de la Ripelle, H. Fiedeldey, and S.A. Sofianos , Phys. Rev. **C38**, 449(1988).

[24] W. Oehm, S.A. Sofianos, H. Fiedeldey , and M. Fabre de la Ripelle, Phys. Rev. **C44**, 81(1991); **C43**, 25(1991); **C42**, 2322(1990).

[25] M.M. Giannini end G. Ricco, La Rivista del Nuovo Cim. **8**, 1(1981).

[26] W. Sandhas, Invited Lecture on Few-Body Problems in Nucl. Physics, 13-16 March 1978, ICTP, Trieste.

[27] E.O Alt, P. Grassberger, and W. Sandhas, Phys. Rev. **C1**, 85(1970).

[28] E.A Bartnik, H. Haberzettl, and W. Sandhas, Phys. Rev. **C34**, 85(1986); T.A Frank, H. Haberzettl, Th. Januschke, U. Kerwath, and W. Sandhas, Phys. Rev. **C38**, 1112(1988).

[29] S.A. Sofianos,N.J.Mc Gurk, and H. Fiedeldey ,Nucl. Phys. **A318**, 295(1979)

[30] S.A. Sofianos, H. Fiedeldey, H. Haberzettl, and W. Sandhas, Phys. Rev. **C26**, 228(1982).

[31] S.A. Sofianos, H. Fiedeldey , and W. Sandhas, Phys. Rev. **C32**, 400(1985).

[32] R.A. Adam, H. Fiedeldey, and S.A. Sofianos, J.Phys. **G17**, 1157(1991).

[33] N.W. Schellingerhout, J.J. Schut, L.P. Kok, Phys. Rev. **C46**, 1192(1992).

[34] P.S. Podmore and H.S. Sherif, in *Few-Body Problems in Particle Physics*, ed. by R.J.Slobodrian et. al. (Laval Univ. Press., Quebec, Canada, 1975), p517; H.S. Sherif, Phys.Rev. **C19**, 1649(1979).

[35] T.A. Tombrello, Phys. Rev. **138**, B40(1965); **143**, 772(1966).

[36] R.I. Jones, PhD thesis, Virginia Polytechnic Inst. and State Univ., Blacksburg, Virginia, USA, 1988.

[37] A.N.Gorbunov, Proc. of the P.N. Lebedev Physics Institute, **71**, 1(1976).

[38] I. Reichstein, D.R.Thompson, and Y.C. Tang, Phys. Rev. **C3**, 2139(1971).

[39] H. Furutani, H. Horiuchi, and R. Tamayaki, Progr. of Theor. Phys., **62**, 981(1979).

Few-Body Systems, Suppl. 7, 201—203 (1994)

# NUCLEAR TRANSITIONS IN MUONIC MOLECULES

V.B.Belyaev[1], H.Fiedeldey[2], S.A.Rakityansky[1],

and S.A.Sofianos[2]

[1] JINR,Dubna,Russia,   [2] UNISA,Pretoria,South Africa

### ABSTRACT

A scheme, analogous to the Linear Combination of Atomic Orbitals (LCAO), is proposed to calculate rates of reactions of nucleon-group transfer between nuclei confined in muonic molecules. As an example the rates of the reactions $d - \mu -^7 Be \rightarrow p - \mu -^8 Be(2+)$ and $p - \mu -^{10} Be \rightarrow t - \mu -^8 Be$ are estimated.

The existence of muonic molecules gives us an unique opportunity to investigate various nuclear reactions at extremely low energies which can never be achieved in collision experiments. Being confined to a rather small area for relatively long time, nuclei can penetrate through the Coulomb barrier and interact via strong forces.

A question arises : what kind of the reactions is more suitable for to be investigated by means of the muonic systems? Obviously, these are the reactions with small values of the momentum transfer, because they can take place at large distances.

Comparing the spectra of light nuclei, one can find a number of cases where two different reaction thresholds of a same nuclear system are very close on the nuclear energy scale. For example, the thresholds of $d +^7 Be$ and $p +^8 Be(2+)$, $p +^{10} Be$ and $t +^8 Be$ are separated from each other by only 44 Kev and 4.5 Kev respectively. Thus, the reactions $d +^7 Be \rightarrow p +^8 Be(2+)$ and $p +^{10} Be \rightarrow t +^8 Be$ are the possible candidates for investigation in the three-body resonances $d - \mu -^7 Be$ and $p - \mu -^{10} Be$.

In this report we present an estimate of the reaction rates for these nuclear transitions, treating the muonic resonances as stable molecular four- and six-body systems $(n, p,^7 Be, \mu)$ and $(p, n, n, \alpha, \alpha, \mu)$ respectively.

Our calculations are based on a LCAO-like ansatz [1] for the wave function, where as the analogs of the "atomic orbitals" we use the wave functions of the nuclei. Such approximation seems to be justified because the sizes of the "orbitals" (nuclei) of $\sim 10^{-13}$cm are much smaller than the molecular sizes of $\sim 10^{-11}$cm.

These "orbitals" $\Psi_1$ and $\Psi_2$ are eigenfunctions

$$H_i \Psi_i = E_i \Psi_i, \quad i = 1, 2$$

of the Hamiltonians $H_1$ and $H_2$, describing initial and final states of the reactions, with the strong forces between the nuclei being switched off. In terms of $H_i$ the total Hamiltonian can be represented in the two equivalent ways

$$H = H_1 + V_1^s, \qquad H = H_2 + V_2^s,$$

where $V_1^s$ and $V_2^s$ are the strong interactions between the initial and final pairs of nuclei confined in the molecule. In the absence of $V_i^s$ the nuclear reactions $a + b \rightarrow c + d$ could not take place and the initial and final states

$$\Psi_1 = \Psi_1^{mol} \Psi_a^{nucl} \Psi_b^{nucl},$$

$$\Psi_2 = \Psi_2^{mol} \Psi_c^{nucl} \Psi_d^{nucl}$$

could not mix to each other. When $V_i^s$ are switched on, the transitions become possible, and we look for a solution of the full Schroedinger equation

$$H\Psi = E\Psi$$

in the form of a linear combination

$$\Psi = C_1\Psi_1 + C_2\Psi_2$$

in the spirit of the LCAO. As a result our problem is reduced to the simple system of linear equations for $C_1$ and $C_2$:

$$C_1(< \Psi_1, H, \Psi_1 > -E) + C_2(< \Psi_1, H, \Psi_2 > -E < \Psi_1, \Psi_2 >) = 0,$$

$$C_1(< \Psi_2, H, \Psi_1 > -E < \Psi_2, \Psi_1 >) + C_2(< \Psi_2, H, \Psi_2 > -E) = 0.$$

Then , the transition probability $P$ is

$$P = |< \Psi_2, \Psi >|^2 = |C_1 < \Psi_2, \Psi_1 > +C_2|^2$$

To perform the calculations we need the overlap integral $< \Psi_1, \Psi_2 >$ and the matrix elements $< \Psi_i, V_j^s, \Psi_k >$, $i, j, k = 1, 2$. To evaluate them we used some simple wave functions of the nuclei $d$ and $t$, and alpha-cluster form for the wave functions of $^8Be$ and $^{10}Be$ nuclei. In the case of the transition $(d - \mu - ^7 Be) \rightarrow (p - \mu - ^8 Be(2+))$ we have assumed $(n - ^7 Be)$ two-body model for $^8Be^*(2+)$ state. Correspondingly, $V_i^s$ in all the cases were sums of the two-body $N - N$, $N$–alpha, and $N - ^7 Be$ strong potentials which describe the low– energy data [2-6].

These matrix elements also include $\Psi_i^{mol}$ which, by definition, are pure Coulomb three-body $(a - \mu - b)$ wave- functions. In our problem muon serves only to bind the nuclei. Hence, it is reasonable to consider $\Psi_i^{mol}$ as a two- body $(a - b)$ wave-function depending only on distance between nuclei $a$ and $b$ which interact to each other via an effective molecular potential generated by the Coulomb repulsion and attraction to the muon space-charge.

For the system $(p - \mu - {}^{10}Be)$ and $(t - \mu - {}^8Be)$ we have constructed such a potential, using the calculated [7] binding energies of other mesomolecules ( having the same nucleus charges ) to find the potential parameters. Then we performed calculations of the matrix elements with the obtained $\Psi_i^{mol}$.

Meanwhile, for the reaction $(d - \mu - {}^7Be) \rightarrow (p - \mu - {}^8Be)$ we applied [8] an additional approximation consisting in taking $|\Psi^{mol}(0)|^2$ (the zero is internuclear distance) out of the integrals, which is reasonable because overlapping of the nuclear functions vanishes rapidly with increase of the internuclear distance. As value of $|\Psi^{mol}(0)|^2$ for that reaction we used an order-of-magnitude estimate based on knowledge of typical energy gap between the ground and first P-wave states [9].

In this way we have obtained the transition probabilities $P \sim 10^{11} s^{-1}$ for $(d - \mu - {}^7Be) \rightarrow (p - \mu - {}^8Be^*(2+))$ and $P \sim 10^3 s^{-1}$ for $(p - \mu - {}^{10}Be) \rightarrow (t - \mu - {}^8Be)$.

Accurate calculations first of all demand more refined wave functions of the molecules. There is also a perceptible dependence on the model used for the description of the strong forces, particulary on the N-alpha potential.

## References

1. Schutte, C.J.H.: The wave mechanics of atoms,molecules and ions. London, Edward Arnold (Publisher) Ltd.: 1968

2. Malfliet,B.A., Tjon,J.A.: Ann.Phys.**61**, 425(1970)

3. Satchler,G.R., Owen,L.W., Elwyn,A.J., Morgan,G.R., Walter,R.L.: Nucl. Phys.,**A112**, 1(1968)

4. Dubovichenko,S.B., Dzhazairov-Kakhramanov,A.V.: Sov.J. Nucl.Phys., **51**, 971(1990)

5. Danilin,B.V., Zhukov,M.V., Ershov,S.N., Gareev,F.A., Kurmanov,R.S., Vaagen,J.S., Bang,J.M.: Phys.Rev., **C43**, 1541 (1990)

6. Robertson,R.G.H. : Phys.Rev., **C7**, 543(1973)

7. Belyaev,V.B., Decker,M., Haberzettl,H., Khaskilevitch,L.J., Sandhas,W.: Few-Body Systems, Suppl., **6**, 332(1992)

8. Belyaev,V.B., Fiedeldey,H., Sofianos,S.A.: Yad.Fiz., **56**, 61 (1993)

9. Common,A.K., Martin,A., Stubbe,J.: Comm.Math.Physics, **134**, 509(1990)

Few-Body Systems, Suppl. 7, 204—207 (1994)

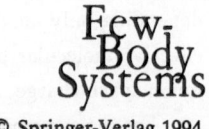

Few-
Body
Systems
© Springer-Verlag 1994
Printed in Austria

# THEORY OF FORMATION AND DECAY OF METASTABLE STATES OF HADRONIC HELIUM ATOMS

G.Ya. Korenman and S.N. Yudin

Institute of Nuclear Physics, Moscow State University,
Moscow 119899, Russia

### Abstract

Theory of mesic atom formation is applied to obtain primary populations of recently-discovered metastable states of hadronic He atoms. The calculated primary populations of the long-lived kaonic states in He are in agreement with the experimental data. Life times of these states exceed the measured value, suggesting a collisional quenching of metastable states. For antiprotonic He, preliminary estimations of population are given.

Metastable states of hadronic He atoms were discovered recently by T. Yamazaki et al.[1-4]. They revealed one group of such $\pi^-$ states, two groups of $K^-$ states, and four groups of $\bar{p}$ states. The populations and life times of these groups were determined. The existence of metastable states of hadronic He atoms has long been conjectured [5,6]. It has been shown, that the states of neutral hadronic He atoms are metastable when the angular momentum $l = n - 1$ (circular orbits) and the principal quantum number $n > (\mu/m)^{1/2}$, where $\mu$ and $m$ are reduced masses of the hadron and the electron. Auger transitions from these states are strongly suppressed due to high multipolarity, while radiative $E1$-transitions are rather slow. However, the available calculations [6,7] did not give the primary level populations, which are of key importance for the experimental study of the yields and the time distributions of the decay and delayed absorption/annihilation products. In this paper we show, that the primary populations of metastable hadronic states can be obtained using the theory of mesic atom formation [8,9].

As follows from [9], the primary populations of hadronic He atoms are given by the equation

$$\rho_{nl} = \frac{1}{I_0}\theta(E_{nl} - E_0)\theta(E_0 + I_0 - E_{nl}) \mid \frac{dE_{nl}}{dn} \mid \theta(l_n - l)\frac{2l + 1}{(l_n + 1)^2}, \tag{1}$$

where $E_{nl}$ is the energy of the $nl$-state of the hadronic atom, $E_0$ and $I_0$ are the energy and the ionization potential of the He atom, $l_n = \min\{L_n, (n - 1)\}$, and $L_n$ is the maximum of all $l$ values satisfying the condition,

$$(l + 1/2)^2 \le 2\mu R_0^2(E_{nl} - E_0 - V_0). \tag{2}$$

Here $R_0$ and $V_0 = -W_0$ are the radius of inelastic hadron-atom interaction and potential energy at the $R_0$ distance. To calculate primary populations of metastable states, it is necessary first to find energy levels and wave functions of states with large angular momenta, to calculate transition rates from the states, and to elucidate which states are metastable. We present the results for kaons and preliminary estimates for antiprotons.

For a neutral hadronic helium atom we use as variational wave function the product of an electron $1s$-function with an effective charge and of a meson function with radial part of the form

$$R_{nl}(r) = A \exp(-\beta r) r^l P_k(r), \tag{3}$$

where $\beta$ is a variational parameter, and $P_k(r)$ is a polynomial of degree $k = n - l - 1$. The energies obtained depend only slightly on the angular momentum.

The life time of the hadronic $nl$-state is determined by radiative and Auger transition rates. Radiative $E1$-transition rates were calculated taking into account the change of the electron wave function with hadron transition. Auger transition rates were calculated in the framework of perturbation theory. For the initial states we use the variational wave functions, while the final states were described by H-like wave functions of a hadron in the field of charge $Z = 2$ and Coulomb continuum functions of an electron in the field of charge $(Z - 1)$. Auger transitions are allowed energetically between the states, for which the final electron energy is non-negative,

$$\varepsilon = E_{nl} + \mu Z^2 / (2n_f^2) \geq 0 \tag{4}$$

The minimal change of the principal quantum number ($\Delta n = n - n_f$) of the kaon is equal to unity for small $n$, but it grows rapidly with increasing $n$. The minimal multipolarity of an Auger transition is $L_0 = \Delta n - k$, if $L_0 < l$.

The calculated rates of radiative and Auger transitions of kaonic helium are shown in Figure 1 for $l = n - 1$ (solid lines) and for $l = n - 2$ (dashed lines). Radiative transition rates fall monotonously with increasing $n$. The Auger transition rates behave non-monotonously. If the minimal multipolarity $L_0$ stays invariable, then the Auger transition rate changes weakly with increasing $n$, but the rate decreases sharply (by two orders at least) when $L_0$ increases one unit.

The criterion for the metastability of a state is connected with the possibility to observe decay and/or delayed absorption of a hadron. It is seen from Fig.1 that circular orbits with $n \geq 27$ ($L_0 \geq 4$) and near-circular orbits with $n \geq 29$ ($L_0 \geq 4$) are metastable. Auger transition rates for these states are below that of radiative transitions, and total transition rates are less than the kaon decay rate $\lambda_0 = 0.808 \cdot 10^8 \ s^{-1}$. For the lower orbits the transition rates are more than $\lambda_0$.

For the calculation of primary populations $\rho_{nl}$ of metastable states we use Eqs. (1) and (2), together with the energy levels obtained above. The derivative in Eq. (1) was estimated, neglecting by a weak dependence of energy on $l$. The calculations were carried out with two sets of parameters [9] : (a) $R_0 = 1.0$, $W_0 = 0.184$ and (b) $R_0 = 0.88$, $W_0 = 0.325$ (both values in a.u.). Results for kaonic helium metastable states are given in Table 1 (upper and lower lines for each $l$ correspond to "a" and "b" parameter sets). Non-populated metastable states are not included in the table.

Table 1. Primary populations of metastable
states of kaonic helium ($10^{-3}$)

| n | 31 | 32 | 33 | 34 | 35 | 36 | 37 | 38 |
|---|---|---|---|---|---|---|---|---|
| l=n-1 | 4.8 | 3.4 | 2.5 | 1.9 | 1.6 | 1.3 | 0 | 0 |
| | 0 | 3.6 | 2.6 | 0 | 0 | 0 | 0 | 0 |
| l=n-2 | 5.1 | 3.7 | 2.8 | 2.1 | 1.7 | 1.4 | 1.2 | 1.1 |
| | 5.2 | 3.9 | 2.9 | 2.3 | 1.8 | 0 | 0 | 0 |

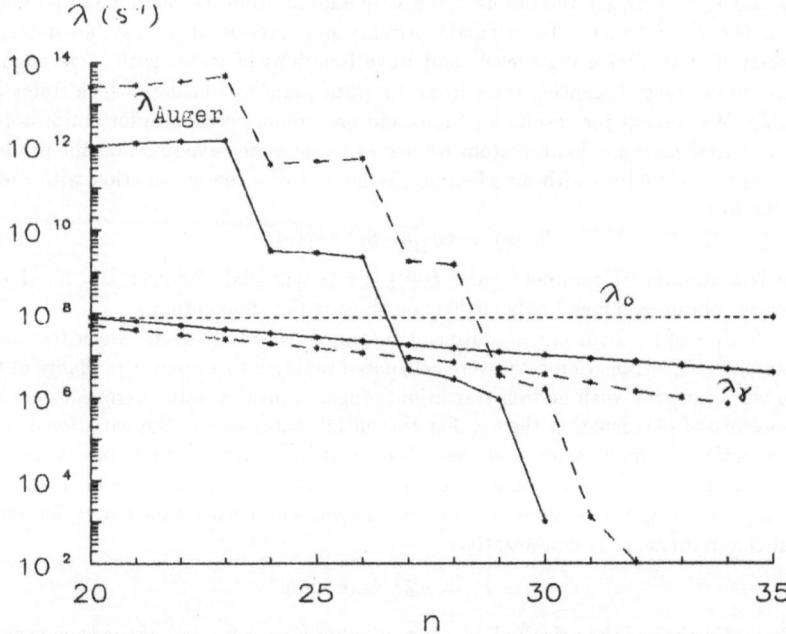

Fig.1. Kaonic helium transition rates for $l = n - 1$ (solid lines) and $l = n - 2$ (dashed lines).

The total population of the states account for 3.5% (per stopped kaon) in the variant "a", and 2.2% in the variant "b". The first figure is in agreement with the value of the population of the longest-lived kaonic state group [4] 4.1 ± 0.3 %. The second variant of the calculations is in better agreement with the initially published value [1] of the stopped $K^-$ decay fraction $f = 1.9 \pm 0.3$ %. The calculated individual life times of these states are more than 100 ns, whereas the measured life time of the group is about 60 ns. It can be considered as evidence for a quenching of metastable kaonic states due to collisions with He atoms.

According to the paper [4], there is another group of metastable kaonic states, which has life time 1.8±0.2 ns and population 1.3±0.3%. One could suppose from an inspection of Fig. 1, that this component is related to the circular states with $n$=24, 25, 26 and to the near-circular states with $n$=27 and 28. However, according to our calculations these states are not populated during the atomic capture of a kaon. To explain the experimental data on this component, there is a need to examine states with lower angular momenta ($l < n - 2$) and, possibly, to take into account the diffuseness of the population boundary on $E$ and $l$ in eq. (1).

Most interesting experimental results are obtained for $\bar{p}$He metastable states [3, 10]. The revealing of four groups of states with very different life times calls for extensive calculations including a wide range of highly-excited states. The calculations for antiprotons in the framework of the foregoing approach are in progress. Today we have an estimate of the total populations of metastable states with $l = n - 1$ and $n - 2$, which account for

1.0 % and 1.5 %, respectively, with the parameter set "a". The experimental population of the longest-lived group of $\bar{p}$-states is 1.85±0.14 %. The full results of the calculations for $\bar{p}$He metastable states will be published elsewhere.

In conclusion, the theory gives a satisfactory picture of the formation of the longest-lived groups of hadronic helium metastable states. Yet, problems of the more short-lived components as well as of collisional quenching of metastable states are still open.

## REFERENCES

1. Yamazaki, T., Aoki, M., Iwasaki, M. et al.: Phys. Rev. Lett. **63**, 1590 (1989).

2. Nakamura, S.N., Iwasaki, M., Outa, H., et al.: Phys. Rev. **A 45**, 6202 (1992).

3. Iwasaki, M., Nakamura, S., Shigaki, K., et al.: Phys. Rev. Lett. **67**, 1246 (1991).

4. Yamazaki, T.: preprint INS-Rep.-952, Univ. of Tokyo, 1992.

5. Condo, G.T.: Phys. Lett. **9**, 65 (1964).

6. Russell, J.E.: Phys.Rev.Lett. **23**, 63 (1969); Phys.Rev. **188**, 187 (1969); Phys.Rev. **A 1**, 721 (1970); Phys.Rev. **A 6**, 2488 (1972).

7. Shimamura, I.: Phys. Rev. **A 46**, 3776 (1992).

8. Cohen, J.S., Martin, R.L., Wadt, W.R.: Phys. Rev. **A 27**, 1821 (1983).

9. Dolinov, V.K., Korenman, G.Ya., Moskalenko, I.V., et al.: Muon Cat. Fusion **4**, 169 (1989).

10. Yamazaki, T., Widmann, E., Hayano, R.S., et al.: Nature **361**, 238 (1993).

Few-Body Systems, Suppl. 7, 208—212 (1994)

# SIGNATURE OF A NARROW πNN-RESONANCE IN THE ENERGY DEPENDENCE OF THE PIONIC DOUBLE CHARGE EXCHANGE*

R. Bilger, H. Clement, K. Föhl, K. Heitlinger, G.J. Wagner
Physikalisches Institut der Universität Tübingen
Auf der Morgenstelle 14, D-72076 Tübingen, Germany

C. Joram, W. Kluge, R. Wieser
Institut für Kernphysik der Universität Karlsruhe
Postfach 3640, D-76021 Karlsruhe, Germany

M.G. Schepkin
Institute of Theoretical and Experimental Physics
B. Cheremushkinskaja 25, Moscow 117259, Russia

R. Abela, F. Foroughi, D. Renker
Paul Scherrer Institut
CH-5232 Villigen PSI, Switzerland

### Abstract

All presently available data on the pionic double charge exchange reaction $(\pi^+, \pi^-)$ on nuclei exhibit a very peculiar, resonance-like energy dependence near $T_\pi \approx 50$ MeV, while the angular distributions behave quite regularly. These features are in accordance with the signature of a narrow resonance in the $\pi$NN-subsystem having $T(J^P) =$ even $(0^-)$ and a mass of 2.065 GeV.

### Introduction

It is well-known that in consequence of the substructure of the nucleon there should be nontrivial, i.e. non-nucleonic resonances in the $B = 2$ system. Although such dibaryon resonances have been predicted [1,2] since long, no unambiguous evidence for their existence has been found up to now, despite a vast number of dedicated experiments.

*This work has been funded by the German Federal Minister for Research and Technology (BMFT) under the contract numbers 06 TÜ 656 and 06 KA 266, and by the DFG (Mu 705/3, Graduiertenkolleg).

Here we present evidence for the existence of a low-lying isoscalar $0^-$ dibaryon resonance embedded in the nuclear medium. Due to its quantum numbers this resonance does not couple to the nucleon-nucleon (NN) channel, where most of the searches have been done, but nearly exclusively to the $\pi$NN-channel [1,2]. Therefore an ideal reaction to look for resonances in this channel is the pionic double charge exchange (DCX) reaction $(\pi^+, \pi^-)$ or $(\pi^-, \pi^+)$ on nuclei. Also, since charge conservation ensures that the reaction takes place on at least two nucleons, the conventional DCX mechanism is basically a 2-step process with correspondingly very small cross sections, which, however, are extraordinarily sensitive to small NN distances [3-5]. The largest DCX cross sections are expected for transitions to double isobaric analog states (DIAT), since in this case there is maximum overlap between initial and final states. On the other hand ground state transitions (GST), in particular on $N = Z$ nuclei, should be strongly suppressed [3] in a conventional purely nucleonic mechanism.

### Experimental Results

Experimentally the situation is quite different in this respect. Moreover all transitions show a very peculiar energy dependence. As an example Fig. 1 shows the energy de-

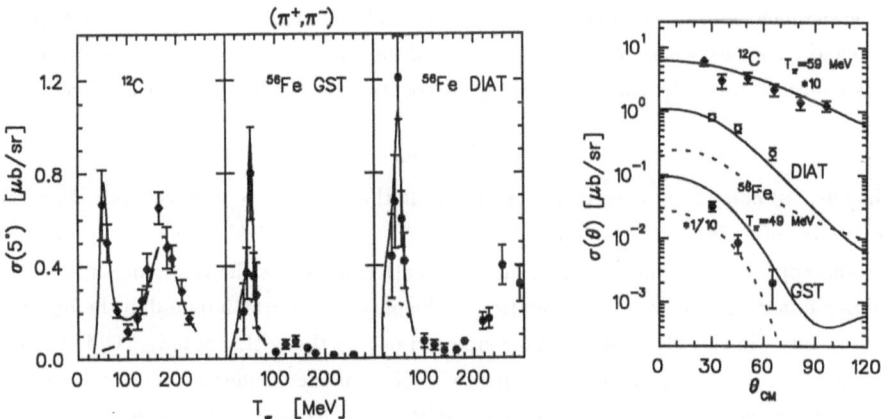

Fig. 1. Excitation function of forward angle ($\Theta = 5°$) cross sections (left) and angular distributions (right) near $T_\pi \approx 50$ MeV for DCX transitions on $^{12}$C and $^{56}$Fe. The solid lines give the full calculations including the $d'$ resonance; the non-resonant part is indicated by the dotted lines and the $\Delta$-excitation in case of $^{12}$C is shown by the dashed lines.

pendence of the forward angle cross sections of the DCX measurements on $^{12}$C [6] and $^{56}$Fe [7,8]. For the latter reaction both GST leading to the doubly closed shell nucleus $^{56}$Ni and DIAT are presented; the data below 100 MeV are from this work, for details see ref. [7]. The data on $^{12}$C exhibit a broad structure near the energy of the $\Delta$, and then again a strong increase in cross section towards $T_\pi = 50$ MeV. For $^{56}$Fe the energy dependence is markedly different at high energies. In the $\Delta$ region the cross section is lower by an order of magnitude relative to the $^{12}$C results, whereas at low energies the cross section again exhibits a clear resonance-like energy dependence, reaching at $T_\pi \approx 50$ MeV a cross section even larger than in case of $^{12}$C. The behaviour in the $\Delta$ region is well understood as a consequence of strong pion absorption and successive double $\Delta$ mechanism [9], the latter showing up markedly only for non-analog transitions.

The strong energy-dependence at low energies is common to all DCX transitions measured so far and clearly signals another resonance. For a discussion of the transitions in $T = 1$ nuclei, as well as in $^{44}$Ca and $^{48}$Ca see ref. [10]. The observed energy behaviour at low energies has missed any satisfactory explanation so far. The dotted lines in Fig. 1 show our calculations within the AGGK concept [3]. The failure to account properly for the observed energy dependence is common to other current theoretical investigations. Plane wave calculations [3,5,11] even predict a cross section minimum around $T_\pi \approx 50$ MeV which is related to the deep minimum in the forward angle single charge exchange cross section caused by the s-p interference at this particular energy.

In contrast to the resonance-like energy dependence the angular distributions at low energies behave quite regularly (see Fig. 1) and thus immediately rule out a pion-nucleus resonance with some spin $J$ as origin of this phenomenon, since then the angular distributions would be characterized by $P_J^2(\cos\Omega)$, which actually is not observed.

### Analysis

In refs. 10 and 11 we have demonstrated that the data are reproduced assuming the existence of a $J^P = 0^-$ resonance — called $d'$ — in the $\pi$NN-subsystem. In analogy to the appearance of the $\Delta$ in pion-nucleus scattering, where the resonance in the $\pi$N-subsystem is smeared out by the motion of the particular nucleon bound in the nucleus, a resonance in the $\pi$NN-subsystem is smeared out by the centre-of-mass (CM) motion of the NN-pair within the nucleus. Since a $\pi$NN-resonance demands two closely spaced identical nucleons, we have to assume that these nucleons have to be in a relative s-state with spin $S = 0$. Calculations with realistic wave functions show that most of the observed widths (FHWM = 20 - 30 MeV) of the resonance-like structure in low-energy DCX can be associated with the CM-motion of the NN-system, which gives a smearing of about 18 (12) MeV in light (heavy) nuclei. Hence $\Gamma$ is small, in the order of a few MeV.

For a detailed analysis of the $d'$ resonance we have to account properly for the interfering nonresonant "background"(dotted lines in Fig. 1), and in case of $^{12}$C for the low-energy tail of the $\Delta$-resonance.

The measured angular distributions uniquely determine the spin of $d'$ to be zero, for $J > 0$ much steeper angular distributions are predicted [10]. The solid lines in Fig. 1

as well as in refs. [10,12] show, without any arbitrary normalization, the final results of our analysis, which gives $\Gamma_+ = \Gamma_- \approx 0.2$ MeV, $\Gamma \approx 5$ MeV, $M_R \approx 2.065$ GeV as well as $J^P = 0^-$ for $d'$. We emphasize that with the same set of resonance parameters we are able to provide a good description [10,12] for all presently available low-energy DCX-data.

<u>Discussion</u>

The resonance parameters as given above reflect the $d'$ resonance embedded in the nuclear medium. Hence both the total resonance energy $M_R$ and the total width $\Gamma$ may be affected by medium effects giving rise to binding energy and spreading width of $d'$. Since $\Gamma$ is in the order of a few MeV, $d'$ must have even isospin, otherwise decay into NN would be allowed, causing a much larger width for $d'$. If existing in vacuum, $d'$ can decay therefore only into the three $\pi$NN channels, and with a tiny probability also by $\gamma$ emission. Thus we have $\Gamma_{d'} = 3\Gamma_+$ and most of the observed width for $d'$ within the nuclear medium has to be attributed to spreading. Such a decay mechanism in the medium would be, e.g., $Nd' \to 3N$.

A low-mass dibaryon state with $J^P = 0^-$ would, in fact, be in agreement with predictions based on QCD-string models, if $d'$ has isospin $T = 0$. These models predict an isoscalar triplet of states $0^-, 1^-$ and $2^-$ in this very energy range, all other NN-decoupled $6q$-states being substantially heavier[1,2]. In those models $d'$ is basically composed of a diquark with angular momentum $l = 1$ relative to a four-quark cluster with $S = 1, T = 0$ (Fig. 2). Since the mass of $d'$ is close to the $\pi$NN-threshold, also the tiny width $\Gamma_{d'}$ appears to be very reasonable.

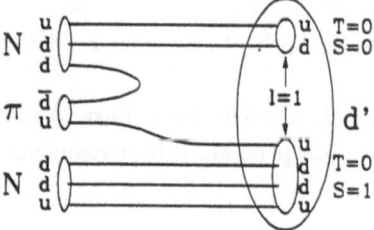

Fig. 2. Interpretation of the $\pi$NN resonance as a dibaryon state with $T(J^P) = 0(0^-)$.

Not many possibilities seem to be left for an independent test of the existence of $d'$. One possibility would be, in principle, $d + \gamma \to d' \to pp\pi^-$. However, the $\gamma$-branch is tiny. Alternatively we suggest [13] to search for $d'$ in the reaction $pp \to d'\pi^+ \to pp\pi^-\pi^+$, which experimentally gives the favorable situation of four charged particles in the exit channel. Estimates for the resonance contribution near threshold to this reaction give [13] cross sections in the order of nb-$\mu$b, which render measurements

212

feasible. Experimental studies of this reaction are currently being planned at COSY. First preliminary results from such an experiment at ITEP are encouraging [14].

## References

1. Mulders, P.G., Aerts, A.T., de Swarts, J.J.: Phys. Rev. **D21**, 2653 (1980)
2. Kondratyuk, L.A., Martemyanov, B.V., Schepkin, M.G.: Sov. J. Nucl. Phys. **45**, 776 (1987)
3. Auerbach, N., Gibbs, W.E., Ginocchio, J.N., Kaufmann, W.B.: Phys. Rev. **C38**, 1277 (1988); Comm. Nucl. Part. Phys. **20**, 141 (1991)
4. Bleszynski, E., Bleszynski, M., Glauber, R.J.: Phys. Rev. Lett **60**, 1483 (1988)
5. For a survey see Clement, H.: Prog. Part. Nucl. Phys. **29**, 175 (1992) and references therein
6. Faucett, J.A. Rawool, M.W., Dhuga, K.S., Zumbro, J.D. Gilman, R., Fortune, H.T., Morris, C.L., Plum, M.A.: Phys. Rev. **C35**, 1570 (1987)
7. Bilger, R., Barnett, B.M., Clement, H., Krell, S., Wagner, G.J., Jaki, J., Joram, C., Kirchner, T., Kluge, W., Metzler, M., Wieser, R., Renker, D.: Phys. Lett. **269**, 247 (1991)
8. Seidl, P.A., Bryan, M.A., Burlein, M., Burleson, G.R. Dhuga, K.S., Fortune, H.T., Gilman, R., Green, S.J., Machuca, M.A., Moore, C.F., Mordechai, S., Morris, C.L., Oakley, D.S., Plum, M.A., Rai, G., Smithson, M.J., Wang, Z.F., Watson, D.L., Zumbro, D.J.: Phys. Rev. **C42**, 1929 (1990) and references therein
9. Gilman, R., Fortune, H.T., Johnson, M.B., Siciliano, E.R., Toki, H., Wirzba, A., Brown, B.A.: Phys. Rev. **C34**, 1895 (1986) and references therein
10. Bilger, R., Clement, H.A., Schepkin, M.G.: Phys. Rev. Lett. **71**, 42 (1993)
11. Siciliano, E.R., Johnson, M.B., Sarafian, H.: Ann. Phys. **203**, 1 (1990)
12. Bilger, R., Clement, H., Föhl, K., Heitlinger, K., Joram, C., Kluge, W., Schepkin, M., Wagner, G.J., Wieser, R., Abela, R., Foroughi, F., Renker, D.: Z. Phys. **A343**, 491 (1992)
13. Schepkin, M., Zaboronsky, O., Clement, H.: Z. Phys. **A345**, 417 (1993)
14. Vorobyev, L., Pivnuyk, N., Schepkin, M.: private communication

Few-Body Systems, Suppl. 7, 213—216 (1994)

© Springer-Verlag 1994
Printed in Austria

# EXOTIC DIBARYONS CORRELATED WITH EXPERIMENT

*Earle L. Lomon*
*Center for Theoretical Physics*
*Laboratory for Nuclear Science*
*and Department of Physics*
*Massachusetts Institute of Technology*
*Cambridge, MA 02139-4307*

**Abstract**

Recent data shows a structure in $C_{nn}$ closely resembling the effect of an exotic $^1S_0$ resonance at 2.7 GeV/$c^2$ predicted in 1987. The hybrid model used in the prediction joins the confined and deconfined regions by the $R$-matrix method. Odd parity exotic resonances are also predicted. The preliminary $A_y$ data indicates the presence of such resonances, at the predicted mass range or lower.

It has been predicted that exotic states of six $u$ and $d$ quarks are observable in nucleon-nucleon scattering at nucleon-nucleon masses of about 2.6 GeV/$c^2$ and more [1–3], although some earlier calculations implied a lower mass [4]. Those calculations which yield lower masses may not adequately treat the repulsive confining effects [2] or the clustering into hadrons at longer range. In reference [1] these effects are taken into account by the $R$-matrix boundary condition connecting the deconfined region to the outer region of hadron exchange interactions. Recent preliminary data [5] indicates such structure exists in $A_{00nn}(\equiv C_{nn})$ as shown in Fig. 1, and perhaps also in $A_{00n0} = A_{000n}(\equiv P$ or $A_y)$ as shown in Fig. 2 ($T_{kin} = 2.0$ GeV corresponds to 2.7 GeV/$c^2$ mass). The curves in Fig. 1 are the predictions made in 1987 [1] for c.m. angles of 63° and 90°. The hybrid model of Ref. 1 uses the R-matrix method to connect the Cloudy Bag Model $[q(1s_{\frac{1}{2}})]^6$ quark interior region with the coupled channel, meson-exchange potential exterior region, separated at 1.05 fm. At 90°, where the data shows strong structure, the prediction is in good agreement with the width and size of the structure, and is only 30 MeV/$c^2$ low in mass and about 15% high in absolute value compared with the data. $A_{00nn}$ is purely singlet at 90° consistent with the $^1S_0$ character of the exotic state. A phase shift analysis based on earlier data [6] finds a $^1S_0$ resonance at this energy. The data at 63° is more scattered, but statistically requires a substantial drop over a small energy range. The prediction fits the position and absolute magnitude as well as at 90°, but has a larger drop than the data. The poorer agreement at 63° may be due to the effect of interference with triplet partial waves. The 1987 model [1] did not include

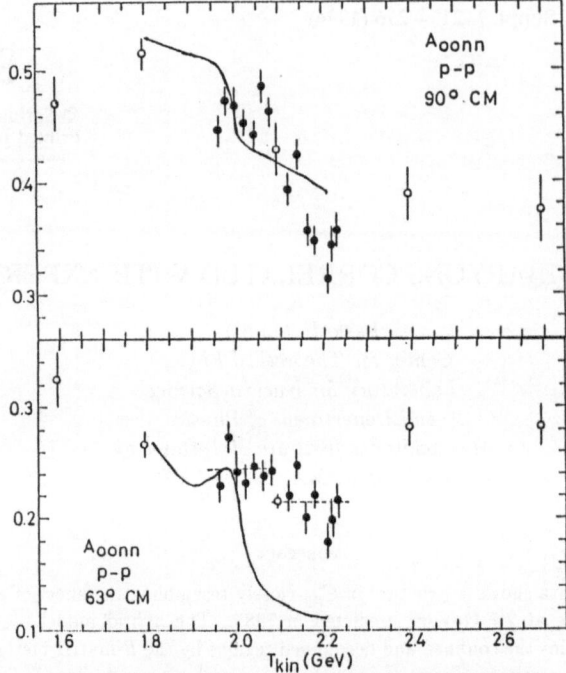

Fig. 1. $A_{00nn}(pp)$. The filled circles are recent preliminary data [5] and the open circles are 1987 Saturne results. The dashed curves indicate the drop in average value from one region to another. The solid curves are the predictions of the $R$-matrix hybrid model [1].

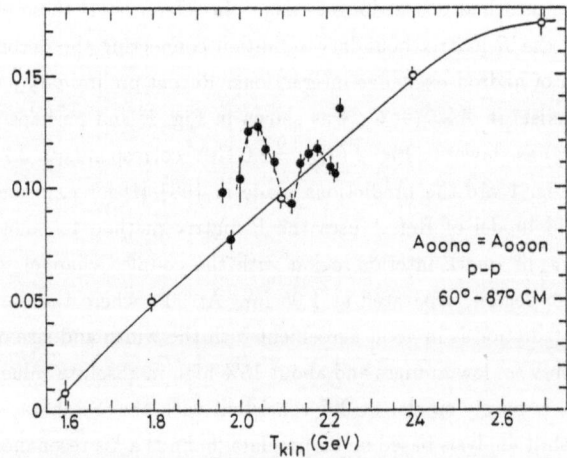

Fig. 2. $A_{00n0} = A_{000n}(pp)$ data. Filled circles are recent preliminary data [5] and the open circles are 1987 Saturne results. The solid and dashed curves are smooth fits to the old and new data respectively.

the nearby odd-parity exotic states [7] which will affect the behavior of the triplet phase parameters in this region.

Fig. 3 displays the spectrum of two-nucleon exotics predicted by the $R$-matrix method [7] for both the above $[q(1s_\frac{1}{2})]^6$ even parity quark states and the $[q(1s_\frac{1}{2})]^5 q(1p_\frac{1}{2})$ odd parity states. The observable $\sigma P$ depends only on triplet amplitudes. Unless the structures in $P$ indicated by the preliminary results of Fig. 2 are only reflections of structure in $\sigma$, odd parity resonances are required. As seen in Fig. 3 the lowest predicted odd parity resonances are about 200 MeV/$c^2$ higher than the first energy structure in $P$ indicated by the data. However an exotic of the $[q(1s_\frac{1}{2})]^5 q(1p_\frac{3}{2})$ quark configuration may be closer to the first experimental structure in $P$ because the energy eigenvalue of a $1p_\frac{3}{2}$ Dirac quark in a 1.05 fm cavity is about 100 MeV less than that of $1p_\frac{1}{2}$ quark. This configuration is expected to produce exotics as low as approximately 2.74 GeV/$c^2$ ($T_{\text{kin}} = 2.12$ GeV). This is still about 100 MeV above the first experimentally indicated structure, but it, and the lowest energy exotics due to $1p_\frac{1}{2}$ state quark, may correspond to the higher energy experimental structures. As indicated in Fig. 3 the odd parity exotics are 10-90 MeV/$c^2$ apart, consistent with the oscillations in the preliminary $P$ data. The narrow width of the structures is implied by the small phase space for decay into a nucleon and an odd-parity nucleon isobar, because of the higher masses of those isobars.

Amplitude analyses [8] have indicated the possibility of very rapid changes of the phase of some of the five complex phases at the angle and energy of the apparent structure in $P$. This implies that the trajectories of those amplitudes pass near the origin at that energy. It has been suggested that this effect, rather than a resonance, is the cause of at least some of the structure in $P$. However for a smooth trajectory the magnitude and phase of an amplitude are related so as to cancel the structure in $P$. Whether or not the amplitude phases have the large variations indicated in Ref. 8, if the structures indicated by Fig. 2 exist they require resonant-like structures in the amplitudes. The effects of exotic odd parity resonances are produced by the interference of a resonant triplet partial wave with other triplet partial waves. For instance, an inelastically resonant $^3P_1$ interfering with an elastic $^3P_2$ gives rise to a contribution to $\sigma p = \text{Re}\,(a^* e)$ at $\theta_{cm} = 45°$ of

$$-\frac{6}{k^2}\,\text{Im}\left[\left(\eta_1\, e^{-2i\delta_1} - 1\right) e^{i\delta_2}\sin\delta_2\right]$$

and the resonant structure of $\eta_1$ and/or $\delta_1$ will appear in the observable.

The singlet $^1S_0$ and $^1D_2$ structures have also shown up, with less statistical significance, in $\Delta\sigma_L(pp)$ [9]. Further $pp$ and $np$ scattering experiments can do much to determine the existence and character of the states involved. In addition, these exotic states can be produced by the absorption of photons or pions on deuterium or heavier nuclei. An energy transfer of 750 MeV or more is required. The well understood electromagnetic probe has the advantage over pions of exciting $I = 0$ as well as $I = 1$ states. In fact the inverse reaction, $pp \rightarrow d\pi^-$ has already produced evidence of structure at 2.7 GeV/$c^2$ dibaryon mass [10].

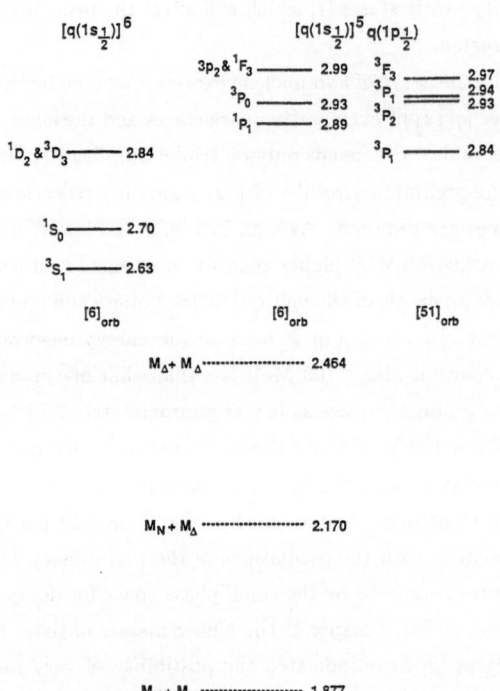

Fig. 3. Nucleon-nucleon exotic states predicted by the $R$-matrix hybrid model [1]. The masses of even parity states are displayed in the left column and those of odd parity states to the right, by solid lines. Two baryon thresholds are denoted by centered, dashed lines. Masses are given in units of GeV/$c^2$.

[1] P. Gonzalez, P. LaFrance and E. Lomon, *Phys. Rev.* **D35**, 2142 (1987).

[2] Yu. S. Kalashnikova, I. M. Narodetskii and Yu. A. Simonov, *S. J. Nucl. Phys.* **46**, 689 (1987).

[3] C. W. Wong, *Prog. Part. Nucl Phys.* **8**, 223 (1982).

[4] A. Th. M. Aerts, P. J. G. Mulders and J. J. de Swart, *Phys. Rev.* **D17**, 260 (1978); M. Oka and K. Yazaki, *Phys. Lett.* **90B** 41 (1980); T. Goldman et. al., *Phys. Rev.* **C39** 1889 (1989).

[5] J. Ball et. al., CEN/Saclay report, Feb. 1993.

[6] N. Hoshizaki et. al., *Proc. of Workshop on Experiments by KEK Polarized Proton and Electron Beams* [KEK, Iburaki, Japan, Oct. 1988].

[7] E. Lomon, *J. Physique* **58** Supple. Colloque C6, 363 (1990).

[8] C. D. Lac et. al., *J. Phys. France* **51**, 2689 (1990).

[9] I. P. Auer et. al., *Phys. Rev. Lett.* **62**, 2649 (1989).

[10] R. Bertini et. al., *Phys. Lett.* **203**, 18 (1988).

Few-Body Systems, Suppl. 7, 217—220 (1994)

Few-
Body
Systems
© Springer-Verlag 1994
Printed in Austria

# ALPHA-PARTICLE BINDING ENERGIES FOR REALISTIC NUCLEON-NUCLEON INTERACTIONS

H. Kamada and W. Glöckle

Institut für theoretische Physik II
Ruhr-Universität Bochum, D-44780 Bochum, Germany

**Abstract** The 4N-Yakubovsky equations for the $\alpha$-particle ground state have been solved rigorously for various realistic NN potentials.

How well describe realistic NN forces the motion of nucleons in nuclei and nuclear reaction? The triton is underbound by about 0.4-0.9 MeV (taking charge-independence-breaking (CIB) into account). That spread reflects the off-shell uncertainty in the present day NN potentials. This underbinding is only about 1-2% in relation to the typical potential energy of $\approx 50$MeV. Three-nucleon scattering based on the same realistic NN potentials shows an overall good agreement with the data [1]. It appears natural to test that simple picture further in the $\alpha$-particle, whose ground state is especially dense. Such a study has been undertaken with great care by the Argonne-Urbana group[2] using variational Monte Carlo techniques and also by the exact Greens Function Monte Carlo method[3]. We attack that 4-body problem by directly solving the Schödinger equation in the form of 4N-Yakubovsky [4](Y) equations. To that aim we developed techniques[5] in momentum space and in a partial wave decomposition. The Y-equations read

$$\psi_1 = -G_0 \ T \ P \ P_{34} \ \psi_1 + G_0 \ T \ P \ \psi_2 \tag{1}$$

$$\psi_2 = G_0 \tilde{T} \tilde{P} \ (1 - P_{34}) \ \psi_1 \tag{2}$$

In that set occur permutation operators $P_{34}$, $P \equiv P_{12} \ P_{23} + P_{13} \ P_{23}$ and $\tilde{P} \equiv P_{13} \ P_{24}$, the free 4-body propagator $G_0$ and the operators T and $\tilde{T}$ describing the full dynamics within a 3-body subcluster and for two noninteracting 2-body clusters, respectively. They obey

$$T = t + T \ P \ G_0 \ t \tag{3}$$

$$\tilde{T} = t + \tilde{T}\,\tilde{P}\,G_0\,t \qquad (4)$$

where t is the off shell- t-matrix. The 4-body wavefunction composed of 18 Y-components is

$$\Psi = (1 + P - P_{34}\,P + \tilde{P})(\psi_1 - P_{34}\,\psi_1 + \psi_2) \qquad (5)$$

The Y-components $\psi_1$ and $\psi_2$ stand for the 3+1 and 2+2 partitions of 4 bodies, respectively. In momentum space and partial wave decomposed Equs (1-2) are coupled integral equations in 3 variables ( the 3 relative momenta of two types). We kept the NN force to be different from zero for NN total angular momenta $j \leq 3$. This required for an accurate result to take 190 channels ( different orbital angular momenta, spin and isospin combinations ) into account. Since the high force components with $j \geq 4$ contribute to the triton binding energy less than 20 keV, one estimates from the correlation between the $^4He$ and $^3H$ binding energies, discussed below, that the $j \leq 3$ truncation is accurate within about $5 \times 20 = 100$ keV. We show our results for the current realistic NN forces in table 1. For the sake of comparison we also include $^3H$ binding energies (34 channels results) and the deuteron d-state probabilities $P_d$, demonstrating again the well known fact that binding energies increase with decreasing $P_d$. While the first four potentials underbind $^4He$ by about 4 MeV in relation to the Coulomb corrected experimental value of 29 MeV, the OBE potential predictions of Bonn B and Bonn A come much closer to the data point. The reason appears to be nonlocalities in those OBE potentials. While Bonn A could be criticized because of its wrong $\epsilon_1$-phase-shift behavior at large energies, this does not apply to Bonn B. According to [6] , however, that high energy behavior is not decisive for the binding energies, what matters is $P_d$. The last column displays variational estimates. The most recent one [7] is very close to our result. If one plots the 3N and 4N binding energies against each other, as shown in Fig. 1, one sees that they are strongly correlated in a rather narrow band ( Tjon line). For simple forces such a correlation has been found the first time in [8]. One can consider that correlation to be disappointing, since no independent information can be gained from the $\alpha$ particle binding energy, which is not already known from $^3H$ ( of course only within the group of forces studied up to now ).

**Table 1** :   3N- and 4N- binding energies for various realistic NN potential.

| Potential | $P_d$ [%] | $^3H$ [MeV] | $^4He$ [MeV] | References |
|-----------|-----------|-------------|--------------|------------|
| Reid | 6.47 | 7.34 | 23.45 | |
| $AV_{14}$ | 6.08 | 7.68 | 24.62 | 24.45 [7] |
| Paris | 5.77 | 7.46 | 24.26 | 25.5 [9] |
| Nijmegen | 5.39 | 7.63 | 25.03 | 23.92 [3] |
| Bonn B | 4.99 | 8.14 | 27.04 | |
| Bonn A | 4.38 | 8.32 | 28.11 | |
| exp. | | 8.48 | 28.30 (29.) | |

That band is in reality even narrower if CIB in the state $^1S_0$ is taken into account. For Bonn B, which is a "np potential" this amounts to a reduction of the binding energy of about 500 keV [5], whereas for a "pp potential" an increase of binding energy occurs.

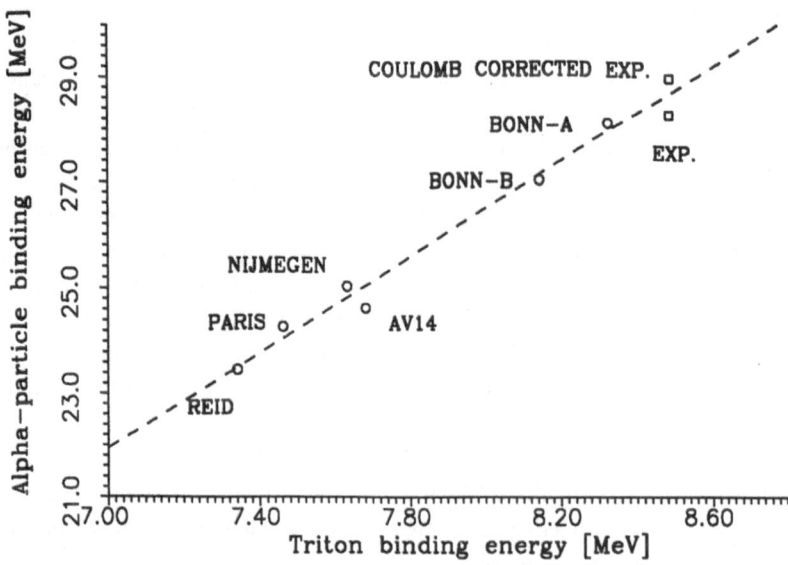

**Fig. 1** : The α-particle against the triton binding energies for various realistic NN forces are correlated in a band, which includes the experimental point.

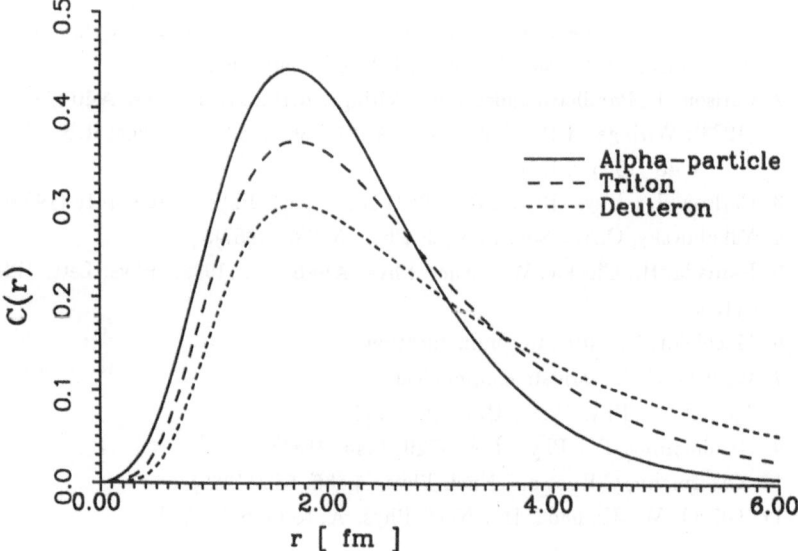

**Fig.2** : The correlation function for the deuteron(dotted curve), triton (dashed curve) and the α-particle(solid curve) based on the Bonn B potential.

As in the triton again the lack of binding energy is of the order of one or a few percent in relation to the total potential energy which is typically 100 MeV.

The charge radius based on the Bonn B potential turns out to be 1.63 fm (extended nucleons), which is close to the experimental value of $1.671 \pm 0.014$fm[10].

Finally we comment on the probability $C(r)$ to find 2 nucleons at a certain distance $r$. It is normalized as $\int_0^\infty drC(r) = 1$. In Fig. 2 we compare that quantity for the deuteron, $^3H$ and $^4He$ using the Bonn B potential. Aside from the kinematical surface effects (different separation energies) the short range behavior is rather similar. For instance if one normalizes the deuteron and $^3H$ correlation functions in the maximum to the one of $^4He$ and subsequently shifts them slightly by $\approx 0.1$fm then the 3 curves coincide nearly perfectly. It is tempting to conjecture that the short range correlation for $r \leq 1.7$ fm is the same for all nuclei. Further studies in that direction are planned.

The inclusion of a 3NF into the 4N Yakubovsky scheme is straightforward and a first feasibility study has already been undertaken [11]. It will be now exciting to investigate low energy scattering in the 4N system to see whether the "experimental" spectrum can be understood on the basis of modern NN forces.

**Acknowledgement**

This work was supported financially by the Deutsche Forschungsgemeinschaft (H.K.). The numerical calculations were performed on the NEC SX3 of the Unversität Köln and the CRAY Y-MP of the Höchstleistungsrechenzentrum in Jülich.

**References**

1. Glöckle, W., Kamada, H., Witała, H. : Second Chinese-German Symposium on Medium Energy Physics, Bochum 1992, to be published.
2. Carlson, J., Pandharipande, V.R., Wiringa, R.B. : Nucl. Phys. **A401**,59 (1983); Wiringa, R.B : Nucl. Phys. **A401** , 86 (1983) ; Wiringa, R.B. : Phys. Rev. **C43** 1585 (1991).
3. Carlson, J. : Phys. Rev. **C38** (1988) 1879; Nucl. Phys. **A508**, 141c (1990).
4. Yakubovsky, O.A. : Sov. J. Nucl. Phys. **5**, 937 (1967).
5. Kamada, H., Glöckle, W. : Nucl. Phys. **A548**, 205 (1992); Phys. Lett. **B292**, 1 (1992).
6. Machleidt, R. : private communication.
7. Wiringa, R.B. : private communication.
8. Tjon, J.A. : Phys. Lett. **B56**, 217 (1975).
9. Goldhammer, P. : Phys. Rev. **C29**, 1444 (1984).
10. Ottermann, C.R. et al : Nucl. Phys. **A436**, 688 (1985).
11. Glöckle,W., Kamada, H. : Nucl. Phys. **A**, to be published.

Few-Body Systems, Suppl. 7, 221–224 (1994)

# RECENT RESULTS FROM THE NUCLEON-NUCLEON PROGRAM AT SATURNE II

R. A. Kunne

Laboratoire National SATURNE, CNRS/IN2P3 et CEA/DSM
CE-Saclay, 91191 Gif-sur-Yvette Cedex, France

(for the Nucléon-Nucléon collaboration)

Abstract

The systematic study of NN spin observables allows the unambiguous determination of the five complex $pp$ and $np$ scattering amplitudes. Recent results from our Saturne II experiment include the determination of the $np$ amplitudes between 0.8 and 1.1 GeV, as well as the measurement of the $pp$ asymmetry and spin correlation $A_{oonn}$ in small energy steps. The latter data are preliminary, but show an intriguing structure. An overview of the future plans concludes this report.

The Nucléon-Nucléon experiment [1] is an ongoing program at the Saturne II accelerator in Saclay (France), whose goal is the unambiguous determination of the five complex $pp$ and $np$ elastic scattering amplitudes at several angles and energies. Such a direct reconstruction requires the measurement of a complete set of observables at each angle and energy. Accounting for statistical uncertainties, this so called *complete experiment* can be done by the measurement of eleven to fifteen spin observables.

The experiment is now in its third phase. Data taking for phase I –$pp$ scattering between 0.8 and 2.7 $GeV$– and phase II –$np$ scattering between 0.8 and 1.1 $GeV$– have been finished. As the polarized neutron beam available at Saturne II is limited to a maximum energy of 1.1 $GeV$, phase III studies quasi-elastic $pn$ scattering on a polarized $^6LiD$ target.

The experimental setup [2] covers about $40^\circ$ in the C.M. system. It consists currently of a polarized target assembly and two detector arms. The forward detector arm is equipped with four MWPC's, a magnet for momentum analysis of protons and a neutron detector. The latter is used for distinguishing neutrons from protons and to provide a time–of–flight measurement. The recoil detector arm is equally comprised of four MWPC's. A carbon plate permits double diffusion, assuring the spin analysis of scattered protons. With an efficiency of a few percent the carbon also allows the detection of backward neutrons by charge exchange. Solenoids in the incoming beam and holding magnets provide the possibility of rotating target and beam particle spin in any desired direction. Two polarimeters, one on the beam line itself and one on Saturne's second extraction, permit on- and off-line monitoring of the beam polarization.

## Recent *pp* results

The amplitude reconstruction of the *pp* data measured in phase I of our program [3] gives below 1.6 GeV unique solutions at almost most angles. However, due to the large errors of some observables at 1.6, 1.8, 2.1, 2.4 and 2.7 GeV there exist two types of solutions, mainly differing in the sign of the relative phases of the amplitudes *a* and *b* [4] (with respect to the phase of the amplitude *e*). At most energies, one type of solution is favoured, but this situation is inverted at 2.1 GeV, where the second type of solution has the better chi-squared.

If real, such a double sign inversion of the phase of amplitude *a* around 2.1 GeV would imply two maxima in the analyzing power energy dependence, at fixed CM angles. A maximum in the higher energy interval was found in the existing data [5] At about 2.2 GeV one set of the ANL-ZGS data shows considerably larger values of the analyzing power in a large angular region.

The structure suggested by the asymmetry data and by the amplitude reconstruction motivated us to remeasure the elastic spin observables $A_{ooon}$, $A_{oono}$ and $A_{oonn}$ with relatively high accuracy at 14 energies from 1.96 to 2.23 GeV (see [6]). The equality $A_{ooon} = A_{oono}$ was used to eliminate the less precisely known beam polarization.

Our preliminary asymmetry results for 65° in the CMS are shown in figure 1 as a function of energy. They are consistent with the already existing data, including our own. At all angles covered the asymmetry shows the double maxima that are expected if the phase of amplitude *a* indeed changes sign twice in the neighbourhood of 2.1 GeV.

There exist several other spin observables that might be used to lift the ambiguity in the amplitude analysis at 2.1 GeV and that can be studied with the present set–up. We plan to measure in november 1993 the triple index spin observables $N_{os"sn}$ and $N_{onsk}$. These observables distinguish clearly between the two solutions: at 55° CM and 2.1 GeV for instance, the calculated values of these two observables are $N_{os"sn} = 0.51$ or -0.21 and $N_{onsk} = -0.44$ or 0.28 for the S1 and S2 solutions, respectively.

Several observables will be measured simultaneously with $N_{os"sn}$ and $N_{onsk}$. Especially the spin correlation $A_{oosk}$ will be measured with a much higher accuracy than was achieved before. All these observables may help to suppress the statistical

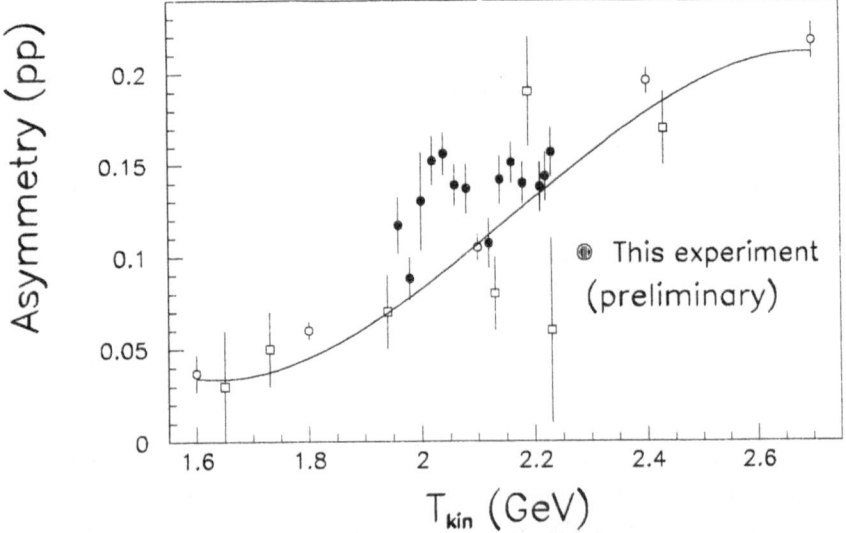

**Figure 1.** *pp asymmetry as a function of kinetic energy around 2.1 GeV, for fixed CM angle. •: preliminary results of this scan; open circles: our phase I [3]; open squares: other results [5].*

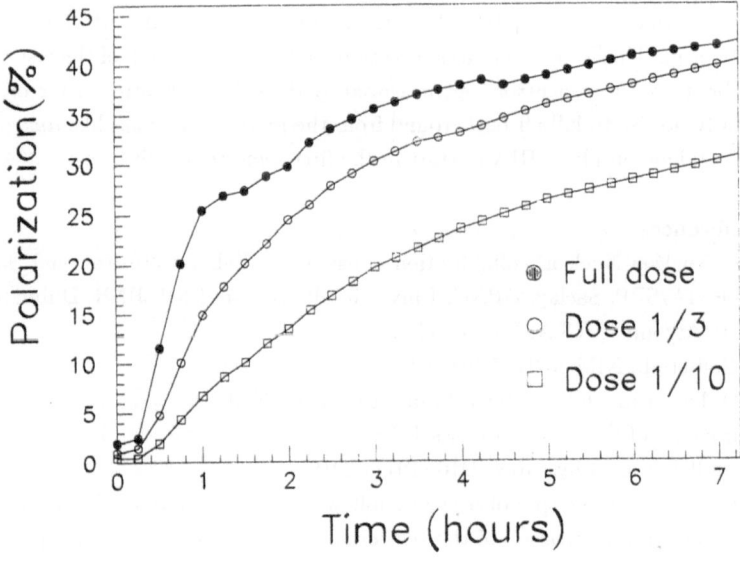

**Figure 2.** *Polarization of the deuterons in the Saclay ⁶LiD target. The curves corresponds to three different irradiation doses of the target material with low energies electrons (see text).*

ambiguity.

## $np$ scattering developments

In phase II of the experiment we used the Saturne II neutron beam to study $np$ interactions at 8 energies between 0.8 and 1.1 GeV. Preliminary results were published in [7].

For phase III of the experiment (quasi-elastic $pn$ scattering on a deuteron target) the Saclay polarized target group is developing a polarized $^6LiD$ target. $^6LiD$ has several advantages over deuterated alcohols:

- the fraction D in the target is larger ($\sim 0.15$ for alcohols; 0.5 for $^6LiD$);
- potentially larger polarization possible (25-40% for alcohols; 50% for $^6LiD$);
- the target polarization is easier to measure from the NMR line shape.

The target material is doped by injecting electrons using a low energy electron accelerator. The final polarization and the relaxation time depend on the total dose, energy and intensity of the electron radiation as well as the holding temperature and field. Our best results are shown in figure 2, obtained with 3 MeV electrons. The three curves correspond to three different radiation doses (in arbitrary units). The target material that had received the highest dose polarized the fastest, but had also the shortest relaxation time: 29 days. The other two samples both had a relaxation time exceeding 65 days.

Another change in phase III will be the addition of a second neutron detector on the recoil arm. This is necessary as simulations show, that the TOF of the fast particle alone –be it proton or neutron– as measured by the existing neutron detector, is not sufficiently precise to kill all background from the inelastic channels like $np\pi$.

Data taking on phase III will start in the first semester of 1994.

## References

[1] The Nucléon-Nucléon collaboration consists of: Laboratoire National Saturne; DAPNIA/SPP, Saclay; DPNC, Université de Genève; LNP–JINR, Dubna; ANL–HEP, Argonne; UCLA, Los Angeles.

[2] J. Ball et al., NIM A327 (1993) 308.

[3] C.D. Lac et al., Journ. Phys. France 51 (1990) 2689.

[4] Our choice of amplitudes follows J. Bystricky et al., Journ. de Phys. 39 (1978) 1.

[5] D. Miller et al., Phys. Rev. D16 (1977) 2016.

[6] Our notation of the spin observables follows [4]: $X_{srbt}$ where the indices stand for the polarization direction of the *scattered, recoiled, beam* and *target* particles.

[7] J.Ball, PhD Thesis, Faculty of Science, University of Paris, Orsay, No. 2132 (1992)

Few-Body Systems, Suppl. 7, 225—230 (1994)

Few-
Body
Systems
© Springer-Verlag 1994
Printed in Austria

# MEASUREMENT OF THE NP→PPπ⁻ REACTION AT 443 MEV

M. G. Bachman, P. J. Riley
*University of Texas at Austin*

A. Amer, A. Berdoz, J. Birchall, J. Campbell, C. A. Davis, N. E. Davison, W. R. Falk,
S. A. Page, W. D. Ramsay, A. Sekulovitch
*University of Manitoba*

D. A. Hutcheon, C. A. Miller
*TRIUMF*

P. Green, E. Korkmaz
*University of Alberta*

B. W. Mayes, L. Pinsky, Y. Tzamouranis
*University of Houston*

D. L. Adams, G. S. Mutchler
*Rice University*

D. J. Margaziotis
*California State University, Los Angeles*

We measured the relative differential cross section and analyzing powers $A_{SO}$, $A_{NO}$ and $A_{LO}$ for the reaction np→ppπ⁻ at an incident neutron energy of 443 MeV. The geometry of our detector allowed us to sample almost all the available phase-space. We reconstructed the full kinematics of each event from the measured trajectories of at least two of the three final state particles. Cuts based on the vertex reconstruction and kinematic $\chi^2$ allowed us to remove most background, primarily nn→npπ⁻ from the target walls and other materials. These cuts gave a signal-to-noise ratio of 22:1, but fewer than 3% of the total data survived the cuts. The resulting ~400,000 np→ppπ⁻ were binned against p-p invariant mass and angles which are most suited to partial waves analysis. Comparisons are made to predictions of the model of Kloet and Lomon. These observables represent the first measurements of their kind at this energy and it is hoped that their determination will allow us to extract the role of isospin–0 amplitudes in the process of pion production.

## Introduction

All NN→NNπ reactions are dominated at medium energies by the production of an intermediate resonant Δ which is the major contributor to cross sections. Nonresonant amplitudes may play a significant role, however, especially in spin observables which are strongly affected by interferences between partial wave amplitudes. The np→ppπ⁻ and np→nnπ⁺ reactions are best suited for studying the role of nonresonant amplitudes in the inelastic process because the $T=0$ initial state is available. The np system is an equal mixture of two isospin channels: $T=0$ and $T=1$. To conserve isospin, the Δ resonance ($T=3/2$) may proceed only from the $T=1$ initial state; at medium energies, other resonances (ΔΔ, N*, *etc.*) are too far off-shell to have any effect on the reaction. Thus, all partial waves available in the $T=0$ channel represent nonresonant amplitudes. Furthermore, the $T=1$ channel can proceed via only the $Δ^0$ ($J=3/2$, $T=3/2$, $T_3=-1/2$) resonance, much weaker than the $Δ^{++}$ which dominates in pp→npπ⁺ and pp→dπ⁺. The threshold for Δ production is 633 MeV, so an np reaction at 443 MeV ought to be particularly sensitive to nonresonant $T=0$ amplitudes.

Due to isospin invariance, the NN→NNπ reactions can be written in terms of three independent cross sections, shown in the table below. The first index refers to the isospin of the NN initial state; the second index is for the isospin of the NN final state. There is no $σ_{00}$ since this transition would violate isospin conservation. The cross section $σ_{01}$ represents contributions from the $T=0$ initial state only. Because the $T=0$ cannot produce a Δ at medium energies, $σ_{01}$ is expected to be small. Early cross section measurements confirmed this suspicion, but it is not clear that $σ_{01}$ is zero.

| Isospin cross sections | Reaction cross sections |
|---|---|
| $σ_{11}$ = | $σ(nn→nnπ^0) = σ(pp→ppπ^0)$ |
| $σ_{11} + σ_{10}$ = | $σ(nn→npπ^-) = σ(pp→npπ^+)$ |
| $\frac{1}{2}(σ_{11} + σ_{01})$ = | $σ(np→nnπ^+) = σ(np→ppπ^-)$ |
| $\frac{1}{2}(σ_{10} + σ_{01})$ = | $σ(np→npπ^0)$ |

Some measurements of np→nnπ⁺ and np→ppπ⁻ searched for $T=0$ effects by subtracting reaction cross sections to deduce the magnitude of $σ_{01}$. Doing this gives conflicting values with large uncertainties. For example, a summary of NN→NNπ in 1982 by VerWest and Ardnt [1] concluded that $σ_{01}$ was essentially zero below 1 GeV, whereas a similar study by Bystricky, *et al.* [2], did not find such a conclusion consistent with isospin invariance. The problem reflects the difficulty in doing high accuracy cross section measurements for these reactions, and the paucity of data available. A plot of this data is shown in figure 1.

Some measurements have sought $T=0$ effects by looking for differences between the differential cross sections of np→NNπ± and the pure $T=1$ reaction, pp→ppπ⁰. The results between 400–600 MeV reveal a significant difference between the shapes of the two differential cross sections, indicating a $T=0$ influence. Other measurements have looked for interference between $T=0$ and $T=1$ by comparing the differential cross sections of np→ppπ⁻ and np→nnπ⁺. The amplitude for np→NNπ± is a linear combination of $T=0$ and $T=1$ amplitudes. Due to charge symmetry, the $T=0$ amplitude changes sign upon interchange of the neutron and proton. This results in a forward-backward asymmetry in the pion CMS cross section, or alternately, a difference between np→ppπ⁻ and np→nnπ⁺ pion yields. Such is a signature of interference between $T=0$ and $T=1$. While some measurements claim to have found evidence for forward-backward interference [3], this evidence for $T=0$ is not strong.

Few measurements of spin observables have been made, owing to the difficulty in obtaining high quality polarized neutron beams. A limited phase-space TRIUMF measurement at 450 MeV was published in 1989 [4]; a full phase-space Saturne measurement for $A_{NO}$ was published in 1992 for energies of 572 MeV to 1134 MeV [5]. Our TRIUMF measurement was done at 443 MeV.

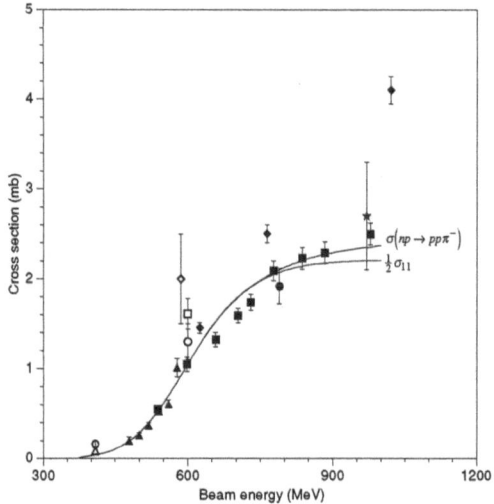

**Figure 1.** Plot of world cross section data for np→NNπ±. Curves reflect fits of $\sigma_{01}$ and $\sigma_{11}$ as determined by VerWest and Ardnt [1].

## Our Experiment

Our experiment (TRIUMF E372) used the polarized neutron facility at the TRIUMF cyclotron. An optically-pumped sodium ion source provided a high intensity polarized proton beam of ~50% polarization. The protons were accelerated to 457 MeV, then directed on to a liquid deuterium target, and the scattered neutrons collimated at 9°. The quasifree reaction p+d→np+(p) at 457 MeV has a spin transfer coefficient of $K_{SS}$~0.8 at a lab angle of 9°; the 443 MeV neutrons had typical polarizations of 32%–35%. Charged particles in the neutron beam were swept away with a magnet; neutral contamination from $\gamma$'s produced in the LD$_2$ were eliminated because they arrived at our LH$_2$ target before the neutrons. Two spin precession magnets allowed us to precess the neutron polarization into S, N and L directions.

The protons from np→ppπ− at 443 MeV are constrained to fall within a cone of 42° in the lab. Since our detector subtended an angle of greater that 42°, we could measure the trajectories of all the protons produced, and thus cover all phase-space. The detector used 18 scintillator bars and two multiwire drift chambers to measure the track angles and velocities of all particles passing through it. These measurements were used to define a kinematic $\chi^2$ which was minimized to reconstruct the events in phase-space. Only events which registered two charged particles in the scintillator bars were triggered. The detector is shown in figure 2.

The $\chi^2$ minimization resulted in good separation between legitimate np→ppπ− events coming from free protons in LH$_2$ and quasifree events from nuclei in the target walls and other materials. Background events were predominately nn→npπ−; most were rejected by the kinematic cuts. The cuts were severe but effective, cutting over 97% of the recorded data and resulting in a signal-to-noise ratio of 22:1. The resulting data set of ~400,000 events for S, N and L polarization was of high quality.

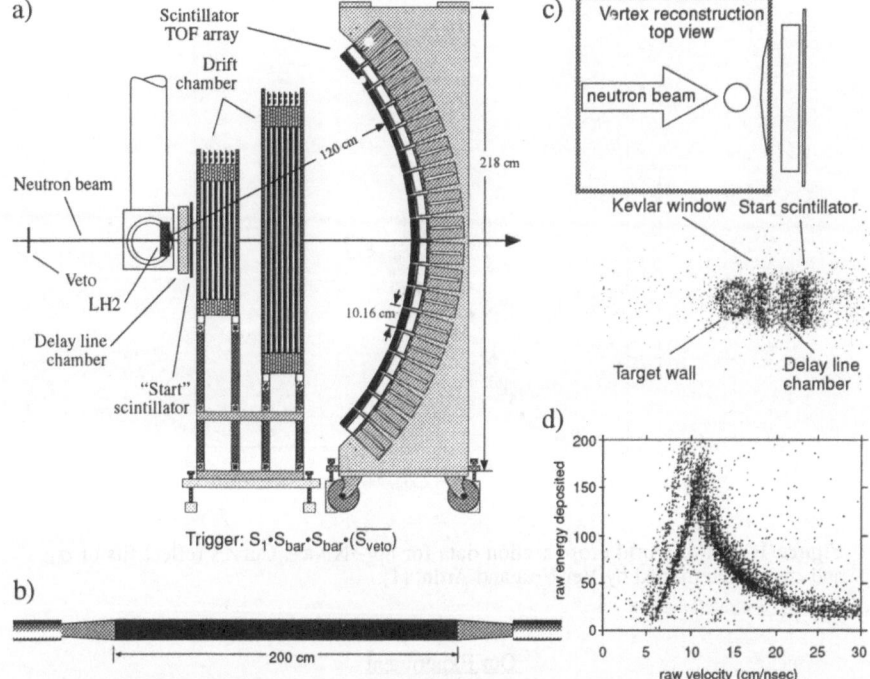

**Figure 2.** Relevant pictures of the detector: a) Side view of target, tracking detector and timing array; b) view of individual scintillator bar; c) top view of vertex traceback for empty target run, showing target walls and vacuum window; d) response of scintillator showing particle velocity *vs.* energy deposited in the bar. This information allowed event reconstruction and particle identification.

The events were binned against angular variables and invariant p-p mass as defined in the coordinate system shown in figure 3. Extracted asymmetries and relative differential cross sections are shown in figure 4. The strong all-negative $A_{NO}$ over pion CMS angle reflects interference of partial waves with the pion in an *s*-state and a *p*-state. We made comparisons to predictions of the model of Kloet and Lomon [6] which are also shown. $A_{SO}$ and $A_{LO}$ are required to be zero by parity conservation for angular variables $\theta$ and $\psi$ (when integrating over the remaining angles), and these were found to indeed be zero.

These measurements represent the first full phase-space spin observables for this reaction below 572 MeV. Projections of the cross sections and analyzing powers against proton-proton angles ($\psi$, $\chi$) are potentially useful in later partial wave analyses. The strong non-zero analyzing powers over all angular variables reveals the interference of several partial waves.

## References

1. B. J. VerWest and R. A. Arndt: *Phys. Rev.* **C25**, 1979 (1982).
2. J. Bystricky, *et al.*: *J. Physique* **48**, 1901 (1987).
3. G. B. Yodh: *Phys. Rev.* **98**, 1330 (1955); R. Handler: *Phys. Rev.* **138,** 1230 (1965); H. Fischer, *et al.*: *SIN Newsletter* **20**, NL15 (1988).
4. C. Ponting, *et al.*: *Phys. Rev. Lett.* **63,** 1792 (1989).
5. Y. Terrien, *et al.*: *Phys. Rev. Lett.* **B294**, 40 (1992).
6. W. M. Kloet and E. L. Lomon: *Phys. Rev.* **C43**, 1575 (1991).

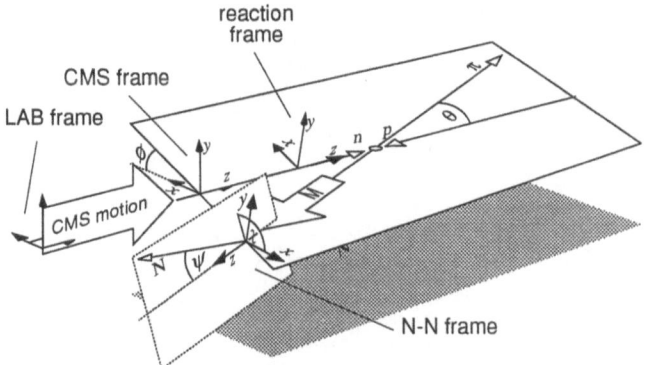

**Figure 3.** Coordinate system defining pion CMS angle and proton angles.

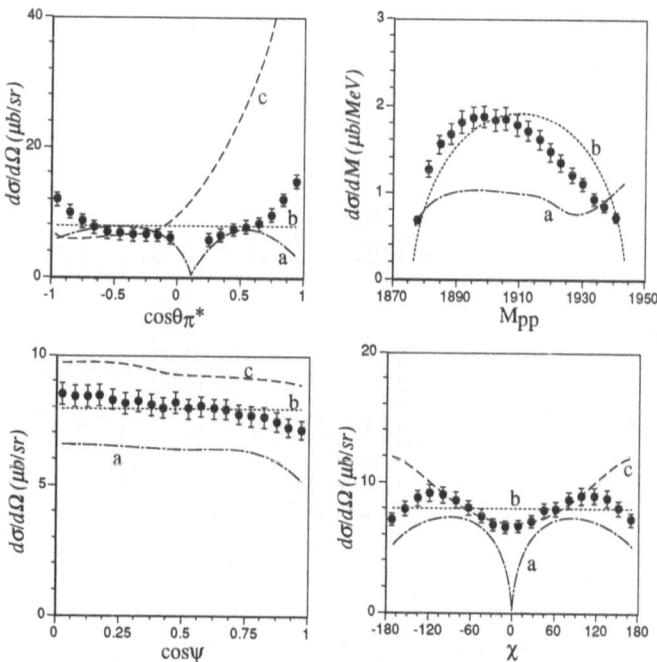

**Figure 4.** Relative differential cross sections. Cross sections are normalized to 100 $\mu$b. Data are corrected for detector acceptance; error bars are estimates of systematic error. Key to curves: a) detector acceptance; b) phase-space prediction; c) Kloet-Lomon prediction.

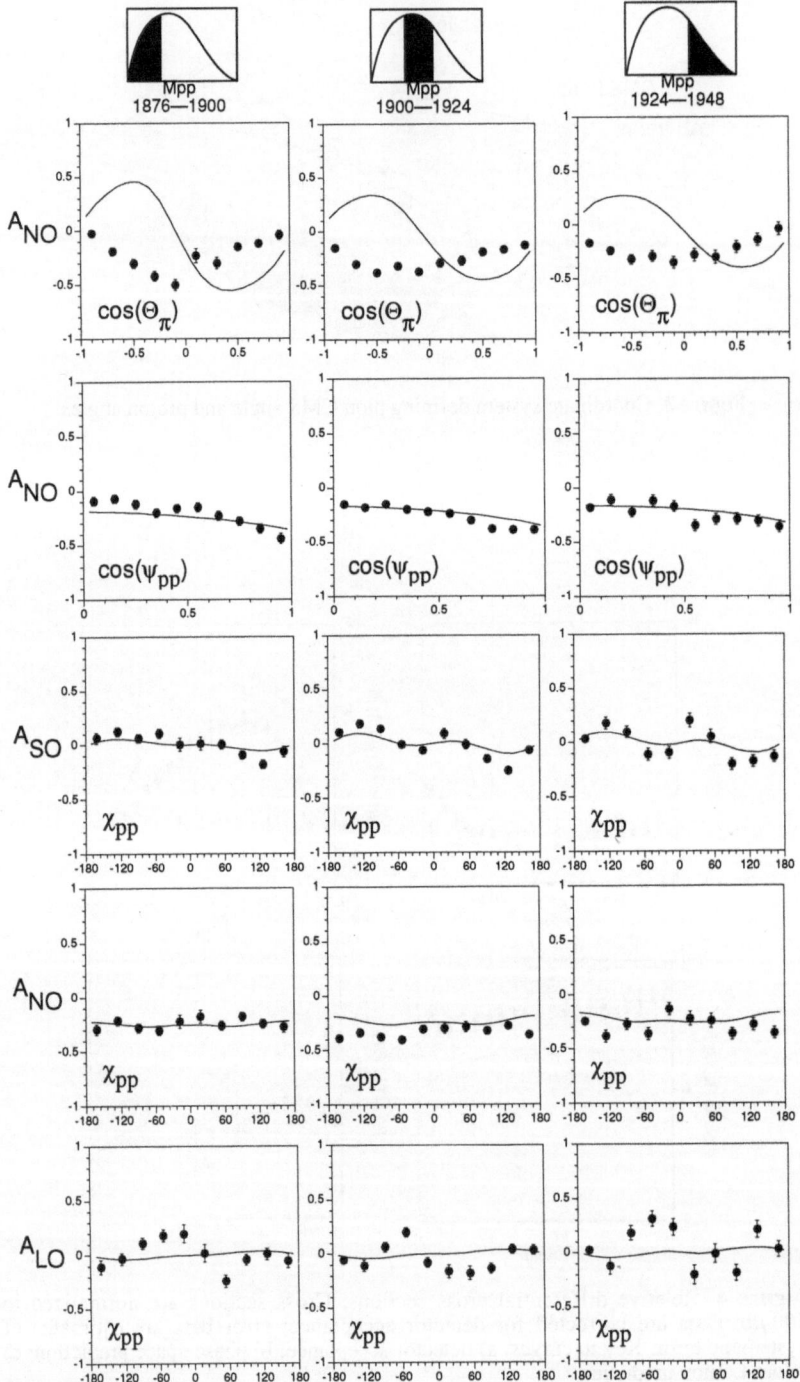

**Figure 5.** Analyzing powers $A_{SO}$, $A_{NO}$, $A_{LO}$ for angular variables defined in figure 3. Also shown are predictions from the model of Kloet and Lomon. Not shown are $A_{SO}$ and $A_{LO}$ for $\cos\theta$ and $\cos\psi$ which are zero everywhere.

Few-Body Systems, Suppl. 7, 231—234 (1994)

Few-
Body
Systems

© Springer-Verlag 1994
Printed in Austria

# A GAUGE INVARIANT UNITARY THEORY FOR PION PHOTOPRODUCTION

C. H. M. van Antwerpen and I. R. Afnan

School of Physical Sciences, The Flinders University of South Australia,

Bedford Park, South Australia 5042, Australia

ABSTRACT: A covariant, unitary and gauge invariant theory for pion photoproduction on a single nucleon is presented. To achieve gauge invariance at the operator level one needs to include both the $\pi N$ and $\gamma\pi N$ thresholds. The final amplitude can be written in terms of a distorted wave in the final $\pi N$ channel provided one includes additional diagrams to the standard Born term in which the photon is coupled to the final state pion and nucleon. These additional diagrams are required in order to satisfy gauge invariance.

Most calculations to date for pion photoproduction on a single nucleon have included two ingredients: (i) A Born term which is taken to be gauge invariant. (ii) A final state interaction or distortion that is needed to satisfy the Watson theorem or unitarity. However, the inclusion of the pionc degrees of freedom into the problem changes the current and charge distribution and thus the coupling of the photon to the nucleon. Hence any consistent theory of pion photoproduction has to include the effect of the pionic degrees of freedom on the charge and current distribution. In other words one needs to satisfy unitarity and gauge invariance at the same time. Clearly, the inclusion of both of these symmetries to all orders requires a full solution to the field theory. Here, we propose to present a formulation that satisfies two-body unitarity and gauge invariance to first order in the electromagnetic coupling.

The starting point of this formulation is the one particle irreducible $\pi NN$ three-point Green's function given by

$$
\begin{aligned}
G(q, p'; p) &= \int d^4 x_1 d^4 x_2 d^4 x_3 \, e^{i(p' \cdot x_3 + q \cdot x_1 - p \cdot x_2)} < 0 | T(\phi(x_1) \psi(x_2) \bar{\psi}(x_3)) | 0 > \\
&= S(p') \Delta(q) \Lambda_5^{(1)\dagger}(q, p'; p) S(p) ,
\end{aligned}
\tag{1}
$$

where the spin, isospin labels have been suppressed. Here, $S$ is the nucleon propagator, $\Delta$ is the pion propagator, and $\Lambda_5^{(1)\dagger}$ is the one pion irreducible $\pi NN$ vertex. Since at this stage we are considering two-body unitarity only, we will restrict the dressing of the nucleon so that the final $\pi N \leftarrow \gamma N$ amplitude has at most one pion in every intermediate state. This allows us to write the $\pi NN$ three-point Green's function as

$$
G(q, p'; p) = S_0(p') \Delta(q) \Lambda_5^{(1)\dagger}(q, p'; p) S(p) ,
\tag{2}
$$

where $S_0(p) = (\not{p} - m)^{-1}$ and $S(p) = (\not{p} - m - \Sigma^{(1)}(\not{p}))^{-1}$, with $\Sigma^{(1)}$ including all contributions to two-body unitarity from mass renormalization.

The photoproduction amplitude is then constructed by gauging Eq. (2) and applying the LSZ reduction to the resulting four-point Green's function. The method of gauging employed was the minimal substitution $p_\mu \rightarrow p_\mu - eA_\mu$, where $p_\mu$ is the four momentum of the particle. This results in the following substitutions [1, 2]

$$
S(\hat{p}) \rightarrow S(\hat{p}) + S(p')\Gamma_\mu(k, p', p)S(p)A^\mu ,
\tag{3}
$$

$$
\Delta(\hat{q}) \rightarrow \Delta(\hat{q}) + \Delta(q')\Gamma_\mu^\pi(k, q', q)\Delta(q)A^\mu ,
\tag{4}
$$

$$
\Lambda_5^{(1)\dagger}(\hat{q}, \hat{p}', \hat{p}) \rightarrow \Lambda_5^{(1)\dagger}(\hat{q}, \hat{p}', \hat{p}) + \Gamma_\mu^{CT}(k, q, p', p)A^\mu .
\tag{5}
$$

i.e. the photon couples to all possible external lines and vertices present. We have followed a procedure developed by Ohta [2] in which (i) a Taylor series expansion is assumed to exist for the form factors present, (ii) perturbation theory can be used to replace various operators by eigenvalues, and (iii) we restrict ourselves to first order in the electromagnetic coupling. Applying the substitutions of Eqs. (3-5) to Eq. (2) leads to the four classes of diagrams [3] describing the photoproduction amplitude,

$$
\Lambda_5^{(1)\dagger} S \Gamma_\mu^{(1)} + \Gamma_\mu^{(2)} S_0 \Lambda_5^{(1)\dagger} + \Gamma_\mu^{\pi(2)} \Delta \Lambda_5^{(1)\dagger} + \Gamma_\mu^{CT(1)} ,
\tag{6}
$$

where $\Gamma_\mu^{(1)}$ is the one-particle irreducible photon nucleon vertex, $\Gamma_\mu^{\pi(2)}$ is the photon pion vertex which is taken to be the bare vertex, and $\Gamma_\mu^{CT(1)}$ is the $\pi N \leftarrow \gamma N$ amplitude resulting from the coupling of the photon to the $\pi NN$ vertex, $\Lambda_5^{(1)\dagger}$. Here the irreducibility is given with respect to the number of pions and nucleons only. This amplitude is by definition gauge invariant with the photon vertices satisfying their corresponding Ward-Takahashi identities [1, 4, 5].

To establish the fact that the amplitude for $\pi N \leftarrow \gamma N$, as given in Eq. (6) satisfies two-body unitarity, we follow the procedure of Araki and Afnan (AA) [6] and classify a diagram's contribution to the amplitude according to it's irreducibility. However, unlike AA who included first the $\pi N$ threshold for two-body unitarity and then the $\pi\pi N$ and $\gamma\pi N$ cuts for three-body unitarity, in this case we include both the $\pi N$ and $\gamma\pi N$ unitarity cuts, since the corresponding branch points are at the same energy, to satisfy two-body unitarity.

Exposing the $\pi N$ unitarity cut results in the following integral equation for the $\pi N$ amplitude [6]

$$t^{(0)} = v(1 + gt^{(0)}) \tag{7}$$

where $v$ is the Born amplitude given by $v = t^{(2)} + \Lambda_5^{(2)\dagger} S_0 \Lambda_5^{(2)}$ , and $g = S_0 \Delta$ is the $\pi N$ propagator. For the $\pi N \leftarrow \gamma N$ amplitude exposing the $\pi N$ cut gives

$$M^{(0)} = \tilde{v} + vgM^{(0)} \quad , \tag{8}$$

with $\tilde{v} = M^{(2)} + \Lambda_5^{(2)\dagger} S_0 \Gamma^{(2)}$. In this case $\tilde{v}$ is not the Born amplitude, since exposing the $\gamma\pi N$ cut in $\Gamma^{(2)}$ gives

$$\Gamma^{(2)} = \Gamma^{(3)} + \Gamma^{3(3)} g \Lambda_5^{(1)\dagger} \quad , \tag{9}$$

where $\Gamma^{3(3)}$ is the $N \leftarrow \gamma\pi N$ amplitude. Similarly, exposing the $\gamma\pi N$ cut in $M^{(2)}$ gives

$$M^{(2)} = \Gamma^{CT(3)} + \tilde{F}_{3;c}^{(3)} g \Lambda_5^{(1)\dagger} + \Gamma^{\pi(2)} \Delta \Lambda_5^{(1)\dagger} + \Gamma^{(3)} S_0 \Lambda_5^{(1)\dagger} \quad , \tag{10}$$

where $\Gamma^{CT(3)}$ is the $\pi N \leftarrow \gamma N$ amplitude, $\tilde{F}_{3;c}^{(3)}$ is the $N\pi \leftarrow \gamma\pi N$ amplitude, and $\Gamma^{\pi(2)}$ is the $\pi \leftarrow \gamma\pi$ amplitude. This results in $\tilde{v}$ being given by

$$\tilde{v} = \Gamma^{CT(3)} + \tilde{F}_{3;c}^{(3)} g \Lambda_5^{(1)\dagger} + \Gamma^{\pi(2)} \Delta \Lambda_5^{(1)\dagger} + \Gamma^{(3)} S_0 \Lambda_5^{(1)\dagger} + \Lambda_5^{(2)\dagger} S_0 \Gamma^{(3)} + \Lambda_5^{(2)\dagger} S_0 \Gamma^{3(3)} g \Lambda_5^{(1)\dagger} \tag{11}$$

which contains more physics than just the Born amplitude. By comparing the above result for the photoproduction amplitude which satisfies two-body unitarity, with the amplitude in Eq. (6), we can establish that they are in fact identical [7].

If we now iterate Eq. (8) and make use of Eq. (7), we find that

$$M^{(0)} = (t^{(0)}g + 1)\tilde{v} = (t^{(0)}g + 1)\tilde{v}_B + (t^{(0)}g + 1)\tilde{v}_R \quad , \tag{12}$$

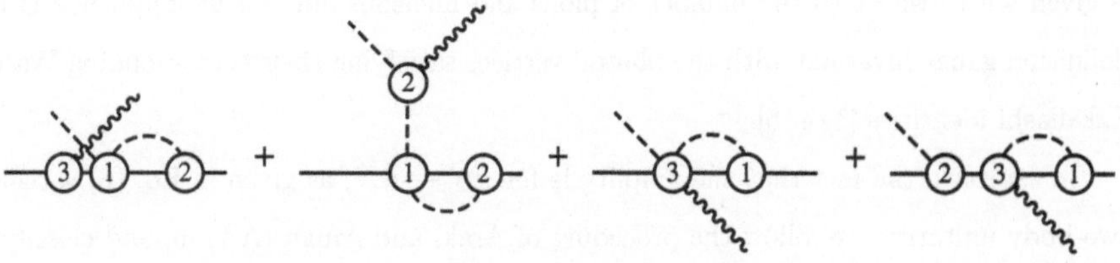

234

Figure 1: These are the non-Born diagrams which contribute to the pion photoproduction amplitude. The numbers in the circle give the irreduciblity of each amplitude.

where $\tilde{v}_B$ is the Born amplitude and $\tilde{v}_R$ contains the additional diagrams required to maintain gauge invariance which are illustrated in Fig. 1.

This result also proves that the commonly used procedure for unitarizing the Born amplitude in terms of the $\pi N$ amplitude, i.e.

$$M^{(0)} = (t^{(0)} g + 1) \tilde{v}_B \quad , \tag{13}$$

does not satisfy gauge invariance. This is due to the fact that the derivation of Eq. (13) involves the inclusion of the $\pi N$ threshold, which is what unitarity requires, and only the lowest order diagrams containing a $\gamma \pi N$ intermediate state, into $\tilde{v}_B$. However, to satisfy gauge invariance, the full $\gamma \pi N$ threshold should also be included, which is what one expects considering the fact that the two thresholds start at the same energy. We should also note that the additional terms resulting from the inclusion of the $\gamma \pi N$ threshold give rise to the dressing of the vertices in $\tilde{v}_B$, and this dressing is required to satisfy gauge invariance.

## REFERENCES

[1] E. Kazes, Nuovo Cimento **13**, 1226 (1959).

[2] K. Ohta, Phys. Rev. C **40**, 1335 (1989); *ibid* **41**, 1213 (1990).

[3] H. W. L. Naus, J. H. Koch, and J. L. Friar, Phys. Rev. C **41**, 2852 (1990).

[4] J. C. Ward, Phys. Rev. **77**, 293 (1950).

[5] Y. Takahashi, Nuovo Cimento **6**, 371 (1957).

[6] M. Araki and I. R. Afnan, Phys. Rev. C **36**, 250 (1987).

[7] C. H. M. van Antwerpen and I. R. Afnan, to be publlished.

Few-Body Systems, Suppl. 7, 235—238 (1994)

Few-
Body
Systems
© Springer-Verlag 1994
Printed in Austria

# THE $\gamma d \to \pi^0 d$ AND $\gamma d \to pn$ REACTIONS IN THE $\Delta$-RESONANCE REGION*

P. Wilhelm and H. Arenhövel

Institut für Kernphysik, Johannes Gutenberg–Universität

D–55099 Mainz, Germany

Coherent pion photoproduction on the deuteron and deuteron photodisintegration in the $\Delta$–resonance region is studied treating the final state interaction within a $NN$–$N\Delta$ coupled channel approach.

The $\gamma d \to \pi^0 d$ and $\gamma d \to pn$ reactions are particulary interesting in order to investigate the $N\Delta$–dynamics, or more precisely the respective $N\Delta - N\Delta$ and $NN - N\Delta$ interactions. Both reactions are strongly linked through the dominant electromagnetic $\Delta$–excitation mechanism. Therefore, we have studied them both in parallel, adopting a $NN$–$N\Delta$ coupled channel approach for the final state interaction [1].

The theoretical concept with respect to the hadronic part is based on [2]. The model includes explicit pion degrees of freedom. Hence its configuration space is build up from $NN$–, $N\Delta$– and $\pi NN$–sectors. The coupling between the latter two is generated by an explicit $\pi N\Delta$–vertex. Its iteration on one hand leads to the dressing of the bare $\Delta$–particle and on the other hand yields a retarded $\pi$–exchange potential diagonal in $N\Delta$–space. Transitions between $NN$ - and $N\Delta$–space are mediated by regularized static $\pi$–exchange.

*Supported by the Deutsche Forschungsgemeinschaft (SFB 201)

The diagonal interaction in $NN$–space contains a renormalized (via subtraction of a $N\Delta$–box graph) realistic $NN$–potential for which the OBEPR version of [3] was used. The diagonal interaction in $\pi NN$–space is switched off in the present calculation. Thus a violation of unitarity is still present. To calculate the required $\pi d$– and $NN$–scattering waves a set of coupled integral equations of Lippmann–Schwinger type has to be solved in momentum space. In order to test our ansatz we have also calculated the pure hadronic reactions $\pi d \rightarrow \pi d$ and $NN \rightarrow NN$. The comparison with experimental $NN$ scattering phase shifts in the $^1D_2$ partial wave has been used to fix the regulator mass in the $NN$–$N\Delta$ transition potential to $\Lambda_\pi = 700\,\text{MeV}$ inserting $(\Lambda_\pi^2 - m_\pi^2)/(\Lambda_\pi^2 + \vec{q}^{\,2})$ at each vertex.

For the electromagnetic interaction, our model includes the usual nucleonic one–body current as well as static $\pi$– and $\rho$–exchange currents. The latter ones were constructed consistently with respect to gauge invariance to the static potentials introduced above. Furthermore, nonresonant background pion production is taken into account. Of course, direct $\Delta$–excitation is the most important photoabsorption mechanism. We consider only the dominant magnetic dipole excitation. It is described by an effective energy dependent and complex coupling to ensure the unitarity of the $M_{1+}(3/2)$ multipole amplitude.

We start the discussion of our results with differential cross sections for $\gamma d \rightarrow \pi^0 d$ shown in Fig. 1. In the resonance region, final state interaction reduces the cross section significantly and thus leads to an improved agreement with experiment at least around $90°$. Nevertheless, a shift in the energy dependence still remains. For extreme angles, the comparison with the available data is less conclusive. In this context we emphasize, that we have not found such a shift when comparing our pion deuteron elastic cross sections to the data. On the contrary, our results, shown in Fig. 2, tend at small angles to overestimate (underestimate) the data for energies below (above) the resonance position around $T_{lab}^\pi = 180\,\text{MeV}$.

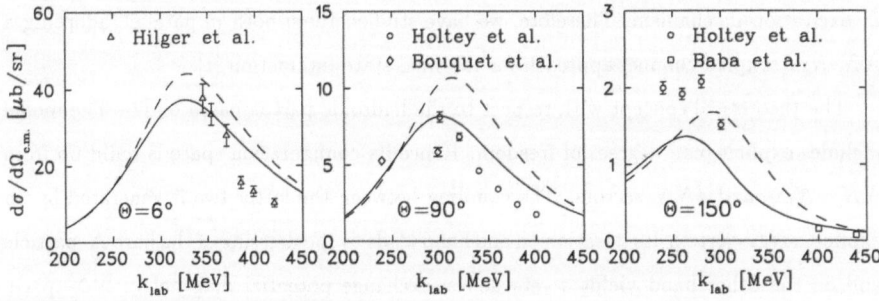

**Fig. 1**: Differential cross sections for $\gamma d \rightarrow \pi^0 d$ compared to experiment [4]: complete calculation (full) and without final state interaction (dashed).

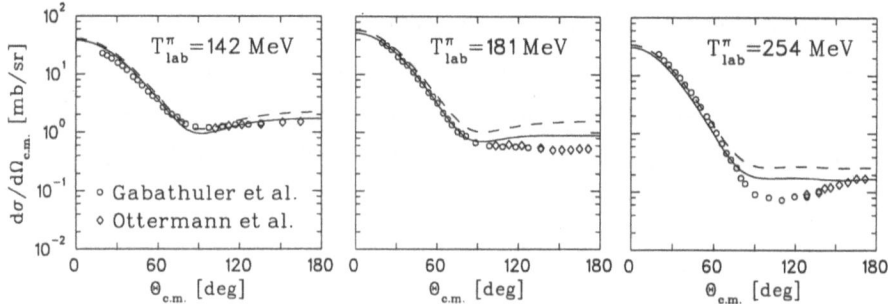

**Fig. 2**: Differential cross sections for $\pi d \to \pi d$ compared to experiment [5]: complete calculation (full) and without final state interaction (dashed).

In Fig. 3 we show the total cross section for $\gamma d \to pn$. The comparison with a treatment of the $\Delta$ in impulse approximation (IA) demonstrates the relevance of final state interaction for this reaction at least above $k_{lab} = 260$ MeV. Obviously, our cross section is definitely too small for energies below 350 MeV. This is in agreement with [7], where the $\gamma N\Delta$-coupling was fixed in a similar way as here. Since this coupling is weaker than the effective one used in [8], we have also considered a modified $\gamma N\Delta$-coupling, which was determined from the elementary amplitude under the assumption of vanishing nonresonant contributions to the $M_{1+}(3/2)$ multipole. Using this coupling, we achieved a good agreement of the total cross section with experiment over the whole energy range as demonstrated in Fig 3. Because we have effectively incorporated the Born terms in the modified coupling we are led to the conclusion that the framework of static $\pi$-exchange currents, which in principle contain the Born terms, gives only a poor description of them.

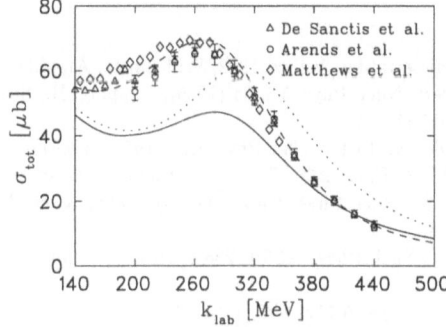

**Fig. 3**: Total cross section for $\gamma d \to pn$ compared to experiment [6]: complete calculation (full), with modified $\gamma N\Delta$-coupling (dashed) as described in the text and IA (dotted).

Finally, differential cross sections and photon asymmetries for $\gamma d \to pn$ are plotted in Fig. 4 and 5. Although the modified $\gamma N\Delta$-coupling gave a good description of the total

238

cross section, problems in these observables still remain. At lower energies for example, our cross sections show a too strong decrease at extreme angles, in particular at forward angles. Furthermore, at 360 MeV our model clearly fails to reproduce the strong negative asymmetry showing instead a relative maximum around 90° which can be traced back to the influence of higher partial waves.

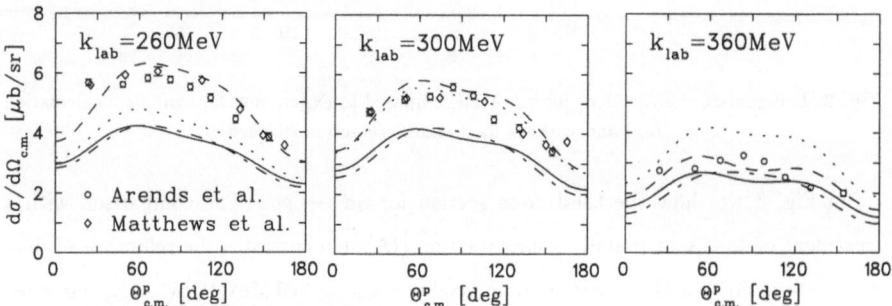

**Fig. 4**: Differential cross sections for $\gamma d \to pn$ compared to experiment [6]: Curves as in Fig. 3 and complete calculation without $N\Delta$ configurations in $j \geq 4$ partial waves (dash–dotted).

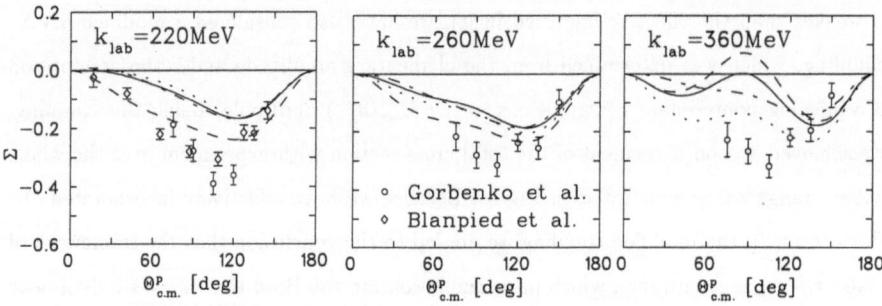

**Fig. 5**: Photon asymmetry $\Sigma$ for $\gamma d \to pn$ compared to experiment [9, 10]: Curves as in Fig. 4.

[1] Wilhelm, P.: Ph.D. dissertation, Mainz, 1992; Wilhelm P., Arenhövel, H.: to be published
[2] Sauer, P.U.: Prog. Part. Nucl. Phys. **16**, 35 (1986); Pöpping, H., Sauer, P.U., Zhang, X.-Z.: Nucl. Phys. **A474**, 557 (1987)
[3] Machleidt, R., Holinde, K., Elster, Ch.: Phys. Rep. **149**, 1 (1987)
[4] von Holtey, G. et al.: Z. Phys. **259**, 51 (1973); Bouquet, B. et al.: Nucl. Phys. **B79**, 45 (1974); Hilger, E. et al.: Nucl. Phys. **B93**, 7 (1975); Baba, K. et al.: Phys. Rev. **C28**, 286 (1983)
[5] Gabathuler, K. et al.: Nucl. Phys. **A350**, 253 (1980); Ottermann, C.R. et al.: Phys. Rev. **C32**, 928 (1985)
[6] Arends, J. et al.: Nucl. Phys. **A412**, 509 (1984); De Sanctis, E. et al.: Phys. Rev. **C34**, 413 (1986); Matthews, J.L. et al.: private communication (1990)
[7] Tanabe, H., Ohta, K.: Phys. Rev. **C40**, 1905 (1989)
[8] Leidemann, W., Arenhövel, H.: Nucl. Phys. **A465**, 573 (1987)
[9] Gorbenko, V.G. et al.: Nucl. Phys. **A381**, 330 (1982)
[10] Blanpied, G.S. et al.: Phys. Rev. Lett. **67**, 1206 (1991); Sandorfi, A.M. et al.: private communication (1991)

Few-Body Systems, Suppl. 7, 239—242 (1994)

# FINAL STATE INTERACTION EFFECTS IN THE
# COUPLED-CHANNEL
# NN-NΔ APPROACH OF THE πNN SYSTEM

M. T. Peña [1], H. Garcilazo [2], and P. U. Sauer [2]

[1] CEBAF, Newport News, Virginia 23606, USA

[2] Institut für Theoretische Physik, Universität Hannover, D-3000 Hannover 1,
Germany

Within a coupled-channel NN-NΔ approach for the πNN system, we study the effects of the NN interaction in the pion-spectator final states (FSI). We found that in particular the elastic πd → πd spin observables are sensitive to this interaction, especially for backward scattering angles. In general, we obtain a better agreement with the data.

## Introduction

Different unified theoretical studies of the unitarity-coupled reactions involving the πNN system at intermediate energies were developed throughout the last decade [1]. As accurate experimental studies of hadron structure by electromagnetic probes become a reality, theorists are urged to improve their predictions for the baryonic systems in general, and, given its role in electro- and photo-induced probing processes, for the πNN system in particular [2]. The final aim is to test our present understanding of the nuclear force from quark and meson degrees of freedom. A first step in this direction was undertaken in Refs. [3,4] where a study of the NΔ interaction on observables of the πNN system was presented.

Above 300 MeV the nucleon-nucleon amplitudes become inelastic. Up to 1 GeV the inelasticities are dominated by single pion production that proceeds, since the nucleon is a composite particle, through the formation and decay of its resonances. Therefore, the extension above the pion production threshold of the conventional meson-exchange based picture of the NN interaction leads naturally to formalisms that consider the $\Delta$ and possibly higher resonances within a coupled-channel approach. We present here a calculation motivated by these underlying ideas. Such an approach has already a history [5,6] suggesting that traditional phenomenological parametrizations of the baryonic dynamics at short distances may fail an overall description of the NN and $\pi$NN data. Neverthless, calculations of this nature are still important as indicators of any signature of the sub-hadronic degrees of freedom, even at intermediate energies.

## Aspects of the Calculation

In this work we used the general formalism of Refs. [3,4]. It is based on a non-covariant, time-ordered NN-N$\Delta$ coupled-channel approach, developed in momentum space where relativistic kinematics for the pion is easily introduced. It is justified by the fact that the $\Delta$ is the most important mode of nucleonic excitation which provides the dominant mechanism for pion scattering, production and absorption. The considered Hamiltonean acts on and couples the baryonic sector of the Hilbert space $\mathcal{H}_N$ and $\mathcal{H}_\Delta$ and the pionic sector $\mathcal{H}_\pi$, where a pion is added to the baryons.

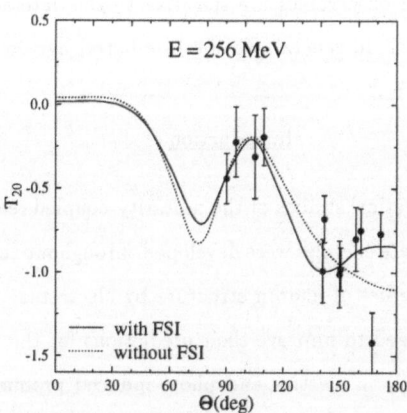

Figure 1: $T_{20}$ observable for $\pi$d elastic scattering at the pion lab energy 256 MeV.

For the first time and in contrast to Refs. [3,4], in this work we lift the restriction of neglecting any interaction in the pionic sector $\mathcal{H}_\pi$. This approximation, made in previous papers, constitutes a severe assumption, since i) it violates two-body unitarity constraints: the reactive pion production content of the NN scattering reaction from the NN $\rightarrow$ $\pi$d inelastic channel is underestimated, ii) it implies an inconsistent description of any $\pi$d process: in the presence of a pion two nucleons could never be bound. The NN interaction in the presence of a spectator pion which we introduce now in

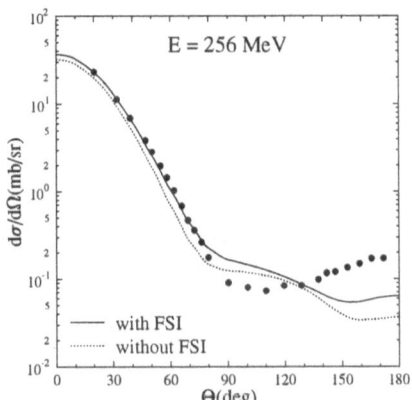

Figure 2: $\pi$d-$\pi$d differential cross section at the pion lab energy 256 MeV.

the $\mathcal{H}_\pi$ sector of the Hilbert space, is taken to be the Graz separable parametrization of the Paris potential [7] in the dominant $^1S_0$ and $^3S_1 - ^3D_1$ channels. The use of this separable form of the NN interaction is convenient for practical reasons, since it simplifies crucially the needed integration in the spectator pion momentum variable, that occurs in the iterated box diagram.

As in Ref. [3], we solve the Alt-Grassberger-Sandhas (AGS) three-particle scattering equations, extended to particle absorption. We obtained the two and three-baryon components of the multichannel transition matrix, from which the physical amplitudes and cross sections follow directly.

## Results

Among the considered $\pi$NN observables, the $\pi$d elastic observables, namely the spin observable $T_{20}$ and the differential cross-section at backward angles, are most sensitive to the improvement in the model that we present here. Figures 1 and 2 show that the

242

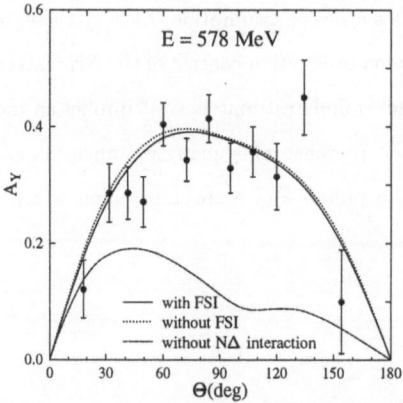

Figure 3: $A_y$ for the reaction $pp \rightarrow n\Delta^{++}(p\pi^+)$ at NN lab energy 578 MeV.

new calculation of these two observables leads to a better agreement with the data. Figure 3 shows the beam asymmetry $A_y$ for the reaction $pp \rightarrow n\Delta^{++}(p\pi^+)$. Being highly sensitive to the N$\Delta$ interaction, it is nearly unchanged by the pion-spectator state NN interaction, as expected. Therefore the failure [3,4], shared with other models, in reproducing the structure of $A_y$ at higher energy persists. Such failure was checked to be due to an incorrect description of the $\Delta$ helicity 3/2 contribution.

Work to incorporate a two term (resonant + non-resonant or background) $\pi$N $P_{33}$ interaction, allowing for a larger value of the $\pi$N$\Delta$ vertex cut-off is under way. This last correction is expected to alter more significantly the $\pi$d $\rightarrow$ NN reactions [2] and constitutes the needed step to avoid the extreme supression of high pion momentum transfer.

[1] H. Garcilazo and T. Mizutani,"$\pi$NN Systems", (World Scientific, Singapore, 1990).

[2] C.Fayard, G.H. Lamot, T. Mizutani, and B. Saghai, Phys. Rev. C46, 118 (1992).

[3] M. T. Peña, H. Garcilazo, U. Oelfke, and P. U. Sauer,Phys. Rev. C45, 1487(1992).

[4] A. Valcarce, F. Fernández, H. Garcilazo, M. T. Peña, and P. U. Sauer, submitted to Phys. Rev. C.

[5] T.-S. H. Lee, A. Matsuyama, Phys. Rev. C36, 1459 (1987)

[6] E. E. Van Faassen and J.A. Tjon, Phys. Rev. C29, 195 (1984)

[7] J. Haidenbauer and W. Plessas, Phys. Rev. C30, 182 (1984).

Few-Body Systems, Suppl. 7, 243—246 (1994)

Few-
Body
Systems

# CROSS SECTION AND ANALYZING POWER $A_y$ IN THE PROTON INDUCED DEUTERON BREAKUP REACTION AT 65 MeV

M. Allet, K. Bodek, W. Hajdas, J. Lang, R. Müller, S. Navert, O. Naviliat-Cuncic,
J. Sromicki, J. Zejma

*Institut für Teilchenphysik, Eidgenössische Technische Hochschule, Zürich,
Switzerland*

L. Jarczyk, St. Kistryn, J. Smyrski, A. Strzałkowski

*Institute of Physics, Jagellonian University, Cracow, Poland*

W. Glöckle, J. Golak, H. Witała

*Institut für Theoretische Physik II, Ruhr Universität, Bochum, Germany*

B. Dechant, J. Krug

*Institut für Experimentalphysik I, Ruhr Universität, Bochum, Germany*

One of the basic questions in nuclear physics deals with the nature of the two-nucleon (2N) interaction. While QCD can not yet be solved in the nonperturbative regime required for an answer, meson theory has achieved some maturity and provides realistic 2N-forces, which are able to describe very well the great amount of 2N data. It is now of interest to see whether those forces can also be used in systems, where more than two nucleons interact. The simplest one, the three-nucleon system (3N) has always been considered as an ideal testing ground for our understanding of the 2N-interactions. Assuming 2N forces only, the Hamiltonian for the 3N system is fixed. Does it describe the experimental 3N observables? Is it necessary to introduce additionally genuine 3N forces in the dynamics of the 3N system? Now, with the advent of supercomputers, the 3N Faddeev equations can be solved in a numerically rigorous way for any local or nonlocal 2N interaction [1,2]. Therefore, the meson-exchange dynamics in nucleon-nucleon forces can be tested reliably in the 3N system by comparing the calculations with precise experimental data. The aim of the reported experiment is to provide accurate continuum 3N observables in those kinematical regions where, according to model calculations, the 3N force effects are enhanced and, simultaneously, the sensitivity to details of the 2N potential is small [3].

The experiment was carried out at the Philips Cyclotron in the Paul Scherrer Institute - Villigen, Switzerland. The cyclotron provided a transversally polarized proton beam with an energy of 65 MeV, an average intensity of about 300 nA and a polarization $|P_y| = 0.75$. The beam was focused on a deuterium gas target mounted in a scattering chamber and cooled down to 77 K. The beam polarization was continuously monitored in a transmission polarimeter by observing the asymmetry in $\vec{p}+{}^{12}C$ elastic scattering at 45.8° with a pair of NaI(Tl) scintillation detectors. Every second the sign of the polarization was reversed by switching the radio frequency transitions at the ion source, accordingly. For the largest angles 3 mm thick surface barrier (SB) detectors were used whereas the other detectors were $\Delta$E-E telescopes built from NE102 plastic scintillators. To reduce systematic errors they were arranged symmetrically on both sides of the beam axis. Standard electronics and an on-line data acquisition system collected the coincidences between each two telescopes and, simultaneously, single events with reduced rate by a factor of 1000. The kinematically complete break-up cross sections were determined from spectra of coincidences and normalized to the p+D elastic scattering cross section [5] extracted from single spectra. The accidental coincidences were measured simultaneously and subtracted in the off-line analysis.

In the first phase of the experiment we have measured cross sections and analyzing powers $A_y$ in the exclusive reaction $D(\vec{p}, pp)n$ at 65 MeV in several kinematical regions with special attention paid to four collinearity configurations: $(\theta_1, \theta_2, \phi_{12}) =$ (20°, 116°, 180°), (30°, 98°, 180°), (45°, 76°, 180°), (60°, 60°, 180°) and two quasi free scattering configurations pp-QFS (30°, 60°, 180°) and (44°, 44°, 180°). The experimental data are compared to the theoretical predictions based on four realistic 2N potentials: BONN-B, PARIS, NIJMEGEN and AV14. In the rigorous Faddeev calculations the charge independence breaking of the 2N-force in the state ${}^1S_o$ was exactly taken into account by admitting an admixture of total isospin T $= 3/2$. By Monte Carlo simulations it has been proven that the averaging due to finite geometry (beam intensity profile, gas target size, solid angles of detectors) and the energetic resolution play a negligible role in the configurations studied, so the point geometry calculations are directly compared to the experimental data. It is gratifying to notice a satisfactory agreement between measured and calculated observables. Some examples are presented in Fig. 1. The gross structures of the cross sections and their absolute magnitudes are well described by the theory. A similar statement holds for the analyzing powers. In this study, special attention was paid to the exact collinearity points. We do not find any intriguing structures neither in the cross sections nor in the analyzing powers. Moreover, the smooth behaviour of the observables around those points is well reproduced by the calculations based on pure 2N interactions. The deviations between theory and experiment do not surpass the differences which exist among the predictions by the different potentials. There is therefore not much

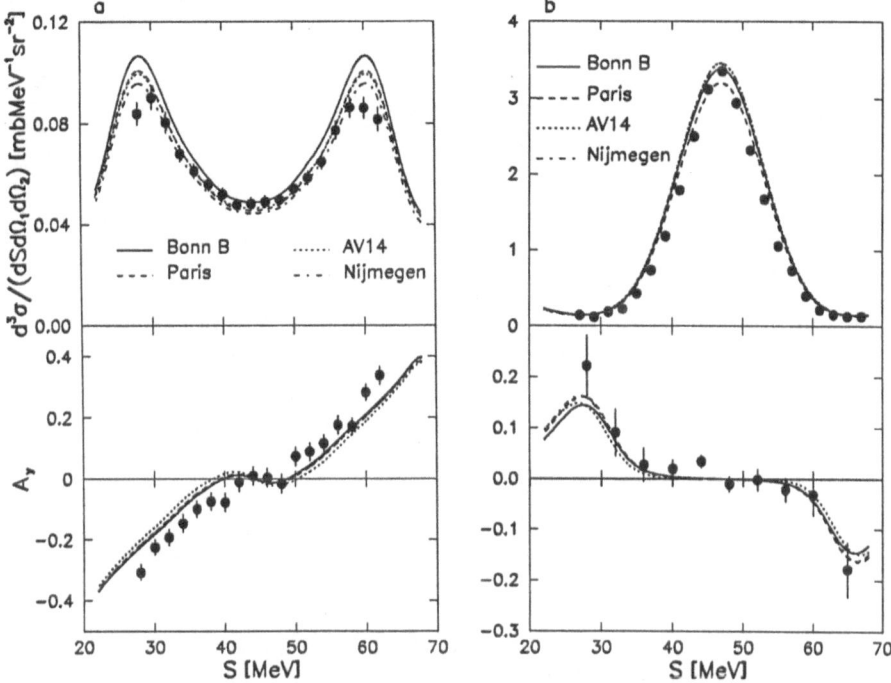

Figure 1: Cross section and analyzing power for the configurations: (a) collinearity (60°, 60°, 180°), and (b) pp-QFS (44°, 44°, 180°)

room for big contributions coming from genuine 3N forces. On the other hand, in some parts of the phase space the discrepancies are greater than the variations due to different potentials used in the Faddeev calculations. The reason for that is unclear and it will be interesting to see the outcome of future calculations including a 3N force. Coulomb effects have been neglected in this study. However, the good quantitative agreement of the theoretical predictions for the n+n+p system with the experimental data measured in the p+p+n system poses an upper limit for such effects to few percents only.

The experiment is being continued and incorporates now detection of neutrons. The following configurations are studied: symmetric space star SSS (54°, 54°, 120°), forward plane star FPS (35°, 35°, 180°), backward plane star BPS (75°, 75°, 180°), quasi free scattering np-QFS (44°, 44°, 180°) and final state interaction np-FSI (44°, 44°, 0°). They were chosen since in the existing analyses they reveal striking discrepancies between experiment and theory [6]. Fig. 2 shows a preliminary analysis of sample data taken for SSS together with theoretical predictions.

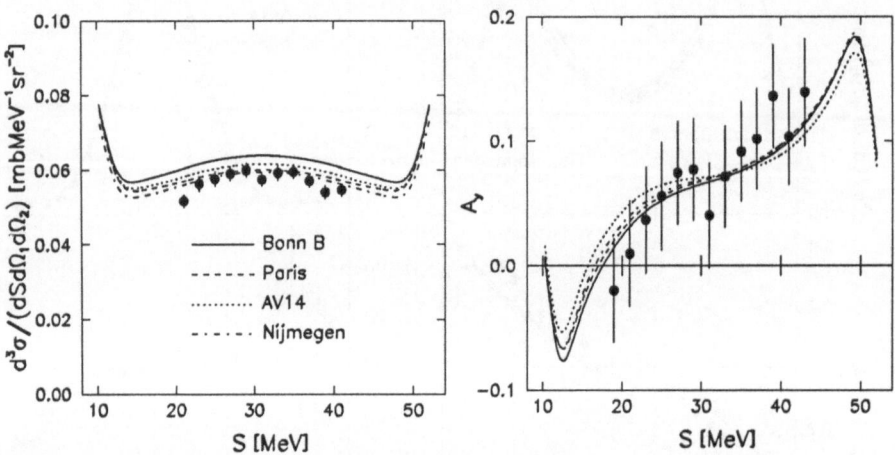

Figure 2: Cross section and analyzing power for the SSS configuration $(54°, 54°, 120°)$-preliminary analysis of sample data.

# References

[1] W. Glöckle, H. Witała, Th. Cornelius, Nucl. Phys. **A508**, 115c (1990);
J. Styczeń and Z. Stachura (eds.) Proceedings of the 25th Zakopane School on Physics, vol. 2: Selected topics in nuclear structure, World Scientific, Singapore 1990, p. 300

[2] H. Witała, Th. Cornelius, W. Glöckle, Few-Body Systems **3**, 123 (1988)

[3] W.Meier, W.Glöckle, Phys. Lett. **128B**, 329 (1984)

[4] H.Witała, Th.Cornelius and W.Glöckle, Few Body System **5**, 89 (1989)

[5] H. Shimizu, K. Imai, N. Tamura, K. Nishimura, K. Hatanaka, T. Saito, Y. Koike, Y. Taniguchi: Nucl. Phys. **A382**, 242 (1982)

[6] G. Rauprich, S. Lemaitre, P. Niessen, K.R. Nyga, R. Reckenfelderbäumer, L. Sydow, H. Patz gen. Schieck, H. Witała, W. Glöckle, Nucl. Phys. **A535**, 313 (1991),
B.Dechant, Ruhr-Universität Bochum, dissertation, 1992

Few-Body Systems, Suppl. 7, 247—250 (1994)

# MESON PRODUCTION NEAR THRESHOLD

### via the reaction $p + d \rightarrow\ ^3He + X$

M.A.Duval, R.Frascaria, F.Roudot [1] and R.Siebert
Institut de Physique Nucléaire, 91406 Orsay, France


J.Bisplinghoff, J.Ernst, F.Hinterberger, R.Jahn, R.Joosten, T.von Oepen and R.Wurzinger
Bonn University, Bonn, Germany


J.Arvieux, F.Plouin and W.Spang
LNS, C.E.Saclay, 91191 Gif-Sur-Yvette, France

*This experimental program focuses on meson production near threshold via the reaction $p + d \rightarrow\ ^3He + X$. The incident beam energy and the $^3He$ momentum are matched in such a way that the mesons are produced at rest in the c.m. system. Clear signals can be observed at the $\omega, \eta', K\bar{K}$ and $\phi$ thresholds.*

The reaction $p + d \rightarrow\ ^3He + X$ is at present extensively studied at LNS (Saclay), the missing mass for X going from the $\omega$ to the $\phi$ meson. The proton beam delivered by the MIMAS/SATURNE facility is focussed on a liquid deuterium target. The outgoing $^3He$ particles are detected at $0°$ with the SPES4 spectrometer. For each chosen incident proton energy the SPES4 is tuned in such a way that the detected $^3He$ and the X system are produced at rest in the c.m. system. Consequently the system $^3He$ and X - which can be a meson or a system of mesons - is studied at very small relative c.m. momentum.

---

[1] *Speaker IPN Orsay France*

Apparatus and Detection.

The synchrotron delivers protons with energies from 1.25 to 1.85 Gev with an intensity of $10^{11}$ p/c. A secondary emission monitor and a scintillator telescopes viewing a thin $CH_2$ film are used to monitor the relative proton flux. Absolute calibration of the monitors were made by activation measurements from $^{12}C(p,pn)^{11}C$.

The proton beam is focussed on a liquid deuterium target at a special position relatively to the SPES4 spectrometer allowing the detection of the scattered $^3He$ particles at $0^\circ$. The experiment utilizes the SPES4 beam line which is a 32 meter long spectrometer allowing the analysis of particles up to 4 GeV/c to analyse the scattered particles. The solide angle of this spectrometer, defined by a lead collimator, is $\Delta\Omega = 2.5 10^{-4}$ and the momentum acceptance is $\Delta p/p = \pm 3.5\%$ with a resolution of $10^{-4}$. The structure of the apparatus is shown in fig.1.

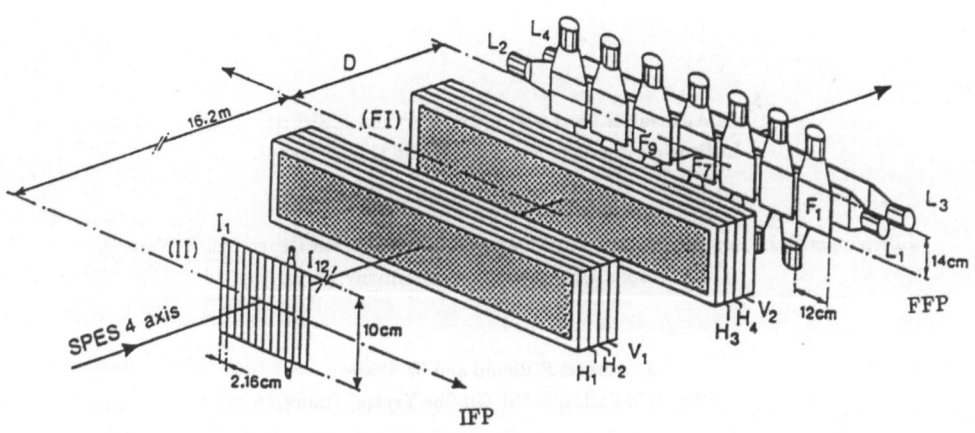

fig.1.

Different time of flight (TOF) measurements are perfomed between the intermediate focal plane (IFP) and the final focal plane region (FFP). The starts of the acquisition are given by 12 scintillator counters of the IFP, the stops are delivered either by 13 scintillators counters of the FFP or by 2 thin scintillator strips covering the whole focal plane. An ultra-fast coincidence circuit allows the rejection of particles with TOF differences of a few nanoseconds.

In addition, two multidrift counters located close to FFP allow the reconstruction of detected particle trajectories in both horizontal and vertical planes. This allows the elimination of any particle that does not come from the target.

Results

The data taken during the last run of June 1992 are presented in fig.2 and fig.3. and give what is called the "Excitation function for meson near threshold". The presented results are normalized on beam monitors. Fig.2 presents clear peaks, beyond an important multipion background, at $\omega$, $\eta\prime$ and $\phi$ thresholds. A blow up of the $\eta\prime$ to $\phi$ region is presented in fig.3 showing small structures at both $K^+K^-$ and $K^0\bar{K}^0$ thresholds. Unfortunately, due to problems coming from beam monitor variation, a change of about 7% in normalization occurred. The dots and the squares in the figures correspond to the two different periods. Small deviations from the general slope which are present in the spectrum are possibly due to these monitor problems. Further evidence for those is required and could be given as a result of a new data taking experiment which is planned next fall.

Total cross sections for $\omega$, $\eta\prime$ and $\phi$ near threshold were extracted from the previous run of 1991. This new run confirms and improves the measurements of the cross sections. A preliminary theoretical analysis of meson production near threshold has been performed by C. Wilkin using an empirical three particle model (ref.1). The comparison with our data is very encouraging; the calculation describes the measured cross section surprisingly well. A 2-pion/2-kaon production in a one pion exchange model (ref.2.) is in progress with the aim to interpret simultaneously the $\pi^- + p \rightarrow n + X$ (ref.3) and the $p + d \rightarrow {}^3He + X$ reactions, X representing two pions or two kaons.

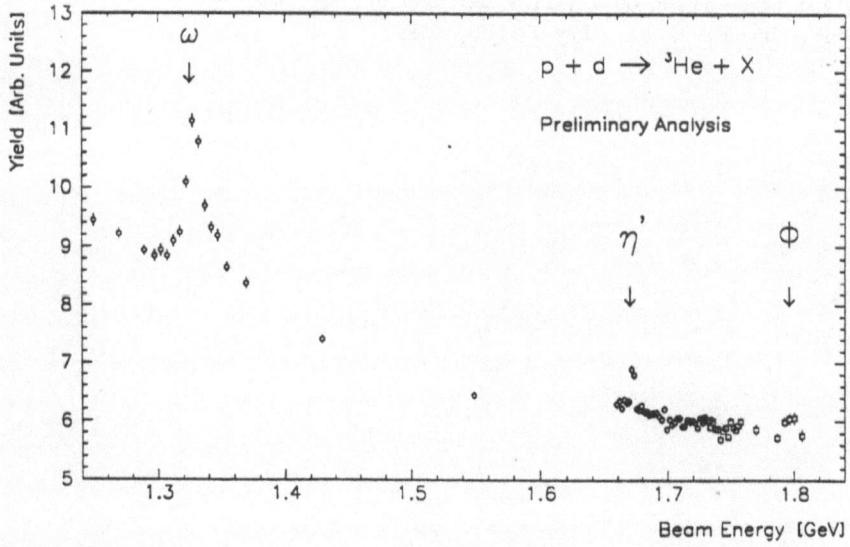

fig.2.

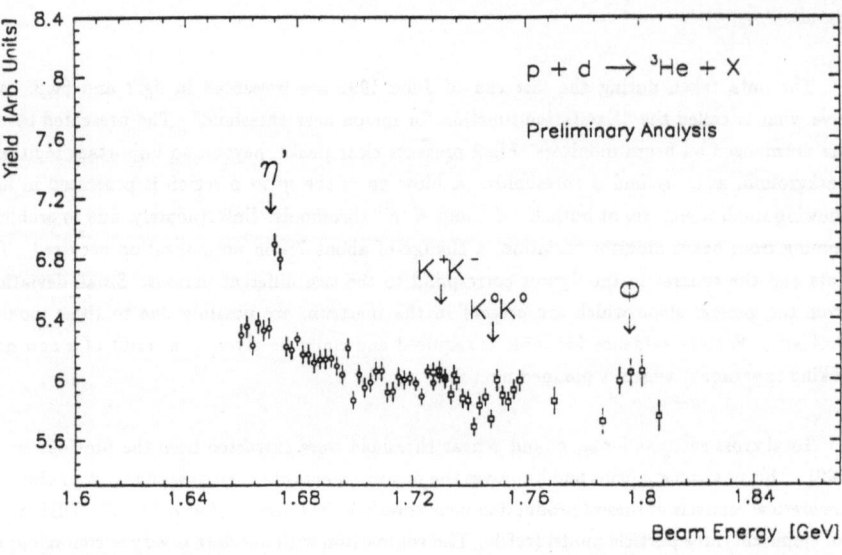

fig.3.

[1] C.Wilkin, *Hadroproduction Of Light Mesons Near Threshold*, Proceedings of the Sixièmes Journées d'Etudes SATURNE (JES6), Mont Sainte-Odile, May 1992, p.121

[2] J.M. Laget and J.F. Lecolley, Phys. Rev. Lett. 61 (1988) 2069.

[3] D.M. Binnie et al., Phys. Rev. D8 (1973) 2789.

Few-Body Systems, Suppl. 7, 251–254 (1994)

Few-
Body
Systems
© Springer-Verlag 1994
Printed in Austria

# INFLUENCE OF ISOBARS ON THE DEUTERON ELECTRIC STRUCTURE FUNCTION A(q²)

W. Plessas, Ch. Brandstätter, S. Cvijetic, J. Haidenbauer,
L. Mathelitsch, P. Obersteiner, J. Pauschenwein, and R. Wagenbrunn
Institute for Theoretical Physics, University of Graz
Universitätsplatz 5, A-8010 Graz, Austria

We have studied the role of the nucleon isobars $\Delta(1232)$ and $N^* = N(1440)$ in deuteron electromagnetic form factors. Here we report on our results for the electric structure function $A(q^2)$ of elastic $e - d$ scattering in the domain $0 \leq q^2 \leq 80$ fm$^{-2}$ of momentum transfers.

It is commonly accepted that beyond the nucleonic configuration $NN(^3S_1 - {}^3D_1)$ also non-nucleonic, but hadronic, components occur in the deuteron, namely $NN^*(^3S_1 - {}^3D_1)$ and $\Delta\Delta(^3S_1 - {}^3D_1 - {}^7D_1 - {}^7G_1)$. Yet, their quantitative influence on various deuteron properties is not well established.

Following previous studies by other groups [1], we are investigating the effects of the "small" $NN^*$ and $\Delta\Delta$ components of the deuteron in observables of $e - d$ scattering. Here we present results for the elastic structure function $A(q^2)$ calculated with three modern dynamical models of the $NN$ interaction:

i) the potential Graz-II Ext3, i.e. a phenomenological separable $NN$ potential [2] (reasonably constructed, though, after the Paris potential [3], and often used in the past for few-nucleon studies) extended to a multi-channel model with phenomenological $NN^*$ and $\Delta\Delta$ components [4,5];

ii) the potential BEST7 Ext, i.e. the Bonn one-boson-exchange $NN$ potential OBEPQ [6], represented in a separable form and extended to a multi-channel model, again with phenomenological $NN^*$ and $\Delta\Delta$ components [7];

iii) the potential CCF [8], i.e. a completely meson-theoretical coupled-channel $NN - \Delta\Delta$ potential derived along the full Bonn model FULLF with the folded-diagram technique [9].

These potentials reproduce all deuteron properties reasonably well in agreement with experimental data but rely on different descriptions, specifically, of the non-nucleonic components (different probabilities of the channels, different shapes of their respective wave functions, etc.).

In the figures we demonstrate the results for $A(q^2)$ from the calculations of elastic $e - d$ scattering along the formalism and ingredients as in ref. [4], i.e. in impulse approximation (IA) with the addition of relativistic (spin-orbit, Darwin-Foldy and nuclear-motion) effects (RC) and the most important meson-exchange currents (MEC) for the nucleonic sector ($\pi$-pair, $\rho\pi\gamma$ and $\pi$-retardation currents). For the CCF potential (and its purely nucleonic partner FULLF), however, no meson-retardation-current contributions must explicitly be added, as their effects are already included in the impulse approximation, due to the instantaneous character of the meson exchanges in the folded-diagram approach. In all cases, RC and MEC were considered only for the $NN$ configurations.

As seen from Fig. 1, the effects of the non-nucleonic components are certainly remarkable, especially at higher momentum transfers. But, as becomes evident from the comparison in Fig. 2, this is even true at lower momentum transfers, in view of the accuracy of the experimental data.

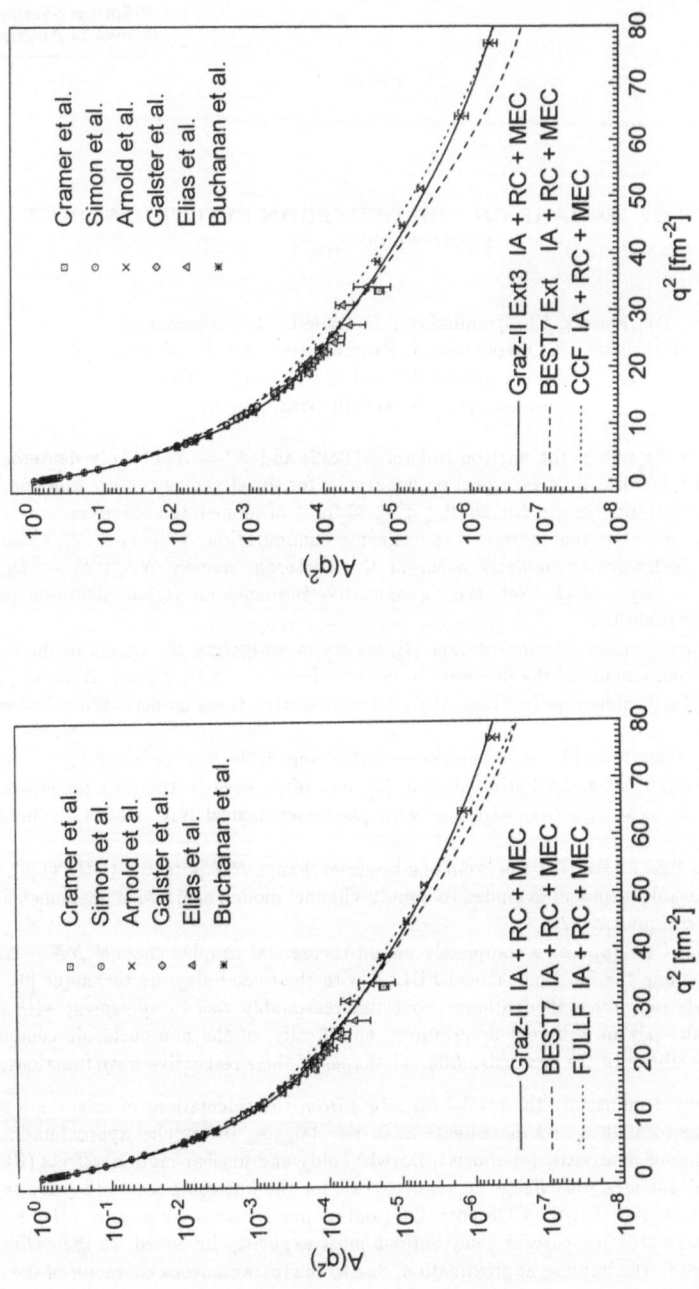

**Fig. 1.** Comparison of electric structure functions as obtained from the IA+RC+MEC calculations. Left: Only nucleonic components for the potentials Graz-II, BEST7, and FULLF. Right: With inclusion of isobar components for the corresponding multi-channel potentials Graz-II Ext3, BEST7 Ext, and CCF. Experimental data as shown in ref. [10].

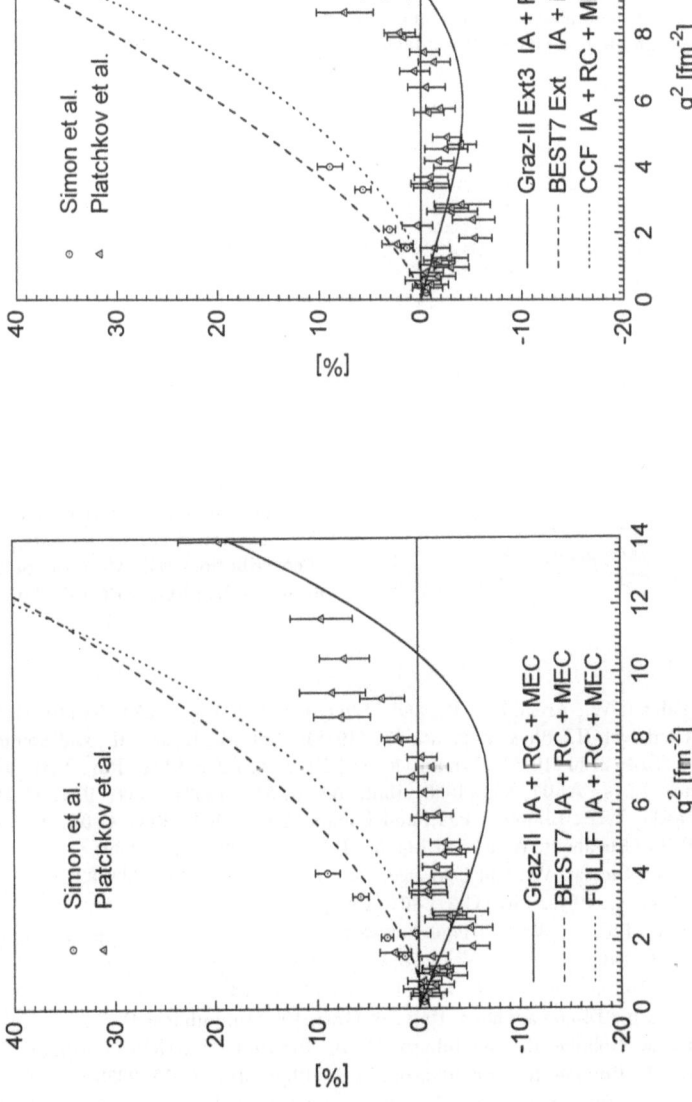

Fig. 2. Same comparison as in Fig. 1 for low momentum transfers. The results and experimental data [11,12] are shown as relative percentage deviations from the IA calculation for Graz-II.

It is surprising that the more realistic models BEST Ext and above all CCF are not capable of reproducing, in particular, the Saclay data [12]. Somehow accidentally, the model Graz-II Ext3 yields the best description over the whole momentum-transfer region. The percentage contributions of the non-nucleonic components are shown in Fig. 3. Their qualitative dependences on $q^2$ are the same for all models, while their magnitudes differ. The influence is biggest in the CCF model, also because it has the highest total probability for the non-nucleonic configurations, namely 1.34, as compared to 0.80 and 1.00 for the interaction models Graz-II Ext3 and BEST7 Ext, respectively.

Of course, the results shown here are also dependent on several ingredients of the calculation (nucleon and isobar form factors, meson-nucleon vertex parametrizations in the MEC, etc.), what we cannot discuss here due to space limitations. Likewise it is interesting to examine the role of non-nucleonic components in the magnetic structure function $B(q^2)$, the charge ($F_c$), quadrupole ($F_Q$), and magnetic ($F_M$) form factors as well as the polarization observables, above all $T_{20}(q)$ [13].

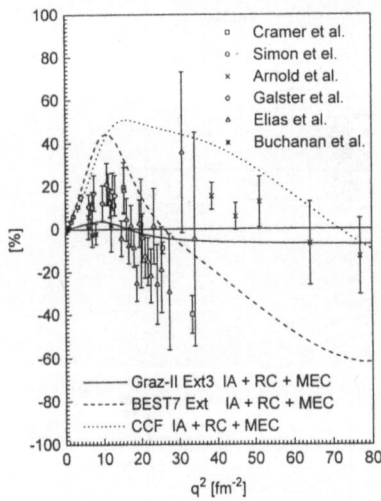

Fig. 3. Influence of non-nucleonic components on the electric structure function shown as percentage contributions relative to the purely nucleonic IA+RC+MEC calculation for Graz-II

1. See, e.g.: Fabian, W., Arenhövel, H., and Miller, H.G.: Z. Phys. **271**, 93 (1974); Weber, H.J. and Arenhövel, H.: Phys. Rep. **36**, 277 (1978); Gari, M., Hyuga, H., and Sommer, B.: Phys. Rev. C 14, 2196 (1976); Dymarz, R. and Khanna, F.C.: Phys. Rev. Lett. **56**, 1448 (1986); Nucl. Phys. **A507**, 560 (1990); ibid. **A516**, 549 (1990); Phys. Rev. C **41**, 2438 (1990); Sitarski, W.P., Blunden, P.G., and Lomon, E.L.: Phys. Rev. C **36**, 2479 (1987); Blunden, P.G., Greenberg, W.R., and Lomon, E.L.: ibid. **40**, 1541 (1989).
2. Mathelitsch, L., Plessas, W., and Schweiger, W.: Phys. Rev. C **26**, 65(1982).
3. Lacombe, M. et al.: Phys. Rev. C **21**, 861 (1980).
4. Obersteiner, P., Plessas, W., and Pauschenwein, J.: Few-Body Syst. Suppl. **5**, 140 (1992).
5. Obersteiner, P.: Doctoral Thesis, Univ. of Graz, 1992 (unpublished).
6. Machleidt, R., Holinde, K., and Elster, Ch.: Phys. Rep. **149**, 1 (1987).
7. Brandstätter, Ch.: Diploma Thesis, Univ. of Graz, 1993 (unpublished).
8. Haidenbauer, J., Holinde, K., and Johnson, M.B.: Preprint KFA-IKP(TH)-1993-12.
9. Haidenbauer, J., Holinde, K., and Johnson, M.B.: Phys. Rev. C **45**, 2055 (1992).
10. Pauschenwein, J., Plessas, W., and Mathelitsch, L.: Few-Body Syst. Suppl. **6**, 195 (1992).
11. Galster, S. et al.: Nucl. Phys. **B32**, 221 (1971).
12. Platchkov, S. et al.: Nucl. Phys. **A510**, 740 (1990).
13. Cvijetic, S., Mathelitsch, L., Pauschenwein, J., and Plessas, W.: In preparation.

Few-Body Systems, Suppl. 7, 255—259 (1994)

# Tensor and Vector Analyzing Powers in the Reaction ²H(e,e'p)

*C.W. de Jager,[a] R. Alarcon,[d] H. Arenhövel,[e] M. Bouwhuis,[a] J.F.J. van den Brand,[b] M. Bucholz,[b] H.J. Bulten,[b] S. Choi,[d] J. Comfort,[d] M. Doets,[a] R. Ent,[g] M. Ferro-Luzzi,[f] W. Haeberli,[b] J. Konijn,[a] J. Lang,[f] D.J.J. de Lange,[a] W. Leidemann,[e] M. Miller,[b] D. Nikolenko,[c] E. Passchier,[a] A.R. Pellegrino,[h] S.G. Popov,[c] P. Quin,[b] I. Rachek,[c] J.J.M. Steijger,[a] O. Unal,[b] H. de Vries,[a] C. Zegers,[a] Z.-L. Zhou[b]*

[a]NIKHEF-K, P.O. Box 41882, 1009 DB Amsterdam, The Netherlands, [b]Department of Physics, University of Wisconsin, Madison, Wisconsin, USA, [c]Budker Institute for Nuclear Physics, Novosibirsk, Russia, [d]Department of Physics, Arizona State University, Tempe, Arizona, USA, [e]Johannes Gutenberg Universität, Mainz, Germany, [f]Institut für Mittelenergiephysik, ETH, CH-8083 Zürich, Switzerland, [g]Massachusetts Institute of Technology, Cambridge, Massachusetts, USA, [h]Free University, Amsterdam, The Netherlands

### Abstract

The first experiment to be performed an the internal target facility of NIKHEF-K will be a study of electron-induced quasi-elastic proton knock-out from tensor-polarized deuterium. Here, we present the first results from the experimental tests as well as the results of a Monte Carlo simulation, which show the feasibility of the proposed experiment, even at modest luminosities.

## Introduction

The structure of the deuteron ground-state wavefunction is a central question in intermediate energy nuclear physics. Spin-dependent scattering experiments allow access to new and crucial observables through asymmetry measurements which depend on the deuteron orientation. In the first internal target experiment[1] to be performed at the AmPS ring electron-induced quasi-elastic proton knock-out from tensor-polarized deuterium will be studied. In this contribution the experimental set-up[2] will be described, and the results of the first test measurements will be presented.

## Instrumentation

Polarized deuterium nuclei from an atomic beam source are fed into an open-ended storage cell, cooled to 80 K, operated in a strong holding field ($\approx$ 30 mT) for maximum tensor polarization or in a weak field ($\approx 1$ mT) for vector polarization. A flux of $1.6 \times 10^{16}$ atoms/s of deuterium, polarized to > 90%, into a 15 mm feed tube has been measured. A future improvement program (increasing the sextupole tipfield by using permanent magnets and replacing the dissociator) is expected to triple the feed rate. The final luminosity goal is $10^{32}$ atoms/cm$^2$s with a 15 mm diameter storage cell and an average stored electron beam intensity of 150 mA.

The scattering chamber consists of an aluminum rectangular box with the feed-through to the atomic beam source in the horizontal plane on the side. At the top and bottom thin windows (0.1 mm stainless steel) give passage to the scattered electron and the ejected proton. Separate magnetic coils allow to orient the magnetic holding field. The complete scattering chamber can be moved vertically over 60 mm in order to allow unrestricted passage of the beam in stretcher mode. Both up- and downstream of the chamber two conductance limiters are mounted to preserve the high vacuum in the rest of the ring.

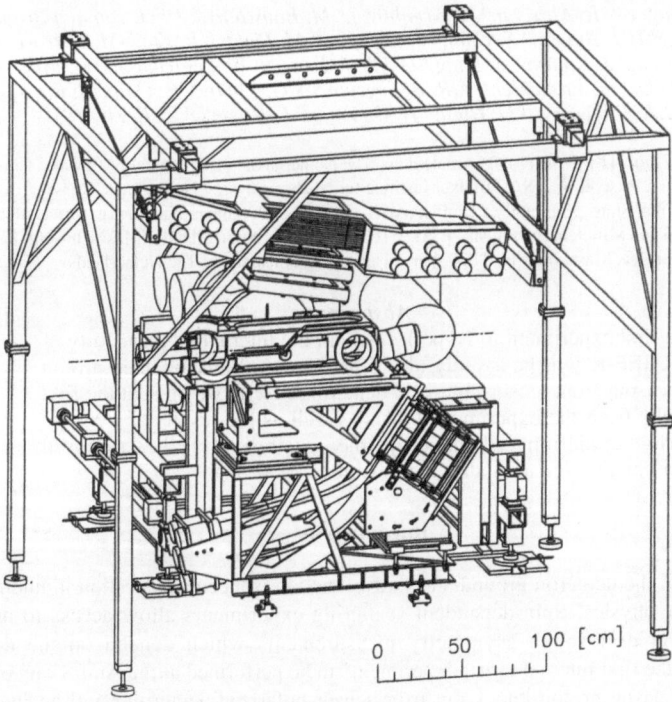

0  50  100 [cm]

Fig. 1. Overview of the NIKHEF internal target set-up

The scattered electrons are detected by a CsI-calorimeter, covering a scattering angle range of 20° to 50°. The calorimeter consists of 60 CsI blocks, arranged in six layers. Each block has dimensions of 15 x 6 x 6 cm$^3$. The detector covers a solid angle of 140 msr and is mounted below the scattering chamber. Only the central quarter of the acceptance will be used in order to limit leakage of the shower out of the side of the calorimeter. Simulation studies predict an energy resolution of 5 %. Two plastic scintillators, 50 and 10 mm thick, are used to define the trigger signal. Vertex reconstruction is done by two sets of wirechambers with a wire spacing of 2 mm.

The knocked-out protons are detected with a range telescope, consisting of 15 layers of plastic scintillator 30 x 50 cm$^2$, each with a thickness of 10 mm, preceded by one layer of 2 mm, so that deuterons can be separated from protons. The instrument is preceded by two sets of wire chambers with a wire spacing of 6 mm. The range telescope covers a solid angle of nearly 300 msr. It will stop protons with an energy of up to 150 MeV and has an energy resolution of about 3 %.

To measure the polarization of the deuterium nuclei in the storage cell, several devices have been or are being constructed: a Breit-Rabi polarimeter, a Balmer polarimeter, a tritium polarimeter (utilizing the reaction $^3H(d,n)^4He$) and a $T_{20}$ polarimeter. The measurement of the average degree of polarization over the storage cell is essential to the determination of the analyzing powers. At least two independent measurements of this property will be used.

## Test results

Several test measurements[3] were performed with a stored electron beam to obtain information on the spatial distribution of the stored beam, so that the diameter of the storage cell could be selected. The first tests were performed with a collimator at the internal target position and two scintillator telescopes. The stored electron beam with an energy of 400 MeV had a lifetime of appr. 100 s. The countrate was measured as a function of the position of each of the scrapers of the collimator. All measurements showed the same behavior, a constant background rate until a position is reached at which the countrate starts to increase drastically. Also the beam lifetime was measured as function of the position of the scrapers, showing that in the domain where the scintillators could be operated, there was no effect whatsoever on the lifetime of the beam. The variation of the lifetime as a function of the scraper position indicates that the tail of the beam distribution is non-gaussian. This is confirmed by the countrate data, which also show that the distribution has tails, extending over 20 times the predicted value of the gaussian-sigma of the beam.

The complete set-up, including a 20 mm diameter storage cell, has been tested with a stored beam at an energy of 400 MeV, an injected intensity of 10 mA and a lifetime of 300 s. All detection systems and the atomic beam source were operated under realistic beam conditions and coincident electron-proton events were detected. Fig. 2 shows the data as a function of the position along the target cell (histogram), together with the results of a Monte Carlo simulation (solid curve). The good agreement between the calculated and observed distribution indicates a good performance of the complete detector set-up.

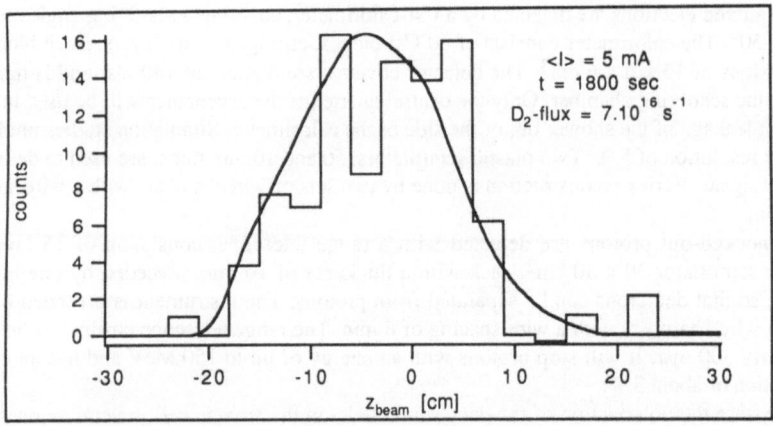

Fig. 2.  Observed (histogram) and calculated (solid curve) distribution of
events along the target cell.

## Model predictions and Monte Carlo simulations

Extensive calculations on the electro-disintegration of the deuteron have been performed by
Arenhövel et al.[4]. They predict sizable asymmetries for quasi-elastic knock-out under
kinematical conditions $E_e$ = 700 MeV and $\vartheta_e$ = 35°. The calculations are sensitive to specific
aspects of the reaction mechanisms (like Plane Wave versus Distorted Wave Impulse
Approximation) and to nuclear structure effects (sensitivity to the NN-potential used, meson
exchange effects and delta-contributions). By an appropriate choice of the kinematics it is
possible to emphasize the relative contribution of specific diagrams.

The deuteron is a spin-one system with the nucleons predominantly in a relative S-state.
However, due to the tensor part in the NN-interaction there is also a considerable D-state
component in the wave function. This component varies significantly (4-7%) for the various
potentials used. In the pure Impulse Approximation there is a straightforward relation
between the tensor analyzing power and the spin structure of the deuteron. Fig. 3a shows the
predicted tensor analyzing power with the target nuclei polarized parallel to the beam. The
FSI-effects are quite small, particularly for values of the p-n angle in the CM system $\Theta < 60°$.
Under such conditions the data would allow a separation of the S-wave and D-wave functions
of the deuteron ground state in momentum space. In the initial phase of the ring and the
internal target facility only modest luminosities of the order of $10^{30}$ cm$^{-2}$s$^{-1}$ are expected. The
error bars indicate the accuracy that can be obtained with 300 beam hours at this luminosity,
demonstrating that even under those modest conditions the asymmetries are measured with a
good accuracy.

In a next stage we also plan to measure the vector analyzing powers. These vector analyzing
powers are especially sensitive to the final state interaction. In PWIA the vector analyzing
power vanish, but in the calculations of ref.[4] the expected analyzing powers are large, and

insensitive to contributions of meson-exchange curves and delta contributions. Fig. 3b shows the vector analyzing power for the above mentioned kinematics. The error bars show the predictions for a luminosity of $4.10^{30}$ cm$^{-2}$s$^{-1}$.

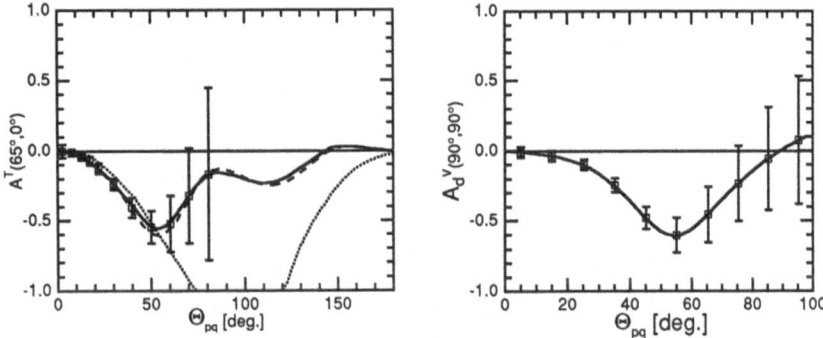

Fig. 3. Asymmetries for quasi-elastic knock-out at $E_e$ = 700 MeV and $\vartheta_e$ =35° as a function of $\Theta$, predicted for the Paris potential by Arenhövel et al.[4]. The error bars indicate the accuracy obtained in 300 hours. Left: Tensor asymmetry $A^T(65°,0°)$ with the target polarization parallel to the beam. Dotted: PWIA; full: PWIA + FSI; dashed: PWIA + FSI + MEC. Right: vector analyzing power $A_d^V(90°,90°)$. The curve corresponds to the prediction of the full calculation.

## References

1 NIKHEF-K proposals 91-12 and 93-04.
2 C.W de Jager, Proc. RCNP-Kikuchi school on "Spin Physics at Intermediate Energies", Osaka, 1992, to be published.
3 C.W. de Jager, Proc.Workshop on "Polarized Ion Sources and Polarized Gas Targets", Madison, 1993, eds. L.W. Anderson and W. Haeberli.
4 H. Arenhövel, W. Leidemann and E.L. Tomusiak, Phys. Rev. **C46** (1992) 455, and references given there.

Few-Body Systems, Suppl. 7, 260–264 (1994)

Few-
Body
Systems
© Springer-Verlag 1994
Printed in Austria

# OUTGOING NUCLEON POLARIZATION IN
# EXCLUSIVE DEUTERON ELECTRODISINTEGRATION

B. Mosconi$^{(+,*)}$, J. Pauschenwein$^{(**)}$ and P. Ricci$^{(*)}$

(+) *Dipartimento di Fisica, Università di Firenze, Firenze, Italy*

(*) *Istituto Nazionale di Fisica Nucleare, Sezione di Firenze, Firenze, Italy*

(**) *Institut für Theoretische Physik, Universität Graz, Graz, Austria*

**Abstract**. Nucleon polarization in deuteron electrodisintegration with polarized beam and target is studied in the standard theory with inclusion of relativistic corrections. The sensitivity of nucleon polarization to the neutron charge form factor $G_E^n$ is discussed at the quasi-elastic peak. Two new components very sensitive to $G_E^n$ models are picked out.

## I. Introduction

We study the outgoing nucleon polarization in the exclusive deuteron electrodisintegration in quasifree scattering in the case of polarized beam and target. In general, the purpose of the study of spin observables is to exploit the enhanced sensitivity of such non-averaged observables in order to test specific points of the theory or to obtain information about badly known quantities such as the neutron charge form factor $G_E^n$. In this note we shall concentrate on the last point. The potential of nucleon polarization in this reaction as a source of information on $G_E^n$ has been pointed out several years ago by Arnold et al.[1] who suggested a measurement of the sideways component of the neutron beam polarization transfer $P_{0x}^h(n)$ in the forward direction. A detailed analysis of the dependence of $P_{0x}^h(n)$ on NN potential models and on two-body effects has been performed by Arenhövel et al. [2]. Unlike previous authors we have considered [3] also the case of deuterons polarized with axial symmetry so that their state of orientation is defined by the vector $P_1^d$ and tensor $P_2^d$ parameters. Defining $P_0^d = 1$, the polarization of the detected nucleon can be written in the compact form

$$\left(\frac{d\sigma}{dE_{e'} d\Omega_{e'} d\Omega_N^{cm}}\right) \mathbf{P} = \left(\frac{d\sigma}{dE_{e'} d\Omega_{e'} d\Omega_N^{cm}}\right)_0 \sum_{I=0}^{2} P_I^d \left[\mathbf{P}_I + h\mathbf{P}_I^h\right] \quad , \qquad (1)$$

where $(d\sigma / dE_{e'} d\Omega_{e'} d\Omega_N^{cm})_0$ is the cross section with unpolarized beam and target and $h$ is the electron beam helicity. We refer to [3] for the explicit expressions of the Cartesian components of the polarization vectors $\mathbf{P}_I$ and $\mathbf{P}_I^h$ in terms of the virtual photon density matrix, the T-transition matrices and the polar angles defining the nucleon momentum and the deuteron symmetry axis. Here we present the two new polarization components as sensitive as $P_{0x}^h(n)$ on $G_E^n$ models picked out in our systematic survey [3]. We also discuss the sensitivity of the results on the choice of the em form factors of the nucleon, essentially the choice of the Dirac $F_1$ or the Sachs $G_E$ form factor as nucleon charge density.

## II. Results

Our results are for the coplanar kinematics in the quasielastic region at $Q^2 = 12 fm^{-2}$ with the deuteron symmetry axis in the reaction plane ($\theta_d = 90°$, $\phi_d = 0°$). In this case only the component of $\mathbf{P}_0$, $\mathbf{P}_2$, and $\mathbf{P}_1^h$ normal to the reaction plane and the two components of $\mathbf{P}_0^h$, $\mathbf{P}_1$, and $\mathbf{P}_2^h$ lying in the reaction plane are non vanishing because of parity conservation. In order to set the scale of the dependence of the nucleon polarization on $G_E^n$ we compare the results obtained with the dipole fit and the Galster model [4] of $G_E^n$ (with p=5.6) and with the same fit and with $G_E^n = 0$. In all the calculations we used the Paris potential [5].

Our calculations are based on the standard nuclear theory with inclusion of meson exchange currents (MEC) and isobar excitation currents (IC) of pionic range and of the most relevant relativistic corrections (RC) evaluated in a perturbative expansion [3,6]. To be explicit, we include the RC's of order $(v/c)^2$ beyond the nonrelativistic one-body charge ($\rho$) and current ($\mathbf{j}$) densities as obtained by a Foldy-Wouthuysen transformation as well as those beyond the pionic $\rho$ in the pseudovector (PV) coupling theory. In addition, we consider the effect of the relativistic transformation of the rest frame wave function which is conveniently translated into effective electromagnetic (em) operators. We recall that only the kinematic boost terms contribute because the pionic ones vanish for PV $\pi$N coupling [7]. Here again we have considered corrections only in the charge density. We would like to note that the neglect of pionic and nuclear motion RC in $\mathbf{j}$ should not be crucial because the knowledge of the corresponding $\rho$ allows us to also approximately evaluate the transverse electric transitions through the use of the Siegert theorem. Such perturbation expansion theory is rather successful at moderate $Q^2$. In fact, our results [8] of the $\phi$-asymmetry of the cross section recently measured at NIKHEF-K [9], are hardly distinguishable from those obtained by Hummel and Tjon [10] in a complete covariant model based on the Bethe-Salpeter equation. The situation may well change at higher $Q^2$ where higher order terms in the perturbative expansion become necessary, because it seems to converge badly.

As said in the Introduction, we have found two new observables as sensitive as $P_{0x}^h(n)$ on $G_E^n$ models. The first one is the longitudinal component of the vector target polarization transfer for emitted neutrons $P_{1z}(n)$, which is reported in Fig. 1 as a function of the center of mass (cm) angle measured from the virtual photon direction. The sign of $\vartheta_{cm}$ has been assumed positive in the half-plane $\phi = 0°$ and negative in the half-plane $\phi = 180°$. Three theoretical approximations ( IA theory, which corresponds to nonrelativistic nucleonic only contributions with final state interaction included; IA+MEC+IC theory, which takes into account MEC and IC ; full theory, which also includes RC ) have been considered in calculations with $G_E^n \neq 0$ in order to show the importance of mesonic and relativistic effects. For clarity, only the full theory results with $G_E^n = 0$ are reported. The strong sensitivity of $P_{1z}(n)$ on $G_E^n$ in the forward direction is manifested by the difference between the solid curve and the dashed curve. The reason for such a large sensitivity of $P_{1z}(n)$ (as well as of $P_{0x}^h(n)$) on $G_E^n$ is that, they are largely determined at $\vartheta_{cm}=0°$ by a longitudinal-transverse structure function, i.e., by the interference between the charge and magnetic neutron form factors. Comparing the two figures it also clearly appears that the full theory results are almost independent of the choice of the Dirac form (Fig. 1a) and the Sachs form (Fig. 1b) of the nucleon em form factors. It is worth noting that such similarity comes out from rather different nonrelativistic and relativistic contributions ( and far from $\vartheta_{cm}=0°$ also from different two-body contributions).Indeed, RC's considerably increase $P_{1z}(n)$ in calculations with $F_1$, causing even a change of sign, while they induce a small lowering in calculations with $G_E$.

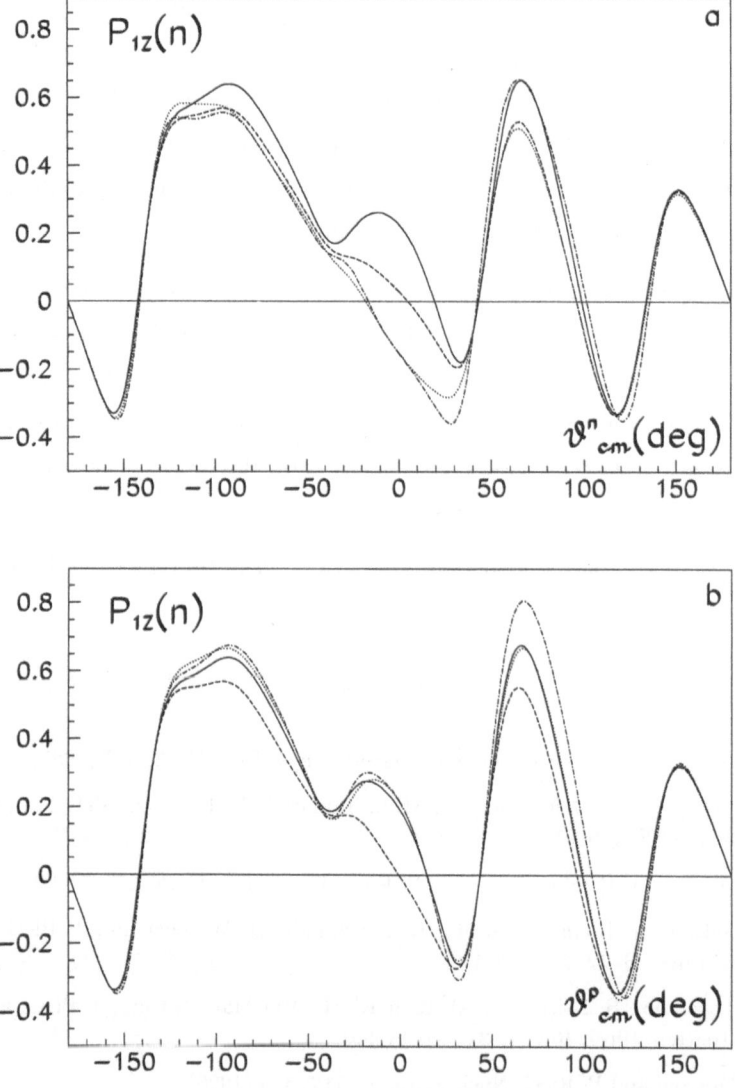

**Fig. 1.** Vector target polarization transfer of the neutron. Calculations with $G_E^n \neq 0$ (Dot-dashed line, IA theory; dotted line, IA+MEC+IC theory; solid line, full theory ). Calculations with $G_E^n = 0$ ( Dashed line, full theory ). The curves in Fig. 1a(1b) are for the Dirac (Sachs) form of the nucleon current.

As is well known, a measurement of a longitudinal polarization presents an additional difficulty. Indeed, it must be turned into a transverse polarization to be measured with the focal plane polarimeters. However, such an experiment seems to be well within the range of the experimental possibilities in the near future in view of the recent advances in the polarized target technique and in the focal plane polarimeters.

The second observable sensitive to $G_E^n$ is the sideways component of tensor target - beam polarization transfer for emitted protons $P_{2x}^h(p)$ at $\vartheta_{cm} = 180°$. Because of lack of space and since the measurement of proton polarization in such conditions is beyond the range of the present experimental possibilities, we do not report the angular distribution of $P_{2x}^h(p)$, which can be found in [3].

In conclusion, we have presented two new components , $P_{1z}(n)$ and $P_{2x}^h(p)$ , of the nucleon polarization in deuteron electrodisintegration which are as sensitive as $P_{0x}^h(n)$ on $G_E^n$ models. We have also shown that the full theory results, i.e. with RC included, are little dependent on the use of the Dirac or the Sachs form of the nucleon current. This fact greatly reduces the ambiguity in the theoretical analysis of deuteron electrodisintegration at the quasi elastic peak.

This work was partly supported by Ministero della Università e della Ricerca Scientifica of Italy.

## References

[1]   R. G. Arnold, C. E. Carlson and F. Gross, Phys. Rev. C **23**, 363 (1981).

[2]   H. Arenhövel, W. Leidemann and E.L. Tomusiak, Z. Phys. **A 331**, 123 (1988); **A334**, 363(E) (1989).

[3]   B. Mosconi, J. Pauschenwein and P. Ricci, Phys. Rev. C **48**, 332 (1993).

[4]   S. Galster, H. Klein, J. Moritz, K.H. Schmidt, D. Wegener and J. Bleckwenn, Nucl. Phys. **B 32**, 221 (1971).

[5]   M. Lacombe , B. Loiseau , J.M. Richard , R. Vin Mau, J. Côté, J. Pirés, and R. De Tourreil, Phys. Rev. C **21**, 861 (1980).

[6]   B. Mosconi and P. Ricci, Nucl. Phys. **A 517**, 483 (1990).

[7]   J. L. Friar , Ann. Phys.(NY) **104**, 380 (1977).

[8]   B. Mosconi, J. Pauschenwein and P. Ricci, Few-Body Systems, Suppl. **6**, 223 (1992).

[9]   M. van der Schaar, H. Arenhövel, H.P. Blok, H. J. Bulten, E. Hummel, E. Jans, L. Lapikás, G. van der Steenhoven, J.A. Tjon, J. Wesseling and P.K.A. de Witt Huberts, Phys. Rev. Lett. **68**, 776 (1992).

[10]  E. Hummel and J.A. Tjon, Phys. Rev. Lett. **63**, 1788 (1989); Phys. Rev. C **42**, 423 (1990).

Few-Body Systems, Suppl. 7, 265—269 (1994)

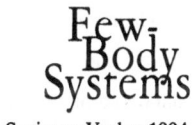

Few-
Body
Systems
© Springer-Verlag 1994
Printed in Austria

# TRINUCLEON THRESHOLD ELECTRODISINTEGRATION

G.A. Retzlaff, H.S. Caplan, E.L. Hallin and D.M. Skopik

Saskatchewan Accelerator Laboratory, University of Saskatchewan,
Saskatoon, Saskatchewan, Canada S7N 0W0

D. Beck, K.I. Blomqvist, G. Dodson, K. Dow, M. Farkhondeh, J. Flanz, S. Kowalski,
W.W. Sapp, C.P. Sargent, D. Tieger, W. Turchinetz and C.F. Williamson

Bates Linear Accelerator Center and Laboratory for Nuclear Science,
Massachusetts Institute of Technology, Cambridge, Massachusetts

W. Dodge, X K. Maruyama and J.W. Lightbody, Jr.

National Institute of Standards and Technology, Gaithersburg, Maryland 20899

R. Goloskie

Department of Physics, Worcester Polytechnic Institute, Worcester, Massachusetts 01601

J. McCarthy, T.S. Ueng and R.R. Whitney

Department of Physics, University of Virginia, Charlottesville, Virginia 22901

B. Quinn

Department of Physics, Carnegie-Mellon University, Pittsburgh, Pennsylvania 15213

S. Dytman and K. Von Reden

Department of Physics, University of Pittsburgh, Pittsburgh, Pennsylvania 15260

R. Schiavilla

Instituto Nazionale di Fisica Nucleare, Lecce, Italy

J.A. Tjon

Institute for Nuclear Physics, University of Utrecht, 3508 TA Utrecht, the Netherlands

## Abstract

Inclusive inelastic electron scattering cross sections for $^3$H and $^3$He were measured for excitation energies below 18 MeV. For six values of the three-momentum transfer q in the range $0.88 < q < 2.87$ fm$^{-1}$, longitudinal and transverse response functions were determined. The experimental data are in good agreement with two recent calculations. One uses variational ground-state wave functions and the orthogonal correlated states method to describe the two- and three-body breakup channels. The other uses bound and continuum Faddeev wave functions for a simple central potential. The inclusion of final-state interactions (FSI) in the Faddeev continuum is found to be very important; inclusion of FSI changes the response functions in the threshold kinematics by a large amount, yielding excellent agreement with the data.

The nucleus is usually described as a collection of non-relativistic nucleons interacting through phenomenological potentials produced by boson exchange. The newest models also contain some non-nucleonic degrees of freedom (DOF) such as meson exchange currents (MEC), nucleon resonances, or three-body potentials. Final state interactions (FSI) are also included in some of these newest models. For the 2- and 3-nucleon systems, exact calculations can be carried out for given nucleon-nucleon potentials, within the framework of this model. Exact, non-relativistic wave functions that include some of the non-nucleonic DOF can be formulated for both two- and three-body bound states and the continuum. Calculating the three-body breakup requires the solution of the three-body equations for the final state, a difficult task. In contrast to the quasielastic region, where relatively low-momentum components of the ground-state wave function are dominant, in the threshold region one can probe the high-momentum parts of the ground-state wave function, due to the mismatch between momentum and energy transfer.

This work deals with the low excitation energy region for inclusive inelastic electron scattering on the $^3$H and $^3$He nuclei. Longitudinal and transverse response functions $R_L$ and $R_T$ were derived from Rosenbluth separations at momentum transfers of 0.88, 1.28, 1.64, 2.08, 2.47 and 2.87 fm$^{-1}$. The targets consisted of cryogenic gas cells filled with $^3$H, $^3$He or $^{1,2}$H. The virial coefficients for $^3$H and $^3$He are not known well enough at our 45 K target temperature. The principle of corresponding states [1] was used to derive densities from other gases with well known coefficients. The subtraction of the radiative tail from elastic scattering was critical in the data analysis, and data with a $^1$H target (which has no inelastic contribution below pion threshold) showed the effectiveness of the subtraction. Last, Rosenbluth separations were performed to derive the longitudinal and transverse response functions from the cross sections. A complete compilation of the data is found in Ref. [2].

Figures 1 and 2 show the response functions, $R_L$ and $R_T$ for each of $^3$He and $^3$H, at momentum transfers of 0.882 and 2.47 fm$^{-1}$ respectively. The zero range approximation (ZRA) model of Ref. [3] was reworked to include $^3$H, and $R_L$ was derived for both nuclei. The ground-state wave function in the ZRA is the solution of a zero range nuclear force whose depth was chosen to fit the ground-state binding energy and radius of the nucleus. The final state was a plane wave model, with its S wave components shifted by the empirical nucleon-deuteron and nucleon-nucleon phase shifts. At our lowest q we have good agreement but as the momentum transfer increases, the model fails to describe the data. A feature of these data is the presence of the enhancement near threshold in the $^3$He but not in the $^3$H, longitudinal response function, more pronounced at lower q. The ZRA model allowed us to examine the multipoles contributing to the response functions. We discovered that the cross-section near threshold for $^3$He was totally dominated by a C0 monopole transition, and this strength was absent in $^3$H and the transverse response.

Two modern calculations that make a serious attempt to treat the three-body continuum states are those of the Urbana [4] and Utrecht [5] groups. Both of these models have been previously applied to the quasielastic peak in the three-nucleon system [5,6].

The Urbana calculation uses a variational ground-state wave function obtained from a realistic Hamiltonian with two- and three-nucleon interactions. The deuteron+nucleon and (nucleon+nucleon)+nucleon continuum wave functions are treated with the Orthogonal Correlated States (OCS) method developed in Ref. [4]. Some of the FSI affecting the knocked-out nucleon in the two-body breakup channel are approximately taken into account by a deuteron+nucleon optical potential. Short-range correlations, orthogonality corrections as well as FSI effects bring the calculated longitudinal and transverse response functions into reasonable agreement with all the data, as shown (OCS, dashed lines) in Fig. 1 and 2. The agreement with the data is somewhat poorer with increasing q

For the Utrecht calculation, the Faddeev equations are solved exactly for the ground and the scattering states using the spin-dependent s-wave Malfliet-Tjon local nucleon-nucleon interaction. No Coulomb effects are included in these calculations, no tensor force is present, and the wave functions are non-relativistic. When this model is compared to the quasielastic data[6], $R_T$ falls below the data, perhaps caused by the lack of MEC in the calculations (not included in either calculation). The thin solid lines in Fig. 1 and 2 correspond to a plane wave impulse approximation final state (PWIA), the heavy lines correspond to the full calculation. Only for the $^3$H transverse response at higher q, where the calculation is ~75% of the data, is there a significant disagreement, otherwise the agreement is excellent. There is a very large difference between the PWIA and the full calculation in some cases, pointing out the importance of FSI. At the QE peak the responses change little (no more than 30% at most), but the threshold region shows much greater differences.

The shape of the transverse response is well described by the Utrecht model. Contributions from MEC to the response functions, which mainly effect the transverse response, have not been included in either model. One nucleus is well described, but not the other. The $^3$He results imply that MEC are not very important or are cancelled out by other effects, while the $^3$H results imply otherwise. Increasing q increases the importance of MEC; this effect is seen as a worsening of the agreement for higher q.

Threshold electron scattering response functions for $^3$He and $^3$H have been measured and compared to several models of the bound and unbound three-nucleon system. The analytic ZRA model yields fair agreement only in the expected range of applicability (low q), but do shed light on the characteristics of the C0 dominance in the two-body breakup near threshold. The modern calculations give good agreement with the data, the Utrecht calculation agrees almost exactly for $R_L$ and the $R_T$ for $^3$He and show most of the observed strength for the $R_T$ for $^3$H. The data not presented in this paper also agree well with the Utrecht work. The Utrecht theory demonstrates the tremendous influence of FSI in the threshold kinematics, which are much larger than in the QE region. Since MEC are not included in the models, it is somewhat surprising (and perhaps fortuitous) that the calculation agreed so well with the measured transverse response functions, and why the $^3$He calculation agrees with the data while the $^3$H response does not. The agreement for the transverse responses worsens as q increases, implying that the MEC are more important as the momentum transfer increases.

# REFERENCES

[1] J.O. Hirschfelder et al., *Molecular Theory of Liquids and Gases* (John Wiley and Sons, 1954), Chapter 4; J.H. Dymond and E.B. Smith, *The Virial Coefficients of Pure Gases and Mixtures* (Clarendon Press, 1980); R.D. McCarty et al., *Selected Properties of Hydrogen*, NBS Monograph 168 (1981).

[2] G. A. Retzlaff, Ph.D. thesis, University of Saskatchewan, 1988.

[3] P.T. Kan et al., Phys. Rev. C **12**, 1118 (1975).

[4] R. Schiavilla and V. Pandharipande, Phys. Rev. C **36**, 2221 (1987); R. Schiavilla, Phys. Lett. **218B**, 1 (1989).

[5] E. van Meijgaard and J.A. Tjon, Phys. Rev. C **45**, 1463 (1992); E. van Meijgaard, Ph.D thesis, Rijksuniversiteit Utrecht, 1989.

[6] K. Dow et al., Phys. Rev. Lett. **61**, 1706 (1988); K. Dow, Ph.D. thesis, MIT, 1987; T. S. Ueng, Ph.D thesis, University of Virginia, 1988.

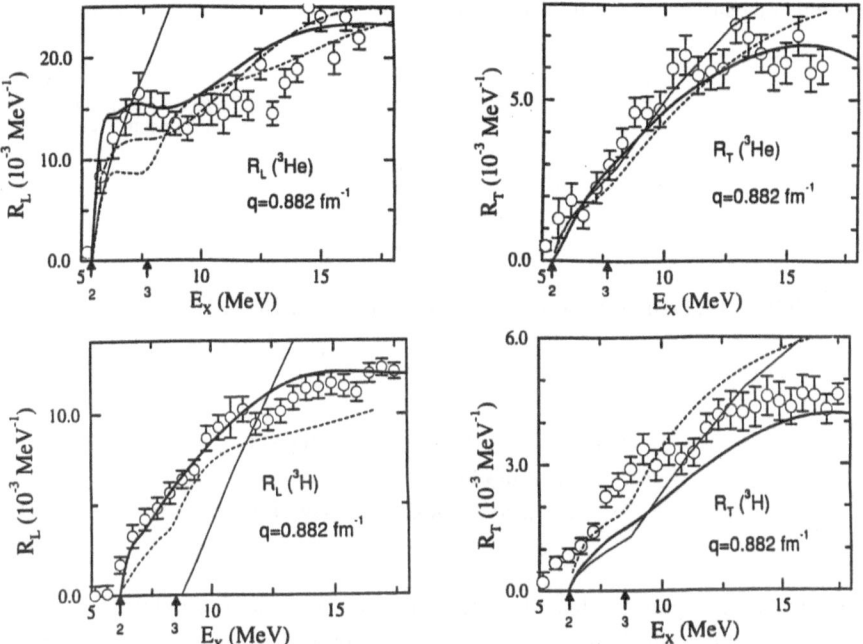

Figure 1: (a) $^3$He longitudinal, (b) $^3$H longitudinal, (c) $^3$He transverse and (d) $^3$H transverse response functions respectively at $q=0.882$ fm$^{-1}$. Circles: data; Dot-dash line: ZRA ($^3$He longitudinal response only); Dashed line: OCS calculation; Thin solid line: Utrecht PWIA; Thick solid line: Utrecht total.

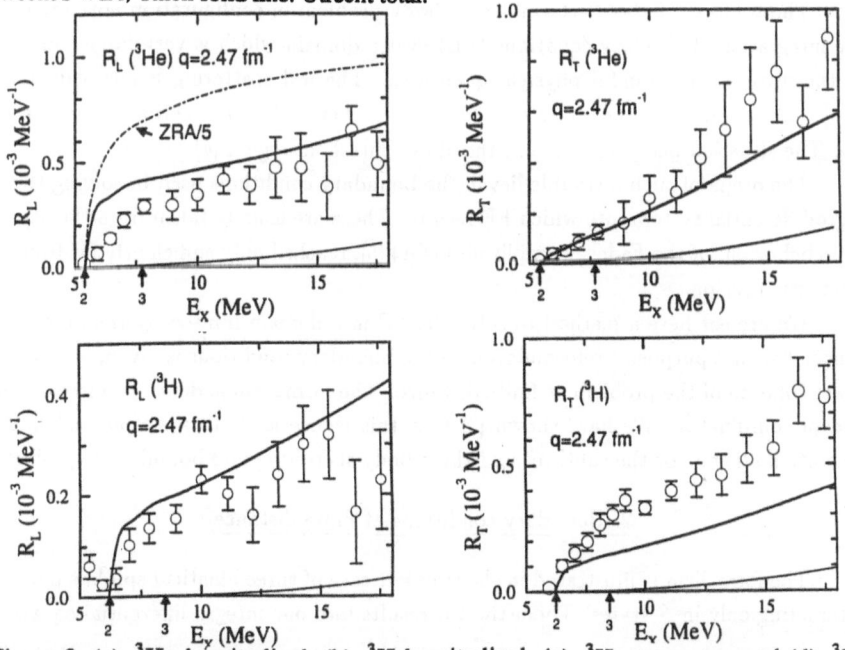

Figure 2: (a) $^3$He longitudinal, (b) $^3$H longitudinal, (c) $^3$He transverse and (d) $^3$H transverse response functions respectively at $q=1.64$ fm$^{-1}$. Circles: data; Thin solid line: Utrecht PWIA; Thick solid line: Utrecht total.

Few-Body Systems, Suppl. 7, 270—273 (1994)

# SOLVING FADDEEV EQUATIONS IN THE INTERACTION DOMAIN

Jaume Carbonell and Claude Gignoux
Institut des Sciences Nucléaires
53, Av. des Martyrs, 38026 Grenoble, France

Abstract: By modifying the boundary conditions we show the possibility to solve the below threshold three-body problem in a domain restricted by the interaction range.

## 1. Introduction

The solution of the scattering three-body problem in configuration space requires the integration of Faddeev equations (FE) over a domain which is very large compared to the involved meaningful physical quantities. The n-d scattering length calculation [1-3] , for instance, extends over a $\sim 70$ fm grid whereas the interaction range is a few fm. The situation becomes worse in the above threshold region [4]

The origin of such a trouble lies in the boundary conditions used in solving the integrodifferential system into which FE results. These are usually taken from the asymptotic behaviour of the Fadeeev amplitudes (FA) [5], reached only sufficiently far from the interaction region.

We present here a method to solve the FE in a domain limited by the interaction range. For that purpose a reformulation of the boundary conditions is given, allowing an exact solution of the problem at finite distance. The formalism is developed in cartesian Jacobi coordinates. We have shown [3] that this choice leads to a simple and unified resolution scheme for the subthreshold three body scattering and bound state problems.

## 2. Boundary conditions at finite distance

The formalism is illustrated in the simplest case of three identical spinless particles interacting only in $S$-states. Then, the FE results into one integrodifferential equation:

$$\left[k^2 + \partial_x^2 + \partial_y^2 - V(x)\right]\varphi(x,y) = V(x)\int_{-1}^{1} du \frac{xy}{x'(u)y'(u)}\varphi(x'(u),y'(u)) \qquad (1)$$

where $x', y'$ are known functions and $k^2 = q^2 - \alpha^2$ the total three body energy.

Equation (1) is usually solved in a compact domain $\Omega = [0, x_{max}] \times [0, y_{max}]$ with the Dirichlet boundary conditions (Fig. 1a):

$$\varphi(0, y) = \varphi(x, 0) = \varphi(x_{max}, y) = 0 \qquad (2)$$

$$\varphi(x, y_{max}) = u_b(x) \qquad (3)$$

$u_b(x)$ being the two body asymptotic wavefunction and $\alpha^2$ its binding energy. For the bound state problem condition (3) is simply replaced by $\varphi(x, y_{max}) = 0$

The integral term in the right hand side of (1) creates an effective long range potential coupling equal-$\rho$ grid points up to very large values of the hyperradius $\rho$. Below that, the boundary conditions (2)-(3) are not justified and lead to unstable numerical results. However, if the interaction has finite range $a$, the inhomogeneous term in (1) vanishes for all $x \in \Omega_E = \{(x, y) \mid x > a\}$ and the FA is in fact the solution of a much simpler problem, the Helmholtz equation,

$$\left[k^2 + \partial_x^2 + \partial_y^2\right] \varphi(x, y) = 0 \qquad (4)$$

with its panoply of specific resolution methods.

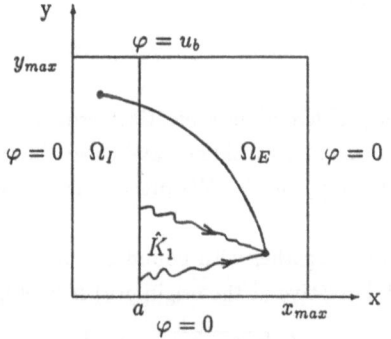
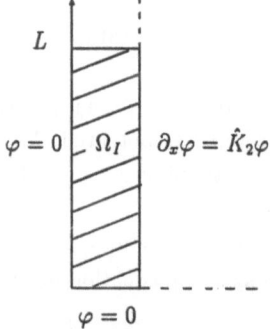

Figure 1a: Local boundary conditons      1b: Integral boundary conditions

Taking advantadge of this fact we used the two following results [6]:

1. The FA in $\Omega_E$ can be obtained from its value along the boundary x=a

$$\varphi(x', y') = \int_0^\infty K_1(x', y'; a, y)\varphi(a, y)\, dy \quad (x', y') \in \Omega_E \qquad (5)$$

This equation maps $\Omega_E$ into its boundary $x = a$. However, in order to solve the problem inside the interaction region $\Omega_I = \{(x, y) \mid x \le a\}$, an additional boundary condition along $x = a$ is required. It is provided by:

2. There exists a relation between the FA and its normal derivative in x=a

$$\partial_x \varphi(a, y') = \int_0^\infty K_2(y'; a, y)\varphi(a, y)\, dy \qquad (6)$$

where the kernels $K_n$ are known analytical functions [6], are independent of the interaction and select among the solutions of (4) those compatible with the proper boundary conditions (in this case outgoing wave functions).

Thus, though the coupling $\Omega_I - \Omega_E$ is unavoidable and the FA in $\Omega_E$ not negligible at all, the last two equations make possible the resolution of the three-body problem in a domain limited by the interaction range. Some remarks are in order:

i)  The main change of this method consists in replacing the local Dirichlet condition $\varphi = 0$ by equation (6) (Fig. 1b). This integral boundary condition is exact for a finite range interaction. Its validity is not based on any asymptotic expansion.

ii)  There exists an equivalent formulation of equations (5-6) which summarizes them in a single relation:

$$\varphi(x',y') \;=\; \int_0^\infty K_3(x',y';a,y)\partial_x\varphi(a,y)\,dy \quad (x',y') \in \Omega_E \tag{7}$$

iii)  In practical calculations $\Omega_I(a)$ is replaced by a compact domain $\Omega_I(a,L) = [0,a] \times [0,L]$ whereas (5-6) requires to integrate the solution $\varphi$ over $[0,\infty]$. The values $\varphi(a,y > L)$ are obtained from the asymptotic behaviour $\varphi(a,y) \sim u_b(a) \sin(qy + \delta)$

### 3. Some applications

#### The zero range potential

A first illustration is provided by the problem of three identical fermions interacting via a zero range potential. This problem was first considered and solved in momentum space for the limiting case of zero total energy in [7]. We present here a solution in configuration space.

For a total spin $J = 3/2$, equation (1) together with boundary condition (7) lead to the following equation for the normal derivative at the origin $g(y) \equiv \partial_x\varphi(0,y)$:

$$g(y) + \alpha \int_0^\infty dy' \left[ K_3(|y - y'|) - \frac{4}{y\sqrt{3}}K_3\left(\sqrt{\frac{3y^2}{4} + (\frac{y}{2} - y')^2}\right)\right] g(y') = 0$$

Its numerical solution gives $A_{\frac{3}{2}} = 5.09\,fm$ (5.1 in [7])

#### The n-d scattering length

The following example is the n-d scattering length in the frame of the MT I-III NN potential model. The quartet scattering length is $A_{\frac{3}{2}} = 6.441 \pm .001\ fm$, obtained by several authors [1-3], by integrating FE over a $\sim 60\ fm$ integration domain.

In Table I are displayed the results provided by the integral boundary condition (6) as a function of the interaction range $a$. Together with the scattering length are given the deuteron binding energy calculated consistently and the potential value at $x = a$. The x-grid is the same up to 4 $fm$ and continued by equally spaced points $x_i$ up to $x = a$. Results displayed in Table I show the applicability of our method for realistic

| $a$(fm) | $B_d$(MeV) | $A$(fm) | V(MeV) |
|---------|-----------|---------|---------|
| 60 | 2.230 | 6.441 | - |
| 10 | 2.230 | 6.441 | -.00001 |
| 8 | 2.230 | 6.441 | -.00032 |
| 6 | 2.230 | 6.441 | -.00955 |
| 5 | 2.228 | 6.440 | -.05395 |
| 4 | 2.212 | 6.444 | -.31663 |

Table 1: n-d quartet scattering length $A$ as a function of the interaction range $a$

nuclear potentials. We remark the high stability of the results down to values of $a$ for which the potential is not negligible. The same quality is found in the doublet case.

A weakly bound system: the $^4He$ atomic trimer

A last example is given by the $^4He$ atomic trimer. It is a very weakly bound, and so very extended, system. A binding energy of 0.096 $°K$ was found in [3] with only S wave component in the FA and requiring an integration over 80 Å. The integral boundary conditions method has been also applied for bound states and provides the same binding energy with $a = 8$ Å. The main interest of this result is the possible application to the resarch of Efimov states.

### 3. Conclusion

We have solved the below threshold three body problem in a domain restricted by the interaction range. The usual boundary condition, based on the asymptotic behaviour at infinity of the FA, is replaced by an integral boundary condition which is exact when the interaction vanishes. This allows a one order of magnitude reduction in the integration domain. The method should be generalized to the above threshold region and to the four body problem.

### References

1. J.J. Benayoun, C.Gignoux, J. Chauvin, Phys. Rev. C **23** (1981) 1854

2. G.L. Payne, J.L. Friar, B.F. Gibson, Phys. Rev. C **26** (1982) 1385

3. J. Carbonell, C. Gignoux, S. P. Merkuriev, Few-Body Systems (1993)

4. W. Glockle, G.L. Payne, Phys. Rev. C45 (1992) 974

5. S.P. Merkuriev, C. Gignoux, A. Laverne, Ann. of Phys. (N.Y.) **99** (1976) 33

6. C. Gignoux, J. Carbonell, to be publisehd

7. G. V. Skorniakov, K. A. Ter-Martirosian, Sov. Phys. JETP **4, 5** (1957) 648

Few-Body Systems, Suppl. 7, 274—277 (1994)

# CLUSTER-DYNAMICAL TREATMENT OF THREE-NUCLEON FORCES

H. Haberzettl and W.C. Parke
Center for Nuclear Studies, Department of Physics
The George Washington University, Washington, D.C. 20052, U.S.A.

Three-body-force contributions entering three-nucleon equations are discussed within the relativistic cluster-dynamical scattering formalism recently proposed by Haberzettl. It is argued that, in addition to the three-body forces usually taken into account, there is an effective enhancement of two-nucleon-force contributions in the medium of three nucleons. The dominant enhancement of this type, a one-pion-exchange three-body force, is included in a five-channel triton binding-energy calculation based on the Paris potential. It is found to provide an increased binding of about 0.6 MeV, a value which closely corresponds to the amount by which the binding energy obtained in coupled-channel calculations with explicit $\Delta$ degrees of freedom differs from experiment.

Triton binding energies calculated with realistic nucleon-nucleon ($NN$) interactions typically fall short of the experimental value of 8.48 MeV by about 0.5-1 MeV. Attempts to resolve this discrepancy in terms of meson-exchange-based three-body forces provide encouraging but, at present, not entirely satisfactory results. (For a review of different three-body-force ansätze, see [1].) The dynamically most detailed description of meson contributions in the three-nucleon system is provided by the coupled-channel approach pioneered by the Hannover group [2]. Using the Paris potential, they find that employing explicit $\Delta$ channels raises the binding energy by 0.4 MeV to 7.85 MeV, which still is about 0.6 MeV short of the experimental value. Extending the investigation of the Hannover group by including multi-$\Delta$ effects, the authors of Ref. [3] recently concluded that $\Delta$ effects alone cannot resolve the discrepancy between theoretical results and experimental findings.

The equations employed in Refs. [2,3] are modeled after nonrelativistic Faddeev-type formulations and as such cannot account for a large class of reaction mechanisms associated with crossed contributions. This drawback is due to the sequential nature of the intermediate meson exchanges which implicitly describes the problem as if the number of particles were a good order parameter when, in fact, it is not.

To consistently account for all meson absorption and production mechanisms in a many-baryon system, including crossed terms, Haberzettl [4] recently put forward a scattering formalism based on clusters rather than individual particles. In its full implementation, this approach provides a fully covariant relativistic, nonlinear description of collision processes for arbitrarily large systems of baryons and mesons.

The starting point of the cluster-dynamical approach of Ref. [4] is the decomposition of the asymptotic Hilbert space of a scattering event into a direct sum of two-cluster sectors, i.e., $\mathcal{H} = \bigoplus_\tau \mathcal{H}_\tau = \bigoplus_\tau (\mathcal{H}_\alpha \otimes \mathcal{H}_\beta)$, where $\tau = (\alpha,\beta)$ labels the partitions consisting of two subclusters $\alpha$ and $\beta$. Each initial or final two-cluster state of the collision process is an element of a particular product subspace $\mathcal{H}_\tau = \mathcal{H}_\alpha \otimes \mathcal{H}_\beta$ of the total Hilbert space. For the three-nucleon problem, these asymptotic clusters consist of a two-nucleon subsystem and a single spectator nucleon, i.e., the cluster structure is (2+1). (It is not necessary to explicitly consider asymptotic configurations consisting of three noninteracting nucleons.) Rather than in terms of individual nucleons, three-body forces are then defined in terms of this (asymptotically measurable) cluster structure.

Restricting, for simplicity, the present discussion to a single pion, in addition to the three nucleons, and generically denoting the interacting $\pi N$ system by $\Delta$, the covariant equations for $(2+1) \rightarrow (2+1)$ transitions then take the diagrammatic form summarized in Fig. 1 [single lines: nucleons (solid), pions (dashed), $\Delta$'s (bold); hatched pair of lines: fully interacting two-nucleon system]. The driving term, Fig. 1(a), is given by all contributions which are irreducible with respect to the asymptotic (2+1) Hilbert-space structure. The corresponding scattering equation is shown in Fig. 1(b); its homo-

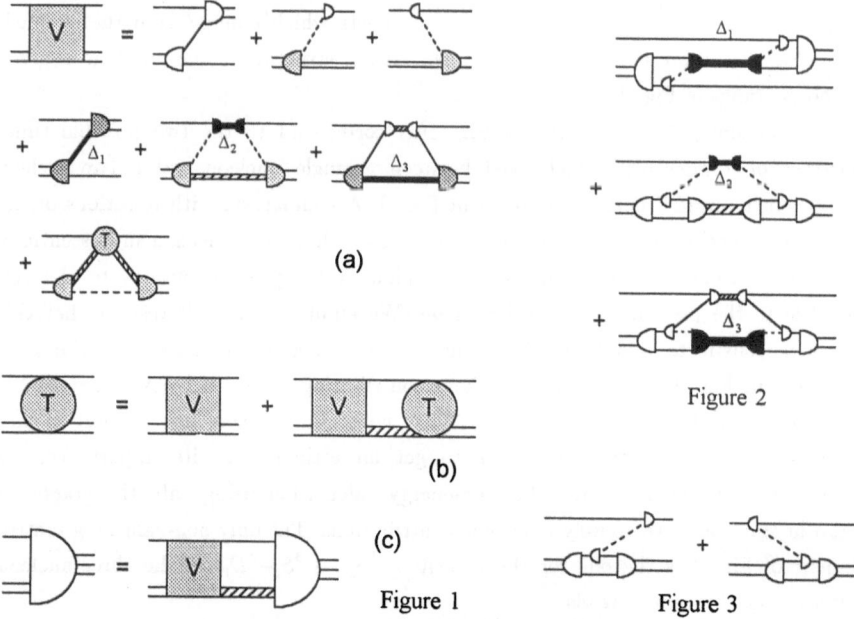

(a)

(b)

(c)

Figure 1

Figure 2

Figure 3

geneous version describing a bound state is depicted in Fig. 1(c). The usual nucleons-only three-body problem is obtained upon neglecting all graphs in Fig. 1(a) containing explicit pions or $\Delta$'s and retaining only the first nucleon-exchange term; in the nonrelativistic limit, the equation of Fig. 1(b) then reduces to the usual Faddeev-type formulation of the three-nucleon problem. The six additional terms here, with explicit pions or $\Delta$'s, represent the three-body-force contributions. For the present discussion, we will ignore the last graph of Fig. 1(a) where the complete $T$-matrix of Fig. 1(b) couples back onto itself nonlinearly.

Figure 2 provides the internal details of the three $\Delta$ graphs of Fig. 1(a). The usual $\Delta$-type three-body force [1-3] is contained in the $\Delta_1$ graph and the $\Delta_2$ and $\Delta_3$ graphs describe $\Delta$ propagation in the presence of fully interacting two-nucleon systems. In this context, we mention that there is no need here, as was done in Refs. [2,3], to subtract an energy-independent $\Delta$-loop piece from the underlying $NN$ potential. The nucleons in a two-body subsystem are fully interacting physical particles forming an asymptotically measurable physical compound system. The $T$-matrix describing their interaction, therefore, must correspond to the one used for the two-nucleon problem, without any subtraction [4], or else the scattering problem of Fig. 1(b) would have unphysical (i.e., unmeasurable) asymptotic states. However, we also note that the particular subtraction procedure adopted in [2,3] in effect provides an approximate treatment of the $\Delta_2$ and $\Delta_3$ graphs of Fig. 2 when the internal degrees of freedom of the spectator two-nucleon systems are frozen [4]. Therefore, we consider the binding-energy increase of about 0.4 MeV reported in Ref. [2] to provide a good indication of what to expect from a fully consistent incorporation of the $\Delta$ graphs within the present framework. The results found in [3] for multi-$\Delta$ contributions, by contrast, provide an inconclusive answer since the higher-order $\Delta$ effects which emerge from the coupled-channel approach are not consistent with the cluster structure of the problem as given by the $\Delta$ graphs of Fig. 2.

The remaining two diagrams in Fig. 1(a) correspond to the two possible time-orderings of a pion being exchanged between a single nucleon and a two-nucleon cluster; their internal details are given in Fig. 3. A comparison with iterations of the one-nucleon exchange graph in Fig. 1(a) shows that this mechanism essentially provides an enhancement of the basic one-pion exchange contributions to the $NN$ potential in the presence of a third nucleon. We emphasize in this respect that this does not constitute double-counting; the detailed dynamical context of the pion exchanges in Fig. 3 is different from those contributing to the full $NN$ $T$-matrix. To our knowledge, the mechanism of Fig. 3 has never been taken into account in any three-body-force calculation. In order to get an estimate for its importance, we performed a five-channel triton binding-energy calculation using only the graphs of Fig. 3 in addition to the purely nucleonic contributions. The only non-vanishing matrix element of Fig. 3 is the one for the transition $^1S_0 \rightarrow {}^3S_1 - {}^3D_1$; in the three-nucleon center-of-mass system, it reads

$$U_{10}^{ij}(\vec{q}\,',\vec{q};E) = \frac{F^2(Q^2)}{(2\pi)^3}\,D_{10}(\vec{q}\,',\vec{q};E)\,\langle d^i|\,G_0\big(E-\tfrac{3}{4m}q'^2\big)\frac{\vec{\sigma}_s\cdot\vec{Q}}{2m}\,\frac{\vec{\sigma}_b\cdot\vec{Q}}{2m}\,\vec{\tau}_s\cdot\vec{\tau}_b\,G_0\big(E-\tfrac{3}{4m}q^2\big)\,|s^j\rangle\,.$$

The Paris potential is employed here in a separable representation, according to the PEST prescription of the Graz group [5], and the states $|s^j\rangle$ and $|d^i\rangle = |g_0^i\rangle + (S_T/\sqrt{8})\,|g_2^i\rangle$ are the PEST form factors for $^1S_0$ and $^3S_1-{}^3D_1$, respectively; we took five terms for $|s^j\rangle$ and six for $|d^i\rangle$ [5]. $S_T$ is the usual tensor operator. The Green's function $G_0(e) = (e-h_0)^{-1}$ describes the free motion of the constituents inside the two-nucleon subsystem with $h_0$ being the kinetic-energy operator of their relative motion. The pion-nucleon couplings here are taken in the static limit: the resulting spin-$\frac{1}{2}$ Pauli matrices $\vec{\sigma}_s$ and $\vec{\sigma}_b$, respectively, pertain to the spectator nucleon or to the nucleon inside the bubble to which the pion attaches itself [see Fig. 3]; $\vec{\tau}_s$ and $\vec{\tau}_b$ are the corresponding isospin operators; the momentum transferred to the pion is $\vec{Q} = \vec{q}\,' - \vec{q}$. The static limit of the pion propagator $D_{10}(\vec{q}\,',\vec{q};E)$ reduces to the usual $-1/(\mu^2 + Q^2)$; in our calculation, we also employed a different form which accounts for retardation according to the two pion-exchange time-orderings in Fig. 3. The $N$-$N\pi$ vertex function $F(Q^2) = g_{\pi N}(\Lambda^2-\mu^2)/(\Lambda^2 + Q^2)$ is the same as in [2,3]. The parameters used here are $m = 938.92$ MeV, $\mu = 138$ MeV, and $\Lambda = 1200$ MeV for the nucleon, pion, and cutoff masses, respectively, and $g_{\pi N}^2/4\pi = 14.4$ for the coupling strength.

We find that the purely nucleonic value of the binding energy is increased by about 0.6 MeV. (Including meson retardation produces essentially the same result.) As mentioned above, this value closely corresponds to the amount by which the binding energy obtained in coupled-channel calculations with explicit $\Delta$ degrees of freedom [2] differs from experiment.

Let us emphasize that we do not claim that the present result is the definitive solution of the triton puzzle. For a complete answer, one should perform a comprehensive calculation within the cluster-dynamical framework outlined in Fig. 1, and the binding-energy increase reported here may then be altered by the presence of other, competing mechanisms. However, the present finding shows that the one-pion-exchange three-body force is an important mechanism which must be taken into account in a realistic calculation.

This work was supported in part by the U.S. Dept. of Energy, Grant No. DE-FG05-86-ER40270.

## REFERENCES

[1] *The Three-Body Force in the Three-Nucleon System* (Lecture Notes in Physics, Vol. **260**), ed. by B.L. Berman and B.F. Gibson (Springer, Berlin, 1986).

[2] Ch. Hajduk, P.U. Sauer, and W. Strueve, Nucl. Phys. **A405**, 581 (1983); Ch. Hajduk, P.U. Sauer, and S.N. Yang, *ibid.* **A405**, 605 (1983); M.T. Peña, H. Henning, and P.U. Sauer, Phys. Rev. C. **42**, 855 (1990).

[3] A. Picklesimer, R.A. Rice, and R. Brandenburg, Phys. Rev. C **44**, 1359 (1991); *ibid.* **45**, 547 (1992); *ibid.* **45**, 2045 (1992); *ibid.* **45**, 2624 (1992); *ibid.* **46**, 1178 (1992).

[4] H. Haberzettl, Phys. Rev. C **46**, 687 (1992); *ibid.* **47**, 1237 (1993); and submitted to Phys. Rev. C. See also contribution to this conference.

[5] J. Haidenbauer and W. Plessas, Phys. Rev. C **30**, 1822 (1984); *ibid.* **32**, 1424 (1985); J. Haidenbauer, Y. Koike, and W. Plessas, *ibid.* **33**, 439 (1986); J. Haidenbauer and Y. Koike, *ibid.* **34**, 1187 (1986).

Few-Body Systems, Suppl. 7, 278—281 (1994)

Few-
Body
Systems
© Springer-Verlag 1994
Printed in Austria

# VARIATIONAL CALCULATIONS FOR SCATTERING STATES IN THREE–NUCLEON SYSTEMS

A.Kievsky, M.Viviani and S.Rosati

Dipartimento di Fisica, Universita' di Pisa, Piazza Torricelli 2, 56100 Pisa, Italy
*INFN*, Sezione di Pisa, Piazza Torricelli 2, 56100 Pisa, Italy

## Abstract
The method of correlated hyperspherical basis functions can be succesfully applied to calculate three–nucleon scattering states. The results obtained for N–d elastic scattering compare very well with those available from Faddeev type approaches.

The Pair correlated Hyperspherical Harmonic (PHH) expansion has been succesfully used[1] to accurately evaluate the three–nucleon bound state properties. For instance, the results of the twelve channel calculation for the AV14 nucleon-nucleon (N–N) interaction coincide with those given by Kameyama et al.[2]. The method can be extended to study three–nucleon scattering states, and in this contribution the elastic N–d scattering below the deuteron break–up will be discussed. For such a process, the wave function is written as a sum of two terms, namely

$$\Psi = \Psi_C + \Psi_A .\qquad(1)$$

The first term $\Psi_C$ is sufficiently flexible to guarantee a detailed description of the system when all the particles are close to each other, and it goes to zero when the nucleon-deuteron distance $r_{Nd}$ increases. Following ref.[1], $\Psi_C$ is constructed by using the standard Faddeev decomposition

$$\Psi_C = \psi(\mathbf{x}_i, \mathbf{y}_i) + \psi(\mathbf{x}_j, \mathbf{y}_j) + \psi(\mathbf{x}_k, \mathbf{y}_k) ,\qquad(2)$$

where $\mathbf{x}, \mathbf{y}$ are Jacobi coordinates; the $i$-th amplitude, with total angular momentum $J, J_z$ and total isospin $T, T_z$, is written in the $L - S$ coupling as

$$\psi(\mathbf{x}_i, \mathbf{y}_i) = \sum_{\alpha=1}^{N_c} \Phi_\alpha(x_i, y_i) \mathcal{Y}_\alpha(jk, i)\qquad(3a)$$

$$\mathcal{Y}_\alpha(jk, i) = \left\{ [Y_{l_\alpha}(\hat{x}_i) Y_{L_\alpha}(\hat{y}_i)]_{\Lambda_\alpha} [s_\alpha^{jk} s_\alpha^i]_{S_\alpha} \right\}_{JJ_z} [t_\alpha^{jk} t_\alpha^i]_{TT_z} .\qquad(3b)$$

$N_c$ is the number of channels taken into account, and each channel is specified by the angular momenta $l_\alpha$, $L_\alpha$, coupled to $\Lambda_\alpha$, and the spin (isospin) $s_\alpha^{jk}$ $(t_\alpha^{jk})$ and $s_\alpha^i$ $(t_\alpha^i)$ of the pair $j$, $k$ and of the third particle $i$, coupled to $S_\alpha$ $(T)$. The antisymmetrization of $\Psi_C$ requires $l_\alpha + s_\alpha^{jk} + t_\alpha^{jk}$ to be odd and the parity of the state is $(-)^{l_\alpha + L_\alpha}$.

The radial amplitude $\Phi_\alpha(x_i, y_i)$ is expanded in terms of the PHH basis functions

$$\Phi_\alpha(x_i, y_i) = \rho^{l_\alpha + L_\alpha} f_\alpha(x_i) \sum_k u_k^\alpha(\rho) \,^{(2)}P_k^{l_\alpha, L_\alpha}(\phi_i) , \qquad (4)$$

where the hyperspherical variables are defined by the relations $x_i = \rho \cos\phi_i$ , $y_i = \rho \sin\phi_i$. The correlation function $f_\alpha(x_i)$ is determined by the nucleon interaction of the pair $(j, k)$ in the specified channel, as discussed in [1]; $^{(2)}P_k^{l_\alpha, L_\alpha}$ is an hyperspherical polynomial and the hyperradial functions $u_k^\alpha(\rho)$ can be determined by the variational procedure described below.

The second term $\Psi_A$ of eq.(1) is required to describe the asymptotic configurations of the system, for large $r_{Nd}$ values, where the effect of the nuclear N–d interaction becomes negligible. $\Psi_A$ is solution of the Schroedinger equation in the asymptotic region and it can be written as a linear combination of the functions

$$\Psi_{L,j}^\lambda = \sum_i \sum_{l_\alpha = 0,2} w_{l_\alpha}(x_i) \mathcal{R}_L^\lambda(y_i) \left\{ [Y_{l_\alpha}(\hat{x}_i) s_\alpha^{jk}]_1 [Y_L(\hat{y}_i) s^i]_j \right\}_{JJ_z} [t_\alpha^{jk} t^i]_{TT_z} , \qquad (5)$$

where $w_{l_\alpha}(x_i)$ is the deuteron wave function component in the waves $l_\alpha = 0, 2$ and $L$ is the relative angular momentum of the deuteron and the incident nucleon; for $\lambda \equiv R$, $\mathcal{R}_L^\lambda(y_i)$ coincides with the regular solution $F_L(y_i)$ of the two–body N–d Schroedinger equation, in absence of nuclear interaction, and for $\lambda \equiv I$ it is taken equal to $\tilde{G}_L(y_i)$, namely the product of the irregular solution $G_L(y_i)$ with a regularizing function going to zero when $y_i \to 0$ and healing to unity for large $y_i$ values. In eq.(5) $\Psi_{L,j}^\lambda$ has been built up in the $j - j$ coupling for sake of simplicity; however, it can be re–written in the $L - S$ coupling and, in conclusion, the total w.f. for an incoming nucleon with angular momenta $L$, $j$ can be written as

$$\Psi_{L,j} = \Psi_{C,L,j} + \Psi_{L,j}^R + \sum_{L'j'} R_{LL'}^{jj'} \Psi_{L',j'}^I , \qquad (6)$$

where the coefficients $R_{LL'}^{jj'}$ are the R–matrix elements. The quantities to be determined in eq.(6) are the functions $u_k^\alpha(\rho)$ and the reactance coefficients $R_{LL'}^{jj'}$. To this end, here we will use the Kohn variational principle, stating[3] that the functional $[R_{LL'}^{jj'}]$ given by

$$[R_{LL'}^{jj'}] = R_{LL'}^{jj'} - \langle \Psi_{L'j'} | H - E | \Psi_{Lj} \rangle , \qquad (7)$$

is stationary for the "optimal" choice of all the trial parameters. The condition that the functional derivatives of $[R_{LL'}^{jj'}]$ with respect to the hyperradial functions $u_k^\alpha(\rho)$ be zero, produces a system of inhomogeneous differential equations. This system can be solved by standard procedures and two sets of hyperradial functions, corresponding to the regular and irregular asymptotic w.f. for the relative angular momenta $L, j$, are obtained. As a final step, the diagonal functionals $[R_{LL}^{jj}]$ are varied with respect to the coefficients $R_{LL'}^{jj'}$ to determine their "optimal" choice; as an example, for $J = 1/2$ the possibles values are $L = 0$, $j = 1/2$ and $L = 2$, $j = 3/2$, and the coefficients $R_{00}, R_{02}, R_{20}, R_{22}$ are solutions of the following set of algebraic equations:

$$\begin{cases} A_{00}^{II} R_{00} + A_{20}^{II} R_{02} = A_{00}^{IR} \\ A_{20}^{II} R_{00} + A_{22}^{II} R_{02} = A_{02}^{IR} \end{cases} \qquad (8)$$

$$\begin{cases} A_{22}^{II} R_{22} + A_{02}^{II} R_{20} = A_{22}^{IR} \\ A_{02}^{II} R_{22} + A_{00}^{II} R_{20} = A_{20}^{IR} \,, \end{cases} \qquad (9)$$

where the coefficients $A_{LL'}^{\lambda\lambda'}$ are given by

$$A_{LL'}^{\lambda\lambda'} = 2 < \Psi_{C,L}^{\lambda} - \Psi_L^{\lambda} \mid H - E \mid \Psi_{L'}^{\lambda'} > \,, \qquad (10)$$

and the functions $\Psi_{C,L}^{\lambda}$ are the internal part of the total w.f., corresponding to the regular ($\lambda = R$) or irregular ($\lambda = I$) asymptotic functions. After solving eqs.(8-9), the improved Kohn estimates for the elements $R_{LL'}$ can be obtained using eq.(7). We notice that, since the R–matrix is symmetrical, the solution of eqs.(7–9) must provide $R_{02} = R_{20}$.

The method outlined here, has been applied to calculate the doublet scattering length for an incident neutron or proton. At zero energy the R–matrix is diagonal, because of the centrifugal barrier, and the scattering length is given by

$$a = - \lim_{k \to 0} \frac{R_{00}}{k} \,. \qquad (11)$$

An important aspect of the problem are the convergence properties versus the number $N_c$ of channels taken into account to construct the internal w.f. $\Psi_C$. To stress this point, the results obtained for two different N–N interactions, the Malfliet and Tjon MT(I-III) central potential[4] acting only in the $s$-wave and the Argonne V14 potential[5], are given in Table 1 for several $N_c$ values.

| $N_c$ | MT(I-III) | | AV14 | |
|---|---|---|---|---|
| | $a_{nd}$ | $a_{pd}$ | $a_{nd}$ | $a_{pd}$ |
| 2 | 0.702 | 0.003 | | |
| 8 | | | 1.211 | 0.980 |
| 10 | | | 1.198 | 0.957 |
| 12 | | | 1.196 | 0.954 |
| Faddeev | 0.70±0.01 ref.[6] | 0.003±0.002 ref.[7] | 1.200 ref.[8] | 0.965 ref.[8] |

**Table 1.** Convergence pattern with $N_c$ for the n–d and p–d doublet scattering length (in fm).

In Table 2 we report the results obtained with the MT(I–III) potential for positive energies, where Faddeev calculations are also available; according to us, the differences between the p–d phase shifts given by the two approaches are due to the truncation of the Coulomb potential used in ref.[9].

In Table 3, the phase shifts and the mixing parameter values, as calculated with the AV14 interaction (for $N_c = 12$), are given for different energies of the incident nucleon. The values $E_{cm}$=0.667, 1.333 and 2.0 MeV have been chosen to allow for a comparison with available experimental values[10]; however, it is known that, in order

| $E_{cm}$ | $^2\delta_0$(n–d) | | $^2\delta_0$(p–d) | |
|---|---|---|---|---|
| (MeV) | present work | Faddeev | present work | Faddeev |
| 0.1 | -3.268 | -3.28 | -0.449 | -0.537 |
| 0.2 | -5.672 | -5.68 | -1.786 | -1.96 |
| 0.3 | -7.945 | -7.95 | -3.513 | -3.73 |
| 0.4 | -10.09 | -10.1 | -5.381 | -5.62 |
| 0.5 | -12.12 | -12.1 | -7.277 | -7.53 |
| 1.0 | -20.66 | -20.7 | -16.00 | -16.2 |

**Table 2.** Phase shifts (in degrees) for the n–d and p–d processes. The Faddeev results are from ref.[9], for the difference in the p-d results see the text.

to fit the (three–body) nuclear properties, it is necessary to add a three nucleon force to the AV14 interaction. This modification will be considered in a forthcoming paper.

As a conclusion, the variational approach proposed in this contribution, allows for a detailed description of the three–nucleon bound and scattering states, even in the case of inelastic processes[11]. The extension of the method to four nucleon systems is in progress and the results for d–d or N–$^3$H($^3$He) scattering processes will be presented in the near future.

| $E_{cm}$ | n–d | | | p–d | | |
|---|---|---|---|---|---|---|
| (MeV) | $^2\delta_0$ | $^4\delta_2$ | $\eta$ | $^2\delta_0$ | $^4\delta_2$ | $\eta$ |
| 0.100 | -4.70 | -0.02 | 0.40 | -1.01 | -0.01 | 0.81 |
| 0.500 | -14.5 | -0.62 | 0.95 | -9.32 | -0.46 | 1.14 |
| 0.667 | -17.7 | -1.00 | 1.04 | -12.6 | -0.79 | 1.19 |
| 1.000 | -23.3 | -1.80 | 1.15 | -18.6 | -1.53 | 1.24 |
| 1.333 | -27.9 | -2.58 | 1.21 | -23.6 | -2.28 | 1.26 |
| 2.000 | -34.9 | -3.91 | 1.26 | -31.4 | -3.62 | 1.26 |

**Table 3.** Phase shifts and mixing parameter $\eta$ values (in degrees) for the AV14 interaction as a function of the total energy.

**References**
1. A.Kievsky, M.Viviani and S.Rosati, Nucl.Phys. **A551, 241** (1993)
2. H.Kameyama, M.Kamimura and Y.Fukushima, Phys.Rev. **C40**, 974 (1989)
3. L.M.Delves in Advances in Nuclear Physics (Vol.5), p.126; M.Baranger and E.Vogt (eds.), New York-London: Plenum Press 1972
4. R.A.Malfliet and J.A.Tjon, Nucl.Phys. **A217**, 161 (1969)
5. R.B.Wiringa, R.A.Smith and T.A.Ainsworth, Phys.Rev. **C29**, 1207 (1984)
6. G.L.Payne, J.L.Friar and B.F.Gibson, Phys.Rev. **C26**, 1385 (1982)
7. J.L.Friar, B.F.Gibson and G.L.Payne, Phys.Rev. **C28**, 983 (1983)
8. C.R.Chen, G.L.Payne, J.L.Friar and B.F.Gibson, Phys.Rev. **C44**, 50 (1991)
9. C.R.Chen, G.L.Payne, J.L.Friar and B.F.Gibson, Phys.Rev. **C39**, 1261 (1989)
10. E.Huttel, W.Arnold, H.Baumgart, H.Berg and G.Clausnitzer, Nucl.Phys. **A406**, 443 (1983)
11. M.Viviani, A.Kievsky and S.Rosati to be published

Few-Body Systems, Suppl. 7, 282—285 (1994)

Few-
Body
Systems
© Springer-Verlag 1994
Printed in Austria

# RELATIVISTIC MESON SPECTROSCOPY IN MOMENTUM SPACE

H. Hersbach and Th.W. Ruijgrok

Instituut voor Theoretische Fysica, Universiteit Utrecht,
P.O. Box 80.006, 3508 TA Utrecht, the Netherlands

## Abstract

In this paper the results of a relativistic constituent-quark-model calculation of the meson spectrum are presented. The model used is based on a relativistic formalism introduced by Ruijgrok and the Groot and is defined in momentum space. The formalism is similar to Dirac's point form in the sense that the four operators for the total momentum are the generators of the Poincaré group that contain the interaction. The quark-antiquark potential used consists of a vector Richardson potential and a scalar linear-plus-constant potential. A consistent treatment of the linear potential in momentum space is briefly discussed. Most meson masses were found to be well described by the model.

## 1 Introduction

It is generally believed that the properties of mesons and baryons should be correctly described by Quantum Chromodynamics (QCD). However, apart from lattice gauge calculations, a practical calculation is impossible at the moment. As a simple replacement nonrelativistic constituent-quark models are surprisingly successful. From a theoretical point of view, however, a nonrelativistic (NR) approach is only justified if the typical momenta of the constituent quarks are small compared to their masses. This condition is certainly not satisfied for the lighter mesons. A relativistic description for mesons is therefore highly desirable. Many approaches in this direction are based on the Bethe-Salpeter formalism, containing an integral over relative energy, which is usually eliminated by a quasi-potential reduction. Different reductions exist which lead to different results. One should keep in mind that quark models are mainly phenomenological. Therefore the question arises, whether a simpler theory could equally well serve as a Lorentz covariant quark model. One candidate [1] is the formalism introduced by Ruijgrok and de Groot (RdG) in 1975. It is defined in momentum space, which in our view is the most natural representation for the formulation of a relativistic theory. It was believed for some time, that it is not possible to use quark models in momentum space, because the confining term of the interquark potential leads to singular equations. Recently, however, it has been realized [2,3] that this singularity is only apparent.

The main purpose of this work is to show that a relativistic momentum-space calculation of the meson spectrum is indeed quite possible and that it leads to good results.

## 2 Resumé of the Ruijgrok-de Groot formalism

The formalism [1] originates from scattering theory. The simplest possible relativistic generalization of the Lippmann-Schwinger equation which has the same structure as the NR theory, is postulated. In addition to the correct NR limit, the formalism has proven to give the proper lowest order relativistic effects [4]. In the intermediate states all particles $i$ remain on their mass shell, so $p_i^0 = \sqrt{|\mathbf{p}_i|^2 + m_i^2} \equiv E_i$. The total four-momentum $P_\mu = \sum_i (p_i)_\mu$ is not conserved in the intermediate states, because for a two-particle system this would not leave any dynamics. Instead, only the total three-velocity $\mathbf{v} = \mathbf{P}/P_0$ is conserved, which appears to be a natural generalization of conservation of three-momentum. In the center-of-momentum system (cms) both conservation laws coincide. Like in the NR case, observables such as the total cross-section, only depend on those scattering amplitudes for which in addition to the conservation of $\mathbf{v}$, also the total energy is conserved. Therefore for such amplitudes the total four-momentum *is* conserved. Concerning the Poincaré invariance a similar situation arises as in Dirac's point form [5] of a dynamical system, in which the commutation relations for the six generators of the Lorentz group $M_{\mu\nu}$ are trivially satisfied and the interaction effects must be put into the generators for the space and time translations, *i.e.*, in the momentum operators $P_\mu$.

From now on we will concentrate on the case of a quark-antiquark system in the cms. In the NR case the momentum dependence of a central potential $V$ appears in the form $|\mathbf{q}|^2 = |\mathbf{p}' - \mathbf{p}|^2$. In the relativistic case this expression must be replaced by $|\mathbf{q}|^2 \to -q^2$. For the present work this replacement is of no use because, due to the lack of four-momentum conservation, the loss of momentum $q_1 = p_1 - p_1'$ by the quark, will in general differ from the gain of momentum $q_2 = p_2' - p_2$ by the antiquark. Instead the following symmetrical substitution is made [4,6]:

$$|\mathbf{q}|^2 \to Q^2 \equiv -q_1 \cdot q_2 = |\mathbf{p}' - \mathbf{p}|^2 - \tau(p', p), \qquad (1)$$

where the term $\tau$, given by $\tau(p', p) = (E_1 - E_1')(E_2' - E_2)$ is responsible for retardation effects. The theoretical justification for this replacement is that in the case of the Coulomb potential, $\tau$ automatically generates the correct form for the Breit interaction [4]. A bound-state equation is obtained by looking for the poles of the scattering matrix. In the cms this leads to the following relativistic wave equation (suppressing discrete quantum numbers), from which the mass $M$ of the meson is to be solved ($\hbar = c = 1$):

$$[E_1 + E_2 - M]\Psi(\mathbf{p}) + \int V(\mathbf{p}', \mathbf{p})\Psi(\mathbf{p}') \left[\frac{m_1 m_2}{E_1' E_2'}\right] d\mathbf{p}' = 0. \qquad (2)$$

## 3 The quark-antiquark potential

The quark-antiquark potential $V$ consists of a vector part $V_V$ and a scalar part $V_S$:

$$V(\mathbf{p}', \mathbf{p}) = \bar{u}_{\lambda_1'}(\mathbf{p}_1')\bar{v}_{\lambda_2}(\mathbf{p}_2) \left[\gamma_\mu^{(1)} \cdot \gamma^{(2)\mu} V_V(\mathbf{p}', \mathbf{p}) + 1^{(1)} 1^{(2)} V_S(\mathbf{p}', \mathbf{p})\right] v_{\lambda_2'}(\mathbf{p}_2')u_{\lambda_1}(\mathbf{p}_1), \quad (3)$$

where $u$ and $v$ are the Dirac spinors for the quark and the antiquark and the $\lambda$'s are their helicities. The potential must contain a one-gluon exchange (OGE) to account for the short-range (high $Q$) and a confining part to account for the long-range (small $Q$) interaction. It is generally believed that the OGE should have a vector Lorentz structure, because this follows from perturbative QCD. About the confining part there is no such consensus. We admitted a mixed vector-scalar character of the confining part.

For $V_V$ the Richardson potential [7] was used:

$$V_V(Q^2) = -\frac{\alpha_0}{2\pi^2 Q^2 \log[1 + Q^2/\Lambda^2]}, \tag{4}$$

which for high $Q$ contains a OGE with a running coupling constant and for small $Q$ a confining term $\sim -1/Q^4$, which gives rise to a linear potential with a string tension $\lambda_V = \frac{1}{2}\alpha_0\Lambda^2$. Below it will be shown how the apparently too singular $1/Q^4$ term has to be interpretated properly. For $\alpha_0$ the value 1.750 was chosen, rather than the value $16\pi/27 = 1.862$ which was proposed by Richardson, because $\alpha_0 = 1.750$ gives a better agreement [6] with the high-$Q$ part of the running coupling constant derived from perturbative QCD (see, e.g. , (B.2) of [8]). The quantity $\Lambda$ was used as a fitparameter.

For $V_S$ a relativistic generalization of the linear-plus-constant potential was used, which is defined via the Yukawa potential [3,6]:

$$V_S = \lim_{\eta \downarrow 0} \left[ -\lambda_S \frac{\partial^2}{\partial\eta^2} + C\frac{\partial}{\partial\eta} \right] (V_{\text{Yukawa}})_\eta, \quad \text{where} \quad V_{\text{Yukawa}} = -\frac{1}{2\pi^2}\frac{1}{Q^2 + \eta^2} \tag{5}$$

for the present work. Definition (5) can be applied to a general class of theories. Its theory dependence is totally reflected by the theory dependence of the well-understood Yukawa potential. In NR configuration space $V_{\text{Yukawa}} = -e^{-\eta|\boldsymbol{r}|}/|\boldsymbol{r}|$, so that in this representation (5) indeed gives rise to a linear potential with string tension $\lambda_S$ plus a constant potential $C$: $V_S = \lambda_S|\boldsymbol{r}| + C$. In momentum space the limit should be taken in a distributional sense. This means that one should perform $\lim_{\eta \downarrow 0} [\int V_\eta(\boldsymbol{p}, \boldsymbol{p}')\Psi(\boldsymbol{p}')d\boldsymbol{p}']$ rather than $\int [\lim_{\eta \downarrow 0} V_\eta(\boldsymbol{p}, \boldsymbol{p}')] \Psi(\boldsymbol{p}')d\boldsymbol{p}'$. The constant part leads to a $\delta$-function, as it should. The pointwise limit of the linear part contains a $-1/Q^4$ singularity. If, however, the limit is taken outside the integral, a careful analysis shows [3] that this singularity is exactly canceled by a counter term, giving a regular, well-defined expression.

Note that besides the OGE, both the linear and the constant potential include retardations, which are hidden in $Q^2$.

## 4 Results and conclusions

Eigenvalue equation (2) in combination with the potential (3), (4) and (5) was solved by expanding the wave function $\Psi$ into cubic Hermite splines. The parameters $\Lambda$, $\lambda_S$ and $C$ of the potential, and the constituent masses $m_{u/d}$, $m_s$, $m_c$ and $m_b$ of the up/down, the strange, the charm and bottom quark were fitted by comparing the calculated masses $M^{\text{the}}$ with the experimental data [8] $M^{\text{exp}}$ of 52 well-established mesons. For this purpose the merit function $\chi^2(\Lambda, .., m_b) = \sum_i \left[(M_i^{\text{the}}(\Lambda, .., m_b) - M_i^{\text{exp}})/\sigma_i\right]^2$ was optimized, by the use of a method based on the Levenberg-Marquardt algorithm. In the fit only the self-conjugate light unflavored mesons, such as the $\eta$, were disregarded. A fair description of these mesons should include an annihilation interaction. For the weight $\sigma_i$ of meson $i$ the uncertainty $dM_i^{\text{exp}}$ of the measured mass, with a minimum of 10 MeV for bottonia and 20 MeV for the other mesons, was taken. For the pion $\pi$ and the kaon $K$ a minimum of 200 MeV was taken, because for these mesons one should include features of the breaking of chiral symmetry. The algorithm converged to $\chi^2 = 263$, for the following parameter set: $\alpha_0 = 1.75$ (not fitted), $\Lambda = 0.324$ GeV, $\lambda_S = 0.077$ GeV$^2$, $C = -1.297$ GeV, $m_{u/d} = 0.512$ GeV, $m_s = 0.766$ GeV, $m_c = 2.066$ GeV and $m_b = 5.474$ GeV.

The following remarks on the parameter set can be made. The constituent-quark masses are substantially larger than what is usually found in quark models. This is caused by the large negative constant $C$, which appears to be necessary in order to

Table 1: Some (for a full table see [6]) meson masses (in GeV) calculated from (2). The experimental values are taken from [8], with the exception of the $h_{c1}$, which is taken from [9].

| $J^{PC}$ | $^{s,t}L_J$ | $M^{exp}$ | $n$ | $M^{the}$ |
|---|---|---|---|---|
| **Light unflavored mesons: $u/d$ quarks.** | | | | |
| $\pi$ $\quad$ $0^{-+}$ | $^1S_0$ | 0.135 | 1 | 0.600 |
| $\pi'$ $\quad$ $0^{-+}$ | $^1S_0$ | 1.300 | 2 | 1.243 |
| $\rho$ $\quad$ $1^{--}$ | $^3S_1$ | 0.768 | 1 | 0.754 |
| $\rho'$ $\quad$ $1^{--}$ | $^3D_1$ | 1.465 | 3 | 1.474 |
| $a_0$ $\quad$ $0^{++}$ | $^3P_0$ | 0.983 | 1 | 1.012 |
| $a_1$ $\quad$ $1^{++}$ | $^3P_1$ | 1.260 | 1 | 1.166 |
| $a_2$ $\quad$ $2^{++}$ | $^3P_2$ | 1.318 | 1 | 1.301 |
| **Strange mesons (Kaons): $s$, $u/d$ quarks.** | | | | |
| $K$ $\quad$ $0^-$ | $^1S_0$ | 0.495 | 1 | 0.762 |
| $K^*$ $\quad$ $1^-$ | $^3S_1$ | 0.894 | 1 | 0.891 |
| $K_1$ $\quad$ $1^+$ | $^1P_1$ | 1.270 | 1 | 1.304 |
| $K_4^*$ $\quad$ $4^+$ | $^3F_4$ | 2.045 | 1 | 2.115 |
| **Charmed mesons: $c$, $u/d$ quarks.** | | | | |
| $D$ $\quad$ $0^-$ | $^1S_0$ | 1.867 | 1 | 1.935 |
| $D^*$ $\quad$ $1^-$ | $^3S_1$ | 2.010 | 1 | 2.006 |
| **Charmed strange mesons: $c$, $s$ quarks.** | | | | |
| $D_s$ $\quad$ $0^-$ | $^1S_0$ | 1.969 | 1 | 2.032 |
| $D_s^*$ $\quad$ $1^-$ | $^3S_1$ | 2.110 | 1 | 2.100 |

| $J^{PC}$ | $^{s,t}L_J$ | $M^{exp}$ | $n$ | $M^{the}$ |
|---|---|---|---|---|
| **Charmonium: $c$ quarks.** | | | | |
| $\eta_c$ $\quad$ $0^{-+}$ | $^1S_0$ | 2.979 | 1 | 3.042 |
| $J/\Psi$ $\quad$ $1^{--}$ | $^3S_1$ | 3.097 | 1 | 3.099 |
| $\Psi'$ $\quad$ $1^{--}$ | $^3S_1$ | 3.686 | 2 | 3.655 |
| $\Psi^{iv}$ $\quad$ $1^{--}$ | $^3D_1$ | 4.159 | 5 | 4.124 |
| $\chi_{c0}$ $\quad$ $0^{++}$ | $^3P_0$ | 3.415 | 1 | 3.437 |
| $\chi_{c1}$ $\quad$ $1^{++}$ | $^3P_1$ | 3.511 | 1 | 3.485 |
| $\chi_{c2}$ $\quad$ $2^{++}$ | $^3P_2$ | 3.556 | 1 | 3.523 |
| $h_{c1}$ $\quad$ $1^{+-}$ | $^1P_1$ | 3.526 | 1 | 3.492 |
| **Bottom mesons: $b$, $u/d$ quarks.** | | | | |
| $B$ $\quad$ $0^-$ | $^1S_0$ | 5.279 | 1 | 5.303 |
| $B^*$ $\quad$ $1^-$ | $^3S_1$ | 5.325 | 1 | 5.336 |
| **Bottonium: $b$ quarks.** | | | | |
| $\Upsilon$ $\quad$ $1^{--}$ | $^3S_1$ | 9.460 | 1 | 9.493 |
| $\Upsilon'$ $\quad$ $1^{--}$ | $^3S_1$ | 10.023 | 2 | 10.011 |
| $\Upsilon^v$ $\quad$ $1^{--}$ | $^3S_1$ | 11.019 | 10 | 11.054 |
| $\chi_{b0}$ $\quad$ $0^{++}$ | $^3P_0$ | 9.860 | 1 | 9.859 |
| $\chi_{b1}$ $\quad$ $1^{++}$ | $^3P_1$ | 9.892 | 1 | 9.882 |
| $\chi_{b2}$ $\quad$ $2^{++}$ | $^3P_2$ | 9.913 | 1 | 9.901 |

obtain a good fit for the entire spectrum. If, for instance one only considers bottonium and charmonium, it turns out that the quality of the fit only weakly depends on the value of $C$. The system is overparametrized and one in fact does not even need a constant in the potential. But, when simultaneously also good results for the lighter mesons are required, the large negative constant arises automatically.

The Regge slopes were found to be compatible with the experimental value $\beta \approx 1.2$ GeV$^2$. It can be shown [6] that this slope is almost entirely determined by the vector part of the string tension $\lambda_V = \frac{1}{2}\alpha_0\Lambda^2 = 0.092$ GeV$^2$. The total string tension $\lambda_V + \lambda_S = 0.169$ GeV$^2$ is comparable to the conventional value of $\sim 0.18$ GeV$^2$.

The results are presented in Table I. The masses of the $\pi$ and $K$ were found to be much too high, as was expected. The rest of the spectrum, however, including both the light and the heavy-meson masses, is quite well described by the model.

## References

[1] E.H. de Groot and Th.W. Ruijgrok, Nucl. Phys. B 101, 95 (1975).
[2] J.R. Spence and J.P. Vary, Phys. Rev. D 35, 2191 (1987), C 47, 1282 (1993),
    F. Gross, and J. Milana, Phys. Rev. D 43, 2401 (1991),
    J.W. Norbury, D.E. Kahana, and K. M. Maung, Can. J. Phys. 70, 86 (1992),
    K.M. Maung, D.E. Kahana and J.W. Norbury, Phys. Rev. D 47, 1182 (1993).
[3] H. Hersbach, Phys. Rev. D 47, 3027 (1993).
[4] H. Hersbach, Phys. Rev. A 46, 3657 (1992).
[5] P.A.M. Dirac, Rev. Mod. Phys. 21, 392 (1949).
[6] H. Hersbach, Preprint THU-93/13 (1993).
[7] J.L. Richardson, Phys. Lett. B 82, 272 (1979).
[8] Particle Data Group, K. Hikasa et al., Review of particle properties, Phys. Rev. D 45, Part 2 (1992).
[9] T.A. Armstrong et. al. (E760 Collaboration), Phys. Rev. Lett. 69, 2337 (1992).

Few-Body Systems, Suppl. 7, 286—289 (1994)

# RELATIVISTIC TWO-BODY BOUND-STATE CALCULATIONS BEYOND THE LADDER APPROXIMATION.

Taco Nieuwenhuis, J. A. Tjon

*Institute for Theoretical Physics, University of Utrecht, Princetonplein 5, 3508 TA Utrecht, the Netherlands.*

Yu. A. Simonov

*Institute for Theoretical and Experimental Physics, 117259 Moscow, Russia.*

### Abstract

In this work the Feynman-Schwinger representation for the two-body Greens function is studied. After having given a brief introduction to the formalism, we report on the first calculations based on this formalism. In order to demonstrate the validity of the method, we consider the static case where the mass of one of the particles becomes very large. We show that the heavy particle follows a classical trajectory and we find a good agreement with the Klein-Gordon result.

## 1    Theory

Recently, the Feynman-Schwinger representation (FSR) was presented [1] as a new covariant formalism to calculate the relativistic two-body Greens function. It was shown that the FSR sums up all ladder *and* crossed diagrams and that it has the correct static limit if one of the masses of the two particles becomes very large. In principle the formalism also admits the inclusion of all valence particle loops and self-energy and vertex-correction graphs. Furthermore it was argued that the formulation was well suited for essentially nonperturbative gauge theories such as QCD, since the formalism can be set up in a gauge invariant way and the possibility exists to include a vacuum condensate in the interaction kernel via the cumulant expansion. In this paper we will concentrate on the static limit.

In order to demonstrate the formalism we consider the $\phi^3$-theory for two charged particles $\chi_1$ and $\chi_2$ with masses $m_1$ and $m_2$ and charges $g_1$ and $g_2$, interacting through exchange of a third, neutral particle $\phi$ with mass $\mu$. The Euclidean action for this theory is:

$$S_E = \int d^4x \left[ \chi_i^\dagger \left( m_i^2 - \partial_\mu^2 + g_i\phi \right) \chi_i + \tfrac{1}{2}\phi \left( \mu^2 - \partial_\mu^2 \right) \phi \right] \tag{1}$$

$$\equiv \int d^4x \left[ \chi_i^\dagger \Lambda_i \chi_i + \tfrac{1}{2}\phi \left( \mu^2 - \partial_\mu^2 \right) \phi \right]$$

where summation over the index $i$ is implied. The Greens function is defined as the transition probability from the initial state $\Psi_i = \chi_1^\dagger(x_1)\chi_2(x_2)$ to the final state $\Psi_f = \chi_1(y_1)\chi_2^\dagger(y_2)$:

$$G(y_1, y_2 | x_1, x_2) \propto \int D\chi_1 D\chi_2 D\phi \ \Psi_f \Psi_i \ e^{-S_E}$$

$$\propto \int D\phi \ (\det \Lambda_1(y_1, x_1)\Lambda_2(y_2, x_2))^{-\frac{1}{2}} \ \Lambda_1^{-1}(y_1, x_1)\Lambda_2^{-1}(y_2, x_2) \ e^{-\frac{1}{2}\int d^4x \ \phi(\mu^2 - \partial_\mu^2)\phi} \tag{2}$$

Next, in order to get the sum of all generalized ladder graphs, we neglect the determinant in (2). This corresponds to neglecting all the $\chi_1$- and $\chi_2$-loops and is often called the 'quenched approximation'. We now wish to rewrite (2) in such a way that we can perform the integration over the remaining field $\phi$ as well. To this end we exploit the Feynman-Schwinger representation for $\Lambda^{-1}$:

$$\Lambda^{-1}(y, x) = \int_0^\infty ds \left\langle y \left| e^{-s(m^2 - \partial_\mu^2 + g\phi)} \right| x \right\rangle$$

$$= \int_0^\infty ds \lim_{N\to\infty} \left( \frac{N}{2\pi s} \right)^{2N} \int \prod_{n=1}^{N-1} d^4 z_n \ e^{-m^2 s - \frac{N}{4s} \sum_{i=1}^N (z_i - z_{i-1})^2 - \frac{gs}{N} \sum_{i=1}^N \phi\left(\frac{1}{2}(z_i + z_{i-1})\right)} \Bigg|_{z_0 = x}^{z_N = y}$$

$$= \int_0^\infty ds \int (Dz)_{xy} \ e^{-m^2 s - \frac{1}{4s} \int_0^1 \dot{z}_\mu^2(\tau)d\tau - gs \int_0^1 \phi(z(\tau))d\tau} \tag{3}$$

Integrating over the $\phi$-field is now straightforward and yields:

$$G(y_1, y_2 | x_1, x_2) \propto$$

$$\int_0^\infty ds_1 \int_0^\infty ds_2 \int (Dz_1)_{x_1 y_1} (Dz_2)_{x_2 y_2} \ e^{-m_1^2 s_1 - m_2^2 s_2 - \frac{1}{4s_1}\int_0^1 \dot{z}_{1,\mu}^2(\tau_1)d\tau_1 - \frac{1}{4s_2}\int_0^1 \dot{z}_{2,\mu}^2(\tau_2)d\tau_2} \ \langle W_\phi \rangle \tag{4}$$

In case we leave out the self-energy contribution and the vertex-corrections, the Wilson loop $\langle W_\phi \rangle$ is simply given by:

$$\langle W_\phi \rangle = \exp\left[ \frac{g_1 g_2 s_1 s_2}{N^2} \sum_{j,k=1}^N \Delta\left( \tfrac{1}{2}(z_{1,j} + z_{1,j-1} - z_{2,k} - z_{2,k-1}) \right) \right]$$

$$= \exp\left[ g_1 g_2 s_1 s_2 \int_0^1 \int_0^1 \Delta\left( z_1(\tau_1) - z_2(\tau_2) \right) d\tau_1 d\tau_2 \right] \tag{5}$$

where $\Delta$ is the free scalar two-point function:

$$\Delta(z) = \frac{\mu}{4\pi^2 |z|} K_1(\mu |z|) \tag{6}$$

This formulation for the Greens function has the great advantage that it is essentially a quantummechanical one in which the two valence particles interact via a nonlocal interaction. One can show that (4) and (5) sum up all ladder and crossed diagrams. Furthermore, if $m_2 \to \infty$ then $G$ obeys the following Klein-Gordon equation:

$$\left( \partial_\mu^2 - m_1^2 + \frac{g_1 g_2}{2m_2} V(|\mathbf{r}|) \right) G(r) = \delta^{(4)}(r) \tag{7}$$

with

$$V(|\mathbf{r}|) = \int_{-\infty}^\infty dt \ \Delta\left( \sqrt{|\mathbf{r}|^2 + t^2} \right)$$

which reduces to the instantaneous Coulomb interaction for the case $\mu \to 0$.

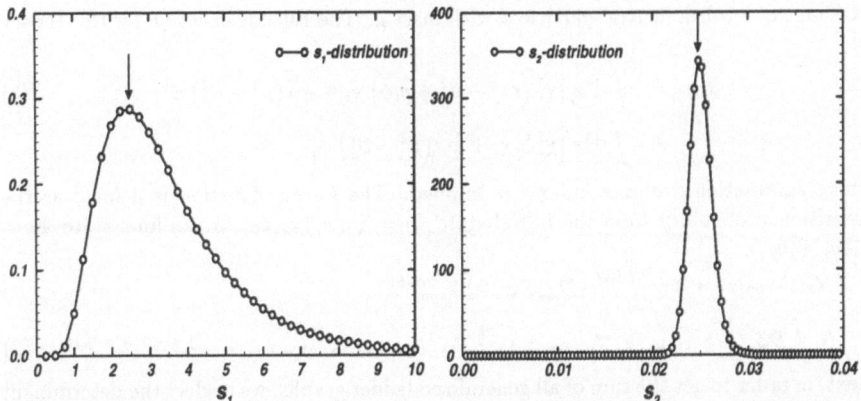

Figure 1: *Normalized distributions of $s_1$ and $s_2$. The arrows indicate the classical values.*

## 2 Results

For a Euclidean theory it is known that for large times $T = \frac{m_1(y_{1,4}-x_{1,4})+m_2(y_{2,4}-x_{2,4})}{m_1+m_2}$ the contributions to $G$ are dominated by the lowest lying states and that they fall off exponentially, $G = \sum_i c_i \exp(-E_i T)$. Hence, we can obtain the ground state energy by observing that:

$$E_0 = -\lim_{T\to\infty} \frac{\mathrm{d}}{\mathrm{d}T} \log G = -\lim_{T\to\infty} \left(\frac{\mathrm{d}}{\mathrm{d}T}G\right)/G \qquad (8)$$

The time derivative with respect to $T$ can be done explicitly.

We project the Greens function on a complete set of total and angular momentum states $|\mathbf{P}, l, m\rangle$, but due to the nonlocal interaction (5) this does *not* imply that the degrees of freedom that are associated with the generators of the symmetries ($\mathbf{P}$, $L^2$, $L_z$), can be integrated out.

For all our calculations we used the Metropolis Monte Carlo algorithm to perform the integrations over $s_1$, $s_2$ and all the coordinates. The value of $N$ in (3) and (5) that we needed to get reasonably stable results was typically 25, so that effectively we had to perform a 200-dimensional integral. Convergence was usually reached after $3 \cdot 10^8$ points which took approximately 30 hours of CPU-time on our fastest workstation. In order to demonstrate that it is indeed possible to determine the actual groundstate within this formalism, we have investigated the case of one light and one heavy particle interacting through exchange of a massless third particle ($m_1 = 1$, $m_2 = 100$ and $\mu = 0$). For the strength of the coupling we took $g^2 \equiv g_1 g_2 = 4500$. The distributions of the coordinates of particle 2 are confined to a very narrow region around their classical values, demonstrating that the heavy particle indeed follows a classical trajectory. Particle 1 shows much more quantum behavior; its distributions have finite widths. The normalized distributions of $s_1$ and $s_2$ are shown in figure 1. Note the difference in scales between both distributions.

In figure 2 we show the average of 5 runs as a function of the number of Monte Carlo points. The exact result for $m_2 \to \infty$, obtained by solving the Klein-Gordon equation (7), is $E_0 = 100.55$ and is indicated by the arrow on the right hand side of the figure. The FSR-result fluctuates around this value and the average over the last $4 \cdot 10^8$ Monte Carlo points is $E_0 = 100.57 \pm 0.02$. This clearly demonstrates the feasibility of the method.

We have also studied in detail the situation where $m_1 = m_2 = 1$ and $\mu = 0$. This case

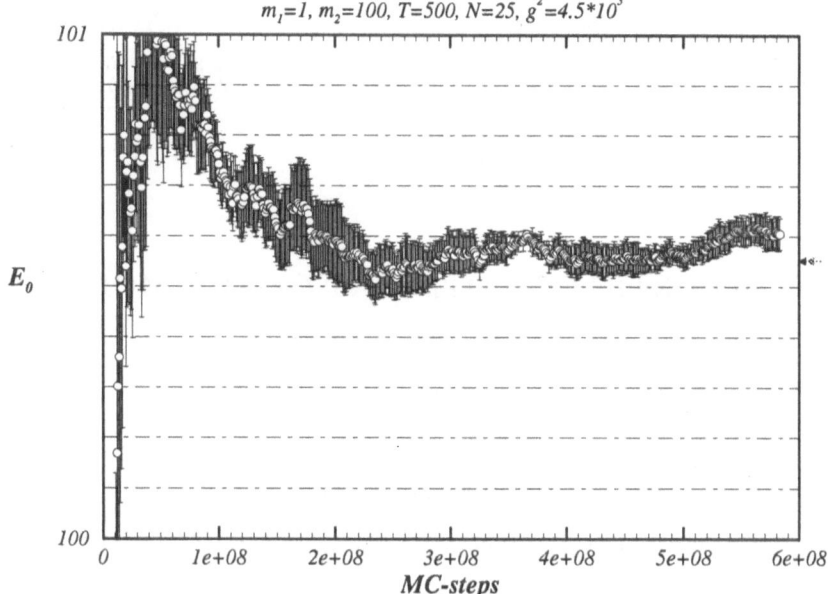

Figure 2: *Bound-state mass in the static limit as a function of the number of Monte Carlo points. T is given in units of the inverse total mass.*

is well suited to test the ladder approximation since the solution of the Bethe-Salpeter equation in the ladder approximation is known exactly [2, 3, 4]. For values of the coupling constant such that there is a binding energy of roughly 10% of the mass of the constituents, we find that the full results lie substantially below the ladder predictions.

In conclusion, we have demonstrated that the FSR is well suited for going beyond the ladder theory in the study of composite systems in field theory. In particular, ground state properties such as the binding energies and bound-state wave functions can readily be determined with this method.

# References

[1] Yu. A. Simonov and J. A. Tjon, *The Feynman-Schwinger Representation for the Relativistic Two-Particle Amplitude in Field Theory*, Ann. Phys., to be published.

[2] N. Nakanishi, *Behavior of the Solutions to the Bethe-Salpeter Equation*, Prog. Theor. Phys. Suppl. **95** (1988) 1-117 and references therein.

[3] G. C. Wick, *Properties of Bethe-Salpeter Wave Functions*, Phys. Rev. **96** (1954) 1124-1134.

[4] R. E. Cutkosky, *Solutions of a Bethe-Salpeter Equation*, Phys. Rev. **96** (1954) 1135-1141.

Few-Body Systems, Suppl. 7, 290—293 (1994)

Few-
Body
Systems
© Springer-Verlag 1994
Printed in Austria

# A RELATIVISTIC CONSTITUENT QUARK MODEL

Felix Schlumpf

Stanford Linear Accelerator Center

Stanford University, Stanford, California 94309, USA

We investigate the predictive power of a relativistic quark model formulated on the light-front. The nucleon electromagnetic form factors, the semileptonic weak decays of the hyperons and the magnetic moments of both baryon octet and decuplet are calculated and found to be in excellent agreement with experiment.

We construct a relativistic constituent quark model consisting of a radial wave function which is spherically symmetric and invariant under permutations times a spin-isospin wave function which is uniquely determined by symmetry requirements [1]. We apply SU(6) symmetry to the rest frame spinors and boost them to the light-front with a Wigner (Melosh) rotation. The three-quark wave functions so constructed are eigenfunctions of mass and spin operators. Eigenfunctions of the four-momentum, which transform irreducibly under the Poincaré group, are obtained from the mass eigenfunctions using light-front symmetry. The current-density operator of the constituent quarks is assumed to be that of Dirac point particles. For the momentum-space wave function a simple function of the invariant mass $M_0$ is assumed. The invariant mass $M_0$

can be written as

$$M_0^2 = \sum_{i=1}^{3} \frac{\vec{k}_{\perp i}^2 + m_i^2}{x_i}, \tag{1}$$

where we used the longitudinal momentum fractions $x_i = p_i^+/P^+$ ($P$ and $p_i$ are the nucleon and quark momenta, respectively, with $P^+ = P_0 + P_z$). The internal momentum variables $k_i$ are given by $k_i = p_i - x_i P$ with $\sum \vec{k}_{\perp i} = 0$ and $\sum x_i = 1$. We choose the following momentum wave functions

$$\phi_H \quad \sim \quad \exp(-M_0^2/2\beta^2), \tag{2}$$

$$\phi_P \quad \sim \quad (1 + M_0^2/\beta^2)^{-p}. \tag{3}$$

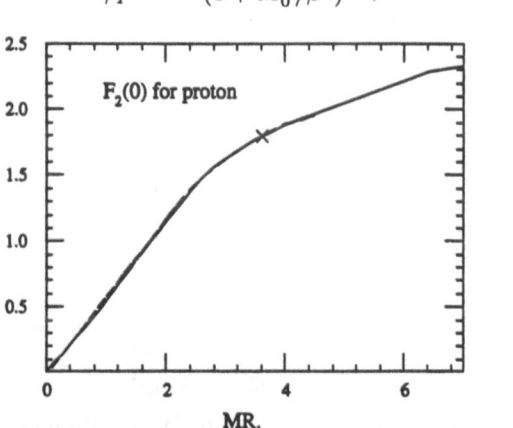

Figure 1: The anomalous magnetic moment $F_2(0)$ of the proton as a function of $M_p R_1$: continuous line, pole type wave function; broken line, gaussian wave function. The experimental value is given by the cross. Our model is independent of the wave function for $Q^2 = 0$.

In Figure 1 the anomalous magnetic moment of the proton $F_2(0)$ is plotted against the radius $R_1 = -6F_1(0)$. Figure 1 shows that the result for $Q^2 = 0$ is *independent* of the wave function chosen. The only parameters are:

- The constituent quark mass $m_i$.

- The scale parameter $\beta$.

For small values of $Q^2$, the two wave functions in Eqs. (2) and (3) still give the same results (Fig. 2). Only for very large momentum transfer $Q^2$ can we see a drastic difference between the different wave functions as shown in Figure 3.

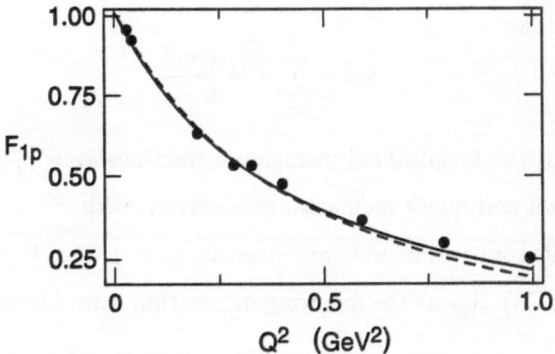

Figure 2: The proton form factor $F_{1p}(Q^2)$. The line code is the same as in the previous figure.

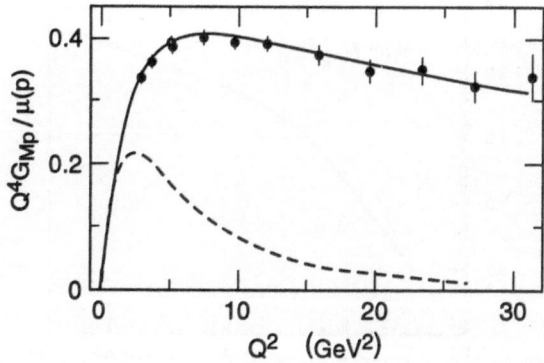

Figure 3: The proton form factor $G_M(Q^2)$: continuous line is the present analysis; broken line gives form factor calculated with a conventionally used wave function. The relativistic constituent quark model does not break down at 2 GeV$^2$, it is even valid up to more than 30 GeV$^2$.

The parameters of the model are fixed by fitting some of the experimental values [1]. The nucleon form factors can be calculated for low, medium and high momentum transfer in excellent agreement with experiment [2]. Figure 3 shows the proton form factor $G_M(Q^2)$ up to more than 30 GeV$^2$. The broken line gives the form factor calculated with a conventionally used wave function. That is the reason why it was believed that the relativistic constituent quark model breaks down at about 2 GeV$^2$. At intermediate energies, $G_M$ and $G_E$ for the neutron and $G_M$ for the proton are in agreement with recent SLAC experiments [3]. A recent pion bremsstrahlung analysis [4] gives a ratio $\mu(\Delta^{++})/\mu(p) = 1.62\pm0.18$, much lower than the nonrelativistic quark model value 2, but in agreement with our value 1.69 [5]. The magnetic moments of the nucleons and hyperons and the semileptonic decays of the baryon octet are also described very well with the same parameters [1].

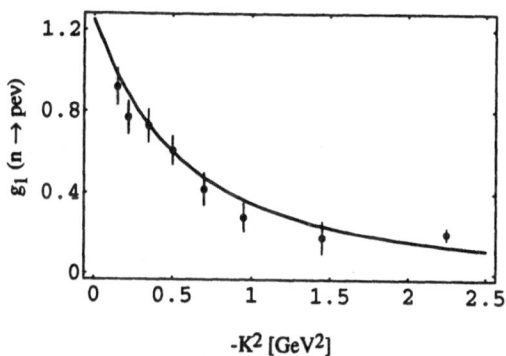

Figure 4: The axial vector form factor $g_1(K^2 = -Q^2)$ for the neutron-proton weak decay.

It is a pleasure to thank Stan Brodsky for stimulating discussions. This work was supported in part by the Schweizerischer Nationalfonds and in part by the Department of Energy, contract DE-AC03-76SF00515.

## References

1. F. Schlumpf, Phys. Rev. D **47**, 4114 (1993).

2. F. Schlumpf, SLAC-PUB-5968 (1992) to be published in J. Phys. G; SLAC-PUB-6050 (1993) to be published in Mod. Phys. Lett. A.

3. P. E. Bosted *et al.*, Phys. Rev. Lett. **68**, 3841 (1992); A. Lung *et al.*, Phys. Rev. Lett. **70**, 718 (1993).

4. A. Bosshard *et al.*, Phys. Rev. D **44**, 1962 (1991).

5. F. Schlumpf, SLAC-PUB-6218 (1993) to be published in Phys. Rev. D.

Few-Body Systems, Suppl. 7, 294—308 (1994)

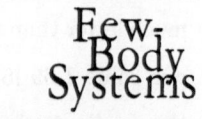

Few-
Body
Systems
© Springer-Verlag 1994
Printed in Austria

# CONVOLUTION APPROACH TO THE $\pi NN$ SYSTEM

B. Blankleider

*School of Physical Sciences, Flinders University of South Australia,
Bedford Park, S.A. 5042, Australia*

A. N. Kvinikhidze

*Mathematical Institute of Georgian Academy of Sciences
Z.Rukhadze 1, 380093 Tbilisi, Georgia*

## ABSTRACT

The unitary $NN-\pi NN$ model contains a serious theoretical flaw: unitarity is obtained at the price of having to use an effective $\pi NN$ coupling constant that is smaller than the experimental one. This is but one aspect of a more general renormalization problem whose origin lies in the truncation of Hilbert space used to derive the equations. Here we present a new theoretical approach to the $\pi NN$ problem where unitary equations are obtained without having to truncate Hilbert space. Indeed, the only approximation made is the neglect of connected three-body forces. As all possible dressings of one-particle propagators and vertices are retained in our model, we overcome the renormalization problems inherent in previous $\pi NN$ theories. The key element of our derivation is the use of convolution integrals that have enabled us to sum all the possible disconnected time-ordered graphs. We also discuss how the convolution method can be extended to sum all the time orderings of a connected graph. This has enabled us to calculate the fully dressed $NN$ one pion exchange potential. We show how such a calculation can be used to estimate the size of the connected three-body forces neglected in the new $\pi NN$ equations. Early indications are that such forces may be negligible.

## INTRODUCTION

Eighteen months ago, at the International Few-Body Conference in Adelaide, one of us (B.B.) summarized the status of our understanding of the $NN-\pi NN$ system as follows [1], "Despite the almost thirty years of effort, we are forced to the conclusion that all current methods have serious deficiencies and that a radical new approach may be needed before this outstanding problem is finally resolved". In particular, it was demonstrated that the so called "unitary $NN-\pi NN$ model" [2-6], which had been the state of the art description for more than 10 years, is theoretically flawed. It is therefore especially pleasing to be able to present, to this conference, just such a radically new approach to the $\pi NN$ system, involving convolution integrals, which resolves the outstanding theoretical

Figure 1: Allowed dressing in the unitary $NN-\pi NN$ model, with associated $Z$ renormalization factors. (a) $\pi N$ nucleon pole graph, (b) $NN$ OPE graph.

problems of the unitary $NN-\pi NN$ model, and which, at this stage, promises to effectively take into account all the diagrams of time-ordered perturbation theory.[1]

As it still may not be well recognized that the unitary $NN-\pi NN$ model is flawed, let us first briefly repeat here the nature of the theoretical error inherent in the model, and how this error can lead to large inaccuracies in predictions of observables.

The origin of the problem lies in the truncation of Hilbert space used to derive the $NN-\pi NN$ equations. Truncation, in this case, means that explicit states with more than one pion are forbidden (certain multi-pion states are, however, included implicitly through the use of two-body, energy-independent potentials). This truncation has serious consequences for the renormalization of both the two-nucleon propagator and the $\pi NN$ vertex. In Fig. 1(a) we show the $\pi N$ nucleon pole diagram where the intermediate state nucleon is dressed by one-pion loops; however, the initial and final state nucleons do not include dressing since two-pion states are neglected in the truncation. Since close to the nucleon pole the dressed one-nucleon propagator is of the form $g(E) \sim Z/(E-m)$, where $Z$ is the residue at the pole, Fig. 1(a) illustrates how each $\pi NN$ vertex $f(E)$ gets effectively renormalized by a factor of $Z^{1/2}$. Thus $f_{\pi NN} = Z^{1/2}f(m)$ is essentially the $\pi NN$ coupling constant, and this fact is used to fix the strength parameter in the form factor $f(E)$. With all other parameters of $f(E)$ fixed to reproduce experimental $\pi N$ phase shifts, this form factor then enters the unitary $NN-\pi NN$ equations as an input. As illustrated in Fig. 1(b), when the $NN$ one pion exchange (OPE) amplitude is calculated in the unitary $NN-\pi NN$ model, the initial and final nucleons are dressed by pions and consequently each external nucleon obtains a renormalization factor of $\tilde{Z}^{1/2}$. The first renormalization problem is the fact that $\tilde{Z} \neq Z$. This arises because two nucleons cannot be dressed at the same time in the truncated Hilbert space; thus, each nucleon in a two-nucleon state cannot obtain its full dressing. This, however, may not be such a serious problem since, in practice, the difference between $Z$ and $\tilde{Z}$ turns out to be quite small. The serious problem, instead, is the size of the effective $\pi NN$ coupling constant in the $NN-\pi NN$ equations. Taking $Z \approx \tilde{Z}$, Fig. 1(b) illustrates that each vertex gets renormalized by a factor of $Z$, so that the effective $\pi NN$ coupling constant here becomes $Zf(m)$; this is a factor $Z^{1/2}$ times the physical coupling constant. With $Z$ being typically between 0.6 and 0.8, we come to the disturbing conclusion that the effective $\pi NN$ coupling constant in the $NN-\pi NN$ equations is smaller than the one used in constructing the input. This observation helps explain why one typically obtains much too small $pp \to \pi^+ d$ cross sections using this model [8-11].

These important observations about the renormalization problem in the unitary $NN-\pi NN$ model were already made in 1985 by Sauer *et al.* [12], yet they seem to have gone largely unnoticed. Perhaps this is partly because one could find less "fatal" reasons for the low cross sections; for example, it was legitimately argued that off-shell effects and the lack of a "backward-going pion" in the $NN \to N\Delta$ amplitude can lead to the underestimation of $pp \to \pi^+ d$ cross sections [10]. However, with the advent of calculations

---

[1]We note that progress has also been made in the four-dimensional relativistic sector where recently derived covariant equations for the $\pi NN$ system resolve overcounting and undercounting problems of previous formulations [7]. In this review talk, however, we shall limit the discussion to the three-dimensional approaches of time-ordered perturbation theory.

where the nucleon and $\Delta$ are treated on an equal footing in the $NN - \pi NN$ model (the so-called $BB - \pi BB$ equations [1]), the effective $\pi NN$ coupling constant is lowered by yet a further factor of $Z^{1/2}$ in the most important $NN \to N\Delta$ amplitude, and it has become very apparent that the renormalization problem is indeed fatal to this type of approach to the $\pi NN$ system.

It may seem that one can fix the renormalization problem "by hand" by strategically including extra $Z^{1/2}$ factors in either $\pi NN$ propagators or $\pi NN$ form factors, or both. But it soon becomes apparent that there is no easy way of doing this without destroying the three-body unitarity of the equations.

Here we report on a completely different formulation of the $\pi NN$ problem where unitary equations are obtained without having to truncate the Hilbert space to some maximum number of pions. Consequently, all possible dressings of one-particle propagators and vertices are retained in our model. In this way we overcome the renormalization problems discussed above. The key element of our derivation is the use of convolution integrals that have enabled us to sum all the possible disconnected time-ordered graphs [13]. Indeed, we basically make only one approximation in deriving the $\pi NN$ convolution equations, namely, we neglect connected three-body forces.

As the details of this new approach have already been published [14], here we shall give only the essential steps of the derivation. A substantial part of our presentation involves an extension of the convolution idea to connected diagrams. In particular, we show how any diagram, connected or disconnected, can be calculated so that all possible dressing contributions are included. We then use the derived formalism to calculate the $NN$ OPE potential where the nucleons are fully dressed. Not only does this calculation indicate that it may be important to include dressing into the popular one boson exchange models of the $NN$ interaction, but it also provides for us a first check on the size of the neglected three-body forces. We find that, in this case, the three-body connected force contribution is negligible.

## THE CONVOLUTION $\pi NN$ EQUATIONS

We consider a time-ordered perturbation theory of nucleons and pions described by a Hamiltonian $H$. The exact form of $H$ need not be specified. The Green function for the $\pi NN \to \pi NN$ process is thus defined by

$$\langle \mathbf{p}'_1, \mathbf{p}'_2, \mathbf{p}'_3 | G(E) | \mathbf{p}_1, \mathbf{p}_2, \mathbf{p}_3 \rangle = \langle \mathbf{p}'_1, \mathbf{p}'_2, \mathbf{p}'_3 | \frac{1}{E^+ - H} | \mathbf{p}_1, \mathbf{p}_2, \mathbf{p}_3 \rangle \tag{1}$$

where $\mathbf{p}_\alpha$ ($\mathbf{p}'_\alpha$) denote initial (final) momenta; here, as below, $\alpha = 1, 2$ label the two nucleons while $\alpha = 3$ denotes the pion. Note that we suppress spin-isospin labels in order to save on notation, and it is assumed that the nucleons are distinguishable since antisymmetrization can be carried out at the end. Note also, that in this section, symbols like $G$ represent operators acting in their appropriate Hilbert space (e.g. $G$ acts in the space of one pion and two nucleons), while in the rest of the paper the same symbols are used for the matrix elements of these operators between momentum eigenstates.

We define the $\pi NN \to \pi NN$ t-matrix $T(E)$ by the equation

$$G(E) = G_0(E) + G_0(E)T(E)G_0(E) \tag{2}$$

where $G_0(E)$ is the free $\pi NN$ propagator with all particles fully dressed (i.e. it is the fully disconnected part of $G(E)$). We can also write

$$T(E) = V(E) + V(E)G_0(E)T(E) \tag{3}$$

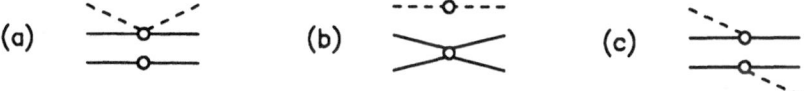

Figure 2: Diagrammatic representation of potentials (a) $V_1$, (b) $V_3$, and (c) $V_4$. The open circles represent all possible graphs excluding those that lead to $\pi NN$ intermediate states. Potentials $V_2$ and $V_5$ are obtained by interchanging the two nucleons in (a) and (c) respectively.

where the potential $V(E)$ is the sum of all possible $\pi NN$- irreducible graphs excluding those consisting of fully disconnected $\pi NN$ states. At this point we shall neglect the connected diagrams of potential $V(E)$; these correspond to connected three-body forces and will be considered again in the next section. It should be stressed that the connected diagrams of $V$ are the only diagrams that are neglected in this model.

For the following discussion we follow the labelling convention where amplitudes that involve the pion interacting with nucleon $i$ are labelled by subscript $i$, while the $NN$ interaction with a pion spectator is labelled by subscript 3. Further, the indices $i$, $j$, are reserved for nucleons 1 and 2 while the Greek indices $\alpha$, $\beta$, $\gamma$ go from 1 to 3 unless otherwise indicated.

Of special relevance to the following discussion is the work of Stelbovics and Stingl (SS) [15], who investigated the disconnectedness problem in the $\pi NN$ system, and then applied it to a model where Hilbert space is chopped at the two-pion level. As discussed by SS, all disconnected $3 \to 3$ diagrams belong to one of five classes of disconnectedness, denoted by $\delta_\alpha$, characterized by an appropriate momentum-space $\delta$-function. Three of these classes, $\delta_1$, $\delta_2$, $\delta_3$, have two of the particles interacting, the third being spectator. The class $\delta_4$ has the pion in initial state being absorbed by nucleon 2 with the final state pion being emitted by nucleon 1. Class $\delta_5$ is the same as $\delta_4$ but with nucleons 1 and 2 interchanged.

According to this classification, the potential $V(E)$ can be written as the sum $V(E) = \sum_{\alpha=1}^{5} V_\alpha(E)$ where $V_\alpha(E)$ consists of diagrams of disconnectedness $\delta_\alpha$. The potentials $V_\alpha(E)$ are represented diagrammatically in Fig. 2. It is worth mentioning that one of the contributions to the potential of Fig. 2(c) is the so called Jennings term which has been shown to be important for calculations of the tensor polarization $t_{20}$ in $\pi d$ scattering [16].

As it stands, Eq. (3) cannot be used directly for calculations as its kernel is disconnected. To derive equations with a compact kernel one proceeds by analogy with the case of Faddeev equations and eliminates the potentials $V_\alpha(E)$ in Eq. (3) in favour of completely summed contributions of disconnectedness $\delta_\alpha$. Let us therefore denote by $\tilde{w}_\alpha$ the Green function consisting of the set of all diagrams, reducible and irreducible, belonging to the disconnectedness class $\delta_\alpha$. We can then define the corresponding amplitudes $w_\alpha$ by

$$\tilde{w}_\alpha = G_0 w_\alpha G_0. \tag{4}$$

Note that $\sum_{\alpha=1}^{5} w_\alpha$ is just the disconnected part of $T$. The resulting compact equations for $T(E)$ can be expressed in a number of ways [15]. Particularly convenient is the expression in terms of the AGS amplitudes $U_{\alpha\beta}$:

$$T = \sum_{\alpha,\beta=1}^{5} (\delta_{\alpha\beta} w_\alpha + w_\alpha G_0 U_{\alpha\beta} G_0 w_\beta) \tag{5}$$

where the $U_{\alpha\beta}$ are found by solving the integral equations

$$U_{\alpha\beta} = \kappa_{\alpha\beta} G_0^{-1} + \sum_{\gamma=1}^{5} \kappa_{\alpha\gamma} w_\gamma G_0 U_{\gamma\beta} \tag{6}$$

where $\kappa_{\alpha\beta}$ is a $5 \times 5$ matrix with elements $\kappa_{11} = \kappa_{22} = \kappa_{33} = \kappa_{51} = \kappa_{42} = \kappa_{14} = \kappa_{25} = \kappa_{45} = \kappa_{54} = 0$, all other elements being equal to 1. Note that the first iteration of Eq. (6) results in a compact kernel.

As our equations and the ones of SS share the same disconnectedness structure, it is especially interesting to compare the actual models used. In the case of the SS model, the simplest possible choice for the potentials $V_\alpha$ was made, namely, that specified by the lowest order graphs. It then turned out that their $w_\alpha$ are not directly related to usual subsystem t-matrices, and must instead be specified by solving very complicated integral equations [17]. By contrast, we take the most complicated possible choice for the $V_\alpha$, namely, the set of all possible contributing graphs of quantum field theory - a choice that cannot even be written down explicitly. The remarkable thing is, that it is just this maximally complicated choice for $V_\alpha$ that enables us to express the needed $w_\alpha$ in terms of simple convolution integrals involving the usual two-body t-matrices $t_\alpha$, the dressed $\pi NN$ vertices $f_i$ (for $N_i \rightarrow \pi N_i$) and $\bar{f}_i$ (for $\pi N_i \rightarrow N_i$), and the dressed one-particle propagators $g_\alpha$.

Introducing the short-hand notation $c = a \otimes b$ to mean the convolution integral

$$c(E) = -\frac{1}{2\pi i} \int_{-\infty}^{\infty} dz\, a(E - z)b(z), \tag{7}$$

the Green functions $\tilde{w}_\alpha$ are specifically given by [13]

$$\tilde{w}_i = \tilde{t}_i \otimes g_j \quad ; \quad \tilde{w}_3 = \tilde{t}_3 \otimes g_3 \quad ; \quad \tilde{w}_4 = \tilde{f}_1 \otimes \tilde{\bar{f}}_2 \quad ; \quad \tilde{w}_5 = \tilde{f}_2 \otimes \tilde{\bar{f}}_1 \tag{8}$$

where $i \neq j$ and, by analogy with Eq. (4), the "tilde" notation denotes the corresponding amplitude with additional initial and final-state propagators; thus, for example, $\tilde{t}_1 = g_{\pi N_1} t_1 g_{\pi N_1}$ where $g_{\pi N}$ is the dressed $\pi N$ propagator. As we have shown in Ref. [13], the convolution integrals effectively sum over all the relative time orderings of one subamplitude of a disconnected diagram with respect to another. In a similar way, as the free $\pi NN$ propagator $G_0$ has all its particles fully dressed, it is given by the double convolution

$$G_0 = g_1 \otimes g_2 \otimes g_3. \tag{9}$$

As the $NN$ channel is hidden in the input terms $w_\alpha$, the relation of Eqs. (5) and (6) to the full $NN$ amplitude is not clear. Neither is it apparent if in the process of iteration, the intermediate $NN$ propagators will obtain their full dressing. For these reasons, we would like to recast Eq. (6) into a form that explicitly exposes the $NN$ channel.

For this purpose, we expose the two-nucleon states in $w_\alpha$ by writing

$$w_\alpha = w_\alpha^0 + w_\alpha^P \tag{10}$$

where $w_\alpha^P$ is the part of $w_\alpha$ that is two-nucleon reducible, and $w_\alpha^0$ is the part of $w_\alpha$ that is two-nucleon irreducible. Since we consider all possible contributions, it is clear that

$$w_i^P = F_i G_{NN} \bar{F}_i \quad ; \quad w_3^P = 0 \quad ; \quad w_4^P = F_1 G_{NN} \bar{F}_2 \quad ; \quad w_5^P = F_2 G_{NN} \bar{F}_1 \tag{11}$$

where $G_{NN}$ is the fully dressed two-nucleon propagator, $F_i$ and $\bar{F}_i$, are the fully dressed $\pi NN$ vertices in the two-nucleon sector. Again we can write these in terms of the one-nucleon propagators $g_i$ and $\pi NN$ vertices $f_i$ and $\bar{f}_i$ through the convolution expressions

$$G_{NN} = g_1 \otimes g_2 \quad ; \quad G_0 F_i G_{NN} = \tilde{f}_i \otimes g_j \quad ; \quad G_{NN} \bar{F}_i G_0 = \tilde{\bar{f}}_i \otimes g_j \quad (i \neq j). \tag{12}$$

With these definitions we can define the amplitudes $U_{\alpha\beta}^0$ corresponding to all possible two-nucleon irreducible contributions, namely

$$U_{\alpha\beta}^0 = \kappa_{\alpha\beta} G_0^{-1} + \sum_{\gamma=1}^{5} \kappa_{\alpha\gamma} w_\gamma^0 G_0 U_{\gamma\beta}^0. \tag{13}$$

Using these definitions in Eq. (6) can then be shown to lead to the following equation for nucleon-nucleon scattering

where
$$T_{NN}(E) = V_{NN}(E) + V_{NN}(E)G_{NN}(E)T_{NN}(E) \qquad (14)$$

$$V_{NN}(E) = \sum_{ij} \bar{F}_i G_0 U_{ij}^0 G_0 F_j \quad ; \quad T_{NN}(E) = \sum_{ij} \bar{F}_i G_0 U_{ij} G_0 F_j. \qquad (15)$$

This result shows explicitly that the $NN$ propagator, which we did not have as an explicit input in Eq. (6), is indeed the fully dressed $G_{NN}$ which is given as the first convolution integral of Eq. (12).

A feature of our formulation is that the input to our derived equations, Eqs. (5) and (6), consists just of usual two-body subsystem off-shell amplitudes. This is an important aspect for practical calculations and was not the case for the formally similar equations of SS. On the other hand, due to the terms $w_4^0$ and $w_5^0$, the kernel of the equations nevertheless is not of the standard type where all interactions are of two-body type. We shall now show how one can rewrite the equations to have standard pair-like interactions in the kernel but at the expense of introducing an extra dimension in the $NN$ sector.

We start by writing an alternative decomposition of $w_\alpha$ to that given in Eq. (10). To do this, we first express the $\pi N$ t-matrix in terms of its nucleon pole and background parts: $t_i = f_i g_i \bar{f}_i + t_i^b$. Then, for $w_i$, we convolute the pole and background parts of $t_i$ separately with the dressed spectator nucleon $g_j$; that is, the decomposition

$$w_i = \bar{w}_i^0 + \bar{w}_i^P \qquad (16)$$

will be defined via the convolutions ($i \neq j$)

$$G_0 \bar{w}_i^0 G_0 = (g_{\pi N} t^b g_{\pi N})_i \otimes g_j \quad ; \quad G_0 \bar{w}_i^P G_0 = (g_{\pi N} f g \bar{f} g_{\pi N})_i \otimes g_j. \qquad (17)$$

We can now write $w_\alpha = \bar{w}_\alpha^0 + \bar{w}_\alpha^P$ generally by defining $\bar{w}_3^P = 0$, $\bar{w}_4^P = w_4$, $\bar{w}_5^P = w_5$, $\bar{w}_3^0 = w_3$, $\bar{w}_4^0 = 0$, and $\bar{w}_5^0 = 0$. By analogy with Eq. (13), we define the amplitudes $\bar{U}_{\alpha\beta}^0$ by

$$\bar{U}_{\alpha\beta}^0 = \kappa_{\alpha\beta} G_0^{-1} + \sum_{\gamma=1}^{5} \kappa_{\alpha\gamma} \bar{w}_\gamma^0 G_0 \bar{U}_{\gamma\beta}^0. \qquad (18)$$

Using these definitions in Eq. (6) and introducing a short-hand notation, illustrated generally by

$$(ab)_{1z} = a_1(E - z)b_1(E - z) \quad ; \quad (ab)_{2z} = a_2(z)b_2(z), \qquad (19)$$

can then be shown to lead to a new kind of $NN$ potential and t-matrix depending on extra variables:

$$V_{NN}(z', z; E) = \sum_{ij} (\bar{f} g_{\pi N})_{iz'} \bar{U}_{ij}^0 (g_{\pi N} f)_{jz}; \quad T_{NN}(z', z; E) = \sum_{ij} (\bar{f} g_{\pi N})_{iz'} U_{ij} (g_{\pi N} f)_{jz}. \qquad (20)$$

In this way the integral equation for $NN$ scattering acquires an extra dimension,

$$T_{NN}(z', z; E) = V_{NN}(z', z; E)$$
$$- \frac{1}{2\pi i} \int_{-\infty}^{\infty} dz'' \, V_{NN}(z', z''; E) g_1(E - z'') g_2(z'') T_{NN}(z'', z; E). \qquad (21)$$

The connection between $T_{NN}(z', z; E)$ and the physical $NN$ t-matrix, $T_{NN}(E)$ follows easily by considering the last two equations of Eqs. (12). By taking the two initial- and final-state nucleons to be on-energy-shell, and setting the $z$ variables to the on-shell energies of the second nucleon, i.e. $z' = \mathbf{p}_2'^2/2m + m$, $z = \mathbf{p}_2^2/2m + m$, one obtains that

$$\langle \mathbf{p}_1' \mathbf{p}_2' | T_{NN}(z', z; E) | \mathbf{p}_1 \mathbf{p}_2 \rangle = \langle \mathbf{p}_1' \mathbf{p}_2' | T_{NN}(E) | \mathbf{p}_1 \mathbf{p}_2 \rangle \qquad (22)$$

where $T_{NN}(E)$ is given in Eq. (15). In this sense $T_{NN}(z', z; E)$ can be considered as an off-shell scattering amplitude with additional energy-like variables $z$ and $z'$.

Our final result for the description of the $\pi NN$ system consists of the coupled equations (18) and (21). The equation for $\bar{U}^0$ is just the Faddeev equation for the $\pi NN$ system with no absorption. In momentum space it is a 6-dimensional integral equation and involves pair-interactions $\pi$-$N$ and $N$-$N$.

The equation for $NN$ scattering, Eq. (21), in momentum space, is a 4-dimensional integral equation. In this sense it is similar to the Bethe-Salpeter equation but with particular forms for the nucleon propagator and $NN$ potential. In the case of Eq. (21), the nucleon propagator contains only positive energy states while the one-pion exchange potential, for example, has $\pi NN$ vertex functions that depend only on one energy variable - in the Bethe-Salpeter case they depend on two energy variables. It is interesting to note that by introducing a fourth dimension into the $NN$ sector, we take into account the $NN$ irreducible diagrams in $w_4$ and $w_5$ while at the same time having only pair-like interactions in the kernel for $\bar{U}^0$.

We conclude by noting that our derivation is model independent. Thus, for example, there was no need to specify either how pions actually couple to nucleons, or what model is used for dressing the pion. We also note that our expressions for the coupled $\pi NN$ equations, Eq. (18) and Eq. (21), or alternatively the coupled equations (13) and (14), express the main features of the physics. Thus, for example, we can easily check that our formulation gives consistent renormalization, simply by substituting the pole term $Z/(E^+ - m)$ wherever the single-nucleon propagator $g(E)$ appears. The unitarity of our equations is also evident from their derivation. For practical calculations, however, either of the coupled set of equations can easily be cast into one set of coupled equations for the reactions $NN \to NN$, $\pi d \to \pi d$, and $NN \to \pi d$. In the case of Eq. (18) and Eq. (21), this would result in coupled equations that are of similar form to the unitary $NN - \pi NN$ equations [5], but with an extra dimension in the $NN$ sector.

## THE DRESSED $NN$ OPE POTENTIAL

The essential feature of the $\pi NN$ equations, presented above, is that they include all possible disconnected $\pi NN$-irreducible graphs. This is the first time that *all* such disconnected diagrams have been included, and has been achieved only after preliminary work showing that convolution integrals can sum all possible time orderings of *disconnected* graphs of time-ordered perturbation theory [13].

With all the disconnected diagrams included, the focus of attention turns naturally to the question of how large the neglected *connected* $\pi NN$-irreducible graphs are. This is not an easy question to answer without a systematic analysis of the whole range of such connected three-body force contributions. Here we shall take the first steps in such an analysis by examining the most basic ingredient of the $NN$ force, namely, the $NN$ one pion exchange potential.

Within the formalism of the $\pi NN$ convolution equations, the OPE potential takes the form appearing in Eq. (15) with $U_{ij}^0$ replaced by $\kappa_{ij} G_0^{-1}$ (the inhomogeneous term of Eq. (13)). It will be sufficient to examine just one time ordering of the exchanged pion, and in this case, the OPE potential of the $\pi NN$ convolution equations is given by

$$V_{12}^{OPE} = \bar{F}_1 G_0 F_2. \tag{23}$$

The consequence of neglecting connected three-body forces becomes immediately apparent upon examination of the dressing contributions to this OPE potential. In Eq. (23), each of the terms $\bar{F}_1$, $G_0$, and $F_2$, is by itself completely dressed. Thus a simple example

Figure 3: (a) Example of a dressing diagram included in the convolution $\pi NN$ equations. (b) Example of a dressing diagram involving a connected three-body force - such diagrams are not included in the convolution $\pi NN$ equations.

of a dressing contribution that is included in Eq. (23) is given by Fig. 3(a). On the other hand, the topologically similar graph of Fig. 3(b) is not included in the formalism of the previous section since it is of the form $\bar{F}_2 V^c F_1$ where $V^c$ is the connected $\pi NN \rightarrow \pi NN$ three-body force.

In order to see the effect of neglecting dressings of the OPE potential like that of Fig. 3(b), it would be useful to have an expression for the fully dressed OPE potential. That is, we need the OPE graph where all possible dressing graphs involving a $\pi NN$ vertex are retained (the pion itself, however, will be assumed to have no dressing contributions). As will be shown in the sections below, such an expression can indeed be easily derived by extending the convolution idea to connected diagrams. Here we just give the result for the corresponding OPE Green function (in the c.m. system) with momentum conserving $\delta$-functions removed:

$$\tilde{V}_{12}^{OPE}(\mathbf{p}', \mathbf{p}; E) = \left(-\frac{1}{2\pi i}\right)^2 \int dz\, dz'\, g(z', \mathbf{p}') \bar{f}(\mathbf{p}', \mathbf{p}, z', z) g(z, \mathbf{p}) \frac{1}{z' - z - \omega_k}$$
$$g(E - z', -\mathbf{p}') f(-\mathbf{p}', -\mathbf{p}, E - z', E - z) g(E - z, -\mathbf{p}). \quad (24)$$

Here $\omega_k$ is the energy of the exchanged pion, and $g(E, \mathbf{p})$ is the dressed nucleon propagator defined as the matrix element of the operator $g$ of the previous section, but with the momentum conserving $\delta$-function removed, i.e. $\delta(\mathbf{p} - \mathbf{p}') g(E, \mathbf{p}) = \langle \mathbf{p}' | \frac{1}{E^+ - H} | \mathbf{p} \rangle$.

A novel feature of Eq. (24) is the need for a $\pi NN$ vertex function $f(\mathbf{p}', \mathbf{p}, E', E)$ that depends on two energy variables. By contrast, in the $\pi NN$ convolution equations, as in most other formulations utilizing time-ordered perturbation theory, the dressed vertex depends only on one energy variable. The relation between one- and two-energy vertices will be given below.

The essential point here is that if one neglects connected three-body forces from Eq. (24), then one will obtain the OPE Green function of the convolution $\pi NN$ equations, which numerically is given as a product of three convolution integrals:

$$\tilde{V}_{12}^{OPE}(\mathbf{p}', \mathbf{p}; E) = \left(-\frac{1}{2\pi i}\right) \left[\int dz\, g(z, \mathbf{p}') \bar{f}(\mathbf{p}', \mathbf{p}, z) g(z - \omega_k, \mathbf{p}) g(E - z, -\mathbf{p}')\right]$$
$$\left[\int dz\, g(z, \mathbf{p}) g(E - \omega_k - z, -\mathbf{p}')\right]^{-1}$$
$$\left[\int dz\, g(z - \omega_k, -\mathbf{p}') f(-\mathbf{p}', -\mathbf{p}, z) g(z, -\mathbf{p}) g(E - z, \mathbf{p})\right]. \quad (25)$$

The fully off-shell OPE potential $V_{12}^{OPE}(\mathbf{p}', \mathbf{p}; E)$ is related to the OPE Green function in the usual way:

$$V_{12}^{OPE}(\mathbf{p}', \mathbf{p}; E) = G_{NN}^{-1}(E, \mathbf{p}', -\mathbf{p}') \tilde{V}_{12}^{OPE}(\mathbf{p}', \mathbf{p}; E) G_{NN}^{-1}(E, \mathbf{p}, -\mathbf{p}) \quad (26)$$

where $G_{NN}$ is the dressed two-nucleon propagator given as in Eq. (12) . For the numerical comparison of Eqs. (24) and (25) we follow our previous work [13] and use the $M1$ $\pi N$ interaction of Ref. [11]. We shall, in each case, calculate the half-off-shell potential $V_{12}^{OPE}(\mathbf{p_0}, \mathbf{p}; E)$, defined by Eq. (26) in the limiting case $\mathbf{p}' \rightarrow \mathbf{p_0}$, where $\mathbf{p_0}$ is the

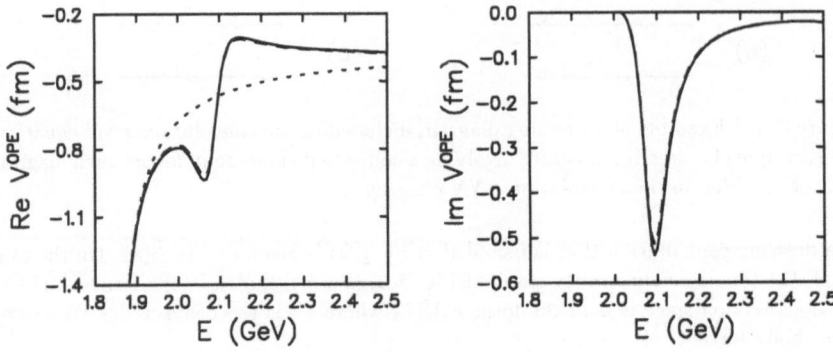

Figure 4: Comparison of the half-off-shell dressed $NN$ OPE potential, $V_{12}^{OPE}(\mathbf{p}_0, \mathbf{p}; E)$ calculated with full dressing (solid curves), and within the convolution $\pi NN$ model (long-dashed curves). The short-dashed curve is for the case where no dressing is included.

on-shell momentum, i.e. $E = p_0^2/m + 2m$. Here we shall consider the simplified case where no dressing is included for the vertices of the exchanged pion; in this case, both the two-energy vertex $f(\mathbf{p}', \mathbf{p}, E', E)$ of Eq. (24) and the single-energy vertex $f(\mathbf{p}', \mathbf{p}, E)$ of Eq. (25), reduce down to the energy-independent bare vertex $f_0(\mathbf{p}', \mathbf{p})$. Note that the restriction to bare vertices is limited to the pion exchanged between the two nucleons, so that the dressed nucleon propagators $g$ are totally unaffected by this simplification. The half-off-shell potential $V_{12}^{OPE}(\mathbf{p}_0, \mathbf{p}; E)$ depends on three variables, the energy $E$, the magnitude $p$ of the off-shell momentum, and the cosine of the angle between $\mathbf{p}_0$ and $\mathbf{p}$. For the numerical comparison, we have examined the potential as a function of energy $E$ for a large range of values of $p$ and $x = \hat{\mathbf{p}}_0 \cdot \hat{\mathbf{p}}$. A typical result is shown in Fig. 4 where, in this case, we have set $p = .1p_0$ and $x = -0.8$. In this figure, the solid curve gives the OPE potential with full dressing, while the long-dashed curve is the result when connected three-body forces are neglected. Also shown is the standard OPE potential where no dressing at all is included (short-dashes). The first observation of note is the significant effect that dressing has on the standard OPE potential. This raises questions about the role such dressing may play in standard descriptions of the $NN$ force. The second observation forms the main result of this section - it is the essential identity of the solid and long-dashed curves, indicating that the contribution of connected three-body forces to the OPE potential is negligible. Although strictly applying to the OPE potential with bare vertices, this is a very encouraging result that gives us hope that connected three-body forces may be small in general.

## DRESSING OF CONNECTED DIAGRAMS - METHOD I

In Ref. [13] we showed how a convolution integral can sum all possible relative time orderings of a *disconnected* diagram of time-ordered perturbation theory. This was then used to demonstrate how two disconnected nucleons can be dressed simultaneously. Here we show that the convolution idea can also be easily extended to *connected* diagrams. The final result shows that the sum of all topologically similar diagrams, connected or disconnected, differing only in the relative time ordering of their vertices, can be expressed through a convolution formula that effectively integrates out initial and final relative energies from the topologically equivalent Feynman diagram. One application of this generalized convolution formula is to the problem of including all possible dressings in connected diagrams.

The starting point of the following discussion is relativistic quantum field theory. Since Feynman diagrams contain, in some sense, all time orderings, it is not surprising that our goal will be to show how to extract the Green function of time-ordered perturbation theory from the Green function of relativistic quantum field theory. Although the procedure we'd like to follow is general, for presentation purposes we specifically consider the $NN \to NN$ process where the interaction is described by a Hamiltonian $H = H_0 + H_I$ involving meson and baryon fields. The explicit form of $H$ need not be specified.

The free fermion field at time $t = 0$ is denoted by $\psi(\mathbf{x})$. At time $t$, we define

$$\psi(\mathbf{x}, t) = e^{iH_0 t} \psi(\mathbf{x}) e^{-iH_0 t} \tag{27}$$

$$\Psi(\mathbf{x}, t) = e^{iHt} \psi(\mathbf{x}) e^{-iHt} \tag{28}$$

being the interaction picture and Heisenberg fields, respectively. The free fermion field $\psi(x) = \psi(\mathbf{x}, t)$ can then be written in terms of its Fourier decomposition as [18]

$$\psi(x) = \sum_s \int \frac{d^3 p}{(2\pi)^{3/2}} \sqrt{\frac{m}{E_p}} [b(\mathbf{p}, s) u(\mathbf{p}, s) e^{-ip \cdot x} + d^\dagger(\mathbf{p}, s) v(\mathbf{p}, s) e^{ip \cdot x}]. \tag{29}$$

We consider the $NN \to NN$ process described by the coordinate space Green function

$$i\mathcal{G}(x, y; x', y') = \langle\langle 0 | T \Psi(x) \Psi(y) \bar{\Psi}(x') \bar{\Psi}(y') | 0 \rangle\rangle, \tag{30}$$

which, because each $\Psi$ is a four-component spinor, can be considered as a $16 \times 16$ matrix. Here $|0\rangle\rangle$ is the dressed vacuum, its relation to the bare vacuum $|0\rangle$ being given by

$$|0\rangle\rangle = \frac{1}{\langle\langle 0 | 0 \rangle} [1 + \frac{1}{0^+ - H} H_I] |0\rangle. \tag{31}$$

The momentum space Green function is defined by

$$\mathcal{G}(p, q; p', q') = \int e^{i(x \cdot p + y \cdot q - x' \cdot p' - y' \cdot q')} \mathcal{G}(x, y; x', y') d^4 x \, d^4 y \, d^4 x' \, d^4 y', \tag{32}$$

which can also be expressed without the momentum conserving $\delta$-function by defining the Green function $G(p, q; p', q')$:

$$\mathcal{G}(p, q; p', q') = (2\pi)^4 \delta^4(p' + q' - p - q) G(p, q; p', q'). \tag{33}$$

We follow Logunov and Tavkhelidze [19] and consider the two-time Green function

$$\mathcal{G}(\mathbf{x}, \mathbf{y}, t; \mathbf{x}', \mathbf{y}', t') \equiv \mathcal{G}(x, y; x'y') \Big|_{\substack{x_0 = y_0 = t \\ x'_0 = y'_0 = t'}} . \tag{34}$$

The two-time Green function in momentum space is defined by

$$\tilde{\mathcal{G}}(\mathbf{p}, \mathbf{q}, E; \mathbf{p}', \mathbf{q}', E') \equiv \int e^{i(-\mathbf{x} \cdot \mathbf{p} - \mathbf{y} \cdot \mathbf{q} + \mathbf{x}' \cdot \mathbf{p}' + \mathbf{y}' \cdot \mathbf{q}' + tE - t'E')}$$
$$\mathcal{G}(\mathbf{x}, \mathbf{y}, t; \mathbf{x}', \mathbf{y}', t') d^3 x \, d^3 y \, d^3 x' \, d^3 y' \, dt \, dt'. \tag{35}$$

A little algebra shows that

$$\tilde{\mathcal{G}}(\mathbf{p}, \mathbf{q}, E; \mathbf{p}', \mathbf{q}', E') = \frac{1}{(2\pi)^2} \int_{-\infty}^{\infty} d\omega \int_{-\infty}^{\infty} d\omega' \, \mathcal{G}(\mathbf{p}, E - \omega, \mathbf{q}, \omega; \mathbf{p}', E' - \omega', \mathbf{q}', \omega'). \tag{36}$$

This result was used by Logunov and Tavkhelidze as the starting point for their study of the two-time Green function. It also constitutes the solution to our problem as we now proceed to show.

It is useful to express $\tilde{\mathcal{G}}$ without momentum conserving $\delta$-functions, thus we define the Green function $\tilde{G}$ by

$$\tilde{\mathcal{G}}(\mathbf{p}, \mathbf{q}, E; \mathbf{p}', \mathbf{q}', E') = (2\pi)^4 \delta(\mathbf{p} + \mathbf{q} - \mathbf{p}' - \mathbf{q}') \delta(E - E') \tilde{G}(E, \mathbf{p}, \mathbf{q}; \mathbf{p}', \mathbf{q}'). \tag{37}$$

Removing the momentum conserving $\delta$-functions from Eq. (36), gives

$$\tilde{G}(E, \mathbf{p}, \mathbf{q}; \mathbf{p}', \mathbf{q}') = \frac{1}{(2\pi)^2} \int_{-\infty}^{\infty} d\omega \int_{-\infty}^{\infty} d\omega' \, G(\mathbf{p}, E - \omega, \mathbf{q}, \omega; \mathbf{p}', E - \omega', \mathbf{q}', \omega'). \tag{38}$$

The rhs of Eq. (38) is a double convolution over initial and final relative energies of Feynman graphs. The goal, therefore, is to show how the lhs, $\tilde{G}(E, \mathbf{p}, \mathbf{q}; \mathbf{p}', \mathbf{q}')$, is related to the Green function of time-ordered perturbation theory. For this we go back to coordinate space and examine the two-time Green function in detail. By the definition of the time-ordered product in Eq. (30),

$$\begin{aligned} i\mathcal{G}(\mathbf{x}, \mathbf{y}, t; \mathbf{x}', \mathbf{y}', t') &= \theta(t - t') \langle\!\langle 0 | \Psi(\mathbf{x}, t) \Psi(\mathbf{y}, t) \bar{\Psi}(\mathbf{x}', t') \bar{\Psi}(\mathbf{y}', t') | 0 \rangle\!\rangle \\ &+ \theta(t' - t) \langle\!\langle 0 | \bar{\Psi}(\mathbf{x}', t') \bar{\Psi}(\mathbf{y}', t') \Psi(\mathbf{x}, t) \Psi(\mathbf{y}, t) | 0 \rangle\!\rangle. \end{aligned} \tag{39}$$

Using Eq. (28) in this equation, and taking the Fourier transform as in Eq. (35) gives the result

$$\begin{aligned} (2\pi)^3 \delta(\mathbf{p} + \mathbf{q} - \mathbf{p}' - \mathbf{q}') \tilde{G}(E, \mathbf{p}, \mathbf{q}; \mathbf{p}', \mathbf{q}') &= \langle\!\langle 0 | \psi(\mathbf{p}) \psi(\mathbf{q}) \frac{1}{E^+ - H} \bar{\psi}(\mathbf{p}') \bar{\psi}(\mathbf{q}') | 0 \rangle\!\rangle \\ &- \langle\!\langle 0 | \bar{\psi}(\mathbf{p}') \bar{\psi}(\mathbf{q}') \frac{1}{E^- + H} \psi(\mathbf{p}) \psi(\mathbf{q}) | 0 \rangle\!\rangle \end{aligned} \tag{40}$$

where we have used Eq. (37), the fact that

$$\int_{-\infty}^{\infty} \theta(t) e^{i\omega t} dt = \frac{i}{\omega + i\epsilon}, \tag{41}$$

and where we have introduced the momentum space fields

$$\psi(\mathbf{p}) \equiv \int d\mathbf{x}^3 \, e^{-i\mathbf{x} \cdot \mathbf{P}} \psi(\mathbf{x}). \tag{42}$$

Using Eq. (29) at $t = 0$ in Eq. (42), one obtains that

$$\psi(\mathbf{p}) = \sum_s \sqrt{\frac{(2\pi)^3 m}{E_p}} [b(\mathbf{p}, s) u(\mathbf{p}, s) + d^{\dagger}(-\mathbf{p}, s) v(-\mathbf{p}, s)]. \tag{43}$$

Eq. (40) is basically the result that we seek. It expresses $\tilde{G}$ in terms of two terms corresponding to the retarded and advanced parts of the Green function of Eq. (39); we shall correspondingly name the first and second terms on the rhs of Eq. (40) as retarded and advanced, respectively.

One may compare Eq. (40) with the Green function of time-ordered perturbation theory, which for the process in question is given by

$$(2\pi)^3 \delta(\mathbf{p} + \mathbf{q} - \mathbf{p}' - \mathbf{q}') G_{\alpha\beta, \alpha'\beta'}(E, \mathbf{p}, \mathbf{q}; \mathbf{p}', \mathbf{q}') = \langle \mathbf{p}, \alpha, \mathbf{q}, \beta | \frac{1}{E^+ - H} | \mathbf{p}', \alpha', \mathbf{q}', \beta' \rangle \tag{44}$$

where the spin components $\alpha, \beta, \alpha', \beta'$ are shown explicitly. Taking into account Eq. (43), we see that the retarded part of $\tilde{G}$ in Eq. (40) is very similar to the Green function of Eq. (44). This similarity suggests a simple transformation of Eq. (40) defined by the equation

$$G_{\alpha\beta, \alpha'\beta'} = \left[ \frac{1}{(2\pi)^3 m} \sqrt{E_p E_q} \, \bar{u}(\mathbf{p}, \alpha) \bar{u}(\mathbf{q}, \beta) \right] G \left[ u(\mathbf{p}', \alpha') u(\mathbf{q}', \beta') \frac{1}{(2\pi)^3 m} \sqrt{E_{p'} E_{q'}} \right] \tag{45}$$

where here $G = G(\mathbf{p}, \mathbf{q}, \mathbf{p}', \mathbf{q}')$ represents any appropriate Green function. Note that Eq. (45) transforms a $16 \times 16$ matrix in spinor space to a $4 \times 4$ matrix in spin space. Under this transformation, Eq. (40) becomes

$$(2\pi)^3 \delta(\mathbf{p} + \mathbf{q} - \mathbf{p}' - \mathbf{q}') \tilde{G}_{\alpha\beta,\alpha'\beta'}(E, \mathbf{p}, \mathbf{q}; \mathbf{p}', \mathbf{q}') =$$
$$\langle\!\langle 0 | b(\mathbf{p}, \alpha) b(\mathbf{q}, \beta) \frac{1}{E^+ - H} b^\dagger(\mathbf{p}', \alpha') b^\dagger(\mathbf{q}', \beta') | 0 \rangle\!\rangle$$
$$- \langle\!\langle 0 | b^\dagger(\mathbf{p}', \alpha') b^\dagger(\mathbf{q}', \beta') \frac{1}{E^- + H} b(\mathbf{p}, \alpha) b(\mathbf{q}, \beta) | 0 \rangle\!\rangle. \qquad (46)$$

If $|0\rangle\!\rangle = |0\rangle$, the advanced term of Eq. (46) disappears, and the retarded term becomes identical with the Green function of time ordered perturbation theory, Eq. (44). Thus the only difference between the transformed two-time Green function of Eq. (46) and the one of time-ordered perturbation theory, is in the type of vacua used to take the matrix element: the transformed version of $\tilde{G}$ uses the dressed vacuum while standard time-ordered perturbation theory, as in Eq. (44), uses the bare vacuum.

In general, $|0\rangle\!\rangle \neq |0\rangle$, however, in the special case where the dressed and bare vacua are equal, Eqs. (44) and (46) become identical. In this case, Eq. (38), after the transformation of Eq. (45), provides us with a formula that expresses the full Green function of time-ordered perturbation theory in terms of a double convolution integral of the full Feynman Green function. This result can be easily generalized to hold for perturbation diagrams of a given order in the interaction, or even for perturbation diagrams of a given topology (the argument is similar to the one used in Ref. [13]). It is in this latter form that the method can give particularly useful formulas, like that of Eq. (24), where all possible time-orderings of a perturbation graph are summed by performing convolution integrals of one Feynman diagram.

A similar result to ours has lately been obtained by Phillips and Afnan [20] using a much more involved argument. In their case, however, they neglected antinucleons altogether in order to obtain the connection of Eq. (38) to time-ordered perturbation theory.[2]

We emphasize that all that is needed in our derivation is the condition $|0\rangle\!\rangle = |0\rangle$. For this to be true, it is sufficient to drop terms from the interaction Hamiltonian that connect vacuum to vacuum states, otherwise antinucleons can be retained. For example, in the simple case of interaction $H_I \sim \bar{\psi}\phi\psi$, expanding each field in terms of creation and annihilation operators gives eight terms; however, of these eight, only terms $b^\dagger a^\dagger d^\dagger$ and $dab$ contribute to the dressing of the vacuum (in lowest order via the process $0 \to N\bar{N}\pi \to 0$), and just these terms can be dropped while still retaining other terms involving $d$ and $d^\dagger$.

## DRESSING OF CONNECTED DIAGRAMS - METHOD II

The above method using the two-time Green function can be used to derive Eq. (24) for the dressed $NN$ OPE potential, but only in the case where vacuum dressing has been neglected. It turns out, however, that Eq. (24) is more general, holding for *any* Hamiltonian $H$, including ones involving vacuum dressing.

Here we derive Eq. (24) using a totally different method that is close in spirit to the one used to derive the convolution formula for disconnected diagrams [13]. This method

---

[2]This paper claims, incorrectly, to have shown that the convolution formula, derived by us in Ref. [13], is a special case of their two-time approach where antinucleons are neglected. If anything, just the opposite is the case: our convolution formula was derived using a method, akin to Method II, which does not need for its validity any approximations whatsoever. We also do not share these authors' criticism of three-dimensional approaches, even though we have ourselves already presented the first consistent four-dimensional approach [7].

involves the temporary introduction of different types of pions and nucleons into a time-ordered perturbation theory description, and provides a way to derive Eq. (24) without making any approximations. In this sense, this method is more general than the one based on the two-time Green function.

The central idea is to interpret the dressed $NN$ OPE potential as consisting of two distinguishable nucleons $N_1$ and $N_2$, each being dressed by its own pion, $\pi_1$ and $\pi_2$ respectively, and in addition, exchanging a third type of pion $\pi$; of the three types of pion, only $\pi$ can interact with both nucleons. We accordingly define $H_1 = H_0(1) + H_I(1)$ to be the Hamiltonian describing the $\pi_1 N_1$ system, and similarly $H_2 = H_0(2) + H_I(2)$ is the Hamiltonian describing the $\pi_2 N_2$ system. The Hamiltonian describing pion $\pi$ and its interactions with the nucleons can be similarly written as $H^\pi = H_0^\pi + H_I^\pi(1) + H_I^\pi(2)$.

Thus, given any Hamiltonian $H = H_0 + H_I$ for which we would like to calculate the $NN$ OPE potential, we can proceed by replacing $H$ with the sum $H' = H_1 + H_2 + H^\pi$, where each individual Hamiltonian $H_1$, $H_2$, and $H^\pi$ has free and interaction parts that are of the same form as $H_0$ and $H_I$, respectively, but each with its own individual fields replacing the corresponding ones of $H$.

Although it may be possible to provide a general formulation, it is more convenient to illustrate the procedure to follow by taking the usual model where the interaction is given by a three-point $\pi NN$ vertex.

As we shall not explicitly need to use any details of the model for $H_1$ and $H_2$, we specify the model in terms of the Hamiltonian involving the pion $\pi$:

$$H_0^\pi = \int d\mathbf{k}\, \omega_k\, a_\pi^\dagger(\mathbf{k}) a_\pi(\mathbf{k}) \tag{47}$$

$$H_I^\pi = \int d\mathbf{k}\, a_\pi^\dagger(\mathbf{k}) J_N(\mathbf{k}) + H.c. \tag{48}$$

$$J_N(\mathbf{k}) = \int d\mathbf{p}\, d\mathbf{p}'\, \delta(\mathbf{p} + \mathbf{k} - \mathbf{p}') \frac{1}{\sqrt{\omega_k}} F_0(\mathbf{p}, \mathbf{p}') a_N^\dagger(\mathbf{p}) a_N(\mathbf{p}') \tag{49}$$

where $N$ can be either $N_1$ or $N_2$, in which case $H_I^\pi$ needs also to be labelled accordingly. Note the relations

$$[a_\pi(\mathbf{k}), H_0^\pi] = \omega_k a_\pi(\mathbf{k}) \quad ; \quad [a_\pi(\mathbf{k}), H_I^\pi] = J_N(\mathbf{k}). \tag{50}$$

Consider now the perturbation expansion of the full $NN \to NN$ Green function with respect to the interactions $H_I^\pi(1)$ and $H_I^\pi(2)$:

$$\langle \mathbf{p}_1', \mathbf{p}_2' | \frac{1}{E^+ - H'} | \mathbf{p}_1, \mathbf{p}_2 \rangle = \langle \mathbf{p}_1', \mathbf{p}_2' | \frac{1}{E^+ - H_1 - H_2 - H_0^\pi} H_I^\pi(1)$$
$$\frac{1}{E^+ - H_1 - H_2 - H_0^\pi} H_I^\pi(2) \frac{1}{E^+ - H_1 - H_2 - H_0^\pi} | \mathbf{p}_1, \mathbf{p}_2 \rangle + \cdots \tag{51}$$

where the term of order $H_I^\pi(1) H_I^\pi(2)$ has been singled out of the complete perturbation series, as it is just this term that coincides with the exact OPE potential specified by the original Hamiltonian $H$. Note how the introduction of the three Hamiltonians $H_1$, $H_2$, and $H^\pi$, enables us to treat meson exchange perturbatively, while nucleon dressing is treated non-perturbatively.

Now consider only this OPE term. Replacing $H_I^\pi(1)$ and $H_I^\pi(2)$ by the integral of Eq. (48), and then using the first of Eqs. (50), we obtain that

$$G_{12}^{OPE} = \int d\mathbf{k}\, \langle \mathbf{p}_1', \mathbf{p}_2' | \frac{1}{E^+ - H_1 - H_2} J_{N_1}^\dagger(\mathbf{k}) \frac{1}{E^+ - H_1 - H_2 - \omega_k} J_{N_2}(\mathbf{k})$$
$$\frac{1}{E^+ - H_1 - H_2} | \mathbf{p}_1, \mathbf{p}_2 \rangle \tag{52}$$

where we have used that $H_0^\pi$ acting on two-nucleon states gives zero, and where one $\mathbf{k}'$-integral has been eliminated using $[a_\pi(\mathbf{k}'), a_\pi^\dagger(\mathbf{k})] = \delta(\mathbf{k}' - \mathbf{k})$. The essential step comes at this stage when we recognize that $[H_1, H_2] = [H_1, J_{N_2}] = [H_2, J_{N_1}] = 0$ which enables us to write Eq. (52) in terms of two contour integrals:

$$G_{12}^{OPE} = \left(-\frac{1}{2\pi i}\right)^2 \int d\mathbf{k}\, dz\, dz' \langle \mathbf{p}_1'| \frac{1}{z'^+ - H_1} J_{N_1}^\dagger(\mathbf{k}) \frac{1}{z^+ - H_1} |\mathbf{p}_1\rangle \frac{1}{z' - z - \omega_k}$$
$$\langle \mathbf{p}_2'| \frac{1}{E^+ - z' - H_2} J_{N_2}(\mathbf{k}) \frac{1}{E^+ - z - H_2} |\mathbf{p}_2\rangle \qquad (53)$$

where the matrix element factors into two, one factor for each nucleon. Because of this factorization, and because all Hamiltonians have the same form as $H$, we may now drop the nucleon labels in Eq. (53).

The matrix elements in Eq. (53) define the two-energy vertices

$$\delta(\mathbf{p}' + \mathbf{k} - \mathbf{p}) f(\mathbf{p}', \mathbf{p}, z', z) = g^{-1}(z', \mathbf{p}') \langle \mathbf{p}'| \frac{1}{z'^+ - H} J_N(\mathbf{k}) \frac{1}{z^+ - H} |\mathbf{p}\rangle g^{-1}(z, \mathbf{p}). \qquad (54)$$

Substituting this definition into Eq. (53) results in the expression of Eq. (24).

Using Eqs. (50), it is straightforward to show that

$$\langle \mathbf{p}'| \frac{1}{z'^+ - H} J_N(\mathbf{k}) \frac{1}{z^+ - H} |\mathbf{p}\rangle$$
$$= \langle \mathbf{p}'\mathbf{k}| \frac{1}{z^+ - H} |\mathbf{p}\rangle + (z - z' - \omega_k) \langle \mathbf{p}'| \frac{1}{z'^+ - H} a_\pi(\mathbf{k}) \frac{1}{z^+ - H} |\mathbf{p}\rangle. \qquad (55)$$

Recognizing that the first term on the rhs is the usual one-energy vertex function, Eq. (55) shows that the one- and two-energy vertex functions coincide for on-mass-shell pions.

## SUMMARY

In order to resolve the renormalization problems suffered by previous formulations of the $\pi NN$ system, we have presented a new approach where the Hilbert space is not truncated to some maximum number of pions. Instead, the guiding principle has been to work only with fully dressed vertices and propagators. What this means in practice, is that all possible disconnected diagrams of quantum field theory have been retained in our model. To achieve this, we have used convolution integrals to sum over all relative time orderings of disconnected graphs. As convolution integrals have often been used in nuclear physics, especially in the four-nucleon problem [21], it may appear surprising why the idea has not been applied sooner to the $\pi NN$ problem. This, however, can be understood when it is realized that the convolutions that we have used are applied to field theory Hamiltonians [13], and result in convolution expressions for *energy-dependent* "potentials". This is quite different from the more usual convolutions of nuclear physics which apply *only* for *energy-independent* potentials.

The convolution $\pi NN$ equations are unitary and do not suffer from renormalization problems. These are the two crucial attributes that are necessary for a theoretically consistent description. Indeed, we should emphasize that the downfall of the $NN - \pi NN$ equations is *not* because the two-nucleon propagator is underdressed, rather, it is simply because the equations themselves do not have both the properties of unitarity and correct renormalization.

The only approximation made in deriving the convolution $\pi NN$ equations is the neglect of connected three-body forces. This is, of course, also an assumption in essentially every model in nuclear physics. Drawing on the bulk experience of the field, such three-body forces are very likely to be small. However, to obtain a more quantitative assessment, explicit calculations of connected three-body force contributions are needed. With this goal in mind, we have extended the convolution idea also to connected diagrams.

Two methods have been presented to derive convolution integrals that sum all the time-orderings of a connected graph. The first method, involving the two-time Green function of relativistic field theory, is useful to obtain the form of the convolution integral in a quick and straightforward manner. However, this method works only in the case when dressed and bare vaccua are equal. The second method, is an extension of the one used by us to derive the convolution formula for disconnected diagrams [13], and involves replacing the true Hamiltonian $H$ by a sum of Hamiltonians, each having the same form as $H$, but applying only to singled out particles. This method works without any assumptions and is therefore more general than the one involving the two-time Green function.

With the convolution formula for connected diagrams derived, we have calculated the fully dressed $NN$ OPE potential, since part of this potential includes connected three-body forces. For this dressed OPE potential, we find that the corresponding connected three-body force is negligible. Thus we can conclude, that at this stage, the convolution $\pi NN$ model is a candidate for effectively summing the whole of time-ordered perturbation theory for the $\pi NN$ system.

# References

[1] Blankleider, B.: Nucl. Phys. **A543**, 163c (1992)

[2] Avishai, Y., Mizutani, T.: Nucl. Phys. **A326**, 352 (1979); **A338**, 377 (1980); **A352**, 399 (1981); Phys. Rev. C **27**, 312 (1983).

[3] Rinat, A.S.: Nucl. Phys. **A287**, 399 (1977).

[4] Thomas, A.W., Rinat, A.S.: Phys. Rev. C **20**, 216 (1979).

[5] Afnan, I.R., Blankleider, B.: Phys. Rev. C **22**, 1638 (1980).

[6] Afnan, I.R., Blankleider, B.: Phys. Rev. C **32**, 2006 (1985).

[7] Kvinikhidze, A.N., Blankleider, B.: Flinders University preprint, FIAS-R-224, September, 1993.

[8] Rinat, A.S., Starkand, Y.: Nucl. Phys. **A397**, 381 (1983); Rinat, A.S., Bhalerao, R.S.: Weizmann Institute of Science Report, WIS-82/55 Nov-Ph, 1982.

[9] Lamot, G.H., Perrot, J.L., Fayard, C., Mizutani, T.: Phys. Rev. C **35**, 239 (1987).

[10] Fayard, C., Lamot, G.H., Mizutani, T., Saghai, B.: Phys. Rev. C **46**, 118 (1992).

[11] Afnan, I.R., McLeod, R.J.: Phys. Rev. C **31**, 1821 (1985).

[12] Sauer, P.U., Sawicki, M., Furui, S.: Prog. Theor. Phys. **74**, 1290 (1985).

[13] Kvinikhidze, A.N., Blankleider, B.: Phys. Rev. C **48**, 25 (1993).

[14] Kvinikhidze, A.N., Blankleider, B.: Phys. Lett. **B307**, 7 (1993).

[15] Stelbovics, A.T., Stingl, M.: Nucl. Phys. **A294**, 391 (1978); J. Phys. G **9**, 1371 (1978); *ibid.* 1389.

[16] Jennings, B.K.: Phys. Lett. **B205**, 187 (1988); Jennings, B.K., Rinat, A.S.: Nucl. Phys. **A485**, 421 (1988).

[17] Afnan, I.R., Stelbovics, A.T.: Phys. Rev. C **23**, 1384 (1981).

[18] Bjorken, J.D., Drell, S.D.: *Relativistic Quantum Fields* (McGraw-Hill, New York, 1965).

[19] Logunov, A.A., Tavkhelidze, A.N.: Nuovo Cimento **29**, 380 (1963).

[20] Phillips, D.R., Afnan, I.R.: Flinders University preprint, FIAS-R-223, September, 1993.

[21] Haberzettl, H., Sandhas, W.: Phys. Rev. C **24**, 359 (1981); Fonseca, A.C.: private communication.

Few-Body Systems, Suppl. 7, 309—316 (1994)

Few-
Body
Systems

# PIONIC HYDROGEN AND THE LOW ENERGY $\pi N$-INTERACTION

P.F.A. Goudsmit

Institute for Intermediate Energy Physics of ETH Zürich

CH-5232 Villigen PSI, Switzerland

## Abstract

A description is given of a direct determination of the $\pi$N s-wave scattering lengths from a measurement of the 1s-level shift and width in pionic hydrogen. The relation with a recently proposed simple dynamical model for the low-energy $\pi N$-interaction is discussed.

## 1. Introduction

An important topic of today's elementary particle physics is the low energy structure of QCD. In this contex the $\pi$N-system plays an important role since it represents the best studied hadronic interaction and since the pion is the lightest hadron.

In the low energy region the $\pi$N-amplitude is dominated by the $s$ and $p$ wave contributions:

$$f(\varepsilon) = b_0(\varepsilon) + b_1(\varepsilon)\vec{\tau}\cdot\vec{t} + (c_0(\varepsilon) + c_1(\varepsilon)\vec{\tau}\cdot\vec{t})\vec{k}_f\cdot\vec{k}_i + (d_0(\varepsilon) + d_1(\varepsilon)\vec{\tau}\cdot\vec{t})i\vec{\sigma}\cdot(\vec{k}_f\times\vec{k}_i)$$

where $\varepsilon$ is the CM kinetic energy of the pion and $\vec{k}_f$ and $\vec{k}_i$ are the final and initial pion momenta. At $\varepsilon = 0$ the parameters $b_0 \equiv b_0(0)$, .....,$d_1 \equiv d_1(0)$ are called "scattering lengths/volumes". Experimental values for these quantities [1] are obtained from $\pi$N scattering cross sections through the use of dispersion relation techniques.

This contribution consists of two parts; in the first part an experiment on pionic hydrogen is described. In the second a recently proposed [2] $\pi$N-interaction model is discussed.

## I. Measurement of the strong shift and width of the 1s-level

This experiment is performed as a combined effort by groups [3] from IMP, the University of Neuchâtel and PSI.

In determinations of $b_0$ and $b_1$ from scattering experiments one has to rely on the extrapolation procedure to $\varepsilon = 0$. This is the case since these experiments get increasingly more difficult with decreasing energy. An additional complication is that several of the more recent $\pi$N-scattering experiments at low energies yield contradictory results. In order to circumvent these problems in the present experiment we determine the $s$ wave scattering lengths experimentally in a direct measurement at $\varepsilon \approx 0$ from the strong shift $\varepsilon(1s)$ and the strong width $\Gamma(1s)$ of the 1s level in pionic hydrogen (see fig. 1).

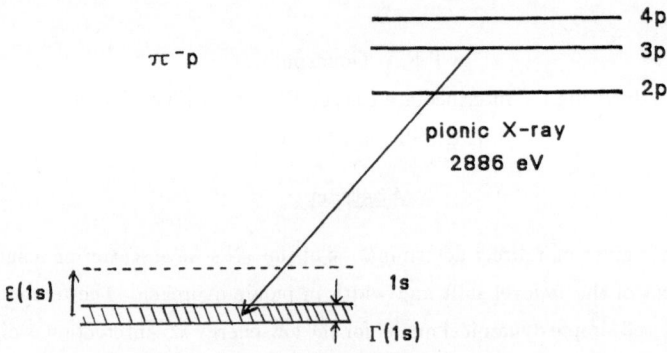

Figure 1

Partial level scheme of pionic hydrogen. The dotted line represents (qualitatively) the electromagnetic position of the 1s level.

$\varepsilon(1s)$ and $\Gamma(1s)$ are related in a known way [4] to $b_0$ and $b_1$. An experimental determination of $\varepsilon$ and $\Gamma$ with relative accuracies if 1% and 10% respectively - being the aim of the present experiment - yields an accuracy of $\pm 0.005\ m_\pi^{-1}$ for both $b_0$ and $b_1$. The reason that such a direct determination has not been possible until now is the smallness of $\varepsilon(1s)$ and $\Gamma(1s)$. Using the values of $b_0$ and $b_1$ from [1] we obtain $\varepsilon(1s) \approx 7$ eV, $\Gamma(1s) \approx 0.9$ eV. The experiment can only be performed with a crystal spectrometer; this calls for the use of special techniques to obtain sufficiently high $\pi$-stopping rates.

A first determination of $\varepsilon(1s)$ in pionic hydrogen was published by Bovet et al. [5] who used a spectrometer equipped with graphite crystals for a measurement of the 2p-1s transition. The instrumental resolution ($\approx 20$ eV at 2.4 keV) did not permit a determination of $\Gamma(1s)$.

In the present experiment both $\varepsilon(1s)$ and $\Gamma(1s)$ are measured for the first time. As a calibration line for the energy of the pionic 3p-1s transition we use the close lying $K_\alpha$ line in argon ($E_{K_{\alpha 1}} = 2957.7$ eV). For the determination of the instrumental line shape the 4f-3d transition in pionic beryllium at 2844 eV is measured. In this transition the strong broadening is negligibly small.

In order to obtain a sufficiently high $\pi$-stopping rate pions from the new $\pi$E5 channel at PSI are injected into a cyclotron trap. In this device the pions are held together in the focusing field of a magnet while being slowed down by a series of degraders until they are stopped in the central cryogenic hydrogen target. The target density equals 15 bar (STP). Measurements of the pionic X-rays at much higher densities or even in liquid hydrogen do not yield an improvement in count rate since here the Stark effect causes an increased absorption in the outer orbits.

The pionic X-rays from the target now fall onto the heart of the apparatus: a unit of six cylindrically bent (110) quartz crystals of a reflecting crystal spectrometer [6]. The instrument has point-to-line focusing properties. In the focal plane the X-ray line is detected with a set of four position sensitive X-ray CCD detectors. Each detector consists of a matrix of $22 \cdot 22$ $\mu$m$^2$ "pixels" each having an energy resolution of $\approx 150$ eV. The total CCD surface used in our experiment equals $1.7 \cdot 10.1$ cm$^2$. In fig. 2 the image seen by a $\approx 1$ cm$^2$ part of one CCD of the argon reflex is shown. The separation (2.1 eV) between the $K_{\alpha_1}$ and $K_{\alpha_2}$ components is clearly visible.

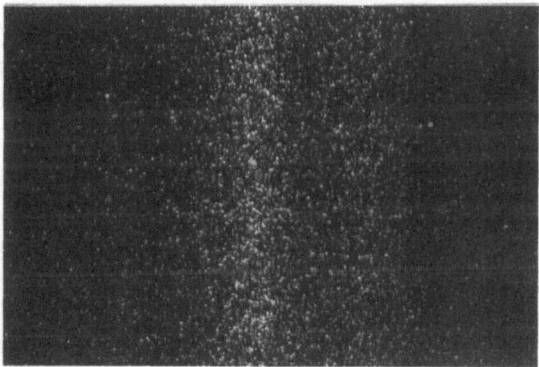

Figure 2

Part of the image of the argon $K_\alpha$-doublet on one CCD detector. The exposure time was 3 minutes.

The projection onto an axis in the dispersion direction (all six crystals, all CCDs) is shown in fig. 3. Taking into account the intrinsic width of the $K_\alpha$ lines of approximately 0.8 eV one finds that the instrumental resolution is indeed below 1 eV as required for the determination of $\Gamma(1s)$.

Argon K-alpha

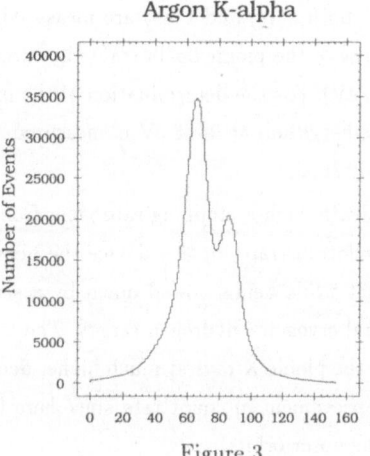

Figure 3

Total spectrum of the argon $K_\alpha$-line (six crystals, four CCD-detectors). The exposure time was 15 minutes.

In addition to the advantage of the position sensitivity it turned out that the use of CCD detectors is vital for our experiment because of their excellent background rejection property. The crystal spectrometer and the detectors are located in a very "hot" area. In spite of heavy shielding the background detected by the CCDs is extremely high in comparison to the X-ray signal which amounts to only $\approx$ 5 events/hr. The background rejection procedure used here is based on the fact that low energy X-rays will in almost all cases deposit their total energy ($\approx$ 3 keV) in a single pixel whereas the background consists, for the vast majority of the cases, of events where a cluster of pixels is involved. A detailed analysis [7] permits us therefore to obtain an important background rejection. This selectively is illustrated in a series of figures. In fig. 4 a detail is shown of the argon X-ray reflex of fig. 2. Indeed, almost all events are seen to be single-pixel events.

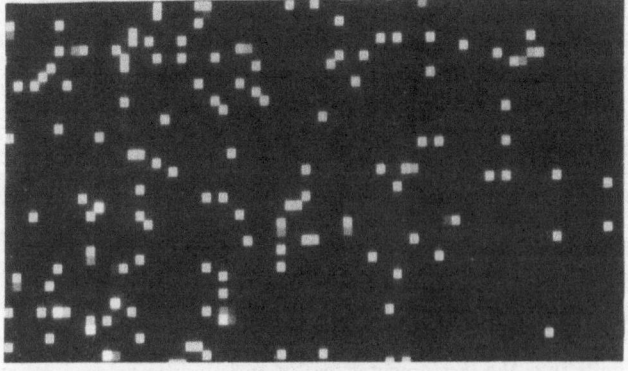

Figure 4

Individual one -pixel events of the argon $K_\alpha$-line of figure 2.

In fig. 5 a $\approx 1$ cm$^2$ large section of one CCD-exposure of 30 minutes at the pionic hydrogen position is shown. Contrary to the argon case the image is now swamped with background events consisting of cluster of charged pixels.

This is seen more clearly in the detail shown in fig. 6. Note that the probability of finding one real 2886-eV X-ray event on an exposure like that of figure 5 is 20% only.

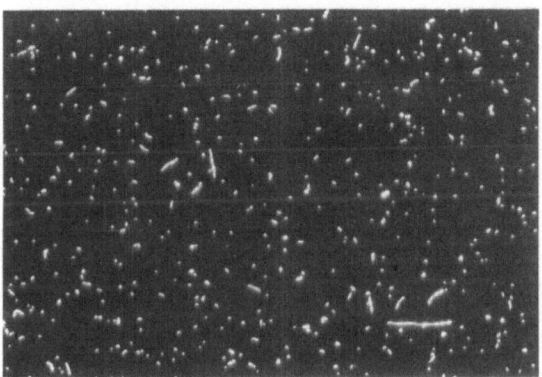

Figure 5
Part of the image of the events on a $\approx$ 1cm$^2$ CCD-section recorded in the pionic hydrogen position. Exposure time: 30 minutes.

In fig. 7 preliminary results are given for the $\pi^-$p(3p-1s) and the $\pi^-$Be(4f-3d) X-ray lines. The broadening of the pionic 3p-1s line is clearly visible. The excellent peak-to-background ratio obtained - in view of the large background in the untreated exposures clearly shows the advantage of the CCD detectors for this measurement.

An improvement of the present measurement with the aim of reaching a final precision in $b_1$ of approximately 2% is under investigation.

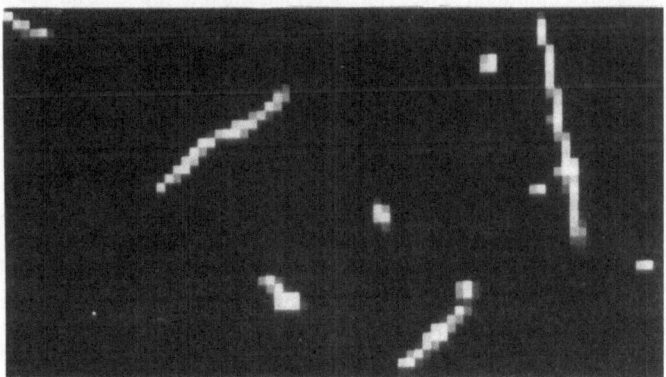

Figure 6
Detail of fig. 5 showing the multiple-pixel events of the background.

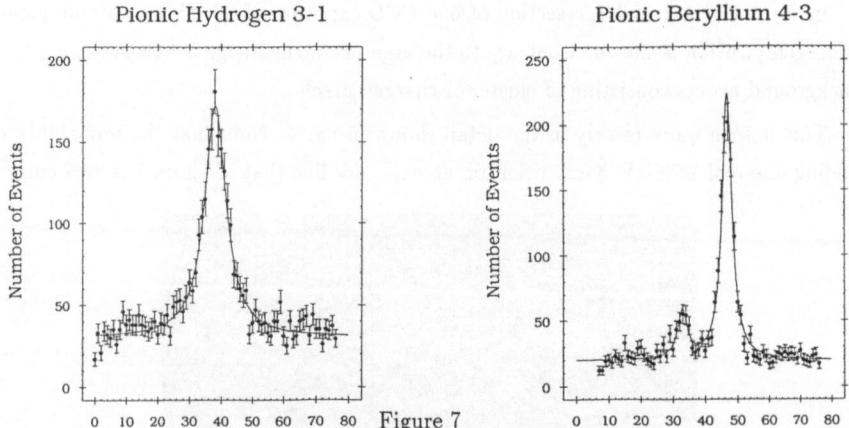

Figure 7

Preliminary results of the pionic hydrogen (3p - 1s) transition and the pionic beryllium (4f - 3d) line shape calibration peak. The energy scales of the two parts (eV/mm) are very similar.

## II. A simple dynamical model of the low-energy $\pi$N-interaction

This part of the research was performed in collaboration with a Gattchina (St. Petersburg) group.

As described in [2,8] we have developed a simple dynamical model for the description of the low energy $\pi$N-interaction. The model is based on the tree level approximation. It includes the exchange of scalar-isoscalar ($\sigma$) and vector-isovector ($\varrho$) mesons in the t-channel and nucleon and $\Delta$ intermediate states in the s and u channels.

Although it is not the first time that such a model is suggested, it seems, however, to be the first one to actually describe both s- and p-wave phase shifts up to the resonance and it does so with a minimum of free (physical) parameters. Specific for our model is the complete relativistic treatment of the $\Delta$-isobar – including the parameter $Z$ associated with the non-pole $\Delta$-isobar contribution, which can be shown to result in important contributions to the $s$-wave. Moreover, a pseudoscalar/pseudovector mixing in the $\pi$NN vertex (mixing parameter $x$) is included. Our model was chosen deliberately to include chiral symmetry violating parts in order to be also capable of describing the $\pi N - \Sigma$ term. A fit to the phase shifts of Koch and Pietarinen for 15 MeV $< \varepsilon <$ 75 MeV [9] and to our value $(b_0 - b_1) = 0.086(4)$ $m_\pi^{-1}$ [4] was used to fix the seven model parameters. A good description of all $s$- and $p$-wave phase shifts was obtained also outside the fitting region, up to about the energy of the $\Delta$-resonance – see fig. 8.

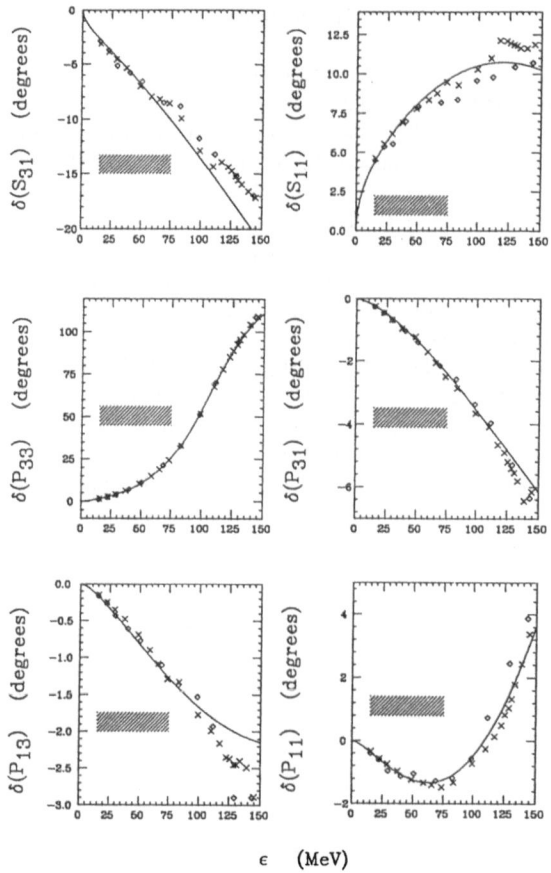

Figure 8

Results of our fit to the phase shifts from [9], $\epsilon$ is the CM pion kinetic energy. The shaded interval indicates the fitting region. For details see [2].

The numerical values of the $s$ and $p$-wave scattering lengths/volumes obtained with our model agree generally well with those of [1]. These exists, however, a small discrepancy in the value of $b_1$. Whereas our model predicts $b_1 = -0.083(1)$ $m_\pi^{-1}$ Koch obtains $b_1 = -0.092(2)$ $m_\pi^{-1}$. This corresponds to a difference in $\Gamma(1s)$ in pionic hydrogen of about 20%. Our present measurement might lead to a solution of this discrepancy.

We expect that our model could play an important role in the investigation of the low energy structure of QCD (chiral symmetry, isospin symmetry) since it provides a link between the $\pi N$-data and, for example, the chiral-perturbation -theory predictions of QCD. The present value deduced from our model for the $\pi N - \Sigma$ term equals $62.5 \pm 5.5$ MeV, this agrees with values deduced through the use of dispersion relation techniques [10].

316

## References

1. R. Koch, Nucl. Phys. **A 448**, 707 (1986).
2. P.F.A. Goudsmit, H.J. Leisi and E. Matsinos, Phys. Lett. **B 299**, 6 (1993).
3. E.C. Aschenauer, A. Badertscher, M. Bogdan, D. Chatellard, J.-P. Egger, K. Gabathuler, P.F.A. Goudsmit, E. Jeannet, L. Knecht, H.J. Leisi, E. Matsinos, A.J. Rusi El Hassani, H.-Ch. Schröder, D. Sigg, L.M. Simons and Z.G. Zhao.
4. W. Beer et al., Phys. Lett **B 261**, 16 (1991).
5. E. Bovet et al., Phys. Lett. **B 153**, 231 (1985).
6. A version with three quartz crystals is described in: A. Badertscher et al., Nucl. Instr. and Methods in Phys. Res. (1993), in print.
7. D. Sigg, to be published.
8. P.F.A. Goudsmit, H.J. Leisi, B.L. Birbrair and A.B. Gridnev, ETHZ-IMP PR/93-2, May 1993, submitted for publication to Nuclear Physics.
9. R. Koch and E. Pietarinen, Nucl. Phys. **A 336**, 331 (1980).
10. R. Koch, Z. Phys. **C 15**, 161 (1982); J. Gasser, H. Leutwyler and M. Sainio, Phys. Lett **B 253** 252 (1991)

Few-Body Systems, Suppl. 7, 317—324 (1994)

Few-
Body
Systems
© Springer-Verlag 1994
Printed in Austria

# THE $E2/M1$ MIXING RATIO IN THE EXCITATION OF THE Δ FROM POLARIZED PHOTO-REACTIONS

A.M. Sandorfi[1], G. Blanpied[4], M. Blecher[6], A. Caracappa[1], C. Djalali[4], G. Giordano[2],
K. Hicks[7], S. Hoblit[5], M.A. Khandaker[6], O.C. Kistner[1], G. Matone[2], L. Miceli[1],
B.M. Preedom[4], D. Rebreyend[4], C. Schaerf[3], R.M. Sealock[5], H. Ströher[8],
A. Tam[4], C.E. Thorn[1], S.T. Thornton[5], C.S. Whisnant[4], and X. Zhao[6]
(The LEGS Collaboration)

[1] *Physics Department, Brookhaven National Laboratory, Upton, NY, 11973*
[2] *INFN-Laboratori Nazionali di Frascati, Frascati, Italy*
[3] *Universita di Roma and INFN-Sezione di Roma, Rome, Italy*
[4] *Department of Physics, University of South Carolina, Columbia, SC, 29208*
[5] *Department of Physics, University of Virginia, Charlottesville, VA, 22903*
[6] *Physics Department, Virginia Polytechnic Institute & SU, Blacksburg, VA, 24061*
[7] *Department of Physics, Ohio University, Athens, OH 45701*
[8] *II Physikalisches Institut, Universität Gießen, Gießen, Germany*

In constituent quark models, a tensor interaction, mixing quark spins with their relative motion, is introduced to reproduce the observed baryon spectrum. This has a consequence completely analogous to the nuclear tensor force between the $n$ and $p$ in deuterium. A $D$ state component is mixed into what would otherwise be a purely $S$-wave object. The $D$-wave component breaks spherical symmetry, resulting in a non-vanishing $<r^2 Y_2>$ matrix element for the nucleon and a static quadrupole moment and deformation for its first excited state, the Δ resonance, at ~325 MeV. The magnitude and sign of this $D$-state component are quite sensitive to the internal structure of the proton and have been of great interest in recent years [1].

The intrinsic deformation of the spin $\frac{1}{2}$ nucleon cannot be observed directly; it must be inferred from transition amplitudes such as $N \to \Delta$. In a spherical bag model, the Δ is viewed as a pure quark-spin-flip transition proceeding only through $M1$ excitation. If there are $D$-state admixtures in the ground state of the nucleon and/or Δ, quadrupole excitation, in addition to spin-flip $M1$, is also allowed. The problem is to evaluate the relative magnitude of this $E2$ excitation in the presence of the dominant $M1$ transition. A variety of models predict this mixing ratio to be quite small, anywhere from –0.9% to –6% [2], so that a high degree of precision is demanded of experiment.

The isospin $(I)$ 3/2 Δ decays with a 99.4% branch to a pion-nucleon $(\pi N)$ final state. An $E2$ photon will produce a $p$-wave pion so that the mixing ratio of interest is written in terms of photo-pion multipoles as $E_{1+}/M_{1+}$. There have been many determinations of the

$I=3/2$ parts of these multipoles from existing pion photoproduction data. For the most part, these agree on the dominant $M_{1+}$ amplitude but differ on smaller components such as the $E_{1+}$. To extract the part of the $E_{1+}^{(I=3/2)}$ multipole associated with the $\Delta$ requires a further decomposition of this amplitude into resonant and background components. This decomposition is not unique, and various models have quoted values ranging from $+4\%$ to $-4\%$ for the ratio of the resonant parts of the $I=3/2$ $E_{1+}$ and $M_{1+}$ amplitudes [3–5].

Of the two decay channels, $\Delta^+ \rightarrow \pi^+ n$ and $\Delta^+ \rightarrow \pi^\circ p$, $\Delta$ excitation in charged-pion production interferes with a large non-resonant $E_{0+}$ background which obscures the presence of $E_{1+}$ components. However, the situation is much more favorable in $\pi^\circ$ production, where backgrounds are greatly reduced. Since the tensor interaction mixes intrinsic spin and orbital motion, its effects appear largest in polarization observables. The pion-photoproduction observable that is most sensitive to the $E_{1+}$ multipole is associated with the $p(\vec{\gamma}, \pi^\circ)$ reaction. The particular observable of interest is the ratio of cross sections measured with linearly polarized photons parallel and perpendicular to the reaction plane $(d\sigma_\parallel/d\sigma_\perp)$ [8]. The sensitivities in reactions not involving a polarization observable are extremely small.

The cross section for $\gamma p \rightarrow \pi^\circ p$ can be measured by detecting either the recoil proton or the two photons from the decay of the $\pi^\circ$. The efficiency of the latter changes with both angle and incident $\gamma$ energy, which is not desirable when studying small effects. Detecting the recoil protons avoids this problem, although at forward angles the proton energy becomes quite low.

# I. The $p(\vec{\gamma}, p)\pi^\circ$ Reaction

We have studied the $\pi^\circ$ photoproduction using linearly-polarized photons between 240 and 330 MeV from the Laser Electron Gamma Source (LEGS) facility located at the National Synchrotron Light Source of Brookhaven National Laboratory [6]. The linearly-polarized $\gamma$ rays were produced by backscattering ultraviolet laser light from 2.6-GeV electrons. The $\gamma$-ray energy was determined, typically to 5 MeV, by detecting the scattered electrons in a tagging spectrometer [7]. Recoil protons were detected at a number of angles.

To test the sensitivity to systematic uncertainties that may survive the $d\sigma_\parallel/d\sigma_\perp$ cross section ratio, four independent experiments have been conducted at a $\pi^\circ$ center-of-mass (c.m.) angle of 105° with different detectors, different methods of determining the $\gamma$-ray energy and monitoring the $\gamma$-ray flux, different polarizations, and using three targets of liquid hydrogen having different cell configurations. The agreement between these four data sets is excellent. Results from three $p(\vec{\gamma}, p)\pi^\circ$ measurements, Expt. numbers L2s, L2p, and L5, have been reported earlier [8]. In the most recent measurement, Expt. L8, recoil protons were detected in coincidence with one of the $\pi^\circ$ decay photons. Details of these measurements are given in the next section on Compton scattering. During all of the experiments, the polarization of the incident $\gamma$-ray was randomly flipped between directions parallel and perpendicular to the reaction plane at a frequency averaging once every 180 sec. The contribution from unpolarized bremsstrahlung in the residual gas of the electron-beam vacuum chamber $(< 1\%)$ was also monitored every 180 sec. All of the data in various energy intervals from 240 to 330 MeV were collected simultaneously, but angle measurements were made sequentially.

The weighted mean of the $d\sigma_\parallel/d\sigma_\perp$ cross section ratios measured at 105° in the three earlier experiments (Expts. L2s, L2p, and L5) are plotted in Fig. 1 (crosses) [8] together with the most recent measurements from Expt. L8 (solid circles). The error bars reflect the combined statistical and polarization-dependent systematic uncertainties, which are smallest for Expt. L8. The data plotted as crosses have been corrected for contamination from $p(\vec{\gamma}, p)\gamma$ events, using the Compton-partial-wave amplitudes of Ref. [9]. In the recent $p(\vec{\gamma}, p\gamma_{\pi^0})\gamma_{\pi^0}$ experiment (L8–solid circles), Compton scattering was resolved and kinematically eliminated. Also shown in Fig. 1 are previously published data (open diamonds and squares) where available [10,11].

Plotted with the data of Fig. 1 are the results of two recent model calculations. The curves lying generally above the data (labeled as NBL) are the work of Nozawa, Blankleider and Lee [4], and result from explicit evaluation of the various diagrams for photo-pion production, including final-state interactions (FSI). The curves lying generally below the data (labeled as DMW) are the work of Davidson, Mukhopadhyay and Wittman [5], in which photoproduction is evaluated in terms of effective Lagrangians, with FSI implicitly included through a unitarization procedure. Both models calculate observables in terms of a few free parameters, most notably the electric ($G_E$) and magnetic ($G_M$) coupling constants of the $\gamma N\Delta$ vertex. These arise in the decomposition of the amplitudes into resonant and background components. In the calculations of Fig. 1, the parameters of both models have been determined by fitting the amplitudes to the Berends and Donnachie (BD) photo-pion multipoles [12]. From this procedure, NBL deduced a resonant $E_{1+}^{(3/2)}/M_{1+}^{(3/2)}$ mixing ratio for $\Delta$ excitation ($EMR$) of –3.1%, while DMW obtain about half that, –1.4%. However, it should be noted that the NBL value of –3.1% reflects the "bare" $\gamma N\Delta$ coupling, without any dressing from FSI, while that of the DMW calculation implicitly included FSI at some level. The predictions of these models for $d\sigma_\parallel/d\sigma_\perp$ are shown in Fig. 1 as dashed-dotted (NBL) and the solid (DMW) curves, respectively. The dotted and dashed curves are obtained by setting the resonant part of the $E_{1+}^{(3/2)}$ multipole to zero ($G_E = 0$).

The sensitivity to a resonant $E2$ component is maximal at 90° c.m. [8]. At 105° the NBL calculations, with model parameters fitted to the BD multipoles, approach the data near the peak of the $\Delta$ (about 325 MeV). However, the energy dependence of the $d\sigma_\parallel/d\sigma_\perp$ ratio provides the crucial test of the resonance-background decomposition, and here their model seems to fail. In contrast, the DMW calculations fall considerably below the data but seem to have an energy dependence similar to the data, at least when fitted to the BD multipoles. However, different energy dependences result when the DMW parameters are fitted to different multipole solutions. Preliminary results of recent measurements from Expt. L8 at 90° c.m. are shown in Fig. 2. The solid and dashed-dotted curves result from fitting the DMW model parameters to the BD and to the recent SP92 [14] multipole sets. At larger angles the comparisons with Nozawa et al. become dramatically worse, while those with Davidson et al. become generally more reasonable, although there the sensitivity to the $E_{1+}$ amplitude is reduced [8].

Recently, Davidson and Mukhopadhyay (DM) have pointed out [13] that the apparent discrepancies between the NBL and DWM predictions for $d\sigma_\parallel/d\sigma_\perp$ nearly disappear when the $u$-channel $\Delta$ (1232) exchange contribution is dropped from their analysis. This component was not included in the model of NBL, although it is not yet clear if this is the only significant difference between these two calculations. It has also been suggested, Refs. [8,13], that the multipoles of the BD analysis may not be reliable for small amplitudes such as the $E_{1+}^{(3/2)}$. In view of this uncertainty, Davidson and Mukhopadhyay have attempted to modify the existing $(\gamma, \pi)$ multipoles and have found that scaling the $E_{1+}^{\pi^0}$ amplitude by a large factor ($\times 2.1$), and refitting their model parameters to this modified amplitude set, gives an $EMR$ of –3.3% and a much better agreement (dashed curve in Fig. 2) with our new polarization results [13]. However, when this prescription is applied to the newer SP92 multipole set the resulting DM predictions ($EMR$=–4.7%) are rather unsatisfactory (dotted curve in Fig. 2). The $(\gamma, \pi)$ multipoles very likely contain significant errors. Evidently a correct description is going to require modifications that are much more complicated than scaling.

The accuracy of our full data set would be sufficient to distinguish differences equivalent to $\sim \frac{1}{5}$ of the separation between the full and 0%-$EMR$ calculations of Fig. 1. However, the discrepancies between the measured $d\sigma_{\parallel}/d\sigma_{\perp}$ ratios and the various calculations described above must be resolved before attempting to confront QCD hadron models with a resonant $E2$ component of $\Delta$ excitation.

## II. The $p(\vec{\gamma}, \gamma p)$ Reaction

We have seen in the previous section that the extraction of a small $E2$ amplitude in the excitation of the $\Delta$ is complicated by the presence of non-resonant background contributions. In the study of $\pi N$ decay channels, polarization has been used to enhance the $E2$ signal relative to the backgrounds. A potentially more sensitive channel is elastic (Compton) scattering of polarized photons [9]. Here, most background contributions cancel in the polarization asymmetry. This has not been previously exploited, since such measurements are technically demanding.

The $p(\vec{\gamma}, \gamma p)$ reaction has been studied at LEGS in Expt. L8 using linearly-polarized tagged gamma rays between 210 and 330 MeV. In this reaction the chief experimental background comes from the $\gamma p \rightarrow \pi^{\circ} p$ channel, where one high energy photon from $\pi^{\circ}$-decay is detected. The cross section for this process is $\sim$200 times that expected from Compton scattering. In this measurement, these channels were separated by detecting the photons in a high resolution NaI spectrometer, together with the recoil protons whose trajectories were tracked through wire chambers and whose energies were measured, both by energy deposition and by time-of-flight (TOF), in an array of plastic scintillators. This arrangement is shown in the Fig. 3. The $\pi^{\circ}$ decay photons closest in energy to Compton scattering are accompanied by low energy photons travelling in nearly the opposite direction. These were suppressed by aditional NaI detectors placed opposite the large high energy $\gamma$-ray spectrometer.

A typical $\gamma/\pi^{\circ}$ separation is shown in Fig. 4 where the $\gamma$-ray energy is plotted against proton TOF. Here, the energy and TOF expected for Compton scattering, as calculated from the tagged beam energy and the proton recoil angles, have been subtracted. Compton scattering is clearly resolved.

Preliminary results for the 90° c.m. photon beam asymmetry are shown in Fig. 5 (solid circles). Also shown in Fig. 5 is the only published datum for this observable (open diamond) [15]. If only $M1$ scattering contributed, the asymmetry would simply reduce to the value of 3/7, independent of energy. This is clearly not the case.

The imaginary part of the Compton scattering amplitude is derivable from $\pi$-production via the Optical theorem and Unitarity, assuming that the $\pi$-production multipoles are known with sufficient accuracy. Dispersion relations can be used to calculate the real part, although there is a fair amount of freedom which comes in through the choice of the subtraction functions that can accompany the dispersion integrals. There have been many speculations that the real part is negligible, at least near the resonance energy of $\sim$325 MeV and possibly over a larger range [16,17]. The result of using $(\gamma, \pi)$ multipoles and Unitarity to calculate the imaginary part of the Compton amplitudes, while leaving the real parts set to zero, is shown as the dashed curve in the Fig. 5 [18]. The real parts are clearly important, even 20 MeV away from the resonance! Pfeil, Rollnik and Stankowski (PRS) have reported the results of two different calculations of the real parts of the Compton amplitudes [16]. The first used fixed-$t$ dispersion relations with the subtraction functions set to zero. The asymmetry predicted by the resulting Compton multipoles is shown as the solid curve in the figure. Although this is in good agreement with the new polarization data presented here, the unpolarized cross sections calculated with these multipoles are significantly higher than measured values [19]. (Preliminary analyses of the cross sections from the present experiment yield similarly higher values near the resonance energy. Although the fixed-$t$ calculation is near these new data at the peak of the $\Delta$, it remains considerably higher than the data at energies off-resonance.) In an effort to reduce the value for the predicted cross section, PRS have adjusted the real parts of the Compton amplitudes. This they refer to as a simultaneous fit to both $(\gamma, \pi)$ and $(\gamma, \gamma)$ [16]. The only significant changes from the fixed-$t$ solution occured in the $f_{mm}^{(1+)}$ and $f_{me}^{(1+)}$ multipoles. Although this solution resulted in a lower unpolarized cross section, the predicted asymmetry (dashed-dot curve) is far from the new data shown in Fig 5.

An alternate approach to calculating Compton scattering has been persued by a group from Tokyo [20]. These authors basically ignored the connection with $(\gamma, \pi)$ and fitted the $(\gamma, \gamma)$ cross section to Breit-Wigner contributions from the spectrum of known baryon resonances. The Compton Born terms were included in the real parts of the amplitudes, but these were scaled down in magnitude during the fitting proceedure. Their predicted cross sections are close to unpolarized data, but the corresponding asymmetry (dotted curve in Fig. 5) does not even have the correct slope in the region of the $\Delta$ resonance.

Sensitivity to the $E2$ excitation of the $\Delta$ comes mainly through the $f_{me}^{(1+)}$ multipole [9,18]. The extraction of this term requires polarization data at several angles. The analysis of such data is presently underway.

This work was supported by the US DOE under contract DE-AC02-76CH00016.

# REFERENCES

[1] S.L. Glashow, Physica (Amsterdam) **96A**, 27 (1979); M. Giannini, Rep. Prog. Phys. **54**, 453 (1991), and references contained therein.

[2] N.C. Mukhopadhyay, in *Topical Workshop on Excited Baryons-1988, Troy, NY*, edited by G. Adams, N. Mukhopadhyay, and P. Stoler (World Scientific, Singapore, 1989), p. 205.

[3] See, for example H. Tanabe and K. Ohta, Phys. Rev. **C31**, 1876 (1985); S.N. Yang, J. Phys. G 11, L205 (1985); A.M. Bernstein, S. Nozawa, and M.A. Moinester, Phys. Rev. **C47**, 1274 (1993).

[4] Private communication; extensions of calculations described in S. Nozawa, B. Blankleider and T.-S.H. Lee, Nucl. Phys. **A513**, 459 (1990).

[5] Private communication; extensions of calculations described in R. Davidson, N. Mukhopadhyay and R. Whittman, Phys. Rev. **D43**, 71 (1991).

[6] A.M. Sandorfi *et al.*, IEEE Trans. Nucl. Sci. **30**, 3083 (1983).

[7] C.E. Thorn *et al.*, Nucl. Instrum. Methods Phys. Res., Sect. **A285**, 447 (1989).

[8] The LEGS Collaboration, G.S. Blanpied *et al.*, Phys. Rev. Lett. **69**, 1880 (1992).

[9] A.M. Sandorfi *et al.*, in *Topical Workshop on Excited Baryons* (Ref. [2]), p. 256; W. Pfeil *et al.*, Nucl. Phys. **B73**, 166 (1974).

[10] V.B. Ganenko *et al.*, Yad. Fiz. **23**, 310 (1976) [Sov. J. Nucl. Phys., **23**, 162 (1976)].

[11] A.A. Belyaev *et al.*, Yad. Fiz. **35**, 693 (1982) [Sov. J. Nucl. Phys., **35**, 401 (1982)].

[12] F. Berends and A. Donnachie, Nucl. Phys. **B84**, 342 (1975).

322

[13] R. Davidson and N. Mukhopadhyay, Phys. Rev. Lett. **70**, 3834 (1993);
     A.M. Sandorfi and M. Khandaker, Phys. Rev. Lett. **70**, 3835 (1993).
[14] R.A. Arndt *et al.*, Phys. Rev. **D42**, 1853 (1990); SP92 solution calculated with the
     code SAID.
[15] G. Barbiellini *et al.*, Phys. Rev. **174**, 1665 (1968).
[16] W. Pfeil, H. Rollnik and S. Stankowski, Nucl. Phys. **B73**, 166 (1974).
[17] M. Benmerrouche and N.C. Mukhopadhyay, Phys. Rev. **D46**, 101 (1992).
[18] A.M. Sandorfi, to be published.
[19] H. Genzel *et al.*, Z. Physik **A279**, 399 (1976).
[20] T. Ishii *et al.*, Nucl. Phys. **B165**, 189 (1980).

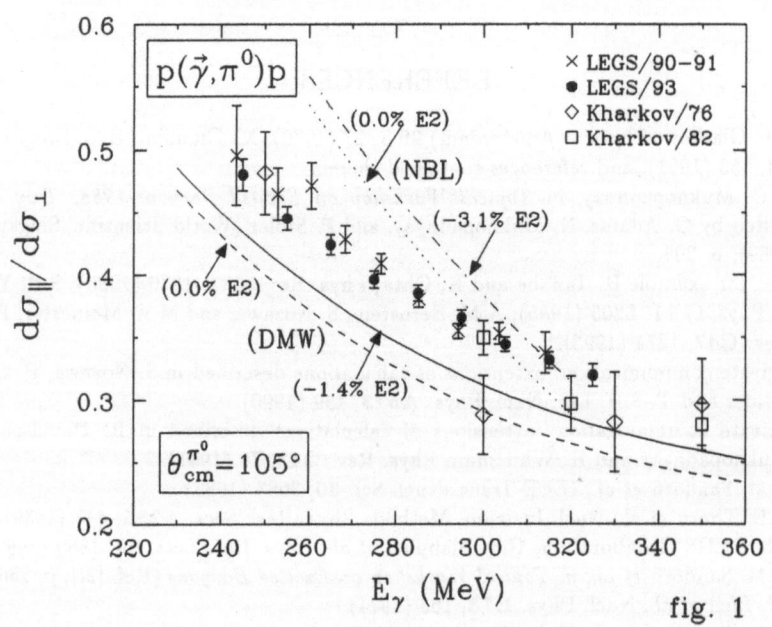

fig. 1

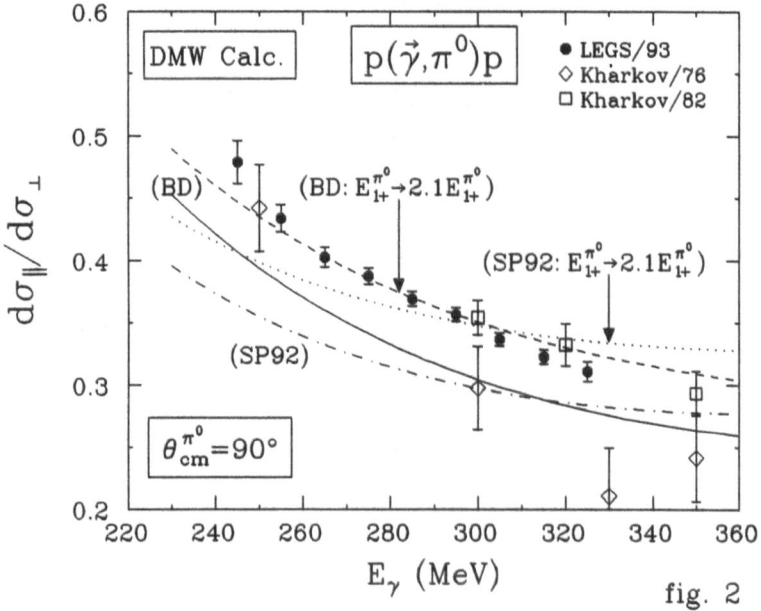

fig. 2

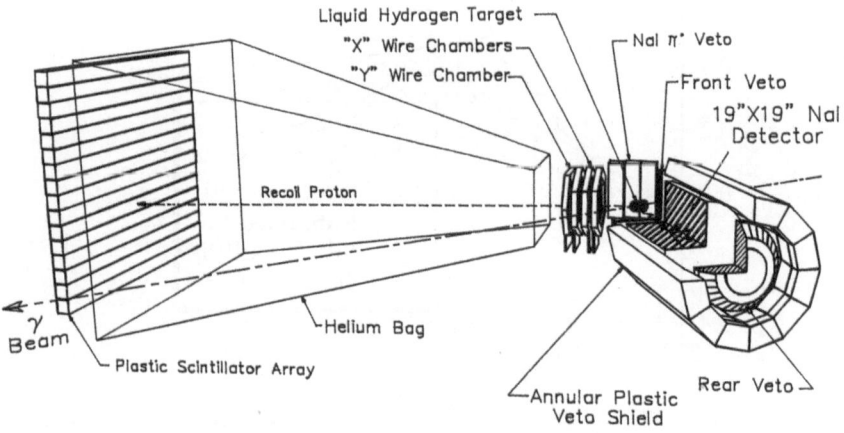

fig. 3

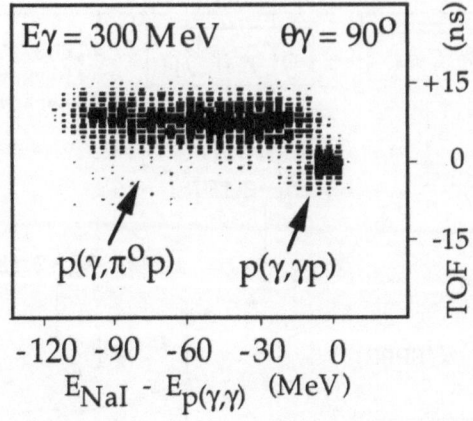

fig. 4

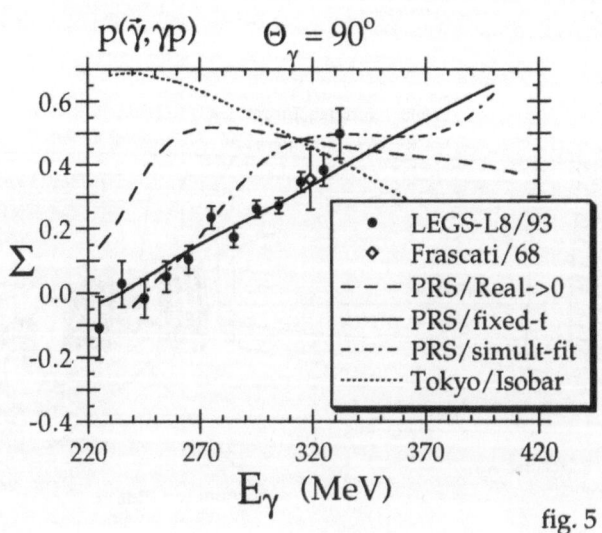

fig. 5

Few-Body Systems, Suppl. 7, 325—338 (1994)

Few-
Body
Systems
© Springer-Verlag 1994
Printed in Austria

# Pion Photoproduction on the Nucleon and Light Nuclei

D. Drechsel

*Institut für Kernphysik, Johannes Gutenberg-Universität, Mainz,*

*D-55099 Mainz, Germany*

### Abstract

Photo- and electroproduction of mesons are well suited tools to investigate the structure of hadrons. With the advent of electron accelerators with high current and large duty-factor, new classes of experiments including polarization degrees of freedom have become possible. Such investigations range from threshold production of mesons to detailed studies of nucleon resonances. In addition, there is the large field of meson production and propagation in the nuclear medium.

## 1  Introduction

Nucleons and nuclei in natural habitat are complex systems, manifestations of the confinement phase of $QCD$. The many-body wave function of such hadrons involves configurations of valence quarks, quark-antiquark pairs (or mesons) and explicit gluon degrees of freedom. The existence of internal degrees of freedom results in a finite size of the nucleon, described by a form factor of the Dirac current and an anomalous magnetic moment multiplied by the Pauli form factor, and in a finite size of the pion. By the same token a spectrum of excited states appears, a series of resonances in the mass region of $1 - 2 GeV$ and a flat continuum at the higher energies, logarithmically rising at the highest observed energies between $200 - 300 GeV$. Finite size effects in the ground state and the existence of an excitation spectrum are not at all independent phenomena, but closely intertwined by sum rules and low energy theorems. These relations reflect the basic symmetries of nature, such as Lorentz, gauge and chiral invariance. Guided by such principles a series of 'model- independent' results has been obtained at threshold within the framework of chiral perturbation theory. Unfortunately, such calculations from first principles are not (yet) possible in the resonance region.

On the experimental side, photo- and electronuclear reactions are a particularly clean instrument to investigate the resonance region and to analyze the multipole content of individual resonance contributions. However, previous experiments have been limited by small currents and low duty-factors. As a consequence the statistics for small amplitudes has been bad and the signal to noise ratio has been small. This situation has changed at the new electron facilities where new classes of coincidence experiments have become possible, and polarization degrees of freedom will play an important role. With a beam polarization of 40 % and more, polarized electrons promise to provide a new capability to measure some of the most wanted observables, in particular in combination with target and recoil polarization.

In the following we will report on some recent developments and future investigations in the field of meson production. For an introduction into the field, in particular for the kinematics and the response functions involved, we have to refer the reader to a recent review [1]. The status of pion production at threshold is discussed in section 2, and section 3 reviews of the field of meson production in the resonance region. In section 4 we present some results for meson production in nuclei. Finally, we give a short summary and an outlook in section 5.

## 2 Threshold production of pions

In 1965 Fubini et al. extended earlier predictions of low energy theorems ($LET$) for pion photoproduction by including the hypothesis of a partially conserved axial current ($PCAC$). In this way they succeeded in describing the threshold multipole $E_{0+}$ as a power series in $\mu = m_\pi/m$ up to terms of relative order $\mu^2$ [1],

$$E_{0+} = \frac{eg_{\pi N}}{8\pi W}\chi_f^\dagger \left(\frac{1}{2}[\tau_\alpha, \tau_0]A^{(-)} + \tau_\alpha A^{(0)} + \delta_{\alpha 0}A^{(+)}\right)\chi_i \tag{1}$$

$$A^{(-)} = 1 + 0(\mu^2), \qquad A^{(+,0)} = -\frac{1}{2}\mu + \frac{1}{2}\mu_N^{(V,S)}\mu^2 + 0(\mu^3), \tag{2}$$

where $g_{\pi N}$ is the pseudoscalar $\pi N$ coupling constant and $\mu_N^{(S,V)} = (\mu_p \pm \mu_n)/2$ are the isoscalar and isovector magnetic moments of the nucleon. The threshold multipoles for the different physical channels are obtained from eq. (1) by evaluating the Pauli matrices $\tau_\alpha$ between nucleon isospinors $\chi_f$ and $\chi_i$, where $\alpha$ is related to the pion charge.

The experimental data for charged pion production, smoothly extrapolated to threshold, agree reasonably well with these predictions. In units of $10^{-3}/m_\pi$, the following threshold values have been derived: $E_{0+}(n\pi^+) = 28.6 \pm 0.2(LET : 27.5)$, $E_{0+}(p\pi^-) = -31.5 \pm 1.0\,(LET : -32.0)$. However, the analysis of the recent Saclay [2] and Mainz [3] experiments has shown a strong fluctuation of the $E_0^+$-amplitude in the case of $p(\gamma, \pi^0)p$.

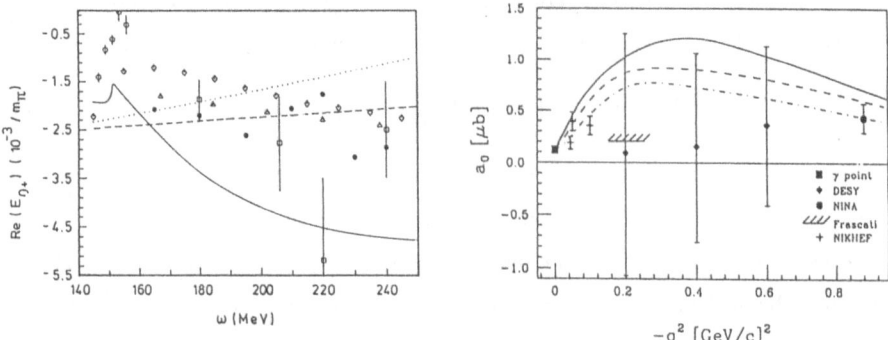

Figure 1: Left: Real part of the $E_{0+}$ amplitude for the reaction $\gamma p \rightarrow \pi^0 p$ between threshold and $250 MeV$ photon energy. Broken curve: Born terms with $PV$ coupling; dotted curve: Born terms plus vector meson exchange and $\Delta$ contribution; full curve: rescattering calculations [4]. The open circles near threshold are the new Mainz data [3], the "data points" at higher energies refer to various multipole analyses [1]. Right: The threshold amplitude $a_0$ for the electroproduction of pions on the nucleon. The new NIKHEF data at small $Q^2 = -q^2$ compared to older experiments and some theoretical predictions [5].

Though the amplitude can be extrapolated to $E_{0+} \simeq -2 \pm 0.2$ at threshold ($LET$ : $-2.4$), its rapid increase over the first few $MeV$ to values around 0 is still not well understood (see fig. 1). Instead, the new data have eventually led to exciting new developments concerning the validity of $LET$ itself. Within the framework of chiral perturbation theory ($\chi PT$), it was shown that certain classes of loop diagrams modifiy the predictions of $LET$ [6] to order $\mu^2$, leading to an additional term

$$\delta A^{(+)}(\chi PT) = \frac{m^2}{16 f_\pi^2}\mu^2, \tag{3}$$

where $f_\pi \approx 93 MeV$ is the pion decay constant. The origin of these corrections may be traced back to so-called triangle diagrams, which are non-analytic and develop a logarithmic singularity in the chiral limit ($\mu \rightarrow 0$), while all previous derivations of $LET$ had assumed a Taylor series expansion in $\mu$. The additional term appears at the same order as the anomalous magnetic moment (another one-loop effect!) and cancels the leading order term in the case of the proton. Strictly speaking, the prediction of the "new $LET$" gives a positive value if expanded to $0(\mu^2)$. However, the full one-loop calculation leads to $E_{0+}(p\pi^\circ) \approx -0.8$, and the extrapolation of the Mainz data to threshold can be obtained by including a counterterm $(-0.5)$ and some isospin breaking effects [6]. Similar results have been obtained in the linear $\sigma$-model [7].

Along the same lines $\chi PT$ gives important corrections for the electroproduction of pions at threshold [6]. As a function of $Q^2$ the amplitude for charged pion production is

$$A^{(-)}(Q^2) = 1 - \frac{Q^2}{6}\left(r_A^2 + \frac{3(\mu_N^V - \frac{1}{2})}{m^2} + \frac{3}{64f_\pi^2}(1 - \frac{12}{\pi^2})\right) + \dots \qquad (4)$$

While the first terms in the bracket, involving the axial radius $r_A$ and the isovector magnetic moment $\mu_N^V$, have been given by Nambu et al. before, the last term is due to triangle and tadpole diagrams at the one-loop level. In a comparison of the data determined by neutrino scattering and electroproduction, it gives a 10% correction to the axial radius. Similarly, there are corrections for the electroproduction of neutral pions involving functions of the ratio $(Q^2/m_\pi^2)$. Obviously these additional terms cannot be expanded in $Q^2$ and $m_\pi^2$ separately. It should be noted that the eventual decrease of the theoretical predictions shown in fig. 1 for the electroproduction amplitude $a_0$ is due to phenomenological form factors, while $\chi PT$ describes the data of the NIKHEF experiment [5] without further modifications. The role of off-shell form factors in threshold electroproduction has also been discussed by ref. [8].

Another interesting consequence of chiral Lagrangians is the existence of nonlinear terms in the pion-nucleon interaction leading to vertices with two or more pions. These contact interactions reduce the threshold cross section for the production of two charged pions by a factor 2-3 in comparison with a sequential pion production as described by PV pion-nucleon coupling. Unfortunately, the overall cross section is small and contributions of intermediate $\Delta$ resonances give a substantial background.

# 3  Resonance Production of Mesons

## 3.1  Photoabsorption

The total cross section for absorption of real photons on the proton is shown in fig. 2. The only contribution in this case is the transverse structure function $R_T$, integrated over all angles, $\sigma_{abs} \sim \bar{R}_T$. Since alle interference terms have vanished, only the most prominent resonances are visible. These are the $P_{33}(1232)$ or $\Delta$ resonance, the dipole excitation

$D_{13}(1520)$ in the second resonance region, the $F_{15}(1680)$ and some other states in the third resonance region. For photon energies $\omega > 2GeV$, the absorption cross section is roughly constant with a slight logarithmic increase at very high photon energies between $200 - 300GeV$. Of particular interest among the double-polarization observables is the response function $\bar{R}^{t}_{TT'}$, with target polarization in the direction of the incoming photon. It measures the difference of the helicity structure functions and is sensitive to the spin structure of the nucleon, $\sigma^{\frac{3}{2}}_{abs} - \sigma^{\frac{1}{2}}_{abs} \sim \bar{R}^{q}_{TT'}$. The Drell-Hearn-Gerasimov sum rule (DHG) connects this absorption cross section with the spin-flip amplitude for forward scattering at threshold, which is determined by the anomalous moment $\kappa$ of the nucleon [9, 10]. The resulting expression shows that the existence of an anomalous magnetic moment requires the dominance of $\sigma^{\frac{3}{2}}$ (circular polarization of the photon in the same direction as polarization of the target proton, overall helicity $\frac{1}{2}$) over $\sigma^{\frac{1}{2}}$ (polarizations in opposite direction, helicity $\frac{3}{2}$).

The DHG result is related to a more general function $I$ $(Q^2)$ describing the spin distribution inside the nucleon [11]. Assuming the validity of the DHG, in particular the convergence of the integral, the value at $Q^2 = 0$ is well known. There are experimental indications that $I(Q^2)$ changes sign in the region of $0.5GeV^2 \leq Q^2 \leq 1GeV^2$. Finally, for $Q^2 > 2GeV^2$, there exist data from deep inelastic lepton scattering measured at CERN and SLAC. The qualitative behaviour of $I(Q^2)$ may be understood in terms of vector dominance models. Quark models, also in relativized versions, seem to miss the DHG value completely. New experiments are being designed to bridge the gap between the DHG and deep inelastic scattering data, in particular at CEBAF [12] and, for lower $Q^2$, by a joint Bonn/Mainz project.

Figure 2: The total photoabsorption cross section for the proton in the resonance region. Also shown are the main decay channels, note in particular the small $\eta$-branch at the energy of the $S_{11}(1535)$.

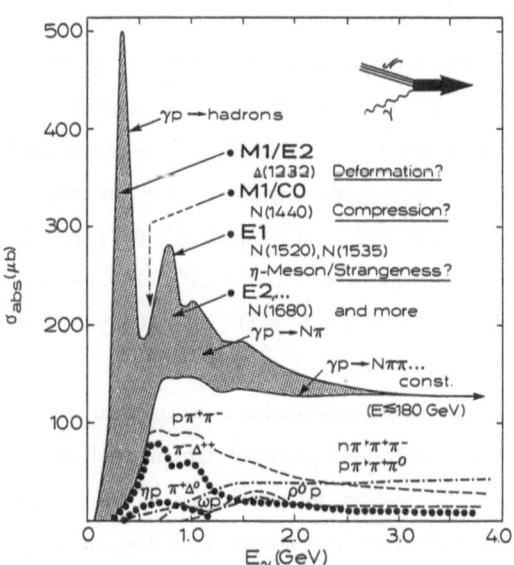

A calculation within the framework of the CQM [13] points out some of the problems of a theoretical evaluation of the DHG. With the usual choice of parameters, there appear large "recoil corrections" requiring a consistent relativistic calculation of the many-body system. Furthermore, the isoscalar-isovector interference term is completely due to $D$-state admixture ("bag deformation"). We note that present evaluations of this term by the dispersion integral disagree with the DHG even in sign. An exact decomposition into multipole contributions indicates a large amount of cancellations between various multipoles. As a consequence the proposed new measurements require a very careful multipole analysis over the whole resonance regions.

## 3.2 The delta resonance, $P_{33}(1232)$

The region between threshold and about 400 $MeV$ excitation energy is dominated by the $P_{33}(1232)$ or $\Delta(3,3)$ resonance, clearly visible in fig. 2 on top of a broad background of mostly S-wave pions. Within the harmonic oscillator quark model, the $\Delta$ and the nucleon are partners with configuration $\{56, 0^+\}_0$, i.e. members of the symmetrical 56−plet of $SU(6)$, orbital momentum $L = 0$, positive parity and no radial nodes. In this approximation the $\Delta$ may only be excited by the magnetic dipole ($M1$ oder $M_{1+}$, respectively). The hyperfine interaction between the quarks admixes a small $D$−state component, resulting in a small electric quadrupole transition ($E2$ or $E_{1+}$, respectively) of the order of 1%, and a small quadrupole moment of the $P_{33}$, $Q_\Delta \approx -.09 fm^2$. On the other hand, perturbative $QCD$ predicts that the spin-flip amplitude should vanish. Therefore the ratio $EMR \equiv E2/M1$ should approach unity in the limit $Q^2 \rightarrow \infty$. While $EMR$ vanishes in the harmonic oscillator quark model, the $CQM$ with hyperfine interaction, relativized quark models and chiral bag models predict $-2\% \leq EMR \leq -1\%$ [14]. Considerably larger values, at variance with the existing data, are obtained in Skyrme models. A recent lattice $QCD$ calculation predicts a small value with large error bars ($-6\% \leq EMR \leq 12\%$) [15]. As has been stated before, the $PQCD$ result approaches 100% at large $Q^2$, without any experimental evidence supporting such a large value up to now.

The $\Delta$ deformation has recently been studied by the Brookhaven group at $LEGS$ using transversely polarized photons (polarized photon asymmetry $\Sigma$) and measuring the cross sections for polarization parallel ($d\sigma_\parallel \sim R_T + R_{TT}$) and antiparallel $d\sigma_\perp \sim R_T - R_{TT}$) to the reaction plane [16]. In particular, $d\sigma_\parallel$ is very sensitive to the bag deformation. The experiment follows quite nicely the prediction of Davidson et al. [17], indicating $EMR \approx -1.5\%$.

Careful studies have shown that the polarized photon asymmetry $\Sigma$ is indeed the most sensitive observable for experiments with real photons. Many more choices seem to exist for electroexcitation with polarization degrees of freedom, apparently some of the longitudinal and transverse interference terms are very sensitive to both the $E_{1+}$ and $L_{1+}$ amplitudes. The coincidence $\vec{e} + p \rightarrow e' + \vec{p} + \pi°$, with polarization transfer to the proton, is a particularly well suited experiment. The use of recoil polarization has been suggested by various authors [18, 19], and the rather striking predictions for such experiments are shown in fig. 3 [20].

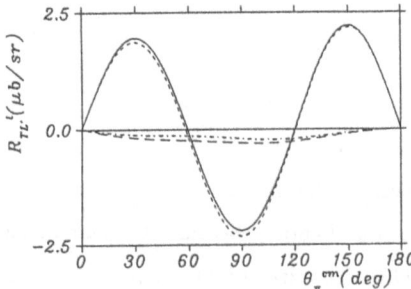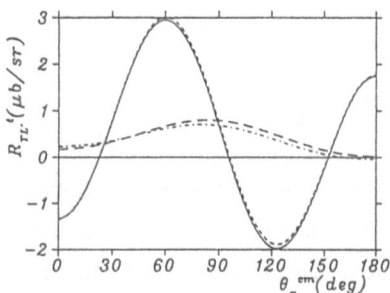

Figure 3: The structure functions $R^t_{TL'}$ and $R^t_{TL'}$ in the region of the $\Delta$($\omega = 1232 \ MeV$, $Q^2 = 0.12 \ GeV^2$). Long-dashed: Born and $M_{1+}$; dashed-dotted: full calculation without $L_{1+}$; short-dashed: Born and all $\Delta$ multipoles; full line: calculation including Born, $\Delta$, and higher multipoles.[20]

## 3.3 The Roper resonance, $P_{11}(1440)$

In the $CQM$ the Roper is a radial excitation of the nucleon occuring at an energy of $2\hbar\omega_0$. The $CQM$ predicts a ratio $A^n_{\frac{1}{2}}/A^p_{\frac{1}{2}} = -\frac{2}{3}$, in reasonable agreement with the data within the large error bars. However, the values for the $CQM$ amplitudes themselves are too small by a factor of 3. The chiral bag model $(CBM)$ predicts a ratio of $-1$ for the pionic contributions and larger values for the amplitudes [14]. As has been pointed out by Burkert and collaborators [21], explicit gluon degrees of freedom might play a role even at low excitation energies. The wave function of such a "hybrid" with a component $|q^3 \times g>$ requires a quark configuration with colour in order to insure an overall colour neutral wave function. As a consequence the quarks can now be in the mixed symmetry

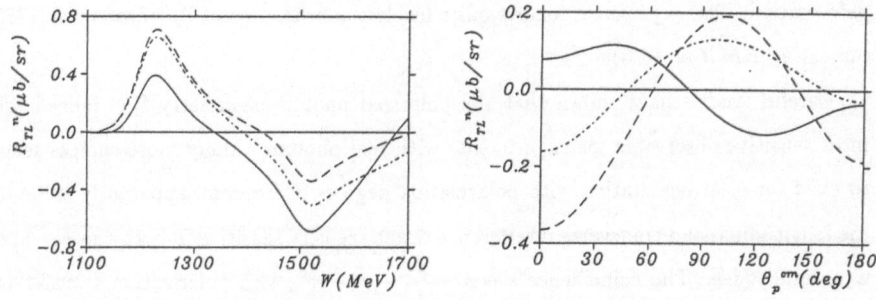

Figure 4: The structure function $R_{TL}^n$ in the region of the Roper resonance at $Q^2 = 0.12 GeV^2$. Left: as function of invariant mass $W$, at $\Theta_p = 90°$; right: at $W = 1350 MeV$ as function of angle. Dashed curve: Born terms $+ \Delta +$ higher resonances; full curve: as in dashed curve $+$ Roper $(q^3)$; dashed- dotted curve: as in dashed curve $+$ Roper $(q^3 \times g)$. The result of the Born approximation vanishes in all cases [20].

representation $\{70, 0^+\}_0$ of $SU(6)$ without orbital or radial nodes. The corresponding Coulomb amplitude vanishes except for relativistic corrections, $S_{\frac{1}{2}}(q^3 \times g) \approx 0$, because the longitudinal photon cannot excite the transverse colourmagnetic field of the gluon. The Roper is certainly a good candidate for such a "hybrid", because it occurs at an extremely low energy for a $2\hbar\omega_0$ state of the $CQM$. The size of its Coulomb excitation $S_{\frac{1}{2}}$ will be quite essential for its classification. While a small or vanishing value will be an indication of a hybrid, very large contributions should be typical of explicit pion degrees of freedom as predicted by the $CBM$.

In this context there are also interesting new data for inelastic scattering of $\alpha$-particles on the proton, showing a broad bump for excitation energies $\omega \approx 0.5 GeV$ at forward angles [22]. According to previous experience with giant resonance studies in nuclei, the data are an indication of a strong Coulomb monopole in the energy range of the Roper.

There are quite a few predictions that polarization degrees of freedom will be useful to unravel the puzzle of the Roper. As has been shown in fig. 4, the structure function $R_{TL}^n$ is sensitive to the Roper over a wide energy range between the $\Delta$ and the second resonance region at $W \approx 1.5 GeV$. Also the angular distribution of the asymmetry is strongly determined by the properties of the Roper [20].

## 3.4   The resonance $S_{11}(1535)$ and eta production

In comparison with its partner $D_{13}(1520)$, the dipole excitation $S_{11}(1535)$ is only weakly seen in pion photoproduction. However, it couples very strongly to the $\eta$ meson, about 50% of its decay width is due to $\eta$ emission. In comparison, the $D_{13}$ has only a $10^{-3}$ branch for $\eta$ decay, and also an excited $S_{11}$ occuring in the third resonance region couples only weakly to the $\eta$. The only other resonance with a sizeable $\eta$ branch is the $P_{11}(1710)$ with 25% $\eta$ decay.

The $\eta$ and its coupling to nucleon resonances is still not well understood. It is relatively strongly mixed with the $\eta'$ (mixing angle between $-10°$ and $-23°$). This $\eta'$ corresponds to the axial singlet current, which is not conserved even in the Goldstone limit of a vanishing $\eta'$ mass. The origin of this behaviour is the axial anomaly of $QCD$, a four-divergence of the axial current related directly to gluonic degrees of freedom. The coupling of $\eta$ and $\eta'$ is one of the reasons that soft-particle theorems do not apply for the $\eta$. Another reason is its large mass, which would make it difficult to extrapolate from a "soft $\eta$" to the real world. Finally, the existence of the $S_{11}$ resonance immediately at threshold overshadows the role of Born terms. This appearance of a resonance immediately at threshold is quite unique. In fact it has been shown recently in elastic pion scattering that the "peak" around $1535 MeV$ may also be described by a cusp at the $\eta$ threshold on top of a nonresonant background [23].

Models of eta photoproduction have been constructed in terms of resonance contributions and Born terms by various authors [24, 25]. Since there is no equivalent to $PCAC$ for the $\eta$, both pseudoscalar $(PS)$ and pseudovector $(PV)$ coupling between $\eta$ and nucleon could exist. Though the predictions for the Born amplitudes differ quite strongly, neither $PS$ nor $PV$ coupling could be ruled out by the old data due to the dominance of the $S_{11}$ resonance.

In order to separate the two isospin amplitudes $A^{(0)}$ and $A^{(1)}$, experiments on a neutron target have to be performed as well. In the case of coherent photoproduction $d(\gamma, \eta)d$, the cross section is determined by the isoscalar amplitude $A^{(0)}$ only. While the theoretical analysis prefers $A^{(0)}/A^{(1)}$ in the range of $15-30\%$, the existing data require at least a ratio of 60%. In particular, the differential cross section for coherent $\eta$ production is larger than the theoretical prediction by about one order of magnitude. In view of these discrepancies, precision data on the deuteron are urgently needed. (This is one of the examples that nuclei are important targets, as filters of spin and isospin, even if one

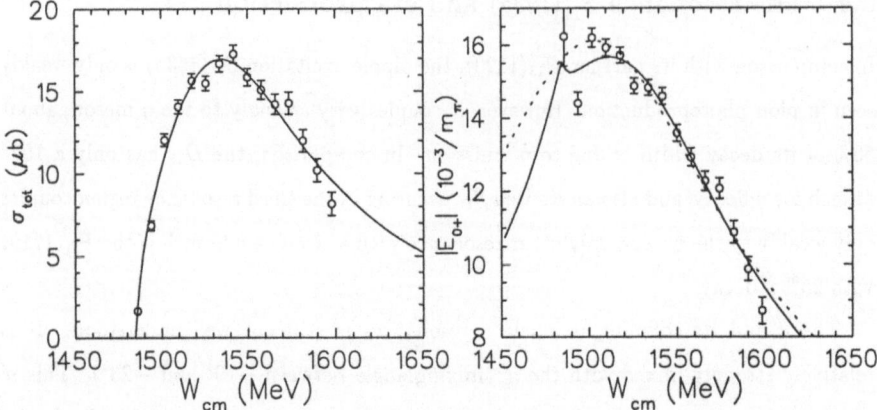

Figure 5: Left: Total cross section for $p(e, e'\eta)$ at $Q^2 = 0.056 GeV^2$. The full line is the result of a Breit-Wigner fit with an energy-dependent width [29]. Right: Threshold amplitude $E_{0+}$ assuming that longitudinal amplitudes and higher partial waves are negligible. The full line is a fit with an energy-dependent width, the dashed line is obtained with a constant width. The preliminary data are from Bonn [29], calculations of ref. [31].

is not interested in the nuclear structure itself!).

Following the first "modern" experiments [26, 27], quite a few new precision data on $\eta$ production will soon be available. At $ELSA$ a Bonn/ UCLA collaboration [28] has measured the reaction $\gamma p \rightarrow \eta p$ with tagged photons in the energy range of $\omega = 710 - 1150 MeV$. In addition, the $ELAN$ group [29] has studied electroproduction at $Q^2 = 0.06 GeV^2$ for energies up to $W = 1590 MeV$. The $TAPS$ collaboration at Mainz [30] has obtained total and differential cross sections for $\eta$ production on the proton, deuterium and some complex nuclei (C, Ca, Nb, Pb). These data range from threshold ($707 MeV$ for the proton, $548 MeV$ for lead) up to $790 MeV$. The best statistics has been obtained from the $2\gamma$ decay branch, but the decay into $3\pi^\circ$ can provide an independent check of systematic errors. As an example for the quality of the new experiments, the preliminary data of the ELAN group have been shown in fig. 5, together with an estimate for the threshold multipole $E_{0+}$.

# 4  Meson Production on Nuclei

The total nuclear absorption cross section for photons has been measured in the $\Delta$ resonance region by various authors using different techniques. The results for the cross section per nucleon show a nearly universal resonance shape and strength over a wide range of nuclei from $^6Li$ to $^{238}U$, indicating an incoherent absorption mechanism. The width of the bound resonance appears to be substantially increased due to Fermi broadening and coupling to many-nucleon channels, but the resonance is still clearly visible. In the energy region above the $\Delta$, such precision data were missing until recently. New absorption experiments have now been performed at Frascati in the energy range of $200MeV \leq \omega \leq 800MeV$ [32] and at Mainz for energies from $50MeV$ to $800MeV$ [33]. In comparison with the proton and the deuteron, where the second and third resonance regions are clearly seen, these authors find no direct evidence for the excitation of the higher resonances in the case of heavier nuclei. While there is still some gap in the region between $1.2GeV \leq \omega \leq 2GeV$, the data extrapolate smoothly to the "asymptotic" region dominated by shadowing effects. Whether the "disappearing" resonances are due to a surprisingly early onset of shadowing or caused by a strong damping mechanism, is an interesting question still under discussion. In this context it should be noted, however, that the Mainz data are consistently higher than the Frascati ones by about 15%. It is also interesting that preliminary results of the DAPHNE group at Mainz [34] do still find a second bump in the cross section for total photoabsorption on $^3He$, in the range of the $D_{13}$ resonance. A detailed analysis of the decay channels in this experiment will certainly shed some light on the nature of the damping of higher nucleon resonances in the nuclear environment.

The photo- and electroproduction of $\eta$ mesons is particularly well suited to study the excitation of the $S_{11}$ in nuclei. While the angular distribution for the proton is relatively flat ($S$ wave production!), we observe a sharp decrease with the scattering angle already for the deuteron and even stronger for heavier nuclei. In particular, the deuteron measures the isoscalar component only and $^4He$ is sensitive to the isoscalar non spin-flip terms [31]. As has been pointed out in section 3.4, there is a strong discrepancy between theory and experiment in the case of the deuteron. For heavier nuclei preliminary data of the TAPS collaboration [30] obtained at Mainz indicate a relatively strong final state interaction of the $\eta$ with nuclear matter in qualitative agreement with the strong damping of the higher resonances seen in photoabsorption.

As a final example for pion production in nuclei we mention an interesting enhancement of the $E_{1+}$ multipole in the polarized photon asymmetry $\Sigma$ for the reaction $^3He(\gamma, \pi^+)^3H$. In Faddeev calculations with realistic nuclear forces, $^3He$ has a $D$-state probability of about 8%. While a simple model with $S$-states only predicts

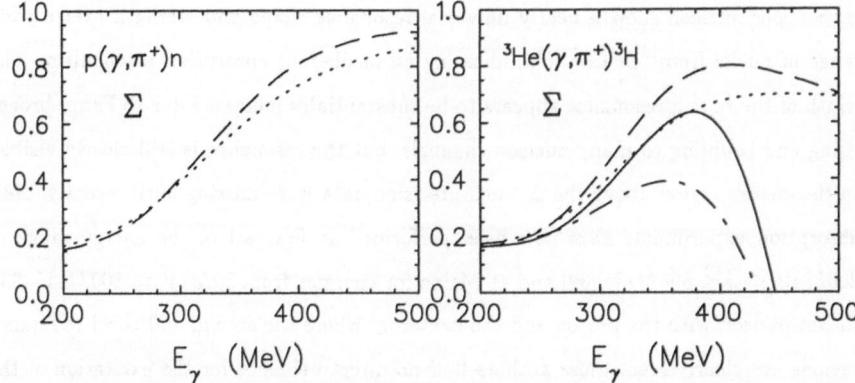

Figure 6: Comparison of the polarized photon asymmetry $\Sigma$ for $(\gamma, \pi^+)$ on the proton and $^3He$ at $\theta_{cm} = 90°$ as function of the photon lab energy. Left: Dashed and dotted lines with and without the $E_{1+}(\Delta)$ multipole for the proton. Right: Dashed and dotted lines with and without the $E_{1+}(\Delta)$ multipole for $^3He$ for a simple S-state wave function; solid and dashed-dotted lines with and without the $E_{1+}(\Delta)$ obtained with a realistic Faddeev wave function including D-state components [31].

$\Sigma(^3He) \approx \Sigma(p)$, the additional D-state allows a strong $E_{0+}/E_{1+}$ interference for the bound nucleon, leading to a considerable increase in sensitivity to the small $E_{1+}$ multipole [35]. As may be seen in fig. 6, the contribution of the $E_{1+}$ is quite small for the proton. In comparison, it gives rise to a huge effect in the case of $^3He$.

## 5   Conclusions

Low energy theorems (LET) have been a powerful tool to determine the threshold amplitudes for reactions involving nucleons and pions. Using covariance, gauge invariance and the partial conservation of the axial current, these amplitudes have been expressed in terms of global properties of the hadrons (masses, coupling constants, magnetic moments). The puzzling experimental data obtained for the production of neutral pions near threshold have led to a series of new theoretical investigations, in particular in the framework of $\chi$PT. It has been shown that the threshold amplitudes are non-analytical in the soft-pion limit, resulting in important corrections to the traditional LET.

With the completion of the new electron accelerators with high duty-factor and high intensity, new types of coincidence experiments with polarization degrees of freedom will be performed. These new experiments will address some of the most challenging questions of intermediate energy physics, related with

- the S-wave amplitude for neutral pions near threshold,

- the electric quadrupole amplitude in the $\Delta$ region ('bag deformation'),

- the structure of the Roper resonance ('breathing mode' vs. 'hybrid'),

- the production of $\eta$-mesons via the $N^*$ (1535),

- the production of two pions and more to test consequences of chiral symmetry,

- radiative pion photoproduction to study the polarizability of the pion,

- the spin structure of pion production and the Drell-Hearn-Gerasimov sum rule,

- the search for 'missing resonances' at excitation energies of about 2 GeV,

- the production of $K$-mesons in order to embed strangeness into the nucleus.

With a better understanding of the elementary amplitude, also meson production on the nucleus can be studied and analyzed with better precision. In some special cases such reactions can serve as filters to project onto certain quantum numbers or to magnify small amplitudes. This is true in particular for the deuteron as a neutron target or for other light nuclei whose structure is reasonably well under control. In heavier nuclei the $A$ dependence of the cross section gives information on global properties of the interaction of mesons with nuclear matter. The new experimental data on $\eta$-production off the proton and a series of nuclei clearly show that such studies can now be performed with an unprecedented accuracy, and similar information will be available for the pion and other mesons within the next few years.

# References

[1] Drechsel, D. and Tiator, L. , J. Phys. G: Nucl. Part. Phys **18**, 449 (1992).

[2] Mazzucato, E. et al., Phys. Rev. Lett. **57**, 3144 (1986).

[3] Beck, R. et al., Phys. Rev. Lett **65**, 1841 (1990).

[4] Nozawa, S. et al., Nucl. Phys. A **513**, 459 (1990).

[5] Welch, T. P. et al., Phys. Rev. Lett. **69**, 2761 (1993).

[6] Bernard, V. et al., Phys. Lett. **B268**, 291 (1991) and Nucl. Phys. **B383**, 442 (1992).

[7] Davidson, R. M., Phys. Rev. **C47**, 2492 (1993).

[8] Scherer, S. and Koch, J. H., Nucl. Phys. A534, 461 (1991).

[9] Drell, S. D. and Hearn, A. C., Phys. Rev. Lett. **16**, 908 (1966).

[10] Gerasimov, S. B., Sov. J. Nucl. Phys. **2**, 430 (1966).

[11] Anselmino, M., Ioffe, B. L. and Leader, E., Sov. J. Nucl. Phys. **49**, 136 (1989).

[12] Burkert, V. , Li, Z., Phys. Rev. **D47**,46 (1993).

[13] Drechsel, D. and Giannini, M. M., to be published.

[14] Drechsel, D., Proceedings of the $6^{th}$ Workshop on Perspectives on Nuclear Physics at Intermediate Energies (1993), and references given therein.

[15] Leinweber, D. B., Draper, T., Woloshyn, R., Contribution to Baryons '92, p. 29 (1992).

[16] Blanpied, G. S. et al., Phys. Rev. Lett. **69**, 1880 ( 1992).

[17] Davidson, R. M., Mukhopadhyay, N. C., Wittmann, R., Phys. Rev. **D43**, 71 (1991).

[18] Lourie, R. W., Nucl. Phys. **A509**, 653 (1990).

[19] Bernstein, A. M., Nozawa, S., Moinester, M. A., Phys. Rev. **C47**, 1274 (1993).

[20] Hanstein, O., diploma thesis, Mainz (1993).

[21] Li, Z. P., Burkert, V., Li, Zh. Phys. Rev. **D46**, 70 (1992).

[22] Morsch, H. P. et al., INPC, Contribution 2.1.17, Wiesbaden (1992).

[23] Hoehler, G. and Schulte, A., $\pi N$ Newsletter No. 7, 94 (1992).

[24] Bennhold, C., Tanabe, H., Nucl. Phys. **A530**, 625 (1991).

[25] Benmerrouche, M., Mukhopadhyay, N. C., Phys. Rev. Lett. **67**, 1070 (1991).

[26] Homma, S. et al., J. Phys. Soc. Japan **57**, 828 (1988).

[27] Dytman, S. A. et al., Bull. Am. Phys. Soc. **35**, 1679 ( 1990).

[28] Anton, G., Proceedings of the $6^{th}$ Workshop on Perspectives on Nuclear Physics at Intermediate Energies (1993).

[29] Wilhelm, M., Ph. D. thesis, Bonn (1993).

[30] Krusche, B., Proceedings of the $6^{th}$ Workshop on Perspectives on Nuclear Physics at Intermediate Energies (1993).

[31] Tiator, L., Proceedings of the 2nd TAPS Workshop, Alicante (1993).

[32] Bianchi, N. et al., Phys. Lett. **B299**, 219 (1993).

[33] Frommhold, Th. et al., Phys. Lett. 295B, 28 (1992).

[34] Tamas, G., private communication (1992).

[35] Kamalov, S. S., Tiator, L. and Bennhold, C., Nucl. Phys. **A547**, 599 (1992).

Few-Body Systems, Suppl. 7, 339—348 (1994)

# PION ABSORPTION IN TRITIUM AND HELIUM

H. J. Weyer

Physics Institute, Basel University, CH-4056 Basel

and

Paul Scherrer Institute, CH-5232 Villigen PSI

Switzerland

## Abstract

The situation of pion absorption in the A=3,4 nuclei is described with emphasis on the reaction modes. The phenomenology of 2N reactions and the evidence for reaction modes involving more than two nucleons are discussed. Similarities between pion and photon absorption are emphasized.

## Introduction

Absorption is one of the most fascinating ways in which a pion interacts with its environment. Generally seen, the absorption of a pion in a nucleus is a manybody problem as in principle all nucleons will be involved. This is true for absorption reactions as for all other reactions in nuclei. To treat such a system in a general way would not be possible. Fortunately, however, the experimental data tell us that the nucleus can be split phenomenologically into participating and spectator nucleons in quite an efficient way. We will see in the following that this approach of a quasifree reaction is a very powerful tool to describe absorption reactions. The way, in which the nucleus is subdivided into actors and spectators can not be determined from first principles. This has to be derived from the experiment.

## 1N Absorption

A reaction, where a real pion (or photon) is absorbed on a single nucleon, is forbidden by momentum conservation. This is not exactly true within a nucleus, as here the rest of the nucleus could help to provide the necessary momentum transfer. In that sense, a one-nucleon absorption (1N absorption) mechanism would be characterized by the genuine absorption taking place on a single nucleon with the rest of the nucleus participating in a minimal way by balancing the total momentum. This approach has been tried in the past by various authors (for a review, see [1]). Because, however, of the large momentum transfer involved, it appeared that the results depend on the very details of the particular models. This contradicts a clean subdivision of the nucleons into participants and spectators. Intuitively, the most obvious candidates for 1N absorption events would be reactions of the type $\pi$ A $\rightarrow$ N (A-1) with (A-1) as a bound system. Experimentally, it is found that the cross section for these reactions is quite low. In addition, investigations (e.g. Fearing [2], Korkmaz et al. [3]) suggest, that even such a reaction can be explained, at least to a large extent, by a 2N absorption mechanism. Therefore, 1N absorption can be ignored in general.

## 2N Absorption

Absorption of a pion on two nucleons (2N absorption) is the dominant mode. In the 'ideal case', the reaction takes place on two nucleons with the rest of the nucleus acting (or better not-acting) as spectator. In a 1N absorption reaction, the momentum transfer to the spectator would have to be several hundred MeV/c. Here, in the 2N absorption case, no momentum transfer to the spectator is needed kinematically and the momentum is just the Fermi momentum between the absorbing pair and the rest of the nucleus. Because this is so low ($\approx$ 100 MeV/c) that it is below usual detection thresholds, there is the typical signature: two high-energy nucleons leaving the reaction vertex essentially back to back.

It suggests itself to relate the absorbing two-nucleon system to a physical deuteron (quasi-deuteron model). In this concept, the measured cross section is factorized into a 'deuteron' part and a 'spectator' part. Such a model has been proposed long ago for pion absorption by Tamor [4] and Heidman [5, 6] and for photon absorption by Levinger [7]. In the simplest case, the influence of the spectator on the reaction reduces just to the distribution function of the momentum between the absorbing pair and the spectator. That means that the momentum distribution measured by 2N absorption is a property of the nucleus rather than of the absorption reaction itself. This relation is most straightforward for the very light nuclei with their simple wave functions (S wave dominance). For $^3$H and $^3$He the momentum distribution of the pair (which in this case is the same as that of the spectating nucleon) has been extracted [8, 9] from the experiment and a good agreement is found with the momentum distribution obtained from $^3$He(e,e'p)pn data. This comparison could even be used to define 2N absorption: it will be shown below that 2N absorption alone is not the only absorption mechanism but that there are strong indications for a genuine absorption mode on three nucleons. This absorption mode exhibits totally different momentum distribution and thus, the comparison to the (e,e'p) data offers a quantitative way to identify the relative strengths. The good agreement between the momentum distribution for the 2N absorption and the (e,e'p) reaction has been confirmed by other experiments [10].

It had been proposed very early [7] that pion and photon absorption could be described by similar models. This aspect has been forgotten somewhat but the phenomenological signatures alone are so similar that a simultaneous discussion is very suggestive. The similarity of 2N absorption of pions and photons is most obvious in the energy range of a few hundred MeV which is dominated by the $\Delta$ resonance. Because the coupling to this resonance is so strong, the first step of an absorption is the creation of this resonance by a $\pi$N$\rightarrow\Delta$ or a $\gamma$N$\rightarrow\Delta$ reaction (if allowed by quantum numbers). Then, for both cases, the $\Delta$ resonance decays by a $\Delta$N$\rightarrow$NN reaction. This picture of a rescattering reaction quite successfully describes the dominant $\Delta$- mediated part of the absorption.

By going from the deuteron as simplest two-nucleon case to two-nucleon systems in heavier nuclei isospin appears as an additional degree of freedom. The nucleon pair in the deuteron has isospin zero. Already in the A=3 nuclei $^3$He and $^3$H there are pp and nn nucleon pairs with isospin 1. It will be discussed below that around the $\Delta$ resonance, absorption on isovector nucleon pairs is strongly suppressed compared to the isoscalar case.

In the sense of a quasifree 2N absorption, the spectator contributes only through the Fermi motion and all the dynamics of the genuine absorption process is contained in the quasideuteron part. It is now very interesting to relate the isoscalar 2N absorption component to the $\pi$d$\rightarrow$pp reaction in a quantitative way. The characteristics of the $\pi$d reaction are the angular dependence for a given pion energy, described by Legendre coefficients, and the dependence of the angle-integrated cross section on the energy of the incoming pion. Angular distributions, for isoscalar 2N absorption in tritium and the

He isotopes have been obtained from different experiments. As an example, Fig. 1(a) shows the data on $^3$H from Salvisberg et al. [9]. Around the $\Delta$ resonance, a fit with

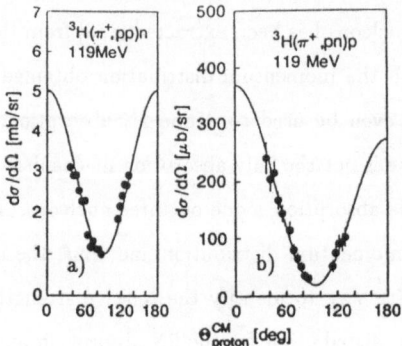

Figure 1: Angular distribution for 2N absorption in $^3$H, (a) for isoscalar and (b) for isovector absorption. The curves are Legendre fits to the data. From Salvisberg et al. [9].

only two Legendre coefficients, $A_0$ and $A_2$ is sufficient to describe the data. Within the errors, the result for $A_2/A_0$ is in very good agreement with the value for $\pi d \rightarrow pp$. Similar results are found in the other experiments on $^3$He and $^4$He [8, 11]. Even more, the absolute value of $A_0$ (i.e. the integrated cross section) can be understood. The tritium and the He nuclei are predominantly in an S state. From that and the Pauli principle it follows that there are 1.5 isoscalar nucleon pairs in the A=3 nuclei and 3 pairs in $^4$He. Fig. 2 shows the integrated isoscalar 2N absorption cross section data available, including the new preliminary data from the $4\pi$ detector LADS at PSI. If one now compares the cross section in the A=3,4 nuclei to that on the deuteron then it appears that, within the errors, they scale just according to the number of isoscalar nucleon pairs. At first sight, this result is somewhat unexpected. It had been discussed since first studies of pion absorption that the cross section should depend on the nuclear density or in other words the nucleon-nucleon correlations. By going from the deuteron to heavier nuclei, this would increase the cross section. On the other hand, a genuine 2N absorption embedded in a nucleus must be expected to be influenced by distortions, i.e. initial and/or final-state interactions. Estimates for distortions come from DWIA calculations which have not much predictive power for such light nuclei as the He isotopes. In order to arrive at the experimental findings which show no large net effect, nucleon-nucleon correlations and distortion effects could be large and of opposite sign or both be small. A detailed

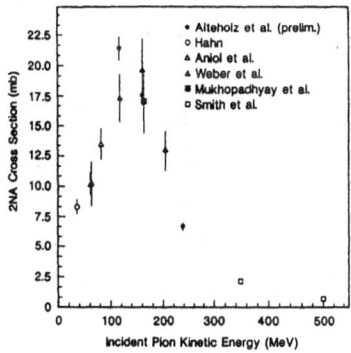

Figure 2: 2N absorption cross section on $^3$He as function of the incident-pion energy. Data are from Alteholz et al. (LADS) [12], Hahn et al. [13], Aniol et al. [14], Weber et al. [8], Mukhopadhyay et al. [15], and Smith et al. [16]. Adapted from Alteholz et al.

investigation in $^3$He [17] did not yield a signature of distortion effects. From that one must conclude that there is also, within the errors, no indication for a density effect in $^3$He. After all, this is not totally unexpected. From the studies on $\pi d \rightarrow pp$ it is known that isoscalar absorption is dominated strongly by the partial wave $^5S_2$ for the intermediate $\Delta N$ system leading to a $^1D_2$ wave in the final pp system. This partial wave is forbidden by quantum numbers for absorption on an isovector nucleon pair. As stated above, the first step in the absorption of a pion is the excitation of a $\Delta$ resonance. Therefore, in isoscalar absorption one is not sensitive to NN correlations but to $\Delta N$ correlations.

The situation is different for isovector absorption. It can be seen already by comparing the angular distributions that different partial waves are involved. The $(\pi^-,pn)$ distribution has a larger $A_2/A_0$ ratio (see Fig. 1(b)) and is not symmetric around 90°. Instead, weaker $\Delta N$ partial waves contribute and also partial waves, which do not proceed via intermediate $\Delta N$ states. Recently, Blankleider et al. [18] have performed a first energy-dependent partial wave analysis for isovector absorption in $^3$He based on the existing experimental data. As an example, Fig. 3 shows the analyzing power from the time-reversed reaction $d(\vec{p}, \pi^-pp)p$ (pp pair in relative S state) measured by Ponting et al. [19] together with the PWA fit from Blankleider et al. [18]. The PWA analysis finds only a weak $^3P_0$ component (intermediate $\Delta N$). The strongest contribution comes from the $^3D_1$ partial wave with about 2/3 followed by the $^3S_1$ partial wave. The asymmetry in the angular distribution around 90° (Fig. 1) is caused by interference terms between

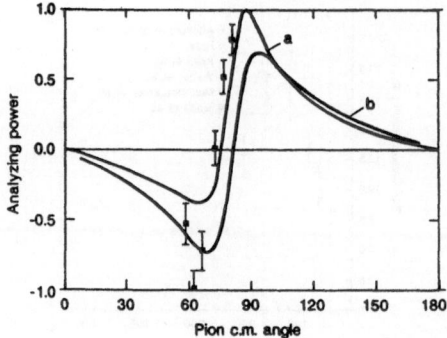

Figure 3: Analyzing power for the reaction d($\vec{p}, \pi^-$pp)p at $T_p$=400 MeV. The data are from Ponting et al. [19]. Curve a) is from the PWA fit of Blankleider et al. [18], b) shows a coupled-channel calculation from Niskanen [20].

these partial waves. Therefore, isovector absorption is sensitive to NN correlations via the $^3S_1$ wave but the dominant partial wave is a D wave which is less sensitive to short-range correlations than a S wave.

## Multistep Mechanisms and Genuine 3N Absorption

If a 2N mechanisms is embedded in a nucleus, effects must be expected due to the nuclear environment. The pion, before arriving at its target pair, may react with other nucleons by a scattering or charge exchange reaction (intial-state interaction, ISI) or, in the final state, an outgoing nucleon may knock on another nucleon (final-state interaction, FSI).

Although caused by the same nucleon-nucleon interaction, two different extreme model cases can be considered for FSI. One case is characterized by two outgoing nucleons having a low relative momentum. Such configurations are known since a long time to correspond to an enhancement in the cross section. This effect has nothing to do with pion absorption but is mainly a function of the relative momentum of the two nucleons. For a pp pair this final state interaction is suppressed by Coulomb repulsion. This Watson-Migdal-type [21, 22, 23] final-state interaction has been seen in many other hadronic reactions with nucleons in the final state. It is a coherent effect, which reflects the increase of the nucleon-nucleon cross section towards zero relative momentum. Therefore, the description is in terms of the low-energy-scattering parameters (effective range and scattering length). Because of the low relative momentum involved this phenomenon is denoted as *soft* final-state interaction. With their quasi-twobody signature, soft-FSI events are collinear in the 3-particle CM system and populate the perimeter of the Dalitz plot. Although it explains the observed structures qualitatively, the quantitative capabilities of the Watson-Migdal model are limited. A more universal approach is provided by Faddeev-type calculations (Witala et al. [24], Bartnik et al. [25]).

Another approach to final-state interactions comes from cascade models. A nucleon on its way out of the nucleus after an absorption reaction may hit on another nucleon but in this case the relative momentum will be high (*hard* final-state interaction). The kinematical signature of hard FSI is totally different from that of soft FSI [17]. If the probability for this effect is not too high, then mostly only one of the nucleons after a 2N absorption will be affected. In a plot of the energies of these two high-energy nucleons (in the A=3 case Dalitz plot) this should lead to band-like structures. Such structure have not been seen in the A=3,4 data [17, 26].

Structures due to initial-state interaction are also predicted. In case of a two-step initial-state interaction, the pion between the first step (scattering or charge exchange) and the second one (2N absorption) could be on its mass shell. It has even be predicted that the cross section for such a mechanism would be maximum in the neighbourhood of the mass shell [27]. It has been pointed out [28] that the effective mass of this exchange pion can be determined from the experimental data and thus a sensitive check on ISI can be performed by studying the distribution of this effective-mass parameter. An analysis of Backenstoss et al. [17] showed no indication of such an enhancement.

Similar multi-step structures as in pion absorption have been proposed also for photon absorption [27]. Currently, for both probes large-solid-angle detectors have been built which will allow to search for such mechanisms with much higher sensitivity. By combining the searches with pions and photons this sensitivity can be enhanced even more. The structures predicted for initial-state interaction originate from the fact that degrees of freedom are frozen resulting in signatures of quasi-two-body kinematics. On top of that, within a two-step picture one expects to see in the measured distributions the characteristic structures of the two-body $\pi$N scattering or the $\gamma$N$\rightarrow \pi$N reaction, respectively. As the first one has maxima at 0° and 180°, the latter at 90°, an unambiguous signal of initial-state interaction should show different structures in pion and photon-induced reactions.

Even, if the question is not finally settled to what extent the 3N absorption cross section behaves exactly according to phase space or not , to our present knowledge the

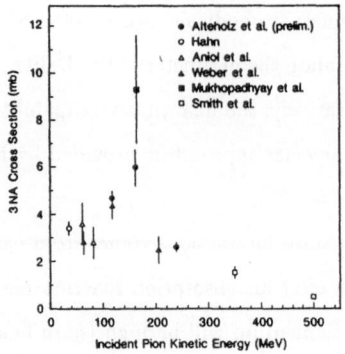

Figure 4: 3N absorption cross section on $^3$He as function of the incident-pion energy. Data are from Alteholz et al. (LADS) [12], Hahn et al. [13], Aniol et al. [14], Weber et al. [8], Mukhopadhyay et al. [15], and Smith et al. [16]. Adapted from Alteholz et al.

deviations are not very large. Thus, the measured data can be extrapolated to the full solid angle and an integrated cross section can be determined. For the previous two-arm experiments this necessitated large extrapolations. The world data on $^3$He are shown in Fig. 4. The figure contains also the preliminary results from the LADS detector. In this case, the systematic uncertainty due to extrapolations is significantly reduced because of the large solid angle acceptance. The comparison of the cross sections for 2N and 3N absorption clearly shows that multi-nucleon absorption is a significant part of the absorption cross section. Similar conclusions can be drawn from the $^3$H and $^4$He measurements and also from photo-absorption data.

Compared to the 2N cross section which populates only a small part, multi-nucleon absorption is present in the whole phase space and, to our current knowledge, with only weak structure. Therefore, the new large-solid-angle detectors which started their operation both, in pion- and photon-absorption experiments, are optimally suited. Interesting new results are beginning to emerge and a unified view of the results with both probes is highly rewarding.

## REFERENCES

1. H.J. Weyer. *Phys. Rep.*, 195:295, 1990.

2. H.W. Fearing. *Progr. Part. Nucl. Phys.*, 7:113, 1981.

3. E. Korkmaz, L.C. Bland, W.W. Jacobs, T.G. Throwe, S.E. Vigdor, M.C. Green, P.L. Jolivette, and J.D. Brown. *Phys. Rev. Lett.*, 58:104, 1987.

4. S. Tamor. *Phys. Rev.*, 77:412, 1950.

5.  J. Heidmann and L. Leprince-Ringuet. *Compt. Rend.*, 226:1716, 1948.

6.  J. Heidmann. *Phys. Rev.*, 80:171, 1950.

7.  J.S. Levinger. *Phys. Rev.*, 84:43, 1951.

8.  P. Weber, G. Backenstoss, M. Izycki, R.J. Powers, P. Salvisberg, M. Steinacher, H.J. Weyer, S. Cierjacks, A. Hoffart, H. Ullrich, M. Furić, T. Petković, and N. Šimićević. *Nucl. Phys.*, A 501:765, 1989.

9.  P. Salvisberg, G. Backenstoss, H. Krause, R.J. Powers, M. Steinacher, H.J. Weyer, M. Wildi, A. Hoffart, B. Rzehorz, H. Ullrich, D. Bosnar, M. Furić, T. Petković, N. Šimićević, H. Zmeskal, A. Janett, and R. H. Sherman. *Phys. Rev.*, C 46:2172, 1992.

10. L.C. Smith, R. Minehart, D. Ashery, E. Piasetzky, M. Moinester, I. Navon, D. Geesaman, J.P. Schiffer, G. Stephens, B. Zeidman, S. Mukhopadhyay, R.E. Segel, M. Sober, B. Anderson, R. Madey, and J. Watson. *Contribution C26 to PANIC Conf.* ,Kyoto, 1987.

11. M. Steinacher, G. Backenstoss, M. Izycki, P. Salvisberg, P. Weber, H.J. Weyer, A. Hoffart, B. Rzehorz, H. Ullrich, M. Dzemidzić, M. Furić, and T. Petković. *Nucl. Phys.*, A 517:413, 1990.

12. T. Alteholz, G. Backenstoss, D. Bosnar, H. Breuer, A. Brković, H. Döbbeling, T. Dooling, W. Fong, M. Furić, P. Gram, N.K. Gregory, A. Hoffart, Q. Ingram, A. Klein, K. Koch, J. Köhler, B. Kotlinski, M. Kroedel, G. Kyle, A. Lehmann, Z. Lin, G. Mahl, K. Michaelian, A.O. Mateos, S. Mukhopadhyay, T. Petković, R.P. Redwine, D. Rowntree, R. Schumacher, U. Sennhauser, N. Šimićević, F.D. Smit, G. van der Steenhoven, D.R. Tieger, R. Trezeciak, H. Ullrich, M. Wang, M.H. Wang, H.J. Weyer, M. Wildi, and K. Wilson. *to be published*, 1993.

13. H. Hahn, D. Ashery, M.A. Moinester, R.R. Johnson, D.R. Gill, A. Altman, M. Sevior, R.P. Trelle, Z. Hermon, E. Cochavi, and R. Natansohn. *Contribution to PANIC90*, 1990.

14. K.A. Aniol, A. Altman, R.R. Johnson, H.W. Roser, R. Tacik, U. Wienands, D. Ashery, J. Alster, M.A. Moinester, E. Piasetzky, D.R. Gill, and J. Vincent. *Phys. Rev.*, C 33:1714, 1986.

15. S. Mukhopadhyay, S. Levenson, R.E. Segel, G. Garino, D. Geesaman, J.P. Schiffer, G. Stgephans, B. Zeidman, U. Engricht. H. Jackson, R. Kowalczyk, D. Ashery, E. Piasetsky, M. Pionester, I. Navon, LC. Smith, R.C. Das, R. R. Whitney, R. Mckeown, B. Anderson, R. Madey, and J. Watson. *Phys. Rev.*, C 43:957, 1991.

16. L.C. Smith, R.C. Minehart, D. Ashery, E. Piasetzky, M. Moinester, I. Navon, D.F. Geesaman, J.P. Schiffer, G. Stephens, B. Zeidman, S. Levinson, S. Mukhopadhyay, R.E. Segel, B. Anderson, R. Madey, J. Watson, and R.R. Whitney. *Phys. Rev.*, C 40:1347, 1989.

17. G. Backenstoss, M. Izycki, R. Powers, P. Salvisberg, M. Steinacher, P. Weber, H.J. Weyer, A. Hoffart, B. Rzehorz, H. Ullrich, D. Bosnar, M. Furić, and T. Petković. *Phys. Lett.*, B 222:7, 1989.

18. B. Blankleider et al. *to be published*, 1993.

19. C. Ponting, D.A. Hutcheon, M.A. Moinester, P.L. Walden, D.R. Gill, R.R. Johnson, F. Duncan, G. Sheffer, P. Weber, V. Sossi, A. Feltham, M. Hanna, R. Olzewski, M. Pavan, F.M. Rozon, M. Sevior, D. Ashery, R.P. Trelle, and B. Mayer. *Phys. Rev. Lett.*, 63:1792, 1989.

20. J.A. Niskanen. *Phys. Rev.*, C 43:36, 1991.

21. K.M. Watson. *Phys. Rev.*, 88:1163, 1952.

22. A.B. Migdal. *JETP (Sov. Phys.)*, 1:2, 1955.

23. M.L. Goldberger and K.M. Watson. *Collision Theory*. Wiley & Sons, 1964.

24. H. Witala, W. Glöckle, and Th. Cornelius. *Phys. Rev.*, C 39:384, 1989.

25. E.A. Bartnik, H. Haberzettl, T Januschke, U. Kerwath, and W. Sandhas. *Preprint* BONN-HE–87–11, 1987.

26. F. Adimi, H. Breuer, B.S. Flanders, M.A. Khandaker, M.G. Khayat, P.G. Roos, D. Zhang, Th.S. Bauer, J. Konijn, C.T.A. de Laat, G.S. Kyle, S. Mukhopadhyay, M. Wang, and R. Tacik. *Phys. Rev.*, C 45:2589, 1992.

27. J.M. Laget and J.F. Lecolley. *Phys. Lett.*, B 194:177, 1987.

28. L.L. Salcedo, E. Oset, D. Strottmann, and E. Hernandez and. *Phys. Lett.*, B 208:339, 1988.

Few-Body Systems, Suppl. 7, 349—360 (1994)

© Springer-Verlag 1994
Printed in Austria

# MONTE CARLO STUDIES OF LIGHT NUCLEI: STRUCTURE AND RESPONSE

J. Carlson
Theory Division
Los Alamos National Laboratory
Los Alamos, NM 87545

R. Schiavilla
Physics Department
Old Dominion University
Norfolk, VA 23529
and
Theory Group
Continuous Electron Beam Accelerator Facility
Newport News, VA 23606

## Abstract

Monte Carlo methods based upon Euclidean path integrals can be used to study both structure and dynamics in few-body nuclei. Traditionally, these methods have been used in the limit of long imaginary time to extract information about low-lying states of light nuclei. Much intriguing dynamics, though, is present in the short time propagators. We briefly review some nuclear structure results, including bound-state results for A=4 and low-energy scattering for A=5. We stress recent developments related to inclusive processes, in particular calculations of the longitudinal and transverse response functions measured in electron scattering experiments. Calculations performed during the last year offer a much clearer picture of nuclear dynamics. In particular, they reproduce the observed quenching in the longitudinal to transverse response ratio, and the relative shift of strength to higher energies between the isoscalar and isovector responses probed in hadronic reactions.

## Introduction

Monte Carlo methods are ideally suited to treat the nuclear few-body problem. Both variational[1] and Green's function Monte Carlo[2] (GFMC) methods are commonly used to study nuclear ground states and low-energy scattering. Here we focus on GFMC or Euclidean path-integral methods. Recently these techniques have been exploited to study a variety of problems in nuclear dynamics, and that will be the main emphasis here. First, though, we will describe the methods briefly and review some low-energy results.

Euclidean methods involve a rotation of the evolution operator to imaginary time: $\exp[iH\tau] \to \exp[-H\tau]$. Since the propagator $\exp[-\tau H]$ is in general unknown, the exponential is broken up into a product of short-time propagators: $\exp[-H\tau] = \prod \exp[-H\Delta\tau]$, where $\Delta\tau = \tau/N$. These short-time or high-temperature propagators are approximated, and Monte Carlo methods used to evaluate the relevant matrix elements. Euclidean formulations are popular in many fields of physics, including lattice QCD and many disciplines of condensed matter physics. Finite-temperature methods have recently been applied to studies of the nuclear shell model.[3]

For few-body problems we are predominantly concerned with ground-state expectation values and transitions, hence it is advantageous to include whatever approximate knowledge we have of the wave function. The ground-state expectation value of an operator $O$ is simply the $\tau \to \infty$ limit of :

$$\langle O \rangle = \frac{\langle \Psi_T| \; \exp[-H\tau/2]O\exp[-H\tau/2] \; |\Psi_T\rangle}{\langle \Psi_T| \; \exp[-H\tau] \; |\Psi_T\rangle}. \tag{1}$$

The better the trial wave function $\Psi_T$, the faster the asymptotic limit is reached. At a formal level, the difference between these zero-temperature calculations and finite-temperature algorithms used in condensed matter and lattice QCD is in the boundary conditions. There are huge practical differences between the algorithms, of course, but in each case one is attempting to exploit the natural description of quantum problems in terms of path integrals. Monte Carlo is simply a tool which makes this possible.

## Nuclear Interaction

We consider a simple non-relativistic formulation of the few-nucleon problem with a Hamiltonian of the form:

$$H = \sum_i T_i + \sum_{i<j} V_{ij} + \sum_{i<j<k} V_{ijk} \tag{2}$$

The NN interaction $V_{ij}$ is fit to two-nucleon scattering data and deuteron properties. The dominant feature of the nuclear interaction is the long-range one-pion-exchange potential (OPEP):

$$V_\pi = [Y_\pi(r)\sigma_i \cdot \sigma_j + T_\pi(r)S_{ij}]\tau_i \cdot \tau_j, \tag{3}$$

where $Y_\pi \propto \exp[-\mu r/r]$, and $T_\pi \propto [1 + 3/(\mu r) + 3/(\mu r)^2]Y_\pi(r)$. The $\sigma$ operators are standard Pauli spin matrices and $S_{ij}$ is the tensor operator. In a simple mean-field approach to nuclear matter, OPEP would give no contribution. However, in realistic calculations of both nuclear matter and light nuclei, it has been found that OPEP contributes roughly two-thirds of the potential energy.[4] Hence, the strong correlations induced by the NN potential and, particularly, by OPEP are an essential feature of any microscopic description of the nucleus. For example, the binding energies of the A=3 and 4 nuclei are only 8.5 and 28.3 MeV, yet the kinetic energies are roughly 50 and 100 MeV, respectively. One primary motivation of few-body nuclear physics is the study of these correlations, both theoretically and experimentally.

All the calculations described here employ a simplified version of the Argonne[5] NN interaction, called the AV8 for the number of operators it contains. Low partial waves are constructed to match those in the V14 model, but higher-order terms in the relative momenta are not included. Momentum-dependent terms are difficult to handle in simple path-integral Monte Carlo schemes, they essentially correspond to different effective masses in each two-body channel. These terms have been constructed to be small in the V14 model, hence we treat the difference between AV14 and AV8 in perturbation theory. The perturbation is about 0.15 MeV in A=3 nuclei and 0.9 MeV in the alpha-particle.

In addition to the NN interaction, we include the Urbana model VIII [6] three-nucleon interaction (TNI) in our calculations. It contains an attractive long-range two-pion exchange term plus a repulsive dispersive term. It is adjusted to reproduce the $^3$H binding energy; the strength parameters are plausible but not derived in any real sense from a microscopic theory. In nuclear matter calculations it nearly reproduces the experimental saturation density but gives too little binding. While it provides a significant fraction of the nuclear binding, the TNI expectation value is still a relatively small fraction ($\approx 10\%$) of the NN interaction one.

Interestingly, these 'reasonable' parametrizations of the TNI also provide good fits to the alpha-particle binding energy.[7] One should not consider this a 'proof' that the TNI is responsible for the missing binding energy in A=3 and 4 nuclei, though, since other effects, including both non-localities in the interaction and relativistic corrections, might have a similar density dependence. Each of these effects is expected to be significantly more important in the A=4 than in the A=3 nuclei, simply because of the fact that there are four triplets in the alpha-particle as compared to one in the A=3 systems. However, no reliable estimate of their contribution is as yet available.

## Ground State Methods and Results

The first step in any path integral calculation is to construct an accurate short-time propagator. The free propagator $G^0$ is defined as:

$$G^0(R, R') = \langle R| \exp[-H\tau] |R'\rangle = N \exp[-(R - R')^2 m/2\tau], \tag{4}$$

it provides a diffusion between $R$ and $R'$. In the limit $\tau \to 0$, $G^0 = \delta(R - R')$. It is useful to construct the approximate interacting propagator from:

$$G(R, R') = G^0(R, R') \; S \prod_{i<j} \frac{g_{ij}(r_{ij}, r'_{ij})}{g^0_{ij}(r_{ij}, r'_{ij})}. \tag{5}$$

Again, in the limit $\tau \to 0$, the ratio $g_{ij}/g^0_{ij} \to \exp[-\Delta\tau(V(r_{ij}) + V(r'_{ij}))/2]$. More accurate expressions for the propagators are used in the actual calculations, but the basic idea is the same.[2] Three-nucleon interactions are included in an obvious way.

Once the short-time propagators are derived, ground-state expectation values are obtained by evolving the system to large $\tau$. The convergence of the calculation with $\tau$ depends upon both the trial wave function and the excited-states spectrum of the Hamiltonian. The binding energy obtained with and without the Urbana TNI models are shown in Table 1. As we noted above, the TNI provides a significant fraction of the alpha particle binding, but a relatively small fraction of the total potential energy. The results obtained here include Coulomb, and are consistent within the error estimates with results obtained by Kamada et al. in the Faddeev-Yakubovsky approach.[8]

Table 1: Binding energies (MeV) for various interaction models

| Interaction | $\langle H \rangle$ | Error |
|---|---|---|
| AV8 | -25.03 | (0.10) |
| AV14 | -24.20 | (0.20) |
| AV14+TNI-VIII | -28.30 | (0.20) |

In addition to the binding energy, one is also interested in other properties such as form factors, momentum distributions, etc. When considering electromagnetic properties of the nucleus, it is necessary to include charge and current operators associated with the nuclear interaction, particularly those due to pion exchange. Their contribution can be quite significant even at relatively small values of the momentum transfer.

In the approach by Schiavilla and Riska [9, 10] the leading two-body charge and current operators are constructed directly from the interaction model. In addition to these, other current operators as well, such as those associated with intermediate $\Delta$-production, are included. These latter terms are parametrized in a plausible way, but are not constrained directly by the interaction model.

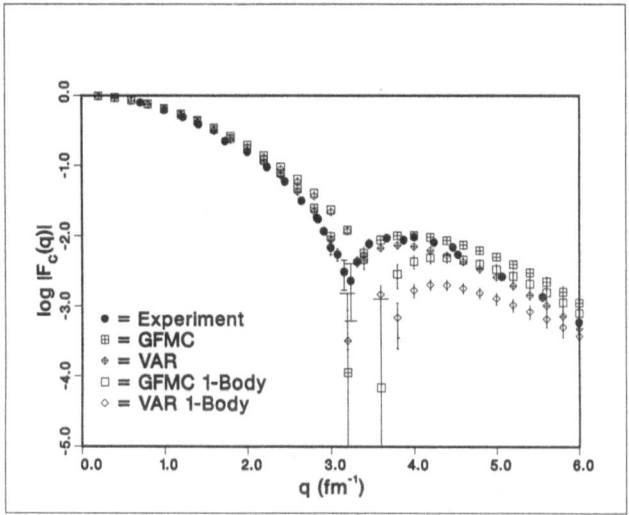

Figure 1) Alpha-particle charge form factor (from Ref. [7])

The alpha-particle charge form factor obtained in this approach is shown in figure 1; similar figures for the trinucleons charge and magnetic form factors are shown in Ref. [6], they are in quite good agreement with experiment up to roughly $q=3$ fm$^{-1}$. As is apparent in the figure, the two-body charge contributions are absolutely crucial in obtaining this agreement.

## Low-Energy Scattering

It is also possible to treat low-energy scattering with Euclidean methods by converting a scattering problem into a bound state one. By enforcing a specific boundary condition and solving for the ground-state energy, the phase shift at that energy can be obtained. Changing the boundary conditions allows one to study the behavior of the phase shift as a function of energy. The method is described in some detail in references [11, 12].

We have, for example, studied the two low-lying P-wave resonances in $^5$He. In figure 2 we show their energy for a nodal radius $R_N$ of 12.5 fm. The $^5$He system is the smallest one where one can directly observe an $L \cdot S$ splitting in the low-lying states. In the figure, the energies of the $J = 3/2^-$ and $1/2^-$ are shown as a function of imaginary time $\tau$ relative to the alpha-particle ground-state energy.

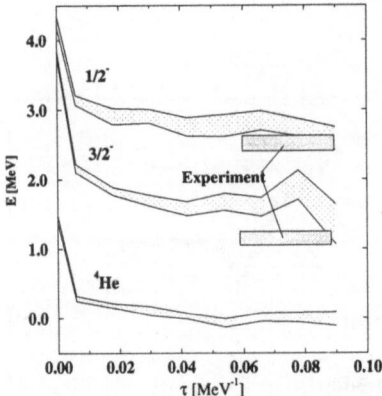

Figure 2) Low-lying P-wave resonances in $^5$He

Neutron-alpha scattering phase shifts are well-known, of course, so that one can easily determine the experimental energy suitable to the boundary condition imposed in the simulation. These energies are shown as the two horizontal bars in the figure. Experimentally, $R_N = 12.5$ fm nearly corresponds to the position of the resonance in the $J = 3/2^-$ channel.

The qualitative agreement between theory and experiment is quite good. In particular, most of the observed $L \cdot S$ splitting is obtained in these calculations. It appears that the TNI gives a significant contribution to this splitting, although it is difficult to assign a precise contribution to individual terms in the Hamiltonian since the problem is very much non-perturbative. The $J = 3/2^-$ and $1/2^-$ relative wave functions are quite different here.

Some quantitative differences remain, however. It appears likely that the present Hamiltonian is unable to give the precise position of the two resonances, both states lie a few tenths of an MeV above the experimental value in our calculations. One obvious shortcoming is the simple TNI model employed, particularly the spin-isospin independent dispersive repulsion. There is little reason to believe that the TNI should be state-independent at short distances. At this detailed level, it will also be fruitful to attempt to improve the treatment of non-local interaction terms, and to study the convergence obtained with different trial wave functions.

## Inclusive Scattering

The applications described above all depend upon the large-$\tau$ or low-energy properties of the propagators. As we have only recently begun to fully appreciate,[13] there is a wealth of information available in the intermediate and short-time propagators as well, information that is directly relevant to inclusive scattering experiments.

Inclusive scattering is traditionally studied in the framework of response functions. The dynamic response function $S(k, \omega)$ represents the response of a system to a weakly-coupled probe $\rho(\mathbf{k})$:

$$S(k, \omega) = \sum_f \langle 0|\rho^\dagger(\mathbf{k})|f\rangle\langle f|\rho(\mathbf{k})|0\rangle\delta(\omega - \omega_f - \omega_0). \tag{6}$$

The Euclidean response is simply the Laplace transform of the standard real-time response:

$$S(k, \tau) = \int d\omega \exp[-(\omega - \omega_0)\tau]S(k, \omega) \tag{7}$$

$$= \langle 0|\rho^\dagger(\mathbf{k}) \exp[-(H - \omega_0)\tau]\rho(\mathbf{k})|0\rangle. \tag{8}$$

It is obvious that $S(k, \tau)$ can be calculated directly in a path integral simulation.

At $\tau = 0$ the Euclidean response is simply the sum rule, while the derivatives with respect to $\tau$ correspond to energy-weighted sum rules. For finite $\tau$, the Euclidean response provides a measure of the energy distribution. As $\tau$ increases, the higher-energy components of the system are gradually suppressed and one is left with the low-energy response. In the $\tau \to \infty$ limit $S(k, \tau)$ is just the square of the relevant nuclear form factor.

Replacing the full interacting propagator with the gaussian free-particle propagator leads to Euclidean equivalents of the standard PWIA response functions obtained from momentum distributions or spectral functions. In these approximations the coupling of the struck nucleon to the rest of the system is ignored. Momentum distribution based calculations assume that the remaining nucleons are at a single energy, while spectral function based calculations take into account the interactions among the residual nucleons. Neither of these approximations is very accurate at momentum transfers of 300-400 MeV/c, however, as we shall see.

We have recently calculated the Euclidean response functions for a variety of simple single-nucleon couplings and also for 'realistic' couplings to longitudinally and transversely polarized virtual photons. The former are mainly of pedagogical interest, while the latter can be directly compared to electron scattering data.

The single-nucleon couplings are simple enough to allow us to provide a physical picture of the final-state interactions. In figure 3, we show the alpha-particle response at $k$=350 MeV/c for the following single-nucleon couplings:

$$\rho_1(\mathbf{k}) = \sum_i \exp[i\mathbf{k} \cdot \mathbf{r}_i] \tag{9}$$

$$\rho_p(\mathbf{k}) = \sum_i \exp[i\mathbf{k} \cdot \mathbf{r}_i] \frac{1 + \tau_z(i)}{2} \tag{10}$$

$$\rho_\tau(\mathbf{k}) = \sum_i \exp[i\mathbf{k} \cdot \mathbf{r}_i] \tau_z(i) \tag{11}$$

$$\rho_{\sigma\tau L}(\mathbf{k}) = \sum_i \exp[i\mathbf{k} \cdot \mathbf{r}_i] \tau_+(i)[\sigma_i \cdot \mathbf{k}] \tag{12}$$

$$\rho_{\sigma\tau T}(\mathbf{k}) = \sum_i \exp[i\mathbf{k} \cdot \mathbf{r}_i] \tau_+(i)[\sigma_i \times \mathbf{k}], \tag{13}$$

For each operator $\rho_O(\mathbf{k})$ there is an associated response $S_O(k,\omega)$, here denoted nucleon, proton, isovector, spin longitudinal, and spin transverse, respectively. The isovector response can be obtained as a weighted average of the spin-longitudinal and transverse responses shown in the figure. The responses have been multiplied by $\exp[k^2/(2m)\tau]$ in order to remove the trivial quasi-elastic kinematics: the response of an isolated nucleon would simply be one on this scale.

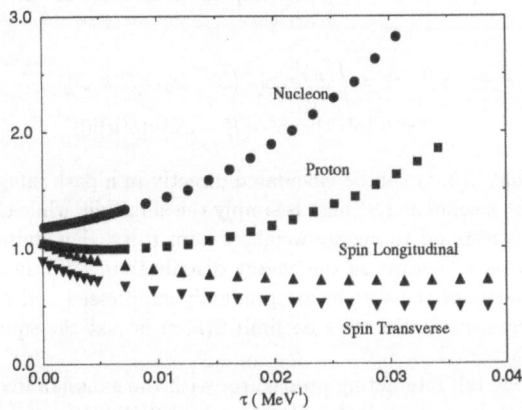

Figure 3) Alpha-particle Euclidean response for simple probes at 350 MeV/c

It is instructive to compare the nucleon $(S_1)$ and proton $(S_p)$ responses. The former is sensitive purely to the average propagation of nucleons. By inserting complete sets of position eigenstates and integrating over the solid angle $d\Omega_k$, $S_1(k,\tau)$ can be calculated as an average over paths of $\sum_{ij} j_0(k|\mathbf{r}_i - \mathbf{r}'_j|)$, where $\mathbf{r}_i$ and $\mathbf{r}'_j$ are the initial and final positions of nucleons i and j. The equivalent expression for the proton response includes only sums over the initial and final proton positions.

The charge-exchange terms in the nuclear interaction play a crucial role in differentiating the proton and nucleon responses. Consider a time $\Delta\tau$ of roughly 0.01 MeV$^{-1}$, corresponding to excitation energies of 100 MeV. In this 'time', the free-particle propagator describes a diffusion on a scale of 1 fm. The relevant charge-exchange matrix element is approximately 40 MeV, implying a $\approx 40\%$ chance of a proton exchanging charge with a neutron during this time. Thus the charge propagates much faster than the nucleons themselves. This rapid propagation in imaginary time implies a shift of strength to higher excitation energies in the proton response, and a consequent quenching in the region of the quasi-elastic peak.

If viewed in a simple one-body picture, this process would have to be described as a 'swelling' of the nucleon charge distribution or as a decrease in the nucleon's effective mass. In fact it is a simple consequence of nuclear dynamics, and therefore the effect depends strongly upon the probe. It is also strongly momentum-dependent; at very high momentum transfers the relevant time scale is very small and the charge-exchange process cannot compete with free-particle propagation.

A comparison of the Euclidean proton response to longitudinal electron scattering is shown in figure 4. Here we present only the inelastic response, the elastic contribution has been subtracted both from theory and experiment. On the left, we have simply performed a Laplace transform of the experimental data. The points with error bars correspond to a 'truncated' response which is assumed to be identically zero beyond the highest measured energy. We have also attempted to extrapolate the measured response to include the effects of the high-energy tails. This extrapolation is given by the dashed lines in the figure and is, by construction, consistent with the Euclidean response at $\tau = 0$. By $\tau = 0.015$ MeV$^{-1}$, however, the extrapolation has no effect and we can directly compare experiment and theory.

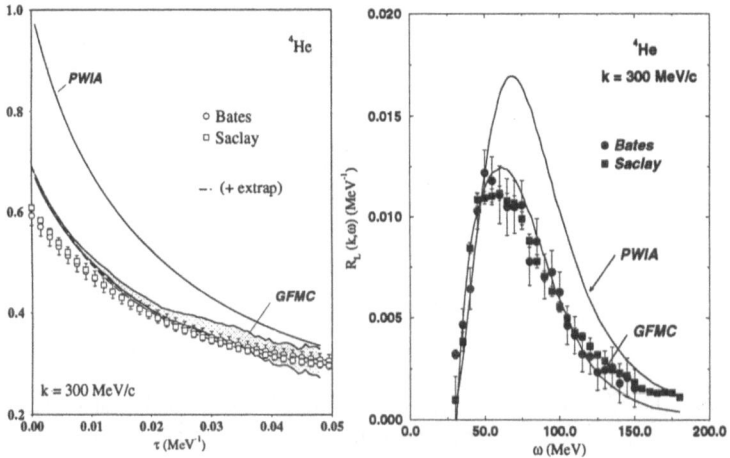

Figure 4) Proton response at 300 MeV/c; imaginary and real-time.

The PWIA and GFMC results are also shown. The latter are represented as a plus or minus one sigma error band. The PWIA response is very different from the observed one, but the full calculation is in good agreement with the data. It is necessary to include the effects of charge-exchange final-state interactions in order to achieve this agreement.

On the right, we have used a fitting proceedure to extract the dynamic response from the Euclidean calculations. This proceedure is described in reference [13], it essentially entails a fit to the difference between the PWIA and full calculations.

This charge-exchange mechanism naturally implies an even stronger shift and quenching in the purely isovector channel. For an isoscalar probe like the alpha-particle, the proton response is just the average of the nucleon and isovector responses. This shift is observed in our calculations, the pure isovector response in figure 3 can be obtained as a weighted average of the spin-longitudinal and spin-transverse isovector responses.

A strong shift is also observed with hadronic probes. Recent (p,n) measurements show a response shifted to energies much greater than $k^2/2m$.[14] The shift is a general feature for all nuclei studied to date; it is also present for a variety of momentum transfers (figure 5). We note that the same shift is not apparent in (p,p′) reactions, confirming the charge-exchange components of the NN interaction as the source of the shift.

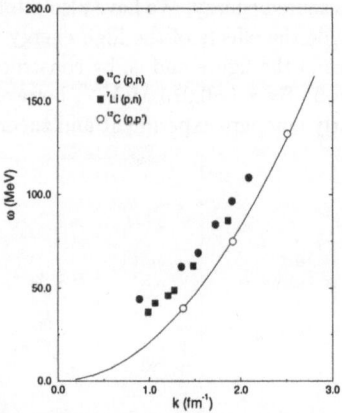

Figure 5) Centroid of (p,n) and (p,p′) quasi-elastic response. The solid line represents the free-particle energy of $k^2/2m$.

Given the predominantly isovector charater of the one-body nuclear current, one would also expect to see a large quenching in the transverse response. Clearly this quenching is not observed in the experiments. In fact, the 'scaled' alpha-particle transverse response is roughly 20 % higher than the longitudinal one in the region of the peak. Using only the one-body current operators, the calculations are far below the experimental data. This is a simple consequence of the charge-exchange mechanism of the nuclear interaction.

In order to obtain a meaningful comparison with the data, the currents associated with this charge-exchange mechanism must also be included. We plot the data and the results of our calculations in figure 6. The data plotted here is truncated at the highest experimental value of the Bates[15] and Saclay [16] experiments, respectively. Since the Saclay experiment extends to higher $\omega$, it will naturally lead to a larger Euclidean response near $\tau = 0$. The difference between Bates and Saclay at high $\tau$ is related to a fairly small relative energy shift in the two experiments.

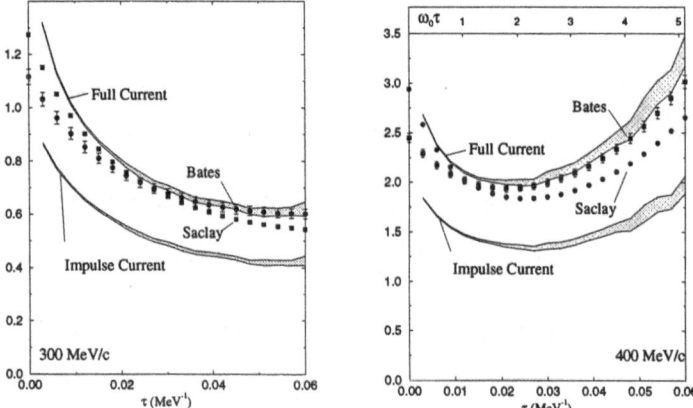

Figure 6) Transverse response at 300 and 400 MeV/c, experiment vs. theory

We have used the current operators described in Ref. [9] for these calculations. The dominant corrections are expected to be those due to the pion currents, but our calculations include the remaining terms as well. In particular, they also contain a static description of the currents associated with delta production. This static model should be adequate at low energies, but obviously is not sufficient for high energies. In the transverse channel, this simplified theory can be directly compared to experiment for $\tau \geq 0.015$ MeV$^{-1}$.

The calculated responses are found to be in good agreement with the experimental results. In order to achieve this agreement, it is absolutely essential to include realistic treatments of both final state interactions and current operators. Only in this way can one simultaneously understand the quenching in the longitudinal channel, the excess strength in the transverse channel, and the large shift observed with hadronic probes.

## Conclusion

Path-Integral Monte Carlo techniques are a useful tool for studying both structure and dynamics in light nuclei. We have employed these techniques to make a convincing case for the consistency between experiment and a microscopic theory based upon realistic nuclear interactions and currents. Both ground-state expectation values and inclusive processes are well described in this framework.

We are currently exploring a variety of extensions to this work, including studies of larger nuclei, extraction of the dynamic response, calculations of polarization observables, and extensions to the higher-energy regime. There remains a large variety of intriguing few-body physics to be pursued with Monte Carlo methods.

# References

[1] J. Carlson, R. B. Wiringa, and V. R. Pandharipande, Nucl. Phys. **A410**, 59 (1983). J. Carlson and R. B. Wiringa, in *Computational Nuclear Physics Vol 1*, K. Langanke, J. A. Maruhn, and S. E. Koonin, eds., Springer-Verlag, Berlin, (1991).

[2] J. Carlson, Phys. Rev. **C38**, 1879 (1988); Nucl. Phys. **A508**, 141c (1990).

[3] C. W. Johnson, S. E. Koonin, G. H. Lang, and W. E. Ormand, Phys. Rev. Lett. **69** 3157 (1992).

[4] S. C. Pieper and V. R. Pandharipande, Phys. Rev. **C 46** 1741 (1992).

[5] R.B. Wiringa, R.A. Smith and T.L. Ainsworth, Phys. Rev. **C29**, 1207 (1984).

[6] R.B. Wiringa, Phys. Rev. **C41**, 1585 (1991).

[7] J. Carlson, Nucl. Phys. **A522**, 185c, (1991).

[8] W. Glöckle and H. Kamada, Phys. Rev. Lett. **71**, 971 (1993).

[9] R. Schiavilla, V.R. Pandharipande and D.O. Riska, Phys. Rev. **C40**, 2294 (1989); Phys. Rev. **C41**, 309 (1990).

[10] J. Carlson, V.R. Pandharipande and R. Schiavilla, 'Modern Topics in Electron Scattering', B. Frois and I. Sick, eds., World Scientific, Singapore (1991), p. 177.

[11] J. Carlson, V. R. Pandharipande, and R. B. Wiringa, Nuclear Physics **A424**, 47 (1984).

[12] J. Carlson, K. E. Schmidt, and M. H. Kalos, Phys. Rev. **C36**, 27 (1987).

[13] J. Carlson and R. Schiavilla, Phys. Rev. Lett **68**, 3682 (1992).

[14] T. N. Taddeucci, 'Spin and Isospin in Nuclear Interactions', S.W. Wissink, C.D. Goodman, and G.E. Walker, eds., Plenum, New York (1991), p. 391; Chrien et al., Phys. Rev. **C21**, 1014 (1980); D. L. Prout, Ph.D. thesis, unpublished, University of Colorado, 1992.

[15] K.F. von Reden et al., Phys. Rev. **C41**, 1084 (1990).

[16] J.F. Danel et al., unpublished; A. Zghiche et al., Saclay preprint (1993).

Few-Body Systems, Suppl. 7, 361–370 (1994)

# NUMERICAL METHODS IN CONFIGURATION-SPACE $A = 3, 4$ BOUND-STATE AND SCATTERING CALCULATIONS

N.W. Schellingerhout

Institute for Theoretical Physics, University of Groningen, Groningen, The Netherlands

*Abstract*

In earlier work [1] we presented an efficient method for solving the three-body Faddeev equations in configuration space, based on the spline method first introduced by Payne *et al.* [2] Recently, we have extended this method so that almost any (nonrelativistic) three-body system imaginable can be handled. We have performed highly accurate calculations for simple model potentials, realistic nucleon-nucleon interactions (including the tensor force), nonlocal potentials, and discontinuous potentials. The efficiency is accomplished by the use of direct product representations where possible, greatly reducing the amount of storage and the amount of computer time required. The flexibility stems from the use of cartesian coordinates, in addition to the traditional polar coordinates, and transforming from one to the other when necessary, thus combining the advantages of both coordinate systems.

# 1 Introduction

Accurate three-body calculations are quite common at present (although a lot of computational effort is still required in most cases). The pioneering work of Payne and coworkers played a decisive role in this. Their trinucleon calculations still stand as an impressive benchmark [3]. The standard of quality of four-body calculations is not yet as high. Green's function Monte Carlo methods give reasonably good binding energies, but there is a need for independent calculations and higher accuracy, especially if one is interested in the scattering problem. The recent work of Glöckle and coworkers [4] represents significant steps towards this goal.

In earlier work we have introduced an efficient solution method for the three-body problem, based on the spline method introduced by Payne. We used this method to solve the bound-state three-body problem accurately with Faddeev equations for the first time. Since then we have focused on generalizing our method to more complicated interactions, and more complicated problems. So far, our method has proven to be effective for all sorts of interactions (including tensor forces, nonlocal, and discontinuous potentials), a wide range of three-body systems (including the trinucleon, $^6$Li in the $\alpha$-$n$-$d$ clustering model [5], atomic systems, and simple scattering below the three-body breakup threshold), and simple four-body systems [6].

In this paper I would like to give an overview of these recent developments. In the following section the numerical methods are discussed in some detail. After this, an overview of results obtained with these methods is given, illustrating the wide applicability of the methods.

## 2 The configuration-space $N$-body problem

A mathematically sound basis for the solution of the three-body problem was first given by Faddeev [7]. After this, Yakubovsky extended the Faddeev framework to the general $N$-body problem [8]. Since we solve the problem in configuration space, I will first briefly discuss the specific difficulties encountered when solving the $N$-body problem in configuration space.

The three-body problem is equivalent to solving the three-body Schrödinger equation in configuration space:

$$(H_0 + V_1 + V_2 + V_3 - E)\Psi = 0 \,, \tag{1}$$

where $V_i$ is the interaction between particles $j$ and $k$, and $H_0$ is the kinetic energy operator. This equation is equivalent to the set of three Faddeev equations

$$(E - H_0 - V_i)\psi_i = V_i(\psi_j + \psi_k) \,.$$

¿From the Faddeev amplitudes $\psi_i$ one can construct a total wave function $\Psi$, which is a solution of the Schrödinger equation, as can be seen by inspection:

$$\Psi = \psi_1 + \psi_2 + \psi_3 \,. \tag{2}$$

The Faddeev equations have a number of advantages over the Schrödinger equation. One is a practical advantage: every equation contains only one potential, enabling the use of a "natural" set of (Jacobi) coordinates for every equation, making treatment of the usually singular potential behavior for small particle separations tractable. (For example, there is no "cusp" problem in the Faddeev equations.) Another practical advantage arises when dealing with identical particles. In this case the equations become dependent, and can be replaced by one equation. The complication of having three potentials in the Schrödinger equation is then replaced by the presence of permutation operators, which are necessary to transform the Faddeev amplitudes from their natural coordinate system into the other coordinate systems. However, the permutation operators are universal to the three-body problem and easy to implement in a numerical method.

The most important advantage of the Faddeev equation over the Schrödinger equation, however, is the way in which it enables us to to specify a unique solution. In momentum space, singularities in the kernel of the Lippman-Schwinger equations, which are related to "unconnected diagrams" (in which one of the particles, the *spectator*, has no interaction

Figure 1: Coordinates in the four-body system.

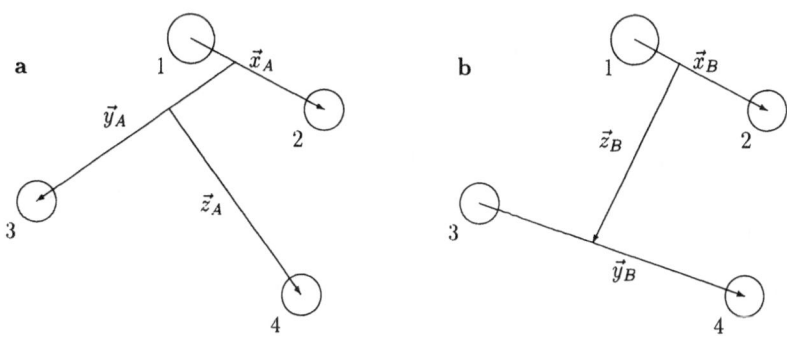

with the other two) make obtaining a unique solution difficult. The kernel of the Faddeev equations has similar singularities, but due to the coupling structure these disappear after one iteration. In configuration space the singularities translate into problems of enforcing the correct asymptotic behavior of the wave function. When, for example, scattering particle 1 off a bound state of particles 2 and 3, it is very difficult to ensure that the ingoing component of the solution only contains a bound state of particles 2 and 3. In the Faddeev formalism, on the other hand, this is a simple matter. Ingoing bound states of other pairs than 23 are excluded by the boundary conditions associated with the equations for $\psi_2$ and $\psi_3$.

For more than three particles, the problems are more involved, and the Faddeev decomposition is not sufficient to obtain a connected kernel. Again it helps to think in terms of interaction diagrams: the problem in a four-body system is that three particles might be interacting, while a fourth remains a spectator, or that there are two interacting pairs, which have no interaction among them. The unconnected clusters must be connected by further decomposition of the wave function. In the general $N$-body problem, $N-2$ levels of nontrivial clustering exist. At every level, the problem of unconnected diagrams will occur, and must be solved. Yakubovsky found a general solution to these problems in Ref. [8].

Faddeev and Yakubovsky components are associated with stepwise fragmentation of the $N$-body cluster into smaller fragments, which can be represented by some tree-like structure, similar to the structure of Jacobi trees. A natural coordinate system can be associated with every component. (Cf. Fig. 1.) In the three-body system there are three Faddeev components, and three coordinate systems, which all have the same structure. In the four-body system there are 18 components and coordinate systems, which can be subdivided into two different structures. The first is associated with the fragmentation $3+1$ (of which there are twelve permutations), the second with $2+2$ (of which there are six permutations). For simplicity, we will only consider identical particles now. This assumption reduces the set of three Faddeev equations to one, and the 18 Yakubovsky equations to two (one associated with the $3+1$ fragmentations and one with the $2+2$ fragmentations).

## 3 The method

We will illustrate our methods for the three-body system in some detail now. The Faddeev equation for three identical particles reads:

$$(H_0 + V - E)\psi = -V(P_1^+ + P_1^-)\psi, \tag{3}$$

where $P_1^\pm$ are the aforementioned permutation operators. Since in this case the three Faddeev components are identical, these operators are just coordinate transformations. Similarly for four particles we can write:

$$(H_0 + V - E)\psi_{3+1} = -V((P_1^+ + P_1^-)(1 + Q_1)\psi_{3+1} + (P_1^+ + P_1^-)T^{-1}X_2\psi_{2+2})$$
$$(H_0 + V - E)\psi_{2+2} = -V(P_2\psi_{2+2} + (P_2X_2 + X_2P_2)T\psi_{3+1}). \tag{4}$$

Here there are two independent components. A larger set of coordinate transformations is needed here due to the more complex coupling structure. The transformations have been written as products, of which the factors are more or less similar to the $P_1^\pm$ of the three-body problem. This allows most techniques discussed for the three-body problem to be used in the four-body problem as well. (In fact the special techniques are even more advantageous in the four-body problem than in the three-body problem.)

Introducing Jacobi coordinates and performing an angular-momentum decomposition in order to retain only the lengths $x_i$ (distance between the constituents of the interacting pair $jk$) and $y_i$ (distance of the spectator $i$ to the interacting pair), we arrive at the following set of coupled equations for the components associated with the angular-momentum channels $\alpha$:

$$\left( \frac{\partial^2}{\partial x^2} - \frac{l_{x_\alpha}(l_{x_\alpha} + 1)}{x^2} + \frac{\partial^2}{\partial y^2} - \frac{l_{y_\alpha}(l_{y_\alpha} + 1)}{y^2} - V(x) + E \right) \psi_\alpha(x, y)$$
$$= V(x)\left( \sum_\beta (P_{\alpha\beta}^+ + P_{\alpha\beta}^-)\psi_\beta \right)(x, y). \tag{5}$$

Note that for simplicity we have assumed the interaction to be central. The infinite set of equations is reduced to a finite set by ignoring very high angular momenta. (Usually it is sufficient to take a very small number of angular momentum states, thus reducing the computational effort enormously.) The operator matrices $P_{\alpha\beta}^\pm$ are integral operators, connecting the point $(x, y)$ in coordinate system $i$ to all points in coordinate system $j$ which are compatible with this point. (Note that because we have absorbed the angle $\hat{x} \cdot \hat{y}$ in the angular momentum states, there is no one-to-one correspondence between points in different coordinate systems.)

We have now arrived in the realm of elliptic partial (integro-) differential equations, for which good solution methods exist. We use the spline method, which consists of approximating the wave function by a linear combination of basis functions, as follows:

$$\psi(x, y) = \sum_{ij} a_{ij} s_i(x) s_j(y). \tag{6}$$

Here $s_i$ and $s_j$ are cubic hermite splines. Splines have two important features, which make them attractive for our problem. First, they are smooth (i.e., $C^1$), which reduces the number of degrees of freedom, and second, they are localized: The domain is divided into a number of smaller (rectangular) intervals, and every spline function is nonzero on at most $2 \times 2$ intervals. This has the advantage that the matrices we will end up with are banded and therefore easier to deal with, but more importantly, that the convergence properties of the basis is much better than for any set of globally defined basis functions. (Discretization errors tend to stay localized.) Now we require that the differential equation is exactly satisfied in a set of (collocation) points only. Taking the four gauss points for each interval is called orthogonal collocation and leads to numerical errors which behave as $h^4$, where $h$ is the typical length of an interval.

After applying orthogonal collocation, the differential equation is reduced to a set of coupled linear equations (for the scattering problem) or a generalized eigenvalue problem (for the bound-state problem), for the unknowns $a_{ij}^{\alpha}$. The matrices appearing in this equation are banded and contain, among others, (derivatives of) the values of the spline functions in the collocation points. Unfortunately this is not the case for the discretized version of the permutation operators. As said earlier, these operators couple a point $(x, y)$ to every point in the permuted coordinate system that is compatible with $(x, y)$. In general these points will form a circle segment.

Since the terms on the left-hand side can be divided into a part which is the identity operator in the $x$ coordinate and one that is the identity in the $y$ coordinate, and the spline basis is separable, we can write the left-hand side as a sum of two direct products:

$$E - H - V = (ES_x - H_{0x} - V_x) \otimes S_y - S_x \otimes H_{0y} . \tag{7}$$

Note that $S_x$ and $S_y$ are numerical representations of the identity operators for the spline method. Since all the matrices in this equation are $N \times N$ instead of $N^2 \times N^2$ ($N^2$ is four times the number of (two-dimensional) intervals), this representation reduces the amount of storage needed considerably, provided we do not have to calculate and store all the matrix elements. Note that a sum of two direct products can be diagonalized very efficiently:

$$A \otimes B + C \otimes D = (A \otimes B)(1 + A^{-1}C \otimes B^{-1}D)$$
$$= (A \otimes B)(U^{-1} \otimes V^{-1})(1 + \Lambda \otimes \Pi)(U \otimes V) \equiv (E \otimes F)\Gamma(U \otimes V), \tag{8}$$

where $U$ and $V$ are diagonalizing matrices for $A^{-1}C$ and $B^{-1}D$, respectively, and $\Lambda$ and $\Pi$ are the diagonal (eigenvalue) matrices. Note that $\Gamma$ is not separable, but since it is diagonal it only contains $N^2$ nonzero elements.

Unfortunately, no factorization can be done for the permutation operators, since the cartesian coordinates are mixed by these operators. Separability can be obtained, however, by introducing polar coordinates:

$$x_i = \rho \cos \theta_i ,$$
$$y_i = \rho \sin \theta_i . \tag{9}$$

Since $\rho$ is invariant under coordinate transformations $P^{\pm}$, the permutation matrices can be separated into a hyperradial and an angular part. Traditionally, one uses polar coordinates rather than cartesian coordinates, and thus ensures separability. We have done this as well, and found a separation of the left-hand side similar to the one shown here. However, the potential can not be factorized using polar coordinates, which is a great disadvantage if one wants to handle nonlocal potentials. Also, this means that the potential term in the left-hand side must be transferred to the right-hand side, which slows down convergence of the iterative solution methods which we use to solve the matrix problem. Finally, the boundary conditions can be expressed more naturally in cartesian coordinates. (The usual hyperspherical formulation is valid only at large distances.) We therefore adopted a sort of hybrid method: we use polar coordinates when doing permutations, and cartesian coordinates at other times. This requires a coordinate transformation which must be efficient if the method is to be of any use at all. Such a transformation will not be separable, but this is not a real problem, since splines are localized functions. The transformation we use is:

$$\psi(x,y) = T(x,y)\vec{a} = \sum_{ij} a_{ij} s_i(\rho(x,y)) s_j(\theta(x,y)), \tag{10}$$

where $\vec{a}$ contains the expansion coefficients for the wave function in the polar representation. The complete transformation from expansion vector to expansion vector is:

$$\vec{b} = (S_x^{-1} \otimes S_y^{-1}) T \vec{a}. \tag{11}$$

Note that $T$ is a sparse matrix, containing at most $16N^2$ nonzero elements. Combining these methods, it turns out that the largest matrix one has to work with has $O(N^2)$ elements, instead of $O(N^3)$, which one would need in conventional methods. (Note that vectors are also $O(N^2)$, so that matrices are roughly of the same size as vectors.)

## 4  Results

In this section we give an overview of various results obtained with the methods described in the previous section. All the results shown here were obtained using relatively small computers (*i.e.*, a small workstation and a Convex C210). The purpose of this section is not to present "new" results, but to show that the methods described above have a very wide range of application. Note that all the three-body bound state results were obtained using the same computer code.

### 4.1  Three-body bound states

In Table 1 results for the trinucleon are shown. (The rows indicated by * are our results. Our results have an uncertainty of less than one in the last digit, unless stated otherwise.) The interaction used in this case was the Reid Soft Core potential. Note that this is a numerically rather difficult potential due to its huge repulsive core and the presence of the tensor force. The results show excellent agreement with Ref. [3]. The first column contains the number of channels and the number of intervals $N$ used. The column denoted by $\langle H \rangle^*$ shows the expectation value of the full hamiltonian (*i.e.*, taking into account all partial waves), which is an absolute upper bound for the three-body ground-state energy.

Table 1: Trinucleon results (Reid Soft Core) for 18 and 34 channels.

| System | $\langle H \rangle$ | $\langle H \rangle^*$ | $E_{\text{coul}}$ | $r_{\text{H}}$ | $P_P$ | $P_D$ |
|---|---|---|---|---|---|---|
| 18, 14 [3] | −7.225 | | 0.647 | 1.68 | 0.08 | 9.43 |
| 18, 20 [3] | −7.231 | | 0.643 | 1.68 | 0.08 | 9.42 |
| 34, 14 [3] | −7.345 | | 0.648 | 1.67 | 0.08 | 9.50 |
| 18, 20 * | −7.233 | −7.346 | 0.643 | 1.68 | 0.08 | 9.41 |
| 34, 20 * | −7.352 | −7.363 | 0.646 | 1.67 | 0.08 | 9.49 |

In the limit of infinitely many channels this number should of course be equal to $\langle H \rangle$. Note that for this system, 34 channels are certainly sufficient for most purposes.

In Table 2 an overview of some results for $^6$Li in the $\alpha$-$n$-$p$ cluster model is given. Since there is a lot of uncertainty regarding the alpha-nucleon interaction, there is a large spread in the results. Also, this is a very difficult system to deal with numerically, because it is so very lightly bound and because of the Pauli exclusion principle which tends to push the nucleons away from the $\alpha$ particle. (This is achieved by the inclusion of a nonlocal term in the potential, which can be handled in a very simple manner thanks to the use of cartesian coordinates.) Note that our result is the only one that uses an alpha-nucleon potential which acts in all partial waves and has the Coulomb effect included exactly. This means that a huge number of channels (i.e., 150) is required to reach an accuracy that is satisfactory. The results obtained by Kukulin are for the same model, but his calculations do not go beyond $d$ waves. These results indicate that including more partial waves raises the quadrupole moment (thus taking it farther from the very slightly negative experimental value), instead of lowering it, as has been a longstanding hope.

Although no efforts have been undertaken to make our program especially suitable for coulombic systems, it turns out that even in this highly specialized field it can produce very accurate results. The numbers shown in Table 3 are upper bounds. More accurate results can be obtained, but require going to quadruple precision. It is probably not worth the effort, considering that excellent specialized methods exist for these problems.

Table 2: Results for $^6$Li in the three-body cluster model.

| Model | $\langle H \rangle$ (MeV) | $\langle r^2 \rangle^{1/2}$ (fm) | $Q$ (e fm$^2$) | $P_P$ (%) | $P_{P'}$ (%) | $P_D$ (%) |
|---|---|---|---|---|---|---|
| Bang [10] | −3.20 | 2.44 | pos. | 0.7 | 5.3 | 1.2 |
| Danilin [11] | ∼ −3.4 | 2.48 | +0.40 | 0.24 | 3.34 | 3.38 |
| proj. (4%) [12, 13] | −3.90 | 2.43 | | 0.48 | 4.65 | 3.40 |
| rep. (4%) [12, 13] | −4.06 | 2.40 | | 0.50 | 4.01 | 3.71 |
| Kukulin [14] | −3.33 | 2.52 | +0.40 | 0.15 | 2.12 | 7.55 |
| * [5] | −3.365(5) | 2.62 | +0.59 | 0.22 | 2.65 | 7.65 |

Table 3: Results for coulombic systems

|  | Helium | $e^- e^+ e^-$ |
|---|---|---|
| variational | −2.90372437705 | −0.2620050702325 |
| * | −2.9037243766 | −0.262005061 |

Table 4: Results for quartet three-body scattering (MT-III potential). Phase shifts are in degrees.

| $E_{cm}$ (MeV) | Ref. [9] | * |
|---|---|---|
| 0 [$a$ (fm)] | $6.442 \pm 0.005$ | $6.443 \pm 0.003$ |
| 0.001 | −2.09 | −2.09 |
| 0.050 | −14.6 | −14.6 |
| 1.000 | −55.8 | −55.9 |
| 1.633 | −66.7 | −66.7 |
| 2.180 | −73.6 | −73.5 |

## 4.2   Three-body scattering

Recently, we have directed our attention to the three-body scattering problem. It turns out that much of the method for the three-body bound-state problem can be retained. However, we are still at a very early stage, and merely show a few results to confirm that our method appears to work for scattering below the three-body breakup threshold as well. (See Table 4.) We expect to be able to extend the method to more complicated interactions and to breakup calculations. The cartesian coordinates will play an essential role since the boundary conditions are more easily expressed in these coordinates. Also, it is much easier to use more intervals in the region where the interaction is significant.

## 4.3   Four-body bound states

In Ref. [6] we have reported on the extension of our methods to the four-body problem ($s$ waves only). We have recently extended our calculations to more realistic systems. Our results are compared with results by H. Kamada et al. [4, 15], which appear to be the most accurate to date. See Table 5. The second column describes the restrictions for the orbital angular momenta, the third column gives the resulting number of channels. As can be seen from this table, the four-body problem is very difficult to solve, not only due to the fact that we must deal with a three-dimensional differential equation, but also due to the number of channels involved. Again, we see near perfect agreement, which for the four-body case appears to be the first time at this level of accuracy.

Table 5: Four-body bound states.

|  | $l_x\,l_y\,l_z$ | $N_{\text{ch}}$ | $-E$ | [4, 15] |
|---|---|---|---|---|
| MT-V | 0 0 0 | $1+1$ | 30.063 | 30.07(3) |
|  | 0 4 4 | $5+3$ | 30.063 | 30.07(3) |
|  | 2 2 2 | $7+5$ | 31.168 |  |
|  | 6 6 6 | $58+34$ | 31.333(2) | 31.36(3) |
| S3 (spin dependent) | 0 0 0 | $2+2$ | 28.785 | 28.80(3) |
|  | 2 2 2 | $83+47$ | 31.00(2) |  |
| SSC(C) (realistic) | 2 0 0 | $3+3$ | 21.220 |  |
|  | 2 2 2e | $34+24$ | 23.63 |  |

# 5 Summary and Outlook

We have developed a highly efficient and flexible method which allows us to solve a wide variety of few-body systems (with, in some cases, unprecedented accuracy). Also, its convergence properties are such that reliable error estimates can be made. This allows excellent separation of numerical and physical uncertainties. These features make it an important tool in few-body physics, with a wide range of applications. (Most notably in nuclear physics, of which the $^6$Li nucleus presents a highly interesting example.)

Our future efforts are directed towards three goals. ($i$) To make the software even more flexible by integrating the three separate programs into a single structure (we now have separate programs for the bound-state and the scattering problem, and for the three- and four-body problem), improving the consistency and quality of the results. This can only be achieved using object-oriented programming techniques (solving physical problems of this complexity poses nontrivial software-maintenance problems). ($ii$) To improve the methods even more. Especially in the three-body scattering problem and the four-body problem there is room for improvement. (Cartesian coordinates will play an important role in this.) This will allow even more complicated systems to be treated with sufficient high accuracy. And finally, ($iii$) to widen the range of problems that can be solved. We are working on solving the scattering problem above the three-body breakup threshold, including three-body forces, and solving the four-body scattering problem. This could open the way to answering a range of interesting physics questions.

# 6 Acknowledgments

This work was performed in collaboration with L.P. Kok. The four-body code was developed by J.J. Schut. The lithium project was suggested by S.A. Coon and performed in collaboration with S.A. Coon and R.M. Adam.

# References

[1] N.W. Schellingerhout, L.P. Kok, and G.D. Bosveld, Phys. Rev. **A40**, 5568 (1989). N.W. Schellingerhout and L.P. Kok, Nucl. Phys. **A508**, 299c (1990).

[2] G.L. Payne, in *Lecture Notes in Physics 273, Models and Methods in Few-Body Physics, Proceedings, Lisboa, Portugal 1986*, ed. L.S. Ferreira and A.C. Fonseca and L. Streit. (Springer-Verlag, Berlin), 1987, pp. 64–99.

[3] C.R. Chen, G.L. Payne, J.L. Friar, and B.F. Gibson, Phys. Rev. **C31**, 2266 (1985).

[4] H. Kamada and W. Glöckle, *Solution of the Yakubovsky equations for four-body model systems*, preprint, Ruhr-Universität Bochum (1992).

[5] N.W. Schellingerhout, L.P. Kok, S.A. Coon, and R.M. Adam, *Nucleon Polarization in Three-Body Models of Polarized* $^6Li$, preprint, University of Groningen (1993). Submitted for publication to Phys. Rev. C.

[6] N.W. Schellingerhout, J.J. Schut, and L.P. Kok, Phys. Rev. **C46**, 1192 (1992).

[7] L.D. Faddeev, Zh. Eksp. Teor. Fiz. **39**, 1459 (1960). [Sov. Phys. JETP **12**, 1014 (1961).]

[8] O.A. Yakubovskiĭ, Yad. Fiz. **5**, 1312 (1967). [Sov. J. Nucl. Phys. **5**, 937 (1967).]

[9] G.L. Payne, J.L. Friar, and B.F. Gibson, Phys. Rev. **C26**, 1385 (1982). C.R. Chen, G.L. Payne, J.L. Friar, and B.F. Gibson, Phys. Rev. **C39**, 1261 (1989).

[10] J. Bang and C. Gignoux, Nucl. Phys. **A313**, 119 (1979).

[11] B.V. Danilin, M.V. Zhukov, A.A. Korsheninnikov, and L.V. Chulkov, Sov. J. Nucl. Phys. **53**, 45 (1991).

[12] A. Eskandarian, D.R. Lehman, and W.C. Parke, Phys. Rev. **C38**, 2341 (1988).

[13] D.R. Lehman and W.C. Parke, Few-Body Systems **1**, 193 (1986).

[14] V.I. Kukulin, V.N. Pomerantsev, Kh.D. Rasikov, V.T. Voronchev, and G.G. Ryzhikh, Australian National University Preprint ANU-ThP-1/92 (1992).

[15] H. Kamada, H. Witała, and W. Glöckle, *Recent progress in the 4N system and pending questions in the 3N system.* Invited talk at the international work shop in Low-Energy Physics, Alma-Ata, Republic of Kazakhstan, September 22–25, 1992.

## Note added after the Conference

It is satisfying to note the close agreement between the numbers of our calculations on atomic systems and those of the calculations by Mandelzweig presented at this conference. Both methods show parallels in that they are direct numerical methods finding *locally* solutions for wave functions, instead of making use of the *global* variational criterium. Moreover, in both methods the wave functions are given in expansion of known and manageable analytical functions. This conference has highlighted that our methods need to be further tested with respect to excited states, and by using local deviation operators as used by Mandelzweig and coworkers.

Few-Body Systems, Suppl. 7, 371—379 (1994)

Few-
Body
Systems

# HYPERSPHERICAL APPROACH TO ULTRA-PRECISE
## NONVARIATIONAL CALCULATIONS IN FEW BODY PROBLEM

V. B. Mandelzweig

Racah Institute of Physics, Hebrew University, Jerusalem 91904, Israel

The correlation function hyperspherical harmonic method (CFHHM) providing extremely precise direct (nonvariational) solution of the few body Schroedinger equation, is reviewed. Given the proper correlation function chosen from physical considerations, the method generates wave functions accurate in the whole range of interparticle distances that lead, in turn, to precise estimates of the expectation values of the Hamiltonian and other operators.

Accurate variational solutions of the three body Schroedinger equation have been available for at least three decades. They have been successfully used in the calculation of relativistic, QED and nuclear size corrections in atomic and molecular three body systems and in fact set a "golden standard" of precision in atomic and molecular physics calculations. However, it is difficult to expect building a complete physical picture of the few body world using only variational methods. Could we imagine contemporary physics achievements if one would be able to solve the two body Schroedinger equation only variationally? A variational wave function coincides with the directly calculated one only on the average, and at least in principle[1] could wildly or even infinitely deviate from it locally. These local discrepancies could lead to wrong estimates of expectation values of operators which have significant contributions from the regions of configuration space where the deviations occur. The availability of a precise direct solution of the few body Schroedinger equation, therefore, is important not only for understanding the analytic structure of the wave function, but also for proper estimation of relativistic, QED and hyperfine effects, as well as the positron annihilation rate in systems like $e^-e^-e^+$, muon sticking probability in mesomolecular ions and parity violation in atoms and nuclei.

During the past two decades a lot of work has been initiated (see refs. 2-5 and references therein) to find accurate direct solutions of three body nonrelativistic equations. The prevailing approach was the numerical solution of the three body equations using one of the modern techniques such as the finite element method[2,3], the hyperspherical coordinate method[4] or the statistical Green function Monte Carlo method[5]. Typical precision obtained was about four significant figures for the energy and the same or less for the wave function. In nuclear physics this approach, based on the Faddeev equations, was a complete success[3] due to limited accuracy requirements in view of the imprecision of our knowledge of the nuclear potentials and of the experiments themselves. However in atomic and molecular physics where current experiments are often probing the tenth or even eleventh significant figure, these numerical methods of solutions are far from offering the necessary precision. The extension to four and more bodies seems very problematic indeed due to the numerical difficulties of dealing with a nine or more dimensional grid and handling the much more complicated Faddeev-Yakubovsky equations[6] describing four and more particles. (For example the Faddeev-Yakubovsky equation for four nonidentical particles has 18 components compared with three components of the usual Faddeev equation ).

Since strictly numerical methods seem not able to go beyond a certain precision, the correct approach should be as analytical as possible, reducing the numerical work to the utmost minimum. Such a possibility is provided by the hyperspherical harmonic method of solution of the N-body Schroedinger equation[7,8], which was introduced into physics in the sixties .

The method is based on changing to 3N-3 Jacobi coordinates (in order to separate center of mass motion) and their parametrization by hyperspherical variables consisting of one radial variable $\rho$ and 3N-4 angular variables $\Omega$. The Schroedinger equation, which in the kinetic energy operator in hyperspherical variables separates into radial and angular parts $T = T_\rho + \dfrac{T_\Omega}{\rho^2}$ is then reduced to the infinite system of coupled ordinary differential equations in the variable $\rho$ by expanding the wave function $\psi$ into eigenfunctions of the angular part of kinetic energy operator $T_\Omega$ called hyperspherical harmonic functions.

These functions which represent the natural generalization of spherical harmonic functions used in the two body problem to the N body case, are characterized by 3N-4 quantum numbers, and are known analytically for any N. The matrix elements of potentials between these functions reduce to integrals containing Legendre and Jacobi polynomials and are calculated analytically as well.

Since the height of the centrifugal potential barrier determined by the eigenvalues $K(K+3N-5)$ of the operator $T_\Omega$ increases quadratically with grand orbital quantum momentum $K = 0,1,2,...$ (which is the analog of orbital quantum number L in the two body case), for bound states the contributions of higher K are expected to be negligible beyond some $K = K_m$ which is called the maximal grand orbital angular momentum. The expansion of the wave function, therefore, may be truncated at this $K_m$, leading to a finite system of the differential equations in $\rho$. This is the only approximation in the hyperspherical harmonics method, which is, in addition, well under control since comparison of calculated values for subsequent $K_m$ allows for an exact estimate of the errors. The convergence in $K_m$, however, is slow due to the following reasons:

i) large degeneracy of the hyperspherical functions set. Each hyperspherical harmonic function is characterized, besides $K_m$, by 3N-5 additional quantum numbers. That means that the number of functions grows as the (3N-5)-th power of $K_m$, and so is the number of coupled differential equations one has to solve.

ii) the singularities of the wave function or its derivatives, due to the singularities of the potential, are difficult to reproduce by a sum of several well behaved hyperspherical harmonic functions. For precise reproduction many hyperspherical harmonic functions are necessary which forces $K_m$ needed for more accurate description of a system to a large value.

iii) clustering, that is the tendency of some particles to form smaller, tighter bound groups of some typical size r with large distances R between them. In analogy with orbital momenta, grand orbital angular momenta with K < ka contribute, that is $K_m$ must be of order ka, where a is a size of the whole system and k is a correspondent wave number. Since a $\approx$ R and since a maximal contribution to a characteristic wave number k is coming from a wave number of a cluster which from the uncertainty relation should be of order 1/r , the maximal global momentum $K_m$ should be of order of R/r, a large number.

To overcome the difficulties of convergence it was suggested few years ago to modify the hyperspherical harmonic (HH) method[7] into the correlation function hyperspherical harmonic (CFHH) method[8-21] by expressing the wave function in the form $\psi = \chi\phi$ (which has been traditionally applied to nuclear bound states),where $\chi$ is a correlation factor which takes singularities (like cusps) and clustering into account, and $\phi$ is a smooth function expanded into hyperspherical harmonics. In case $\chi=1$ the CFHH method reverses to the usual HH method. Proper choice of the correlation factor is expected to facilitate fast convergence both in case of the singular potential and of clustered systems. The solution for $\phi$ proceeds as in the usual HH method, except that the potential V is

replaced by an effective velocity dependent potential $V' = V - \dfrac{\nabla^2\chi}{2\chi} - (\nabla \ln\chi)\cdot\nabla$ , where

$\nabla$ is the six- dimensional gradient operator.

The wave function $\psi$ obtained by the CFHH method converges to an exact solution not on the average as in the variational method, but in an absolute and uniform fashion. The theory and the calculational aspects of the method are summed up in detail in reference 11.

To reach the proper precision, one need first to take a few preparatory steps:

i) find a way to solve hundreds of coupled equations with the proper accuracy.

A sufficiently precise standard numerical integration technique for hundreds of coupled equations is beyond the capabilities of even modern computers owing to the sheer size of the grid. To do that one presents[22] a solution of M coupled radial Schroedinger equations

in a form of a matrix function $\phi$ given by a power series expansion $\phi = A \, \rho^S \rho^\Lambda \sum\limits_{n=0}^{\infty} C_n \rho^n$

in the variable $\rho$ with matrix coefficients $C_n$ determined by recursion relations[22] which are easily handled by computer. Here A is a constant matrix to be found from the boundary condition at infinity, $\Lambda$ is a known diagonal matrix with elements - $(K_m+2)$, $-K_m$ , ........, $K_m$, $(K_m+2)$ and S is a upper triangular matrix whose elements are found from recursion relations. Since by definition $S^M=0$ , $\rho^S \equiv \exp ( S \ln \rho ) = \sum_{n=0}^{M} \frac{S^n}{n!} (\ln \rho)^n$. The expansion contains also power of logarithms as it first was pointed out by Fock[23].

To prevent the exponential growth of $\phi$ at $\rho \to \infty$ , one demands the realization of a matrix equation A B(E)=0 , B(E) is the coefficient in front of the growing exponent which is a matrix function of energy . The energy levels are determined by condition det B(E) = 0. Then the matrix A can be calculated as the solution of the equation A B(E)=0. The current accuracy in solving about 500 coupled equations is of the order $10^{-13}$ for both energies and wave functions and can be further increased, if necessary.

ii)   find analytic expressions for the matrix elements of the potential between hyperspherical states.

The numerical evaluation of the thousands of matrix elements needed would demand an extraordinary amount of computer time. Therefore the knowledge of an analytic expression for matrix elements is extremely important. Since the potentials are usually represented by their power series expansion it is enough to calculate the matrix elements of powers of interparticle distances and to show that the resulting series converge. The analytic expressions for the matrix elements of arbitrary pair potentials in the three body case were calculated for L=0,1 states in refs. 11, 24 and 25, 26 respectively and for arbitrary L in three, four and more particle systems in ref. 27.

A short summary of the main results of the CFHH computations for different atomic and molecular systems is given below:

i) The Helium atom.

Direct solution of the Schroedinger equation was obtained[9,10,11,14,15,19,20] for the $1^1S$, $2^1S$, $3^1S$, $4^1S$ and $5^1S$ states of the Helium atom. The results show that even with the simplest correlation factor the accuracy of the CFHH method (which contains no adjustable parameters) for excited states is comparable to that of the ground state, as well as with the most sophisticated variational calculations involving hundreds of variational parameters. Comparison of our results for $K_m$ = 48 and 56 shows that, even for highly excited states, the expectation values of the Hamiltonian converge to about the eighth significant figure and of the other expectation values to about the fifth significant figure. This accuracy increases to nine significant figures for the energy and to seven significant figures for the wave functions of the ground and first excited states. The CFHH energy values are better than those obtained by the variational calculations of Accad et. al.[28] with Hylleraas type  variational wave functions and by the generator  coordinate computations of Thakkar and Smith[29] and agree well with the most sophisticated variational calculations of Drake[30] and Kono et. al.[31] who use trial functions constructed with two groups of basis functions .They also agree well with the results of Frankowsky[32] and Baker et. al.[33],whose basis functions incorporate logarithmic terms. Both modifications of the Hylleraas expansion yield more rapidly convergent variational energies.

Although accuracy of the directly calculated three body atomic and molecular wave functions is very high, the CFHH calculations are in complete agreement with the best variational calculations which theoretically[1,34] could introduce an error in the expectation values of operators which have significant contributions from the regions of low wave function density. The recently found serious discrepancy between the theoretical and experimental values of the $2^1S$ Lamb shift correction[35] is reported. The discrepancy is, however, reported to have been decreased by more precise calculation of the Bethe logarithm[36] and of quantum-electrodynamic corrections of order $O(mc^2\alpha^6\ln \alpha)$ to the electron-electron interaction[49]. However until now only variational estimates of the relativistic and QED effects in the Helium atom have been available and an independent nonvariational corroboration would be of value. We undertook therefore a careful verification of the currently accepted values of the relativistic and QED corrections[30-33] with the help of the direct CFHH solution of the Schroedinger equation using $K_m=80$. This brought the accuracy of the energy value to 11 significant figures and produced a very high quality $2^1S$ Helium wave function accurate to about 9 digits, for which the average of the absolute value of the local deviation $\Delta = \dfrac{H\psi}{E\psi} - 1$ (see below) is expected to be around 0.0001.

ii) The mesomolecular systems $\mu$dd, $\mu$tt, $\mu$pp and $\mu$dt.

Direct solutions of the Schroedinger equation for the ground states of the $\mu$dd, $\mu$tt, $\mu$pp molecular ions as well as for the ground and excited S-states of the $\mu$dt molecular ion were obtained[13,17,21]. The results clearly illustrate the utility of the CFHH method in the mesomolecular computations. Our energy values are better than those obtained by the adiabatic method[37] or in variational calculations[38] with Hylleraas type variational wave functions. They agree with the most sophisticated variational calculations using Slater type geminals[39] and the generator-coordinate method[40]; the precision of the CFHH calculations converged, for example, for the expectation values of binding energy up to an error in the fifth significant figure.

iii) The negative positronium ion $e^+e^-e^-$.

Direct (nonvariational) solution of the Schroedinger equation for the ground state of the positronium negative ion was obtained[12,18]. The correlation function used was chosen[18] to have proper electron-positron and electron-electron cusps as well as asymptotic behavior. The inclusion of 225 hyperspherical functions yields the nonextrapolated ground state energy value of 0.262005058 atomic units which is slightly lower than the nonextrapolated energy values 0.262004895 and 0.262005056 calculated in the work of

Ho[41] and Bhatia and Drachman[42] which involved hundreds of variational parameters but somewhat higher than the best variational value of 0.262005069 obtained by Petelenz and Smith[43] . This is in agreement with the result of a comparison of values of <H> for $K_m$=56 and for $K_m$=48 which indicates that the accuracy of our present direct calculation of the energy is at least eight significant figures. The previously obtained nonvariational energy values of 0.2620 and 0.2620217 atomic units, calculated respectively by the direct solution of the Schroedinger equation by the hyperspherical coordinates method[44,45] and by the orthogonal collocation method of numerical solution of the Faddeev equations[46], have an accuracy of four significant figures. The accuracy of the CFHH-calculated value[18] of $2.08610 \pm 0.00006$ nsec$^{-1}$ for the two-photon annihilation rate in the latter system was by an order of magnitude higher than obtained in the literature till then. (A very recent variational calculation[47] with eight hundred variational parameters gave the value of 2.086121 nsec$^{-1}$).

Analysis of the Convergence.

In order to properly estimate the accuracy of the method we systematically analyzed[10,13] not only convergence trends of wave functions themselves at many different interparticle distances, but also calculated the local deviation at the same points $\Delta = \dfrac{H\psi}{E\psi} - 1$ and an expectation value of its absolute value $\langle |\Delta| \rangle = \langle |\dfrac{H\psi}{E\psi} - 1| \rangle$. These two quantities were shown[48] to be extremely sensitive measures of the local and overall goodness of the wave function, respectively, and could be used therefore for proper judging of the accuracy of any method of solving the Schroedinger equation. For a true eigenfunction $\psi$, both $\Delta$ and <$\Delta$> are equal strictly to zero. However, $\Delta$ becomes infinite at any of the singularities if they are not properly included in the calculated wave function even when the wave function itself displays very smooth behavior[48].

The calculation of the local deviation $\Delta$ and of the averages of its absolute value <|$\Delta$|> for ground and excited states of the Helium atom and for the ground state of the positronium negative ion indicate that the quality of the CFHH wave functions for $K_m$=48 is very high with <|$\Delta$|> equal to 0.00015 and 0.00033 for the ground and excited states of the Helium, respectively. The last value reduces[20] to 0.00015 and 0.00012 for $K_m$=72 and 80. The proper inclusion of all cusps yields for the ground state of the Helium atom with $K_m$=48 the local wave function accuracy of about $10^{-7}$ for different interparticle distances. The omission of one of the cusps in the excited Helium atom reduces the wave function precision to $10^{-2}$ near the corresponding coalescent point and to $10^{-4}$- $10^{-5}$ away from it. The inclusion of the cusp conditions into the correlation function is shown to be of crucial importance not only near the coalescence points, but also away from them[46].

Graphical Studies.

Graphical representations of three particle CFHH wave functions were obtained and analyzed for the ground and excited states of the Helium atom and for the ground state of the $\mu dd$ mesomolecular ion[16]. The inclusion of adequate singular and cluster correlation behavior is shown to be of crucial importance for a proper description of the wave function. While the hyperspherical expansion by itself is not able to reproduce a correct form of the wave function for finite $K_m$, the inclusion of the correlations results in its proper description even for low values of the maximal grand orbital angular momenta $K_m$. In other words the inclusion of the singular and cluster behavior by properly chosen correlation functions not only accelerates the convergence of the hyperspherical expansion, but also completely changes the form of the calculated wave function and therefore is necessary not only for precise quantitative but also qualitative description of few body systems.

Conclusions.

Summing up, the results of the calculations have shown that the correlation function hyperspherical harmonic (CFHH) method[8-21] of direct solution of the few body Schroedinger equation for bound three-body atomic and molecular systems indeed yields a precision comparable to that obtained by elaborate variational calculations. Our method improves on the precision typical for the numerical (finite element),statistical (Green function Monte Carlo) or adiabatic (Bohr-Oppenheimer or hyperspherical coordinates) methods by four to six orders of magnitude. For $K_m=48$ up to 9 significant figure precision has been obtained for the ground and excited state energies of the Helium atom. This precision could be increased by two more figures by using $K_m=80$, though further increase of $K_m$ will result in the use of a much larger amount of computer time. In addition, the method generates extremely accurate ground and excited state wave functions, whose values for $K_m=48$ at different interparticle distances, including coalescence points, have a precision of up to seven significant figures. Values of the relative local deviation $\Delta = \dfrac{H\psi}{E\psi} - 1$ at different points and of the expectation value of its absolute value, which are known to be extremely sensitive measures of the local and overall goodness of the wave function, respectively, were shown to be very small as well.

The new three body bound state Coulomb calculations reported on this conference by Schellingerhout show the possibility of extremely precise nonvariational solution of the Schroedinger equation also by a direct numerical method. If these results could be confirmed for excited states and for systems with different mass ratio (such as mesomolecules), and if the expectation value of the local deviation operator $\Delta$ will be shown to be small as well, then this numerical method, together with the CFHH method reported here, could be used in the future to verify and improve different expectation values in three body Coulomb problem obtained until now exclusively by the variational method.

In order to further increase the accuracy of the CFHH method in the three body problem and especially to be able to deal with the large degeneracy of the hyperspherical basis for four and more particles one has ideally to be able to choose better and better correlation factors. The simplest way to do this is to make a good guess for the correlation factor and then use a wave function obtained by the CFHH method as a new improved correlation factor for the next round of the CFHH computations, and so on, until one converges to a given precision. In other words, instead of reaching a necessary precision by using the same correlation function and increasing $K_m$ as we do now, one can instead keep $K_m$ constant and iterate the correlation factor. Estimates show that the most time efficient way to do this is to choose a low value of $K_m$ and iterate many times which means, as a fringe benefit, that such calculations could be handled by work stations instead of powerful vectorized machines like the Cray or Convex. If this program can be realized, this would be a breakthrough since in this case accurate calculations of the systems involving four and even more particles can become feasible.

The author thanks Drs. M. Haftel and R. Krivec for many discussions and comments made during years of fruitful collaboration in development of the CFHH method.

REFERENCES

1.B. Klahn,J.D. Morgan, J. Chem. Phys. **81**, 410 (1984).

2.J. Shertzer et. al. Phys. Rev. **A39**, 3833 (1989); **A40**, 4777 (1989); **A43**, 2531 (1991): **A45**, 4393 (1992); **A46**, R1155 (1992);Chem.Phys. Lett. **189**, 287 (1992).

3.G.L.Payne et.al. Phys. Rev. **C22**, 823 (1980); W. Gloeckle et.al. Nucl. Phys. **A381**, 343 (1983); Phys. Rev. **C41**,2538 (1990);Sasakawa et.al., Phys. Rev. Lett. **53**, 1877 (1984); Few Body Syst. **1**, 3 (1986); Few Body Syst. **1**, 143 (1986); for a recent review see J.L. Friar, Few Body Syst. Suppl. **2**, 51 (1987).

4.J.M. Macek, J. Phys. **B1**, 831 (1968) ; J.G. Frey, B.J. Howard, Chem. Phys. **111**, 33 (1987); A. Kuppermann and P.G. Hipes, J. Chem. Phys. **84**, 5962 (1988) ; Chem. Phys. Lett. **133**, 1 (1987) ; D.M. Hood and A. Kuppermann, in "Theory of Chemical Reaction Dynamics", Boston, p. 193, (1986); L. Wolniewiez and J. Hinze, J. Chem. Phys. **85**, 2012 (1986); G.A. Parker et al., Chem. Phys. Lett. **137**, 564 (1987); J. Chem. Phys. **87**, 3888 (1987); J.M. Launay and B. Lepetit, Chem. Phys. Lett. **144**, 346 (1988); J. Linderberg and B. Vessal, Int. J. Quantum Chem. **31**, 65 (1987) .

5.D. Ceperley and B. J. Alder, Phys. Rev. **31A**, 1999 (1985); S.P. Merkur'ev and S.A. Nemnyugin, Sov. J. Nucl. Phys. **53**, 32 (1991).

6. O.A. Yakubovsky , Sov.J. Nucl. Phys. **5**, 937 (1967); L.D. Faddeev in "Three Body Problem in Nuclear and Particle Phys.", J.S.C. McKee and R.M. Rolph eds.(North-Holand,Amsterdam,1970).

7. L.M. Delves, Nucl. Phys. **9**, 391 (1959) ; **20**, 275 (1960) ; F.T. Smith, Phys. Rev. **120**, 1058 (1960); J. Math. Phys. **3**, 735 (1962) ; W. Zickendraht, Ann. Phys. **35**, 18 (1965); **159**, 1448 (1967) ; A.J. Dragt, J. Math. Phys. **6**, 533 (1965); M.A.B. Beg and H. Ruegg, J. Math. Phys. **6**, 677 (1965) ; J.M. Levy-Leblond and M. Levy-Nahas, J. Math. Phys. **6**, 1571 (1965); Yu.A. Simonov,Sov. J. Nucl. Phys. **3**, 461 (1966) ; A.M. Badalyan and Yu. A. Simonov, *ibid.* **3**, 755 (1966).

8. V.B. Mandelzweig , Nucl. Phys. **A508**, 63c (1990).

9. M.I. Haftel, V.B. Mandelzweig, Phys. Lett. **A120**, 232 (1987) .

10. M.I. Haftel, V.B. Mandelzweig, Phys. Rev. **A38**, 5995 (1988).

11.M.I. Haftel,V.B. Mandelzweig, Ann. of Physics **189**, 29 (1989) .

12. M.I. Haftel, V.B. Mandelzweig, Phys. Rev. **A39**, 2813 (1989).

13. M.I. Haftel,V.B. Mandelzweig, Phys. Rev. **A41**, 2339  (1990).

14. M.I. Haftel,V.B. Mandelzweig, Phys. Rev. **A42**, 6324 (1990).

15. R. Krivec, M.I. Haftel,V.B. Mandelzweig, Phys. Rev. **A44**, 7158 (1991).

16. M.I. Haftel,V.B. Mandelzweig, Phys. Rev. **A46**,142 (1992).

17. R.Krivec, M.I.Haftel,V.B.Mandelzweig, Phys. Rev. **A46**, 6903 (1992).

18. R. Krivec, M.I. Haftel,V.B. Mandelzweig, Phys. Rev. **A47**, 911 (1993).

19. M.I.Haftel,V.B.Mandelzweig, Phys. Rev. A , 1993, submitted for publication.

20. M.I.Haftel,V.B.Mandelzweig, "Ultra-precise Nonvariational Calculation of the Relativistic, Finite Size and QED corrections for the $2^1S$ exited state of the Helium atom", Phys. Rev. A, 1993, submitted for publication.

21. R.Krivec, M.I.Haftel,V.B.Mandelzweig,"Precise Nonvariational  Calculation of the mdt  Molecular Ion", Phys. Rev. A , 1993, submitted for publication.

22. V.B.Mandelzweig, Phys. Lett. **A78**, 25 (1980); M.I.Haftel,V.B.Mandelzweig, Ann. of Physics **150**,  48 (1983).

23. V. A. Fock, Isv. Akad. Nauk. **18**, 161 (1954).

24. V.B.Mandelzweig, Phys. Lett. **A80**, 361 (1980);**A82**, 471 (1980).

25. Nir Barnea , V.B. Mandelzweig, Phys. Rev. **A41**, 5209 (1990).

26. Nir Barnea , V.B. Mandelzweig, Phys. Rev. **A44**, 7053 (1991) .

27. R. Krivec, V.B. Mandelzweig, 1990, Phys. Rev. **A42**, 3779 (1990).

28. Y. Accad ,C.L. Pekeris and B. Schiff, Phys. Rev. **A4**,516 (1971)

29. A.J. Thakkar and V.H. Smith, Phys. Rev. **A15**,16 (1977).

30. G.W.F. Drake, Nucl. Instruments and Methods in Phys. Research **B31**,7 (1988).

31. A. Kono and S. Hattori, Phys. Rev. **A34**,1728 (1986).

32. K. Frankowsky, Phys. Rev. **160**,1 (1967).

33. J. Baker,R.N. Hill and J.D. Morgan III, in Proceedings of the AIP Conference "Relativistic,Quantum Electrodynamic,and Weak Interaction Effects in Atoms" ,editors W. Johnson,P. Mohr and J. Sucher ,Santa Barbara, 1988,p.123; J.D. Morgan III (private communication).

34. D.E. Freund,B.D. Huxtable and J.D. Morgan, Phys. Rev. **A29**,980 (1984).

35. T.D. Dinnen et. al., Phys. Rev. Lett. **66**, 2859 (1991); W. Lichten et. al., Phys. Rev. **A43**, 1663 (1991);J. Sansonetti et. al., Phys. Rev. Lett. **65**, 2539 (1990).

36. J.D. Morgan,B.A.P.S. **37**, 864 (1992).

37. S.I. Vinitskii et al, Sov. Phys. JETP **52**,353 (1981).

38. S. Cohen et al,  Phys. Rev. **110**,1471 L (1958) ; W. Kolos et al,  Rev. Mod. Phys. **32**,178 (1960); A. Froman and J.L. Kinsey, Phys. Rev. **123**, 2077 (1961).

39. A.M. Frolov, V.D. Efros, JETP Lett. **39**,544 (1984); A. Yeremin et al, Few Body Systems **4**,111 (1988); S.A. Alexander and H.J. Monkhorst, Phys. Rev. **A38**, 26 (1988); A. Scrinzi, H.J. Monkhorst and S.A. Alexander Phys. Rev. **A38**, 4859 (1988).

40. P. Petelenz and V. H. Smith  Phys. Rev. **A36**, 4078 (1987); **A39**, 1016 (1989).

41. Y.K. Ho,J. Phys. **B16** (1983).

42. A.K. Bhatia and R.H. Drachman, Phys. Rev. **A28**, 2523 (1983); **A32**, 3745 (1985).

43. P. Petelenz and V.H. Smith, Phys. Rev. **A36**, 5125 (1987).

44. J. Botero and C.H. Greene, Phys. Rev. **A32**, 1249 (1985).

45. C.D. Lin, X.H. Liu, Phys. Rev. **A37**, 2749 (1988).

46. C.Y. Hu, A.A. Kvitinsky and S.P. Merkuriev, Phys. Rev. **A45**, 2723 (1992).

47. A.M. Frolov, J. Phys. **B26**, 1031 (1993) .

48. J.H. Bartlett, J.J. Gibson and C.G. Dunn, Phys. Rev. **47**, 679 (1935).

49. G.W.F. Drake, I.B. Khriplovich, A.I. Milstein and A.S. Yelkhovsky, Phys. Rev. **A48**, R15 (1993).

Few-Body Systems, Suppl. 7, 380—387 (1994)

THE ROLE OF TWO-BODY INTERACTIONS
IN THE DESCRIPTION OF FEW AND MANY-NUCLEON SYSTEMS

R. Malfliet
Kernfysisch Versneller Instituut
Zernikelaan 25, 9747 AA Groningen, The Netherlands

Abstract:
Microscopic theories of many-nucleon systems use few-nucleon information
as a starting point. We discuss the sensitivity and the need for a more
accurate knowledge of certain aspects of the elementary interactions.
These include fully relativistic treatments of N-N interaction, the role
of Δ-degrees of freedom and off-shell behaviour in processes like N-N
bremsstrahlung.

The few-body problem[1] and the many-body[2] problem are not such different
issues as many people think. In both cases, microscopic approaches are
structured along the same lines. First of all and as a starting point, a
"realistic" nucleon-nucleon interaction, based on N-N observables (like
phase shifts). Secondly, a specific (parameter-free) prescription to
calculate many-nucleon (A > 2) properties. Examples of dynamical
prescriptions are: for A = 3, non-relativistic Faddeev equations including
or excluding Coulomb forces and/or many-nucleon forces. For A = ∞ (nuclear
matter) the Bethe-Goldstone-Brueckner equations. These prescriptions are,
however, not unique and form a rich field of research in both fields.

In the foregoing we used the term "realistic" in parentheses concerning
the basic N-N interaction, because there is not a single interaction which
in our opinion could claim this unique property[3]. In fact the N-N

KVIKM003

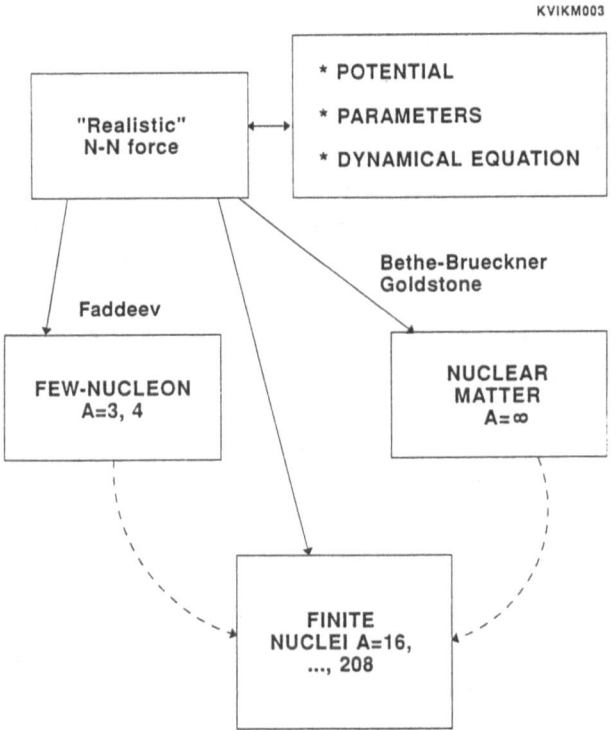

interactions commonly used are based and constructed through the combination of three ingredients. Firstly, a dynamical equation which for instance can be of Schrödinger-type (non-relativistic), Bethe-Salpeter (covariant) or one of its quasi-potential versions (effective relativistic). Secondly, a "driving" term which in field-theoretic language is usually a truncation of all irreducible diagrams, the so-called OBE (One Boson Exchange) model plus possibly phenomenological regularizing terms (form-factors). It may include or exclude meson-retardation and might be restricted to be local or non-local. Thirdly, the data which are needed to fix all parameters. These are usually accurate phase shifts and the agreement or disagreement is expressed through a $\chi^2$-criterion. Here one may choose as data set the phase shifts (and inelasticities) below 200 MeV, 300 MeV, ... any other energy. In this respect it is worthwhile to mention a result from non-relativistic potential-inversion theory: restricting the interaction potential to be local, it will be determined uniquely if one uses all partial wave phase shifts at all energies as input.

382

In this presentation we will focus on two aspects related to the determination of the N-N interaction and elucidate their importance in both few- and many-body physics. The first aspect deals with the question of relativity. The results obtained up to now using non-relativistic approaches with standard modern two-body interactions for A = 3 as well as for A = ∞ all point out to the same conclusion. This is best demonstrated by considering a so-called Coester-diagram where binding energy is plotted versus the inverse of the charge radius (of $^3$He) or versus the Fermi momentum (or density for A = ∞). All results show a linear dependence and if extrapolated, especially in the case A = ∞, miss the empirical point (see fig. 1 and 2). Introducing 3-body forces of the Tucson-Melbourne type might improve the situation for A = 3 but for A = ∞ lead to the disturbing result that even saturation becomes questionable[4]. Moreover the division in 2-body and 3-body forces is model-dependent and needs a more careful and consistent treatment, as we will point out later.

Recently, new results have been obtained by the Utrecht group based on the covariant Bethe-Salpeter equation as the dynamical two-body equation[5,6]. They confirm both for A = 3 and A = ∞ the trend which was obtained using effective relativistic quasi-potential equations (Blankenbeckler-Sugar, Thompson or Gross eqs.). The linear dependence (Coester line) is now shifted as a whole and at least for A = ∞ almost crosses the empirical point (see fig. 2; the points A and B are obtained using the B-S eq.[6] while THM[7] and M[8] are the quasi-potential results). This important

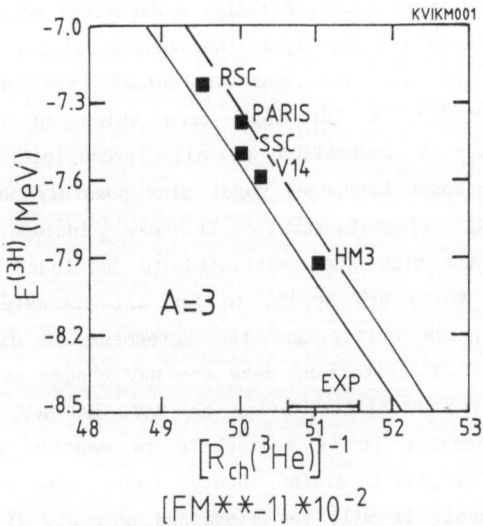

Fig. 1  Triton binding energy versus the inverse of the $^3$He charge radius. Plotted are the non-relativistic results for different N-N potentials. The point Exp. indicates the experimental value.

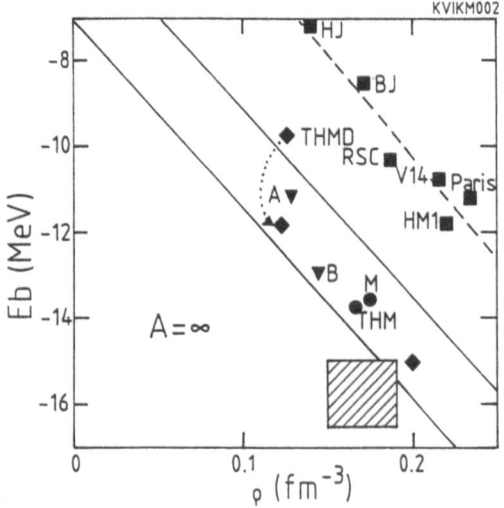

Fig. 2 Binding energy per nucleon in symmetric nuclear matter versus the density. The results (full squares) grouped by the dashed line are non-relativistic results. The remaining points, grouped by the two parallel full lines, are respectively: A and B (ref. 6), THM and THMD (ref. 7), M (ref. 8). The diamond point connected by the dot-ted arrow is from ref. 10 and the unlabeled diamond from ref. 11.

feature needs further quantitative corroboration based on a more "realistic" interaction with a good $\chi^2$-based comparison with data.

The second aspect relating to the N-N interaction is the importance of non-nucleonic degrees of freedom. The pion-production threshold at 280 MeV marks the on-shell manifestation of these degrees of freedom. But it is also clear (remember the potential-inversion result) that the related off-shell manifestation cannot be excluded, *even in the energy region below 280 MeV*. As a first step one could consider the N-Δ coupled channel approach, in its non-relativistic version, applied by the Hanover group to the 3-body problem[9]. They find for the triton an increase in binding, however, more or less along the non-relativistic Coester line. We have applied the N-Δ coupled channel approach in an effective relativistic quasi-potential treatment to nuclear matter. The driving force consists of OBE-type diagrams ($\pi, \rho, \omega, \tau$) both for N-N as for N-Δ (direct and exchange) channel. More details can be found in refs. 7 and 10. The A = ∞ results can be seen in fig. 2. The point THMD labels the earlier results[7] and the point connected with an arrow labels the result of a more accurate and complete calculation[10] (which is quite involved). Anyway the results

384

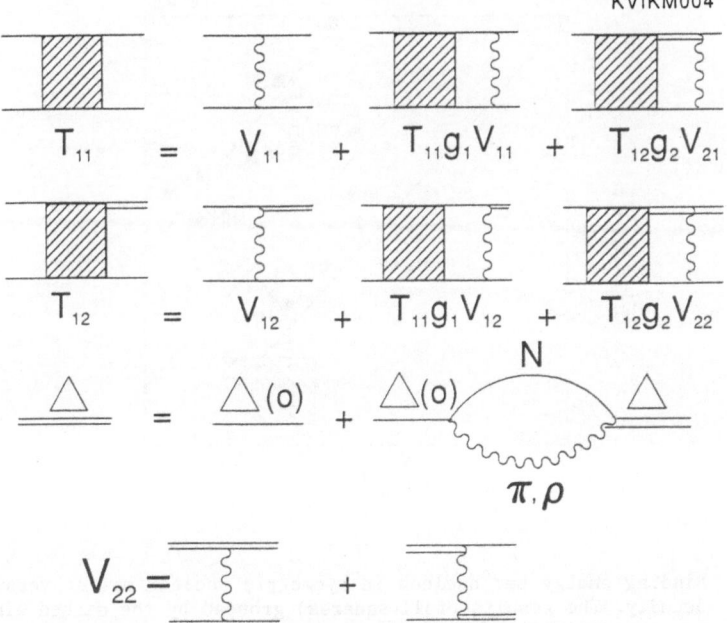

$$T_{11} = V_{11} + T_{11}g_1V_{11} + T_{12}g_2V_{21}$$

$$T_{12} = V_{12} + T_{11}g_1V_{12} + T_{12}g_2V_{22}$$

confirm the existence of a "relativistic" corridor in the Coester diagram indicated in the figure by two parallel lines.

The N-Δ coupled channels approach discussed here is based on single Δ-excitation and thus excludes contributions of Δ-Δ channels. If one wants to take hadronic degrees of freedom seriously these channels should also be included.

In general, the many-body approximations and truncations of large series of diagrams should not only have a physical basis (i.e. ladder type diagrams for low density systems with short-range interactions) but also respect consistency requirements imposed by causality, unitarity, gauge invariance, etc. As an example, thermodynamical consistency is crucial for $A = \infty$ systems[11]. Here, the approximations used should respect thermodynamical relations like the relation between chemical potential and pressure as emphasized by the Hugenholtz-van Hove theorem. Requiring this consistency criterion fulfilled, the saturation point shifts towards the point labeled DJM in fig. 2 as compared to the earlier point labeled THM (in both cases only nucleonic degrees of freedom were considered). In fig. 2 this particular result is the only one which accounts for thermodynamical consistency.

We have discussed a number of qualitative features both in the few- and many-body sectors which demonstrate the influence of the main ingredients of the nucleon-nucleon interaction. Of special importance is the choice of the two-body dynamical equation, which in our discussion has been a relativistic one (either fully covariant or effectively through a quasi-potential approximation) and finally incorporates non-nucleonic degrees of freedom in an effective relativistic approach. These different "choices" for the dynamical 2-body equation, when carried over into the N-body sector (N > 2) lead automatically to specific 3-body type contributions. Also they determine to a large extent the difference in off-shell behaviour which is important for N > 2 systems. The on-shell behaviour is merely fixed by the phase shifts. Therefore interactions which are based on the same phase shifts and use the same dynamical equation have similar off-shell behaviour. The problem of which dynamical equation to prefer is not yet settled and presents a major challenge in N-body systems (few- as well as many-body) since there the off-shell properties play a prominent role. Especially interesting is the difference in off-shell behaviour *below* pion production threshold induced by the coupled N-$\Delta$ approach discussed earlier as compared to the N-N channel only.

The standard way to study the N-N off-shell behaviour is through N-N bremsstrahlung. For example recent experiments on p+p$\rightarrow$p+p+$\gamma$ at TRIUMF have shown very little difference between calculations for the Paris and Bonn potentials[12]. In principle these potentials fit equally well the on-shell data. They use also the same dynamical equation (non-relativistic) and consequently do not show great differences in off-shell behaviour.

This situation is quite different for the N-$\Delta$ coupled system as demonstrated by recent theoretical results from a Groningen-Juelich collaboration[13]: The production of the observed $\gamma$ originates either from a single- or double N-N scattering T-matrix contribution, or from the decay of a $\Delta$ ($\Delta \rightarrow$ N+$\gamma$) from a N-$\Delta$ T-matrix element. Compared to the pure N-N channel situation there are two main differences. First of all, the $\Delta$-decay contribution. This produces a 20-30% increase of the differential cross sections and leaves the analyzing powers relatively unchanged. The latter observables are, however, significantly influenced by the N-N scattering contribution with the inclusion of the intermediate N-$\Delta$ channels. These sizable changes as compared to the old situation are

386

apparent, not only at 280 MeV (above or at pion-threshold) *but also at 200 MeV* (below pion-threshold, see fig. 3). This demonstrates that the off-shell behaviour of a purely N-N scattering T-matrix element is quite different form a N-N T-matrix element which uses also intermediate N-Δ states and is tuned to reproduce similar on-shell properties.

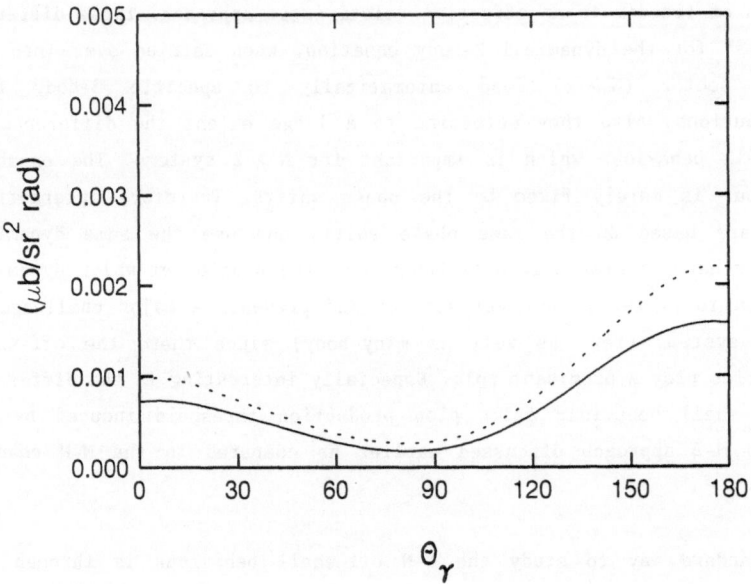

Fig. 3   Differential cross section for p+p→p+p+γ at $E_{lab}$=200 MeV. The two outgoing protons are detected in coplanar geometry at $\theta_1$=12.4° and $\theta_2$=12.0°. $\theta_\gamma$ is the angle of the outgoing photon. The full curve is the calculation based on N-N channels only, while the dotted curve is based on the inclusion of N-Δ channels in a full coupled N-Δ approach. Results based on the work of ref. 13.

This work is part of the research program of the "Stichting voor Fundamenteel Onderzoek der Materie" (FOM), which is financially supported by the "Stichting voor Nederlands Wetenschappelijk Onderzoek" (NWO).

References

1. For a recent overview see the proceedings of the Thirteenth International IUPAP Conference on Few Body Problems in Physics, Adelaide, S.A., Australia, January 5-11: Nucl. Phys. A **543**, 1 (1992)

2. For a recent discussion see "Nuclear Matter and Heavy Ion Collisions": Soyeur, M., Flocard, H., Tamain, B., Porneuf, M. (eds.). Les Houches Winter School (1989)

3. "Realistic" interactions are the Paris, Bonn, Argonne and Nijmegen potentials. The latter has by far the best $\chi^2$-performance

4. Carlson, J., Pandharipande, V., Wiringa, B.: Nucl. Phys. A **401**, 59 (1983)

5. Rupp, G. and Tjon, J.A.: Phys. Rev. C **45**, 2133 (1992)

6. Amorim, A. and Tjon, J.A.: Phys. Rev. Lett. **68**, 772 (1992)

7. Ter Haar, B. and Malfliet, R.: Phys. Rep. **149**, 207 (1987)

8. Machleidt, R.: Adv. Nucl. Phys. **19**, 189 (1989)

9. Sauer, P.U.: Nucl. Phys. A **543**, 291 (1992)

10. De Jong, F. and Malfliet, R.: Phys. Rev. C **46**, 2567 (1992)

11. De Jong, F. and Malfliet, R.: Phys. Rev. C **44**, 998 (1991)

12. Herrmann, V. and Nakayama, K.: Phys. Rev. C **45**, 1450 (1992) and references therein

13. De Jong, F., Nakayama, K., Herrmann, V., Scholten, O.. KVI-Jülich preprint (1993), submitted to Phys. Rev. Lett.

Few-Body Systems, Suppl. 7, 388—394 (1994)

# STABILITY OF HIERARCHICAL TRIPLE STARS

L. G. Kiseleva and P.P.Eggleton

*Institute of Astronomy, Madingley Road, Cambridge CB3 0HA, UK*

### Abstract

We study using a direct N-body simulation the dynamical evolution of hierarchical triple stars in which both orbits are initially circular, and determine the lower limit to the ratio of periods (outer/inner) for which there is dynamical stability. This is in the range 3.0 - 6.3, for the ranges of mass ratios which we consider. We discuss a variety of types of behaviour that we find when the period ratio varies near its critical value.

Newton's low of gravity imprints its signature on a wide range of scales in the Universe. Well-defined gravitational systems are found on scales of increasing size; e.g. multiple stars (2-10 stars), open clusters ($\simeq 10^2 - 10^3$ stars), globular clusters ($\simeq 10^5$ stars), galaxies ($\simeq 10^{11}$ stars), as well as groups ($\simeq 3 - 10$ members) and clusters ($\simeq 10 - 10^3$ members) of galaxies and even superclusters. This hierarchical structure permits a study of nearly isolated systems on different scales. But since Newton until recently, only slow progress was made in undestanding of the gravitational few-body problem. (e.g. Sundman 1912). Because of the complicated nature of the solutions, the few-body orbits in the general cases could not be calculated before the age of computers and the development of appropriate numerical tools (see Aarseth 1988, Valtonen and Mikkola 1991, and references therein). Recently, however, the few-body problem has been revealed to be one involving chaotic dynamics in much of the phase space. Chaos makes the problem unfamiliar and difficult in some way, but it also simplifies the treatment in other respect. The transfer from deterministic to chaotic dynamics takes place when we move from two to three-body systems. Thus, the lower limit to the number of bodies $N$ in a few-body system is $N = 3$. The upper limit of $N$ is less well defined, however when $N > 4$, there usually exist subsystems of $N = 3$ and $N = 4$ that play an important role in the solution of the problem. Nowadays, the few-body problem is recognized as a standard tool in many areas of astonomy and astrophysics. In addition to moddeling real systems, computer N-body simulations may be used for investigating detailed dynamical processes, as well as for checking theoretical predictions.

Most hierarchical triple stars that have two measured orbital periods have a rather large ratio of periods, commonly $\gtrsim 100$, and are probably very stable dynamically. A small but growing number, however, have a period ratio $X$ that is not particularly large. The most extreme examples are $\lambda$ Tau (X=8.3) and CH Cyg (X=7.0). In this paper we investigate the dynamical stability of close triples, using numerical simulations with the code TRIPLE of Aarseth (Aarseth and Zare 1974).

The stability of hierarchical systems in general has been investigated by Harrington (1972, 1975, 1977), Zare (1976, 1977), Szebehely (1976, 1980), Szebehely and Zare (1977), Graziani and Black (1981), Black (1982), Bailyn (1983), Walker and Roy (1981, 1983), and Roy *et al.* (1984). They gave a number of criteria, obtained numerically or analytically, to identify triple systems as stable or unstable. Bailyn (1983), whose methodolgy we follow and extend, showed that various criteria proposed earlier were not reliable to much better than a factor of two in determining the least ratio of periods for stable systems. We have not found other criteria to be much more helpful, especially when ultimately unstable systems are not yet close to disruption and appear fairly stable. This appears to be because these criteria involve not mean but instantaneous orbital parameters: eccentricities, semimajor axes, distances between bodies, relative velocities. But it is clear, both in the numerical experiments described below, and in other works (*e.g.* Mazeh and Shaham 1979, Bailyn 1987, Mazeh *et al.* 1993) that these parameters in triple systems often show periodic or quasiperiodic modulations with significant amplitude. In some cases they may also show a systematic trend to increase or decrease.

In this work we restrict ourselves to the following circumstances:

1. The orbits are coplanar and prograde

2. They start from 'initially circular' conditions, as defined below

3. Only gravitational interaction is considered: we ignore tidal friction, for example

4. The mass ratios in both the 'inner' and the 'outer' pairs are between 1:1 and 100:1.

By 'initially circular' we mean the following. We choose initial conditions such that the inner binary, with masses $m_1, m_2$, would have a circular orbit, in the absence of the third star; and the outer binary, with masses $m_{12}$ ($\equiv m_1 + m_2$) and $m_3$, is also given initial conditions which would imply a circular orbit if the inner binary were replaced by its combined mass at its instantaneous centre of gravity (CG). We always start with the two orbits out of phase with each other by 90°. Thus the only parameter to be specified at the beginning of each simulation, apart from two mass ratios, is the initial period ratio, which we call $X_0$. The mass ratios are given as $\alpha, \beta$, where

$$\alpha = \log_{10}(m_1/m_2) \geq 0.0 \ , \quad \beta = \log_{10}\left(\frac{m_1 + m_2}{m_3}\right). \qquad (1)$$

We restrict ourselves to $\alpha = 0.0, 0.2, \ldots 2.0$ and $\beta = -2.0, -1.8, \ldots 2.0$. Most cases were integrated for 100 outer orbits (strictly speaking, 100 times the *initial* period of the outer orbit), but certain cases were followed for 1000 or 10,000 outer orbits. At several points (20, or less if the step used by the Bulirsch-Stoer integrator was greater than $0.05 P_{in}$) in each *inner* orbit, we compute instantaneous values of inner and outer orbital period, eccentricity, and semi-major axi, determined from the instantaneous positions and velocities of the components. These values were averaged over the whole run. Thus we obtain as functions of $X_0$, $\alpha$ and $\beta$ the quantities $\overline{X}$, $\overline{e}_{in}$, $\overline{e}_{out}$, $\overline{a}_{in}$, and $\overline{a}_{out}$ the first being the 'mean period ratio'. We also computed the variances of these quantities.

Part of our motivation for this investigation is that binary-star interaction, in the form of Roche-lobe overflow, can modify the period of a binary, so that a triple system whose present period ratio implies stability may evolve towards instability as either the inner period increases or the outer period decreases (depending on which of the three components reaches its Roche lobe first, *i.e.* which is initially the most massive). In $\lambda$ Tau, the inner binary is semidetached and can be expected to increase its period, thus decreasing the period ratio towards instability (Bailyn and Eggleton 1983). We should note, however, that in the circumstance that at least one star is close to its Roche lobe it is highly probable that tidal friction cannot be ignored. Undoubtedly this will affect the period ratio at which instability will set in, though we believe that this will be a small effect for most mass ratios of interest. However RLOF is not the only mechanism that can progressively change the period ratio: stellar wind from one or more of the components, if it is strong enough to influence the masses on a nuclear timescale or faster, will also change the period ratio. Such winds are to be expected from any AGB star, for instance, and they may occur when the components are far enough apart for tidal friction to be unimportant.

The dynamical stability of purely gravitating systems with more than two point masses is hard to define in a precise sense. Our initial expectation was the following:

1. for all $X_0$ greater than some value $X_{c1}(\alpha, \beta)$ the system would survive over many orbits, with both the inner orbit and the outer orbit remaining bound while each fluctuates about some well-determined mean orbit. Such a situation we see as 'stable'

2. for all $X_0$ less than some value $X_{c2}(\alpha, \beta)$, either the inner or the outer binary would have its total energy increased to a positive value after a moderate number of orbits. Although the system might not be broken up immediately into a single star and a binary on hyperbolic orbits, this would seem almost inevitable in the long term. Such a situation we see as 'unstable'

3. for $X_{c2} < X_0 < X_{c1}$ the long-term behaviour would be rather unclear

4. $(X_{c1} - X_{c2})/(X_{c1} + X_{c2})$ would be fairly small, perhaps a few per cent, so that a critical period ratio $X_c$ defined as the average of $X_{c1}$ and $X_{c2}$ would be reasonably representative.

In several simulations this expectation was largely borne out. In such cases we find that the mean eccentricities of the inner and outer orbits increase steadily as $X_0$ decreases (usually in steps of 0.01), until abruptly the system disrupts. All smaller values of $X_0$ that we tested also show instability. We must however include within the term 'unstable' a number of cases where a hierarchical triple survived but with the components exchanged, *i.e.* paired differently from the original configuration. It is unlikely that these exchanged systems survive for long; but even if they do they are not relevant to the $(\alpha, \beta)$ pair being investigated.

But there were also many values of $(\alpha, \beta)$ where there were significant departures from this simple behaviour. For example, some cases would show a behaviour which we describe, perhaps somewhat loosely, as 'resonant': as $X_0$ decreases steadily, the mean eccentricity of both orbits would pass through a local maximum before returning to a steady increase until disruption. There might be more than one 'resonance', and some might be quite strong while others might be rather weak. These resonances were sometimes associated with a rational value of the period ratio, in particular the value 4.0, but more commonly were not: for example, with $\alpha = 0.4$, $\beta = 1.8$ the two resonances that we noted had mean period ratios $\overline{X}$ of 4.19 and 3.70.

A more marked departure takes place in a broad band across the $(\alpha, \beta)$ plane, centred roughly on $\beta = 1.2$ at $\alpha = 0.0$ and on $\beta = 0.6$ at $\alpha = 2.0$. Here we find what we call a 'disruptive resonance': the system disrupts in a narrow range of $X_0$, before settling down to the same steady progression as $X_0$ further decreases. Only at some smaller value still for $X_0$ does the system disrupt again, this time for all smaller $X_0$ as far as we could tell.

Table 1 summarises the results we obtained over the whole range of $\alpha$, $\beta$ that we investigated. The Table gives $\overline{X}_{min}$, the lowest value of $\overline{X}$ for which the system was apparently stable. This was not necessarily the value of $\overline{X}$ at the lowest stable value of $X_0$, since $\overline{X}(X_0)$ might reach a local minimum shortly before the first clearly disruptive case. We give $\overline{X}$ rather than $X_0$ because the former is the value that would presumably be measured, at least on average, in a real system. $X_0$ and $\overline{X}$ do not usually differ by more than about 5% for stable cases. Note that as a rule $X_0$ is a little bit greater than $\overline{X}$ for all values of $X_0$ which are not close to such critical values as 'resonance' or disruption. In situations with a disruptive resonance, we give two values for $\overline{X}_{min}$, one being the least value found in the region above the disruptive resonance, and the other in the region below. We have in some cases identified a small 'antiresonance' as a region of stability below a disruptive resonance, but only if we confirmed its stability over 1000 or more outer orbits.

It is not easy to assess the accuracy of these entries, most of which are based on computations lasting 100 outer orbits. In several of the cases which looked most interesting, we explored the boundary between stability and instability using longer integrations, up to 10,000 outer orbits. This usually moved the boundary of stability by $\sim .02 - .04$ in $X_0$, but although we think this is likely to be representative of the Table as a whole we cannot claim this to be established. We give a crude estimate of the uncertainty in the Table by using a colon to indicate values which seem uncertain by more than $\pm.03$, and the absence of a colon to indicate values which seem less uncertain.

It is evident from Table 1 that there is some curiously non-systematic behaviour, for example in the cases where $\alpha=2.0$ and $\beta = 0.2 - 1.0$. The value of $\overline{X}_{min}$, having dropped fairly steadily with $\beta$ until $\beta = 0.2$, first drops rapidly to 3.52, then rises to 3.85 and 3.88, then drops to 3.17, before continuing to decrease fairly steadily. If, in the three cases where there are two values of $\overline{X}_{min}$, we use the lower value instead of the upper, we still find non-systematic behaviour. We have examined these cases in particular detail, but although the values depend somewhat on whether we take 100, 1000 or 10,000 outer orbits, the trend (or lack of it) persists. We do not have a ready explanation for this phenomenon, and hope to investigate it further.

Note that for all sets of $\alpha$ and $\beta$ above the line corresponding to $\beta = -0.4$ in Table 1 we have a direct disruption of the inner binary ($e_{in} > 1$). For all $\alpha$ and $\beta$ in the middle of Table 1 (the area shown in boldface) we have a case which one could call 'a mild exchange': one component of the initial inner binary is replaced by the distant star but the inner eccentricity is never $> 1$ in this case. The new binary (and the new triple system itself) can then survive for some period of time but all such systems are usually unstable for future evolution. For all $\alpha$ and $\beta$ below this area the instability is manifested as the escape of the third body which is the least massive one in this case.

From the point of view of a real stellar triple, in which secular loss or exchange of mass is decreasing the period ratio, it is obviously only the first instability to be reached which should matter. However, if tidal friction is in fact operating on much the same secular timescale, it is likely to keep the orbits more circular. This may then allow the possibility that disruption be avoided, until the lower breakup value of $X_0 = 3.24$ is

Table 1: Lowest values of the average period ratio $\overline{X}$ for stable triple systems

| $\alpha$ $m_2/m_{12}$ | 0.0 .50 | 0.2 .39 | 0.4 .28 | 0.6 .20 | 0.8 .14 | 1.0 .09 | 1.2 .06 | 1.4 .04 | 1.6 .025 | 1.8 .016 | 2.0 .01 |
|---|---|---|---|---|---|---|---|---|---|---|---|
| $\beta\ (m_3/m_{12})$ | | | | | | | | | | | |
| -2.0 (*100*) | 5.90: | 5.90: | 6.03: | 6.10: | 6.15: | 6.17 | 6.19 | 6.21: | 6.21: | 6.23: | 6.20: |
| -1.8 (*63*) | 5.90: | 5.95: | 6.07: | 6.14: | 6.18: | 6.12 | 6.18 | 6.24: | 6.24: | 6.24: | 6.23: |
| -1.6 (*40*) | 5.90: | 5.95: | 6.07: | 6.16: | 6.20 | 6.26 | 6.22 | 6.29 | 6.29 | 6.28 | 6.30: |
| -1.4 (*25*) | 5.90: | 6.00: | 6.08 | 6.18 | 6.23 | 6.26 | 6.28 | 6.29: | 6.29 | 6.32 | 6.32 |
| -1.2 (*16*) | 5.89 | 5.95: | 6.09 | 6.15 | 6.23 | 6.27 | 6.25 | 6.30 | 6.33 | 6.32 | 6.32 |
| -1.0 (*10*) | 5.75 | 5.90: | 5.98 | 6.14 | 6.18 | 6.23 | 6.29 | 6.31 | 6.31 | 6.31 | 6.32 |
| -0.8 (*6.3*) | 5.67 | 5.67: | 5.91 | 6.03 | 6.11 | 6.17 | 6.21 | 6.24 | 6.22 | 6.21 | 6.22 |
| -0.6 (*4.0*) | 5.45 | 5.60: | 5.74 | 5.88 | 5.99 | 6.05 | 6.09 | 6.10 | 6.09 | 6.10 | 6.10 |
| -0.4 (*2.5*) | **5.20** | **5.31** | **5.47** | **5.62** | **5.73** | **5.79** | **5.84** | **5.87** | **5.87** | **5.88** | **5.88** |
| -0.2 (*1.6*) | **4.81** | **4.93** | **5.07** | **5.20** | **5.32** | **5.38** | **5.42** | **5.46** | **5.47** | **5.47** | **5.47** |
| 0.0 (*1.0*) | **4.37:** | **4.40** | **4.50:** | **4.68** | **4.77** | **4.81** | **4.87** | **4.90** | **4.91** | **4.92** | **4.92** |
| 0.2 (*.63*) | **4.27:** | **4.25** | **4.21** | **4.40** | **4.11** | **4.11** | **4.17** | **4.21** | **4.24** | **4.24** | **4.25** |
| | | | | *4.15* | | | | | | | |
| 0.4 (*.40*) | 4.31 | 4.31 | **4.25** | **4.20** | **4.14** | **4.06** | **4.00** | **3.95** | 3.61 | 3.50 | 3.52 |
| | | | | | | | | | | *3.44* | *3.46* |
| 0.6 (*.25*) | 4.34 | 4.33 | 4.30 | **4.25:** | **4.20:** | **4.16** | **3.94** | 3.91 | 3.89 | 3.87 | 3.85 |
| | | | | | | | | | | *3.73* | *3.73* |
| 0.8 (*.16*) | 4.32 | 4.33 | 4.30 | 4.25: | 4.20 | 3.95 | 3.92 | 4.00: | 3.89: | 3.90: | 3.88 |
| | | | | | *3.53* | *3.45* | *3.42* | *3.35* | *3.28* | *3.30:* | *3.22:* |
| 1.0 (*.10* ) | 4.28 | 4.30 | 4.29 | 4.23 | 3.45: | 3.37 | 3.30 | 3.26: | 3.21 | 3.19 | 3.17 |
| | *3.58* | *3.58* | *3.54* | *3.50* | | | | | | | |
| 1.2 (*.063*) | 4.29 | 4.29 | 4.25 | 3.45 | 3.39 | 3.32 | 3.25 | 3.21: | 3.17 | 3.14: | 3.10: |
| | *3.54* | *3.52* | *3.50* | | | | | | | | |
| 1.4 (*.040*) | 4.24 | 3.48 | 3.46 | 3.40 | 3.35 | 3.25 | 3.21 | 3.17 | 3.08 | 3.05 | 3.04: |
| | *3.51* | | | | | | | | | | |
| 1.6 (*.025*) | 3.47: | 3.47 | 3.43 | 3.36 | 3.31 | 3.24 | 3.17 | 3.10: | 3.03 | 3.03 | 3.04 |
| 1.8 (*.016*) | 3.44 | 3.43 | 3.39 | 3.34 | 3.28 | 3.21 | 3.12 | 3.06: | 3.04: | 3.07 | 3.06 |
| 2.0 (*.010*) | 3.41: | 3.41 | 3.37 | 3.31 | 3.25 | 3.18 | 3.10 | 3.02: | 3.05 | 3.06 | 3.04 |

reached. Even there, it is conceivable that disruption might be delayed by tidal friction.

Of course, it is entirely possible that tidal effects might hasten rather than delay disruption. At fairly extreme mass ratios the Darwin instability (Darwin 1879) can occur. We intend to investigate possible effects of tidal friction in future work, by introducing a crude dissipative term into the dynamical equations.

An aim of this work was to find some simple empirical formula to approximate the function $\overline{X}_{min}(\alpha, \beta)$, which might be useful in estimating that point, during the evolution of a triple containing an Algol-like binary, at which the orbit becomes dynamically unstable. For this purpose one probably would be interested in the larger of the two values of $\overline{X}_{min}$, in those situations where there is a disruptive resonance. It seemed possible, for example, that the mean semi-major axis of the inner binary might always be close, at break-up, to a certain fraction of the Roche-lobe radius of that lobe of the outer binary in which the inner binary is situated. However, this turns out not to be a particularly good estimate. Further, since $\overline{X}_{min}$ shows a virtual discontinuity along a curve in the $(\alpha, \beta)$ plane, as seen in Table 1, it seems unlikely that any very simple formula will be a good empirical fit. We hope that in the absence of a simple formula the Table will be reasonably reliable for this purpose. Interestingly, $\lambda$ Tau (see above), at $\alpha \sim 0.6$, $\beta \sim 1.1$, appears to fall very close to this discontinuity, thus rendering slightly less certain the conclusions of Bailyn and Eggleton (1983).

### Acknowlegments

The authors are very grateful to S.J.Aarseth for his TRIPLE code and a number of useful discussions. LGK thanks the Royal Society for a year Research Fellowship at IoA.

# References

[1] Aarseth S.J., Zare K.: Celest. Mech. **10**, 185 (1974)

[2] Aarseth S.J.: Multiple Time Scales, eds. J.U.Brackbill and B.I.Cohen, p.377, New York: Academic Press (1985)

[3] Aarseth S.J.: in The Few Body Problem, IAU Colloq. No.96, ed. M.J.Valtonen, p.287. Dordrecht: Kluwer (1988)

[4] Bailyn C. D., Eggleton P. P.: ApJ, **274**, 763 (1983)

[5] Bailyn C.D., 1983, unpublished CPGS Thesis, Cambridge University

[6] Bailyn C.D.: ApJ, **317**, 733 (1987)

[7] Black D.C.: AJ, **87**, 1333 (1982)

[8] Darwin G., 1879, Proc. Roy. Soc. London, 29, 168

[9] Harrington R.S.: 1975, AJ, **80**, 1081 (1975)

[i0] Harrington R.S.: AJ, **82**, 753 (1977)

[11] Mazeh T., Krymolowski Yu., Latham D.W.: MNRAS, in press

[12] Roy A.E., Carusi A., Valsecchi G.B., Walker I.W.: A&A, **141**, 25 (1984)

[13] Sundman K.F.: Acta Math., **36**, 105 (1912)

[14] Szebehely V.: Celest.Mech., **22**, 7 (1980)

[15] Szebehely V., Zare K.: A&A, **58**, 145 (1977)

[16] Valtonen M., Mikkola S.: Annu. Rev. Astron. Astrophys., **29**, 9 (1991)

[17] Walker I.W., Roy A.E.: Celest.Mech., **24**, 195 (1981)

[18] Walker I.W., Roy A.E.: Celest.Mech., **29**, 117 (1983)

[19] Zare K.: Celest.Mech., **16**, 35 (1977)

Few-Body Systems, Suppl. 7, 395—408 (1994)

# RELATIVISTIC QUASIPOTENTIAL APPROACHES AND ELECTROMAGNETIC FORM FACTORS OF THE DEUTERON

Stephen J. Wallace and Neal K. Devine
University of Maryland
College Park, MD 20742 USA

## ABSTRACT

Quasipotential approaches involve a reduction of four-dimensional dynamics to three dimensions by use of constraints on the relative momenta of particles. They provide covariant forms of dynamics which in principle are equivalent to the four-dimensional dynamics. For each quasipotential reduction it is necessary to define appropriate electromagnetic current operators, which differ from those appropriate to the four-dimensional formalism. Because elastic form factors generally are analyzed in the Breit frame, using wave functions determined in the rest frame of the bound system, it is necessary to have appropriate Lorentz boosts for the three-dimensional dynamics. Several choices for the quasipotential constraint and their implications for electromagnetic scattering from relativistic bound states are reviewed. An 'instant' formalism is presented which has desirable features, such as a conserved electromagnetic current operator and no singularities in the quasipotential. Calculations for elastic electron scattering from the deuteron demonstrate the significance of the relativistic effects and meson-exchange currents.

Introduction

The increasing energies and momentum transfers at which nuclei are probed invite relativistic approaches to nuclear kinematics and dynamics. Approaches based on quantum field theory start from an effective lagrangian for mesons and nucleons. One is able in principle to define the electromagnetic currents in terms of the meson and nucleon degrees of freedom and to achieve some consistency between the currents and the NN interaction [1]. The dynamics is four-dimensional and it incorporates negative-energy states, thus allowing for distortions of the Dirac spinors which describe free particles.

Significant progress has been made by solving the Bethe-Salpeter equation for NN and $D$ using one-boson-exchange interactions [2,3]. It has been established that a quasipotential reduction to three dimensions provides results which are very similar to the full Bethe-Salpeter results provided that meson-nucleon coupling constants in the quasipotential are fit to empirical data, such as NN phases and deuteron properties [4,5]. Such a quasipotential dynamics with effective meson couplings is useful because most meson-nucleon couplings are not known from first principles.

Only a few quasipotential approaches have been considered seriously in the literature but in principle there exists an infinite number of ways to perform a reduction from four dimensions to three dimensions. Blankenbecler and Sugar and Logunov and Tavkhelidze [6] developed a reduction which has zero relative energy in the c.m. frame. The idea is to use the two-nucleon discontinuity of the scattering amplitude, for which both nucleons are on mass shell, together with a dispersion relation, to construct a three-dimensional propagator with exactly the same discontinuity as the Bethe-Salpeter propagator. Hummel and Tjon have recently performed extensive analyses of elastic and inelastic electron scattering from deuterium using the BSLT approach to describe the deuteron vertex functions [7,8]. Gross has developed a reduction based on one particle being on mass shell, the idea being to achieve a good approximation to the EM form factor of a bound state, such as $D$, where a dominant contribution arises from a pole corresponding to the spectator particle being on shell [9,10]. Deuteron form factors have been calculated using the impulse approximation [11].

Another quasipotential possibility for EM interactions recently was proposed based on determining the deuteron vertex functions and currents using an 'instant', or 'equal-time' constraint in the Breit frame [12]. The motivation is to achieve a simple form of EM current conservation in the impulse approximation. This work will be described further on. First we review the Bethe-Salpeter [13] and Mandelstam formalisms [1] because they provide the underlying formalism for quasipotential approaches.

## Bethe-Salpeter-Mandelstam formalism

For analysis of form factors of relativistic bound states, it is appropriate to start from the four-dimensional Bethe-Salpeter formalism for the bound state wave function [13] and the Mandlestam formalism for the EM current [1]. Consider the Bethe-Salpeter t-matrix,

$$T(p, q; P) = K^{BS}(p, q; P) + \int \frac{d^4 p'}{(2\pi)^4} K^{BS}(p, p'; P)$$
$$\times G_0^{BS}(p'; P) T(p', q; P), \tag{1}$$

where $G_0^{BS} = iS_1(\frac{1}{2}P + p')iS_2(\frac{1}{2}P - p')$ is the propagator for two free particles, and $K^{BS}$ is the kernel consisting of two-particle irreducible graphs, which serves to define the dynamics. This equation may be abbreviated as $T^{BS} = K^{BS} + K^{BS}G^{BS}T^{BS}$, with implied four-dimensional integration. Note that $P = p_1 + p_2$ is the total momentum and $p = \frac{1}{2}(p_1 - p_2)$ is the relative momentum.

A bound state with mass M gives rise to a pole at $P^0 = \hat{P}^0 \equiv \sqrt{M^2 + \mathbf{P}^2}$ :

$$T(p, q; P) = \frac{-i}{2\hat{P}^0} \frac{\Gamma(p; \hat{P})\overline{\Gamma}(q; \hat{P})}{P^0 - \hat{P}^0}, \tag{2}$$

where $\Gamma(p; \hat{P})$ is the Bethe-Salpeter vertex function, $\overline{\Gamma} = \Gamma^\dagger \gamma_1^0 \gamma_2^0$ and terms regular when $P^0 = \hat{P}^0$ are omitted.

The electromagnetic current to be used with Bethe-Salpeter wave functions or vertex functions has been developed by Mandelstam [1], who considered a general five-point function, $T_5$, which has a photon coupled in all possible ways to the charges of the two particles and exchanged mesons. The five-point function may be expressed as follows,

$$T_5 = (1 + T'G_0'^{BS})K_5^{BS}(1 + G_0^{BS}T), \tag{3}$$

where four-dimensional integrations are implied and primes denote the fact that total momentum differs in the final state. The five-point kernel $K_5^{BS}$ represents the current and it is given by coupling the photon to particles one and two (lowest order impulse contributions) plus coupling the photon to all possible internal lines of the irreducible two-body kernel $K^{BS}$. Thus $K_5^{BS}$ completely defines the current to be used with a given bound state dynamics, represented by $K^{BS}$. To extract the electromagnetic matrix element for elastic scattering from the bound state, one substitutes Eq. (2) into Eq. (3). The electromagnetic matrix element is proportional to the residue of the double pole term in the resulting expression,

$$\langle J^\mu \rangle = N\, \overline{\Gamma}' G_0'^{BS} K_5^{BS} G_0^{BS} \Gamma, \tag{4}$$

with implied four-dimensional integrations. A normalization factor $N = -1/\sqrt{4P'^0 P^0}$ arises because of the factors in Eq. (2).

In the impulse approximation, $K_5^{BS}$ consists of two terms, $J^{\mu(IA)}(1) = \Gamma_1^\mu S_2^{-1}$, and $J^{\mu(IA)}(2) = \Gamma_2^\mu S_1^{-1}$, where $\Gamma_1^\mu$ is the one-body electromagnetic operator for a particle of charge $e_1$, charge form factor $F_1^1$ and magnetic form factor $F_2^1(q^2)$,

$$\Gamma_1^\mu = e_1 \left( \gamma_1^\mu F_1^1(q^2) + i\sigma_1^{\mu\nu} q_\nu F_2^1(q^2)/(2m_1) \right), \tag{5}$$

with a similar form applying to particle 2. Higher order contributions arise, for example, from the crossed-box graph, from couplings to mesons (the meson-exchange currents) and from vertex graphs representing the nucleon form factor.

Current conservation is a general requirement due to gauge invariance. In the Bethe-Salpeter formalism, conservation of the isoscalar current holds already at the level of the impulse approximation when $K^{BS}$ depends only on the momentum transfer, as is true in one-boson-exchange models of the NN interaction [3,14]. One finds,

$$q_\mu \langle J^\mu \rangle = N\, F_1(q^2) \overline{\Gamma}'(e_1 + e_2)(G'^{BS} - G^{BS})\Gamma, \tag{6}$$

which is equivalent to

$$q_\mu \langle J^\mu \rangle = N\, \overline{\Gamma}'[K'^{BS}(e_1 + e_2) - (e_1 + e_2)K^{BS}]\Gamma, \tag{7}$$

This vanishes provided the Bethe-Salpeter kernel is the same in initial and final states and it commutes with the total charge, $e_1 + e_2$. For the deuteron, the charge is isoscalar and it commutes with one-boson-exchange forms for $K^{BS}$, thus yielding current conservation.

Calculations for the EM interactions of the deuteron based on the BS formalism with the impulse approximation current have been performed by Zuilhof and Tjon [3], who also showed that a quasipotential analysis provides a good approximation to the full BS analysis.

## Quasipotential Approaches

The T-matrix and bound state vertex function of the BS formalism can be produced also with a quasipotential propagator, $G^{QP}(p, P) = \delta_C \hat{g}^{QP}$, where $\delta_C$ denotes a constraint which reduces integrations from four to three dimensions and $\hat{g}^{QP}$ is the three-dimensional propagator.

In the quasipotential formalisms, the T-matrix may be expressed in terms of the propagator $G^{QP}$ and a quasipotential $K^{QP}$ as follows,

$$T = K^{QP} + K^{QP}G^{QP}T. \tag{8}$$

The quasipotential kernel is then defined in terms of the Bethe-Salpeter kernel by:

$$K^{QP} = K^{BS} + K^{BS}(G^{BS} - G^{QP})K^{QP}. \tag{9}$$

For the bound state vertex, one has a homogeneous equation, found by substituting Eq. (2) into Eq. (8) and retaining pole terms:

$$\Gamma(p; \hat{P}) = \int \frac{d^4p'}{(2\pi)^4} K^{QP}(p, p'; \hat{P})\delta_C \hat{g}^{QP}(p'; \hat{P})\Gamma(p'; \hat{P}). \tag{10}$$

A key point is that the quasipotential constraint need not act on momentum $p$ in this expression, showing that the full four-dimensional vertex function on the left-hand side is in principle obtainable from the quasipotential vertex function on the right-hand side. Therefore it is possible to be consistent with an underlying four-dimensional dynamics by working mainly in three dimensions. If one chooses to represent the quasipotential by a set of meson exchanges with effective coupling constants, the corresponding $K^{BS}$ implicitly is defined by Eq. (9).

To write the current matrix element with quasipotential vertex functions, Eq. (10) is used with constraints appropriate to the initial and final states in Eq. (4) to obtain,

$$\langle J^\mu \rangle = N \,\overline{\Gamma}' g'^{\hat{Q}P} \delta_{C'} K_5^{QP} \delta_C \hat{g}^{QP} \Gamma, \tag{11}$$

with

$$K_5^{QP} = K'^{QP} G'^{BS} K_5^{BS} G^{BS} K^{QP}, \tag{12}$$

defining the current appropriate to the quasipotential formalism. Equations (11) and (12) are general and they can be used with any choice of initial and final quasipotential propagators [15–17].

Three quasipotential formalisms have been used in the analysis of deuteron form factors, all of which involve a three-dimensional propagator in the deuteron rest frame of the general form,

$$g_{QP}(\mathbf{p}, P) = \frac{\Lambda_1^+ \Lambda_2^+}{P^0 - 2\epsilon} - \frac{\Lambda_1^- \Lambda_2^+}{D_{-+}} - \frac{\Lambda_1^+ \Lambda_2^-}{D_{+-}} - \frac{\Lambda_1^- \Lambda_2^-}{D_{--}} \tag{13}$$

where $\Lambda_i^\pm$ denotes projection operators to positive or negative-energy states, $P^0$ is the total energy and $\epsilon = \sqrt{M^2 + p^2}$. Differences arise in the treatment of negative-energy states and retardations. The BSLT approach is based on the constraint $\delta_C = 2\pi i \delta(p \cdot P/\sqrt{P^2})$, which reduces in the rest frame to $2\pi i \delta(p_{rest}^0)$. In the rest frame, one has an instant, or equal-time, formalism in the sense that the constraint causes interactions to have zero time-component of momentum transfer. The three dimensional part of the BSLT propagator has $D_{-+} = D_{+-} = P^0 + 2\epsilon \approx 4M$, and $D_{--} = (P^0 + 2\epsilon)^2/(P^0 - 2\epsilon) >> 4M$. The one-particle-on-shell formalism [9,10] is equivalent to the constraint $\delta_C = 2\pi i \delta^{(+)}(\frac{1}{2}P^0 - p^0 - \epsilon)$ and denominators $D_{-+} = P^0$, $D_{+-} = \infty$, and $D_{--} = \infty$. The meaning of $D_{+-} = \infty$ is simply that no propagation takes place in the $+-$ channel. Here we consider the unsymmetrized form of the one-particle-on-shell formalism which has been used in deuteron calculations [11]. It is necessary to symmetrize the propagator with respect to the two particles being on-mass-shell in order to treat identical particles, as in NN scattering [10]. A difficulty then occurs in that unphysical pinching singularities arise when the quasipotential is subject to a constraint where one particle is on shell in the initial state and the other is on shell in the final state. Such singularities do not arise in the equally-off-shell approach because a single, symmetric constraint is used.

An alternative form of the equally-off-shell formalism has been formulated by Mandelzweig and Wallace [18] who include negative-energy states in the same way they arise in box and crossed-box Feynman graphs with an instantaneous interaction. This results in $D_{-+} = D_{+-} = 2\epsilon \approx 2M$ and $D_{--} = P^0 + 2\epsilon \approx 4M$. The Mandelzweig-Wallace propagator exhibits charge-conjugation symmetry as it is unchanged when $P_0 \to -P_0$ and $\Lambda^+$ and $\Lambda^-$ are interchanged.

In general, it is not correct to approximate $K_5^{QP}$ by applying the BS currents directly to quasipotential wave functions. In the BSLT formalism, doing so would be incompatible with momentum conservation for the impulse current. It is this fact which motivates development of appropriate currents and vertex functions in the Breit frame, where $q^0 = 0$ is compatible with the constraints $p^0 = 0$ in initial and final states [12].

## Quasicurrents

For electromagnetic scattering from the deuteron, the requirements of current conservation and the avoidance of unphysical singularities motivate an instant formalism in the Breit frame.

The impulse approximation current is obtained from Eq. (12). Using the one-boson-exchange description of the quasipotential, the quasicurrent involves a box and crossed-box contribution, as well as higher order terms. In principle, the loop integration involved requires the quasipotential to be evaluated for all values of $p^0$. Impulse currents consistent with the quasipotential description of initial and final states are obtained by neglecting the dependence of the quasipotential on $p^0$. The loop integration is then straightforward. The difference between the full loop integration and the one based on $p^0 = 0$ in the quasipotentials represents a meson-exchange correction to the current, which is neglected in the present treatment. The crossed-box contribution to the current is treated in the same fashion as in the analysis leading to the propagator $\hat{g}^{MW}$ of Mandelzweig and Wallace. Considering the box and cross-box contributions to the current of Eq. (12) with the virtual photon attached to particle 1, we find $\hat{K}^{QP} \int dp^0 S_1' \hat{J}^\mu(1) S_1(S_2 + S_2^{cross}) \hat{K}^{QP}$, where $\hat{K}^{QP}$ is the quasipotential evaluated on the constraint $p^0_{Breit} = 0$, $S_1$ and $S_2$ are nucleon propagators in the box graph and $S_2^{cross}$ is the corresponding propagator in the crossed-box graph. A similar expression holds for the contribution to the current where the virtual photon attaches to particle 2. Current conservation similar to that of the BS formalism is obtained with this current provided initial and final states also are treated in a consistent fashion. To show the connection of current conservation to the description of initial and final vertex functions, consider the requirement of $q_\mu J^\mu = 0$. Using the Ward-Takahashi identity, $q_\mu \gamma_1^\mu = S_1^{-1}(p+q) - S_1^{-1}(p)$, it is straightforward to show that

$$q_\mu K^{QP\mu} = (e_1 + e_2) F_1(q^2) \hat{K}^{QP} (\hat{g}'^{MW} - \hat{g}^{MW}) \hat{K}^{QP}, \tag{14}$$

where $\hat{g}^{MW} = \int dp^0 S_1(S_2 + S_2^{cross})$ is the quasipotential propagator of Refs. [18,19].

If one determines the initial state and final states with $p^0 = 0$ *in the Breit frame* using $\hat{g}^{MW}$ in the initial and $\hat{g}'^{MW}$ in the final, then quasipotential factors in the Eq. (11) and (12) may be collapsed by use of $\hat{K}\hat{g}\hat{\Gamma} = \hat{\Gamma}$ in initial and final states. One finds,

$$q_\mu \langle J^\mu \rangle = N F_1(q^2) \overline{\hat{\Gamma}}'(e_1 + e_2)(\hat{g}'^{MW} - \hat{g}^{MW})\hat{\Gamma} \tag{15}$$

It follows in the same fashion as for Eqs. (6) and (7) that using the instant constraint in the Breit frame for both initial and final vertex functions leads to impulse approximation currents obeying isoscalar current conservation.

Current conservation is realized only for positive-energy states in the work of Hummel and Tjon because BSLT vertex functions are used with an ET current derived from the box graph. Moreover a kinematical boost operator is applied to obtain vertex functions in the Breit frame and this is not consistent with current conservation.

A somewhat more complicated form of the current conservation has been shown by Gross and Riska to be necessary for the one-particle-on-shell formalism. The current is conserved by the addition of exchange currents involving the photon interactions with the on-mass-shell particle.

It is possible to obtain isoscalar current conservation in quasipotential approaches similar to that which holds in the Bethe-Salpeter formalism provided that the initial and final vertex functions are consistent with the impulse current. For elastic scattering, the box plus cross graph currents evaluated as described above together with the Mandelzweig-Wallace propagator and the constraint $p^0_{Breit} = 0$ in the initial and final states provide an exactly conserved impulse current.

## Instant Quasipotential Vertex Functions in Breit Frame

It is a straightforward matter in the Bethe-Salpeter formalism to boost the vertex functions to the Breit frame because they are covariant functions [3]. The usual kinematical Lorentz boost operators for Dirac particles are used. The boost is straightforward also for the one-particle-on-shell formalism. However, for the equally-off-shell or instant approaches, one requires a dynamical boost, which is nontrivial.

It is straightforward to derive the following equation which defines in principle the quasipotential which should be used in an instant formalism in the Breit frame such that the mass of the deuteron is independent of the momentum. Assuming that the Bethe-Salpeter kernel has no dependence on total momentum, we find

$$K^{QP}(P_2) = K^{QP}(P_1) + K^{QP}(P_1)(\Delta(P_2) - \Delta(P_1))K^{QP}(P_2), \tag{16}$$

where $\Delta(P) = G^{BS}(P) - G^{QP}(P)$. In this equation, $G^{QP}(P_2)$ is an instant propagator involving $\delta(p^0)$ in a frame where total momentum is $P_2$, e.g., $P_2 = (\sqrt{M_D^2 + \mathbf{q}^2/4}, \mathbf{q}/2)$ for a moving deuteron in the Breit frame, and $G^{QP}(P_1)$ is an instant propagator involving $\delta(p^0)$ in a frame where total momentum is $P_1$, e.g., $P_1 = (M_D, \mathbf{0})$ for the deuteron at rest. Solving Eq. (16) determines the quasipotential which should be used in an instant equation in the Breit frame in terms of the usual quasipotential which is used in an instant equation in the rest frame. Similarly, one may obtain the potential which should be used in the c.m. frame at invariant energy $\sqrt{s}$, corresponding to $P_2 = (\sqrt{s}, \mathbf{0})$, from the potential used in rest frame of the bound state with $M_D$. The implied energy-dependence of the quasipotential is usually ignored in one-boson-exchange models.

Although Eq. (16) is needed here as a 'boost' equation, it is more general and may be used to determine the potential appropriate to an instant form of three-body dynamics, where $P_2 = (\frac{1}{3}M_3, \mathbf{P})$ is the total momentum of two nucleons in the rest frame of the three-body bound state, and $P_1$ is as above the total momentum in the deuteron rest frame. In this application, one does not perform a boost at all but rather a determination of the quasipotential at an an off-shell point. Due to presence of the BS propagator, integration is over four dimensions in Eq. (16), meaning that the four-dimensional extension of the quasipotential is needed in order to carry out the implied loop integration. Some understanding of the boost equation has been obtained in a separable potential analysis.

## Dynamical Boost in a Separable Potential Model

If the instant quasipotential in the rest frame is assumed to have a separable form,

$$K_0^{QP} = |f\rangle \hat{v}_0 \langle f|, \tag{17}$$

then the instant quasipotential of the deuteron moving with momentum $\mathbf{P}$ is a 16 by 16 matrix of the form $K_P^{QP} = |f\rangle \hat{v}_P \langle f|$, with $\hat{v}_P$ determined by solving the matrix equation

$$\hat{v}_{P_2} = \hat{v}_{P_1} + \hat{v}_{P_1}[\langle f|\Delta(P_2) - \Delta(P_1)|f\rangle]\hat{v}_{P_2} \tag{18}$$

The bound state equation of the instant formalism in a frame where the three-momentum is **P** is,

$$\chi_P = S^{QP}(P)\hat{v}_P\chi_P, \tag{19}$$

where $\chi_P \equiv \langle f|\Psi(P)\rangle$ and $S^{QP}(P) = \langle f|G^{QP}(P)|f\rangle$. This equation has been solved using the Mandelzweig-Wallace quasipotential propagator and assuming a scalar potential $\hat{v}_{P_1} = g^2$ with coupling constant $g$ chosen to yield a bound state of mass $M_D$. The resulting energy of the bound state at momentum $\pm\mathbf{q}$ is $\sqrt{M_D^2 + \mathbf{q}^2/4}$, in accord with Lorentz invariance.

When the three-momentum is not very large, one can see that the change of the quasipotential should not be very big. In order to clarify this, consider the forward scattering matrix element of the quasipotential in positive-energy states defined by $\bar{u}_1\bar{u}_2\hat{v}_P u_1 u_2$, with all spinors having momentum $\frac{1}{4}\mathbf{q}$. This yields $g^2$ in the rest frame where $\mathbf{q} = 0$. To estimate the change of the effective coupling constant implied by the boost to a frame where deuteron momentum is $\pm\mathbf{q}/2$, define a parameter $\lambda(\mathbf{q}^2)$ as follows,

$$\bar{u}_1\bar{u}_2\hat{v}_P u_1 u_2 = g^2/\lambda(\mathbf{q}^2). \tag{20}$$

In the separable model, one finds that $\lambda(\mathbf{q}^2)$ varies slowly with momentum $\mathbf{q}^2$ and that simply scaling the rest frame quasipotential by the factor $\lambda(\mathbf{q}^2)$ gives an excellent approximation to the required quasipotential matrix $\hat{v}_{\mathbf{P}=\mathbf{q}/2}$. The rest frame quasipotential changes by 10% in order to realize invariance of the deuteron mass at momentum $\mathbf{q}^2 = 200F^{-2}$.

### Electromagnetic Form Factors of the Deuteron

Calculations of the instant wave functions in the Breit frame have been performed based on solving,

$$g_{MW}^{-1}(\mathbf{p}'; P)\psi(\mathbf{p}'; P) = \int \frac{d^3p}{(2\pi)^3} K^{OBE}(\mathbf{p}', \mathbf{p}; P)\psi(\mathbf{p}; P), \tag{21}$$

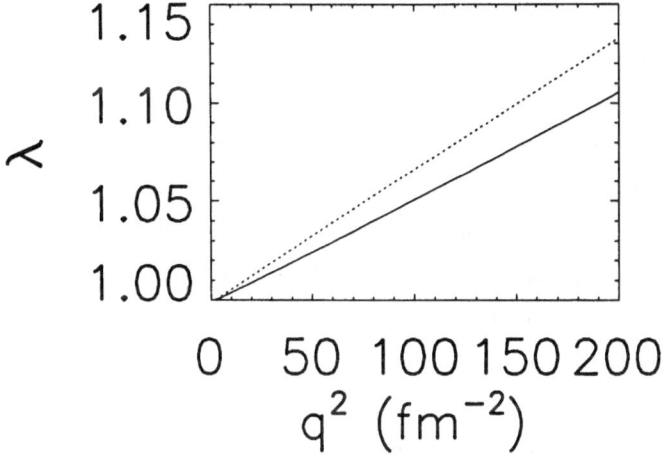

Figure 1: The scaling of the potential, $\hat{K}^{OBE}(p'-p, \mathbf{P} = \pm\frac{1}{2}\mathbf{q}) = \hat{K}^{OBE}(p'-p)/\lambda(\mathbf{q}^2)$, that produces constant deuteron mass, $M_D = (\epsilon_D^2 - \mathbf{P}^2)^{1/2} = 2m_N - 2.22464MeV$: full propagator (solid), $++$ states only (dotted).

for $\mathbf{P} = \pm\mathbf{q}/2$. The one-boson-exchange interaction used has scalar, pseudo-vector, and vector meson exchanges $(\sigma, \delta, \eta, \pi, \omega, \rho)$ using modified Bonn B parameters [16,20].

In the dynamical boost of the deuteron quasipotential to the Breit frame, initial and final vertex functions must be determined for each q. Calculations based on a one-boson-exchange interaction have been performed which show that the latter requirement can be realized in a straightforward manner [12]. However, the variation of the quasipotential has been approximated in a very simple manner in initial calculations. It consists of simply scaling the rest frame quasipotential by a factor $\lambda(\mathbf{q}^2)$ chosen to keep the deuteron mass invariant as follows,

$$K^{QP}(P_2) = K^{QP}(P_1)/\lambda(\mathbf{q}^2), \qquad (22)$$

as opposed to solving Eq. (16). The required factor for the one-boson-exchange model of the NN interaction is shown in Figure 1.

The usual partial-wave analysis is inapplicable due to the nonzero total momentum

**P** and thus the homogeneous equation is solved in three dimensions. See reference [12] for details. We note that with the instant constraint $K^{OBE}$ as well as $g_0$ and $J_{IA}$ are non-singular.

Results for the deuteron magnetic form factor are shown in Figure 2. The solid line shows the result based on the impulse current plus $\rho\pi\gamma$, $\omega\sigma\gamma$, and $\omega\eta\gamma$ meson-exchange-currents calculated with the instant wave functions in the Breit frame. The meson-exchange-current operators of Hummel and Tjon [8,16,17] are used with $g_{\rho\pi\gamma} = .563$, $g_{\omega\sigma\gamma} = -.4$, $g_{\omega\eta\gamma} = -.206$. Results based on the exactly conserved impulse approximation current differ little from use of just $\Gamma_1^\mu \gamma_2^0$. This means that the extra terms required for current conservation play little role in $D$ form factors and the results are very similar to those of Hummel and Tjon, except for boost effects. The dynamical boost produces significant differences. To illustrate this, we compare the impulse approximation contributions based on solving Eq. (21) (dotted line) and two approximations based on using the rest frame wave functions with a kinematical boost as follows,

$$\psi(p, P) = \Lambda_1(\mathcal{L})\Lambda_2(\mathcal{L})\psi(\mathbf{p}_{rest}, P_{rest}), \tag{23}$$

where $P_{rest} = (m_D, 0)$, $P = (\epsilon_D, \pm\mathbf{q}/2)$, and $\mathcal{L}P_{rest} = P$. The kinematical Lorentz transform of the relative momenta, $\mathcal{L}p_{rest} = p$, is ambiguous since it is not possible to simultaneously satisfy both constraints: $(p^0 = 0)$ and $(p^0_{rest} = 0)$. In Figure 2, the dashed line is the result of satisfying $(p^0 = 0)$, while the dash-dotted line is the result of satisfying $(p^0_{rest} = 0)$. The difference of the two kinematical boost prescriptions shown in Figure 2 provides a measure of the ambiguity of kinematical boosts of equally-off-shell wave functions. The dynamical boost does not suffer from the ambiguity associated with assumptions about $p^0$ and it is seen to provide results which differ significantly from those of both of the kinematical boosts. However, a final assessment must await solution of Eq. (16).

Conclusions

Quasipotential approaches are useful for incorporating relativistic effects into analyses of EM form factors in a systematic fashion. For a given quasipotential approach, it is necessary to derive appropriate current operators. Current conservation also requires that vertex functions must be consistent with the current. Three main approaches to

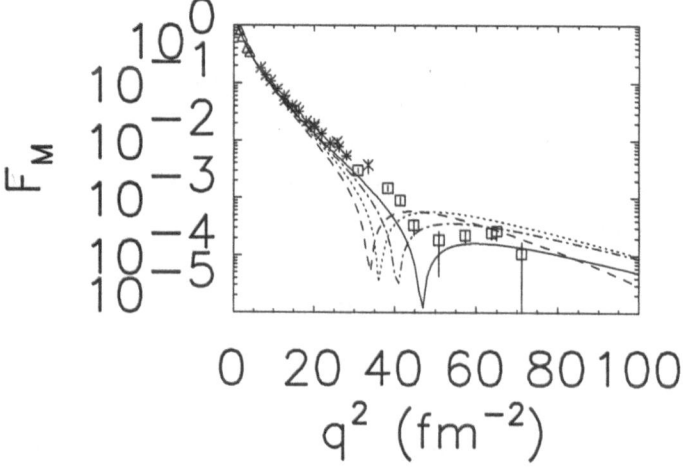

Figure 2: Elastic e-d magnetic form factor: consistent calculation with IA+MEC (solid). Impulse approximation only: consistent calculation (dotted), boost approximations with $p^0(Breit) = 0$ (dashed) and $p^0(cm) = 0$ (dash-dotted). See text.

this problem are the BSLT quasipotential analysis, the spectator-on-shell approach and the instant formalism in the Breit frame. The latter has been shown to have some advantages with respect to current conservation and avoidance of unphysical singularities,

There is reasonable agreement between the corresponding results of Hummel and Tjon and the work of Devine and Wallace based on the instant formalism in the Breit frame. The extra terms in the current which are needed conservation of the impulse are found to cause very small differences. The dynamical boost which is in the instant formalism is an interesting new ingredient. It has been evaluated using a simple approximation which shows significant differences in the deuteron magnetic form factor in comparison with approximate forms of the boost. Comparisons of recent calculations for elastic scattering by $D$, including new results based on the symmetrized spectator-on-shell formalism, [21] shows that the magnetic form factor is the most sensitive observable to differing treatments of relativistic effects and meson-exchange currents.

## ACKNOWLEDGEMENT

Support for this work by the U.S. Department of Energy under grant DE-FG02-93ER-40762 is gratefully acknowledged.

## REFERENCES

1. Mandelstam, S.: Proc. Roy. Soc. London, Ser. **A 233**, 248 (1955)

2. Fleischer, J. and Tjon, J. A.:Phys. Rev. D **21**, 87 (1980)

3. Zuilhof, M. J., and Tjon, J. A.:Phys. Rev. C **22**, 2369 (1980)

4. Zuilhof, M. J., and Tjon, J. A.:Phys. Rev. C **24**, 736 (1981)

5. Tjon, J. A.: in *Hadronic Physics with Multi-GeV Electrons*, B. Desplanques and D. Goutte, eds., Nova Science, Commack, New York, 1990

6. Blankenbecler, R., and Sugar, R. L.: Phys. Rev. **142**, 1051 (1966)

7. Hummel, E., and Tjon, J. A.: Phys. Rev. Lett. **63**, 1788 (1989)

8. Hummel, E., and Tjon, J. A.: Phys. Rev. C **42**, 423 (1990)

9. Gross, F.: Phys. Rev. C **26**, 2203 (1990)

10. Gross, F., Van Orden, J. W. and Holinde, K: Phys. Rev. C **45**, 2094 (1992)

11. Arnold, R. G., Carlson, C. E. and Gross, F.: Phys. Rev. C **21**, 1426 (1990)

12. Devine, N. K. and Wallace, S. J.: Phys. Rev. C **48**, 973 (1993)

13. Salpeter, E. E. and Bethe, H. A.: Phys. Rev. **84**, 1232 (1951)

14. Gross, F. and Riska, D. O.: Phys. Rev. C **36**, 1928 (1987)

15. Jaus, W.: Hel. Phys. Acta **57**, 644 (1984)

16. Devine, N. K.: Ph. D. Thesis, University of Maryland (1992)

17. Hummel, E.: Ph.D. Thesis, University of Utrecht (1991)

18. Wallace, S. J. and Mandelzweig, V. B.: Nuc. Phys. **A503**, 673 (1989)

19. Mandelzweig, V. B. and Wallace, S. J.: Phys. Lett. **B197**, 469 (1987)

20. Machleidt, R. A.: in *Advances in Nucl. Phys.*, J. W. Negele, ed., Plenum, New York, 1989

21. Gross, F.: private communication

Few-Body Systems, Suppl. 7, 409—416 (1994)

© Springer-Verlag 1994
Printed in Austria

# RELATIVISTIC EFFECTS IN qq̄ SYSTEMS

Peter C. Tiemeijer
*Institute for Theoretical Physics, University of Utrecht*
*Box 80.006, 3508 TA Utrecht, The Netherlands*

Extensions of the constituent quark potential model for mesons are formulated and analyzed which include relativistic dynamics and the full Dirac structure of both positive and negative energy states. The meson wave equations are transformed to configuration space where they can easily be solved numerically. Regge-slopes, the pion form factor and radiative, annihilation decay are studied. We find that the projection on the full Dirac structure of the confining potential and of the meson decay vertices gives unexpected large modifications to the nonrelativistic mass spectra and decay vertices. This paper summarizes the main results of [1, 2].

The nonrelativistic constituent quark potential model provides a simple and successful description of meson and baryon properties such as their masses and decay probabilities. In view of its successes it is interesting to study relativistic extensions of this model, in order to understand how much is neglected by making the nonrelativistic (NR) approximation. Relativistic models may also lead to more reliable predictions for processes where high momenta are involved.

Generalization of the ordinary Schrödinger equation to a relativistic covariant form leads to the well-known Bethe-Salpeter equation (BSE). The BSE differs from the NR equation in two respects. First, covariance requires that for fermions the full Dirac structure is taken into account, so for quarks not only the positive energy states are to be considered, but also their negative energy states must be included. Second, the dependence of the bound state wave function and of the potential on the relative three-momentum $p$ becomes a dependence on the four-momentum $(p_0, p)$. This latter feature does not only make the BSE considerably more complex, but it also poses the fundamental problem of defining the confining potential for retarded and advanced times. At present there is no underlying theory which can give a prescription for this. For tehse reasons, the analysis of the relativistic effects has been restricted to quasipotential (QP) approximations to the BSE. Here the $p_0$ dependence of the propagator $S$ of the BSE is eliminated, but the full Dirac structure is kept in the equations.

# 1    The relativistic quark model

Let us briefly describe the relativistic model [1, 2]. In the center of mass (cm) system the momenta $p_1$ and $p_2$ of the two quarks are denoted by

$$p_1 = (\frac{M^2 + m_1^2 - m_2^2}{2M}, \boldsymbol{p}) = (E_1, \boldsymbol{p}) \quad \text{and} \quad p_2 = (\frac{M^2 - m_1^2 + m_2^2}{2M}, -\boldsymbol{p}) = (E_2, -\boldsymbol{p}), \quad (1)$$

with $M$ the total meson mass. The bound state wave equation for the meson wave function $\psi$ reads

$$S^{-1}(\boldsymbol{p})\psi(\boldsymbol{p}) = -\int \frac{d\boldsymbol{p}'}{(2\pi)^3} V(\boldsymbol{p}' - \boldsymbol{p})\psi(\boldsymbol{p}'). \quad (2)$$

Two QP approximations have been studied, namely the Blankenbecler-Sugar-Logunov-Tavkhelidze (BSLT) approximation [3, 4], and an equal-time (ET) approximation [5]. If one defines the relativistic energies $\omega_i = \sqrt{\boldsymbol{p}^2 + m_i^2}$ and the projections

$$\Lambda^{\rho_1\rho_2} = \frac{\rho_1(\boldsymbol{\gamma}^{(1)} \cdot \boldsymbol{p} + m_1) + \omega_1 \gamma_0^{(1)}}{2\omega_1} \frac{\rho_2(-\boldsymbol{\gamma}^{(2)} \cdot \boldsymbol{p} + m_2) + \omega_2 \gamma_0^{(2)}}{2\omega_2}, \quad (3)$$

then the QP propagators can be expressed as

$$S_{\text{BSLT}} = \frac{1}{2(\omega_1 + \omega_2)} \frac{1}{G} \Big[ (\omega_1 + E_1)(\omega_2 + E_2)\Lambda^{++} - (\omega_1 + E_1)(\omega_2 - E_2)\Lambda^{+-}$$
$$- (\omega_1 - E_1)(\omega_2 + E_2)\Lambda^{-+} + (\omega_1 - E_1)(\omega_2 - E_2)\Lambda^{--} \Big], \quad (4)$$

where $G = \omega_1^2 - E_1^2 = \omega_2^2 - E_2^2$, and

$$S_{\text{ET}} = \frac{\Lambda^{++}}{\omega_1 + \omega_2 - E_1 - E_2} - \frac{\Lambda^{-+}}{\omega_1 + \omega_2 + E_1 - E_2} - \frac{\Lambda^{+-}}{\omega_1 + \omega_2 - E_1 + E_2} + \frac{\Lambda^{--}}{\omega_1 + \omega_2 + E_1 + E_2}. \quad (5)$$

Both QP propagators are time-reversal invariant, i.e. invariant for a simultaneous change of the energy labels $\rho_i \to -\rho_i$ and the particle energies $E_i \to -E_i$. Furthermore, both QP equations reduce to the one-body Dirac equation if one of the quarks is taken infinitely heavy. Since no expansion in $p/m$ is made, there is also no need for cut-off parameters to regularize short-distance singularities.

The instantaneous interaction $V$ between the quarks is modeled as the sum of a Coulomb-like part describing the one-gluon-exchange (OGE) interaction and a linearly rising part for the confinement. It takes in coordinate space the form

$$V(x) = -\frac{\alpha(x)}{x} \gamma_\mu^{(1)} \gamma^{\mu(2)} + (\kappa x + c) \left[ (1 - \varepsilon)1^{(1)}1^{(2)} + \varepsilon \gamma_\mu^{(1)} \gamma^{\mu(2)} \right]. \quad (6)$$

The running coupling constant behaves as $\alpha(x) \sim (8/27)\pi / \ln(x_0/x)$ for small distances $x$, and grows to some maximum saturation value $\alpha_{\text{sat}}$ for large separations, according to the interpolation given in [1]. The gauge dependence can be estimated by replacing the Feynman form by the Coulomb form in Eq. (6), $\gamma_\mu^{(1)} \gamma^{\mu(2)} \to \gamma_\mu^{(1)} \gamma^{\mu(2)} + \frac{1}{2}\left[ \boldsymbol{\gamma}^{(1)} \cdot \boldsymbol{\gamma}^{(2)} - (\boldsymbol{\gamma}^{(1)} \cdot \hat{\boldsymbol{x}})(\boldsymbol{\gamma}^{(2)} \cdot \hat{\boldsymbol{x}}) \right]$.

Certain difficulties as found in the one-particle Dirac equation are also present in the QP approach. Let us for a moment consider one fermion in an external potential. If this particle experiences a potential which fluctuates more strongly than $\sim 2m$ over a distance shorter than its Compton length $x_C = 1/m$ new fermion-antifermion pairs can be created. This phenomenon cannot correctly be described by the Dirac equation which describes a

Table 1: *Parameter sets used (in units of* GeV*)*

|  | $m_u$ | $m_c$ | $\alpha$ | $\kappa$ | $c$ | $\varepsilon$ | gauge |
|---|---|---|---|---|---|---|---|
| BSLT-F | 0.20 |  | 0.8 | 0.33 | -0.8 | 0.2 | Feynman |
| BSLT-C | 0.25 | 1.779 | 0.8 | 0.33 | -1.0 | 0.25 | Coulomb |
| ET | 0.25 |  | 0.8 | 0.33 | -1.0 | 0.2 | Coulomb |
| fixed $\alpha$ ET | 0.25 |  | 0.318 | 0.20 | -0.529 | 0 | Feynman |

one-particle theory and thus misses the interactions between the newly created pair and the starting particle. Since the Dirac equation does allow for antifermion components, solutions in this potential can have an unbound number of non-interacting fermions and antifermions thus being unnormalizable and unphysical. This break-down of the Dirac equation is well-known as the Klein paradox. Similar flaws emerge in the QP equations of this work since they also contain negative energy components. Unbound solutions can be expected if the confining strength becomes too strong, $\kappa x_C \gtrsim 2m$ or $\kappa \gtrsim 2m^2$. This domain can be reached in light meson systems. The condition on $\kappa$ depends on the fraction $\varepsilon$ of vector confinement. Similarly, if the OGE potential becomes too strong, $\alpha/x_C \gtrsim 2m$ or $\alpha \gtrsim 2$, irregular solutions may be expected.

The solution of relativistic quark-antiquark equations has long time been considered as difficult due to the singular behavior of the confining potential at small momenta. Recently, various prescriptions for handling this singularity have been published [6–11]. In this work this difficulty has been circumvented by Fourier transforming the QP equations to configuration space. For example, the BSLT equation can be rewritten as

$$(\not{p}_1 + \not{p}_2 - m_1 - m_2)\psi(\boldsymbol{p}) = \frac{1}{2(\omega_1 + \omega_2)}(\not{p}_1 + \not{p}_2 + m_1 + m_2)\int \frac{d\boldsymbol{p}'}{(2\pi)^3}\, V(\boldsymbol{p}' - \boldsymbol{p})\psi(\boldsymbol{p}'), \quad (7)$$

which is easily Fourier transformed to

$$(i\not{\partial}_1 + i\not{\partial}_2 - m_1 - m_2)\psi(\boldsymbol{x}) = \int d\boldsymbol{x}'\, \frac{Z_{\text{BSLT}}(\boldsymbol{x}' - \boldsymbol{x})}{2(m_1 + m_2)}(i\not{\partial}_1 + i\not{\partial}_2 + m_1 + m_2)V(\boldsymbol{x}')\psi(\boldsymbol{x}'). \quad (8)$$

Due to the relativistic phase space factor in the BSLT propagator, a non-locality occurs in the relativistic equation, which is contained in the function

$$Z_{\text{BSLT}}(\boldsymbol{R}) = \int \frac{d\boldsymbol{p}}{(2\pi)^3}\, \frac{m_1 + m_2}{\omega_1 + \omega_2} e^{i\boldsymbol{p} \cdot \boldsymbol{R}} = \frac{1}{2\pi^2 R^2}\, \frac{m_2^2 K_2(m_2 R) - m_1^2 K_2(m_1 R)}{m_1 - m_2}. \quad (9)$$

$K_2$ is the modified Bessel function of the second kind of order two. Next, the QP equations are projected on basis states with definite spin and orbital momentum. The solution to the resulting eight coupled integro-differential equations can be found straightforwardly with spline techniques.

## 2 Relativistic effects in the meson masses

An overview of the differences between the NR Schrödinger results and the relativistic BSLT is given in Fig. 1. The figure shows the masses of the radially unexcited light isovector mesons ($\pi$, $\rho$, $a_0$, etc.). Figure 1(a) shows the results of the Schrödinger equation with the spin-dependent corrections of order $p^2/m^2$. Figure 1(b) gives the masses after the replacement $\boldsymbol{p}^2/(2m) + m \to \omega = \sqrt{\boldsymbol{p}^2 + m^2}$. Both calculations in Figs. 1(a) and 1(b) need cutoff parameters to regularize the short-distance behavior of the $p^2/m^2$ corrections. These cutoffs are no longer present in the calculation for Fig. 1(c) where

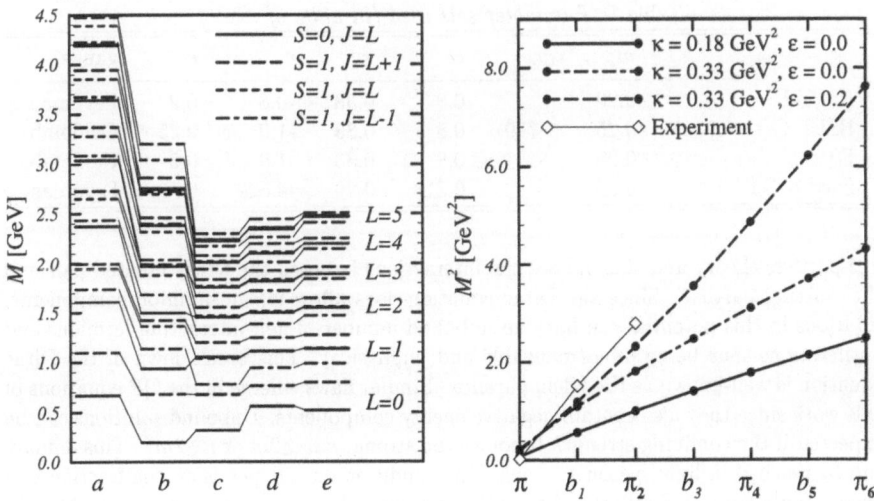

Figure 1: (left) *Light meson spectrum of the radially unexcited states, using the BSLT-C parameters. (a) Schrödinger equation with $p^2/2m$ and $m^{-2}$ corrections. (b) Schrödinger equation with $\sqrt{p^2 + m^2}$ and $m^{-2}$ corrections. (c) Schrödinger equation with $\sqrt{p^2 + m^2}$ and full projection of potential, i.e. ET with $|++\rangle$, (d) BSLT with $|++\rangle$, (e) BSLT with all states.*

Figure 2: (right) *Regge-trajectories of the $\pi$ family using the BSLT-F parameters.*

the full projection $\Lambda^{++} V \gamma_0^{(1)} \gamma_0^{(2)} \Lambda^{++}$ of the potential has been used. In Fig. 1(d) the relativized Schrödinger propagator is replaced by the BSLT propagator for $|++\rangle$ states, $(\omega - E) \rightarrow 2\omega(\omega - E)/(\omega + E)$. Finally, in Fig. 1(e) all energy states are included, leading to the full BSLT results.

The most prominent differences are the reductions of the meson masses caused by the replacement of the NR kinetic energy $p^2/(2m)$ by the relativized expression $\omega - m$, and by the replacement of the NR $V$ by the full projection of $V$ on the positive energy spinors. The last reduction can also qualitatively be seen from the spin-independent corrections of order $m^{-2}$ to a NR scalar potential $V_S$ [12], which read $m^{-2} \left[ \frac{1}{4}(\nabla^2 V_S) + V_S'(d/dx) + V_S \nabla^2 \right]$.

The consequences of these reductions for the Regge-slopes of the light meson systems are large. Experimentally it is found that for light mesons the squares of the masses $M$ of the mesons are proportional to their angular momenta, $M^2 = \beta J + c$, where the Regge-slope $\beta \simeq 1.2$ GeV$^2$ [12]. NR analyses of the heavy mesons indicate that the confining potential should be purely scalar, with a confining strength $\kappa \simeq 0.18$ GeV$^2$. If such a confining potential is employed in the BSLT model for the light mesons, one finds $\beta \sim 0.5$ GeV$^2$, which far too small. This is shown in Fig. 2. For scalar confinement the value $\kappa = 0.18$ GeV$^2$ must roughly be tripled in order to reproduce the experimental $\beta$ for the light mesons. However, if a small admixture $\varepsilon$ of vector confinement is introduced as in Eq. (6), then $\beta$ is greatly increased: for $\varepsilon = 0.2$ the experimental $\beta$ is found at $\kappa \simeq 0.33$ GeV$^2$. The QP models with the full Dirac space do not allow for larger fractions $\varepsilon$ due to Klein's paradox. For $\varepsilon > 0.2(0.25)$ in the Feynman(Coulomb) gauge the net confinement becomes repulsive in some $|+-\rangle$ and $|-+\rangle$ states. Klein's paradox can be circumvented by restricting the QP equations to $|++\rangle$ only. In that case one finds that

Table 2: *Electric dipole decay widths (in keV) calculated from the BSLT-C wave functions, using experimental masses. The first column lists the NR decay widths, and the last columns the full BSLT results. The columns in between show the accumulative effects of the various relativistic corrections as explained in the text. The experimental data are from the particle data group [15].*

| | NR $\langle r \rangle$ | NR $\langle p \rangle$ | BSLT $\vert + \rangle$ | BSLT $+k$ | BSLT $+\Lambda$ | BSLT $+1-$ | BSLT full | Experiment |
|---|---|---|---|---|---|---|---|---|
| $b_1^+ \to \pi^+ + \gamma$ | 170 | 919 | 17 | 0.1 | 0.6 | 0.7 | 0.6 | $230 \pm 60$ |
| $a_1^+ \to \rho^+ + \gamma$ | 57 | 502 | 53 | 57 | 70 | 40 | 41 | |
| $\chi_{c2} \to \psi + \gamma$ | 375 | 514 | 341 | 298 | 281 | 306 | 307 | $270 \pm 30$ |
| $\chi_{c1} \to \psi + \gamma$ | 282 | 479 | 324 | 325 | 345 | 290 | 290 | $240 \pm 40$ |
| $\chi_{c0} \to \psi + \gamma$ | 132 | 381 | 246 | 254 | 269 | 196 | 196 | $90 \pm 40$ |
| $\psi' \to \chi_{c2} + \gamma$ | 19.7 | 82.6 | 45.2 | 45.9 | 46.4 | 41.0 | 41.0 | $22 \pm 3$ |
| $\psi' \to \chi_{c1} + \gamma$ | 24.3 | 85.1 | 47.2 | 46.8 | 46.0 | 43.4 | 43.4 | $24 \pm 4$ |
| $\psi' \to \chi_{c0} + \gamma$ | 23.4 | 55.9 | 28.0 | 26.4 | 24.6 | 25.2 | 25.2 | $26 \pm 4$ |

the popular choice $\varepsilon = 0.5$ (e.g. studied in [10, 13, 14]) reproduces the experimental $\beta$ at $\kappa \simeq 0.20$ GeV$^2$. However, this large amount of vector confinement worsens the description of the fine structure of the mass spectra, e.g. the spin-orbit splittings between the light meson $^3P$ states becomes much too large.

Thus the confining potential favored by the light meson Regge-slopes is considerably stronger than favored by heavy meson spectroscopy. This may indicate that the assumption of flavor independence of the confinement strength is incorrect in the relativistic constituent quark model.

# 3 Relativistic effects in the meson decays

Various processes which are well-understood in terms of simple Feynman diagrams can now be calculated using the relativistic wave functions of the model. Such processes are e.g. $q\bar{q}$ annihilation decay or the interaction between a photon and a meson. Due to the QP approximation ambiguities arise in the evaluation of the latter process; here the initial and final mesons have different cm frames, and hence different prescriptions for treating $p_0$. These prescriptions must be matched in some way. It was found [2] that the differences between the various matching procedures are less than 3% for all reactions considered — except when pions are involved, then differences up to 30% can occur.

## 3.1 Electric dipole decays

The NR expression for the electric dipole decay width between $S$ and $P$ states is given by

$$\Gamma^{E1}(M \to M' + \gamma) = \frac{2J' + 1}{27} \frac{4k^3 e_Q^2}{4\pi} |\langle r \rangle|^2 , \tag{10}$$

where $J'$ is the spin of the final state, $k$ the emitted photon momentum, $e_Q$ the quark charge, and $\langle r \rangle$ the expectation value of the radial distance between the in- and outgoing wave functions. The first column of Table 1 lists some NR $\Gamma^{E1}$ that one obtains from using the positive energy states of the wave functions obtained in the BSLT-C model. It is amusing to note the good agreement with the experimental rates in the last column.

By adding consecutively various relativistic effects one arrives at the full BSLT widths in the column labeled 'BSLT full'.

The NR rates are derived by approximating $2\boldsymbol{p} \simeq im[H, \boldsymbol{r}]$ in the decay current ($H$ is the NR Hamiltonian). The column labeled NR $\langle p \rangle$ shows the decay rate without this approximation, i.e. making the substitution $\langle r \rangle \to 2\langle p \rangle/(mk)$ in Eq. (10). The effect of this replacement is dramatic. Even for the heavy $c\bar{c}$ system the configuration and momentum dipole moments can differ more than a factor of two. This difference is partly due to the neglect of the relativistic expression for the kinetic energy in $H$, i.e. $\omega - m$ instead of $p^2/(2m)$. However, the biggest difference is caused by the neglect in $[H, \boldsymbol{r}]$ of the momentum dependence of the projection on positive energy states of the confining potential. A nice way of explaining these differences is the following. Let us not take the experimental photon momentum $k$ in Eq. (10) but instead the theoretical photon momentum $k_{\text{th}}$ obtained from the mass spectrum of the Schrödinger equation, shown in Fig. 1(a). The Schrödinger spectrum has almost twice as large spacings and hence $k_{\text{th}} \approx 2k$. Using $k_{\text{th}}$ in Eq. (10) one finds more or less consistent configuration and momentum dipole moments. Similarly, if one corrects Eq. (10) for the relativistic expression for the kinetic energy by replacing $p/m \to p/\omega$, then it is possible to use $k_{\text{th}}$ as obtained from the mass spectrum of the Schrödinger equation with relativistic kinetic energy, shown in Fig. 1(b). In general, let us note that all ups and downs in the mass spectra of Fig. 1 have corresponding downs and ups in the decay vertices in Table 2 which tend to compensate each other to give only moderate modifications of the final decay probabilities. This clearly illustrates the importance of having a current consistent with the dynamics.

The next step towards the full BSLT width is to abandon the NR approximation of keeping only the leading terms in $p/m$ and to consider the full projection of the decay vertex on positive energy states. This has several effects; for the light mesons the largest effect is caused by an extra factor $M'/M$ associated with the wave function normalization This full projection decreases the decay rates considerably, as can been read from the column BSLT$|+\rangle$.

The remaining columns show the comparatively small effects of including the momentum shift $k$ between the initial and final wave functions (BSLT+$k$), the Lorentz boosts of the wave functions to their rest frames (BSLT+$\Lambda$), the inclusion of the once negative energy states (BSLT+1-), and the inclusion of the double negative energy states (BSLT full). The full BSLT calculations for the light meson decays give far too low widths. A good description of these decays can be given by using vector meson dominance [16].

## 3.2 Pseudoscalar decay

In previous sections the influence of the relativistic dynamics on the meson masses and the radiative decay widths was investigated by starting from the NR expressions and consecutively adding the various relativistic ingredients. A similar analysis can also be done for the pseudoscalar annihilation decay, and is shown in Fig. 3. Figure 3(a) shows the NR $f_P = \sqrt{6}\langle^1S_0^{++}\rangle_a/\sqrt{M}$, where $^1S_0^{++}$ is the positive energy component of the meson wave function. The subscript indicates that the wave function has been averaged over a small volume of radius $a$ equal to the averaged Compton length of the two quarks. Performing a full projection of the annihilation vertex on plane wave states leads to Fig. 3(b). Figures 3(c) and 3(d) show the pseudoscalar decay constants if the once and double negative energy states are added, respectively. We already pointed out in the previous section that the relativistic corrections to the radiative decay amplitude are partly compensated for by the relativistic corrections to the meson masses. This also happens to the annihilation decay. The meson mass in the denominator of the expression

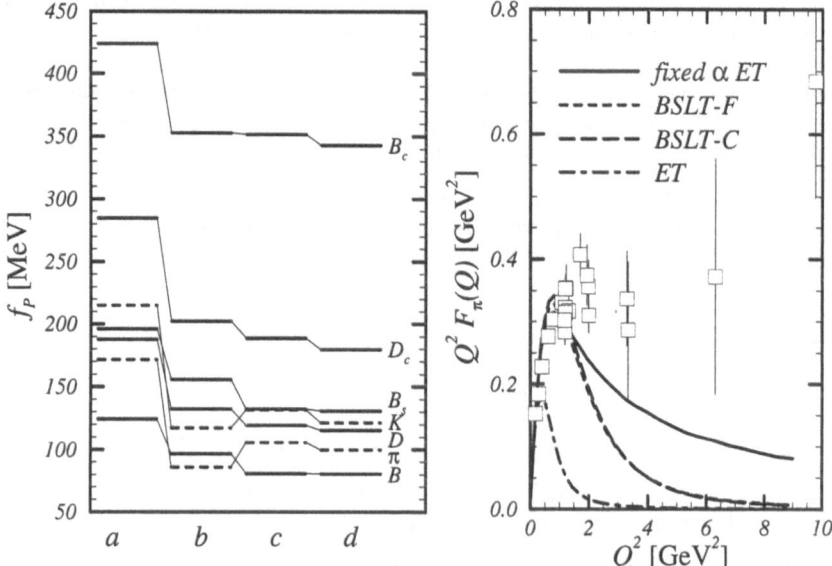

Figure 3: (left) *Pseudoscalar decay constants calculated from the meson wave functions of the BSLT-C model. (a) NR vertex, (b) Full vertex with $|++\rangle$, (c) Full vertex with $|++\rangle$, $|+-\rangle$ and $|-+\rangle$, (d) Full vertex with all states.*

Figure 4: (right) *Pion form factor $F_\pi$ using the fixed $\alpha$ ET model. Experimental points are from [18]. It has been argued that the analysis by which $F_\pi$ is extracted from the data is very model dependent [19].*

for $f_P$ causes the ups and downs in the decay constant to be partly compensated for by the ups and downs in the mass spectrum. This again demonstrates the importance of having a vertex that is consistent with the dynamics.

## 3.3 Pion electromagnetic form factor

The experimental data on the pion form factor $F_\pi$ provide a nice tool for testing simultaneously the low momentum region dominated by the non-perturbative confining potential and the high momentum region dominated by the asymptotic OGE potential. The pion wave function diverges at small distances or large momenta, and one can expect this behavior to play an important role in the large momentum behavior of $F_\pi$. This has indeed been found in [17] for the full BSE. We also see this feature in the QP framework.

The divergences can best be studied when they are maximal. This is obtained with a coupling constant that is fixed and close to its maximum value. This is obtained in the ET model with fixed coupling $\alpha = 0.318$, which is close to the maximum $1/\pi$. The resulting $F_\pi$ for the fixed $\alpha$ ET model and the running coupling models are shown in Fig. 4. Their slopes at zero momentum determine the electromagnetic radius $< r_{\mathrm{e.m.}}^2 >^{1/2}$, which are 0.48, 0.44, 0.44 and 0.65 fm for the fixed $\alpha$ ET, BSLT-F, BSLT-C and ET model, respectively. The experimental radius is $0.663 \pm 0.023$ fm [20]. A striking feature of $F_\pi$ is its large tail at high $Q^2$, especially in the fixed $\alpha$ ET model, which resembles the results of perturbative QCD predicting that due to the one-gluon-exchange the meson form factor will fall off like $Q^{-2}\ln^{-1}(Q^2)$ for large $Q^2$ [21]. In our model the large tail also follows

416

from the OGE potential. It causes the singular behavior at the origin where $\psi$ grows like $x^\gamma$, $-1 < \gamma < 0$. Its momentum wave function $\psi(p)$ consequently falls off like $p^{-2-\gamma}$ for large $p$. Since the behavior of $F_\pi(Q)$ for large $Q$ follows from that of $\psi(p)$ for large $p$, one finds that the smaller $\gamma$ gets the larger the tail of $F_\pi(Q)$ will be. Indeed, the numerical results show that a smaller value of $\alpha$ or a running $\alpha$, which lead to less singular wave functions, give a faster fall-off in the form factors.

This work was partially financially supported by de Stichting voor Fundamenteel Onderzoek der Materie (FOM), which is sponsored by de Nederlandse Organisatie voor Wetenschappelijk Onderzoek (NWO).

# References

[1] P.C. Tiemeijer and J.A. Tjon, Phys. Lett. B **277**, 38 (1992); Phys. Rev. C **48**, 896 (1993); Utrecht preprint THU-92/31, to appear in Phys. Rev. C **49**.

[2] P.C. Tiemeijer, Ph.D. Thesis, University Utrecht (1993).

[3] R. Blankenbecler and R. Sugar, Phys. Rev. **142**, 1051 (1966); A.A. Logunov and A.N. Tavkhelidze, Nuovo Cim. **29**, 380 (1963).

[4] E.D. Cooper and B.K. Jennings, Nucl. Phys. A **500**, 553 (1989).

[5] V.B. Mandelzweig and S.J. Wallace, Phys. Lett. B **197**, 469 (1987); S.J. Wallace and V.B. Mandelzweig, Nucl. Phys. A **503**, 673 (1989); S.J. Wallace in Nuclear and Particle Physics on the Light Cone, Proceedings of the LAMPF Workshop, Los Alamos 1988 (World Scientific, Singapore, 1989).

[6] T. Murota, Prog. Theor. Phys. **69**, 181 (1983).

[7] J.W. Norbury, D.E. Kahana and K.M. Maung, Can. J. Phys. **70**, 86 (1992); K.M. Maung, D.E. Kahana and J.W. Norbury, Phys. Rev. D **47**, 1182 (1993).

[8] F. Gross and J. Milana, Phys. Rev. D **43**, 2401 (1991); *ibid.* D **45**, 969 (1992).

[9] J.-F. Lagaë, Phys. Lett. B **240**, 451 (1990), Phys. Rev. D **45**, 305 (1992), *ibid.* 317.

[10] J.R. Spence and J.P. Vary, Phys. Rev. D **35**, 2191 (1987); Phys. Rev. C **47**, 1282 (1993).

[11] H. Hersbach, Phys. Rev. D **47**, 3027 (1993); Utrecht preprint THU-93/13.

[12] W. Lucha, F.F. Schöberl and D. Gromes, Phys. Rep. **200**, 127 (1991).

[13] M. Chachkhunashvili, T. Kopaleishvili, Few-Body Systems **6**, 1 (1989); A. Archvadze, M. Chachkhunashvili and T. Kopaleishvili, *ibid.* **14**, 53 (1993).

[14] H.W. Crater and P. Van Alstine, Phys. Rev. Lett. **53**, 1527 (1984), Phys. Rev. D **37**, 1982 (1988).

[15] Particle Data Group, K. Hikasa *et al.*, Phys. Rev. D **45**, S1 (1992).

[16] J. Babcock and J.L. Rosner, Phys. Rev. D **14**, 1286 (1976); J.L. Rosner, *ibid.* D **23**, 1127 (1981).

[17] M.L. Goldberger, D.E. Soper and A.H. Guth, Phys. Rev. D **14**, 1117 (1976).

[18] C.J. Bebek *et al.*, Phys. Rev. D **17**, 1693 (1978).

[19] C.E. Carlson and J. Milana, Phys. Rev. Lett. **65**, 1717 (1990); R. Kahler and J. Milana, Phys. Rev. D **47**, 3690 (1993).

[20] E.B. Dally *et al.*, Phys. Rev. Lett. **48**, 375 (1982).

[21] G.P. Lepage and S.J. Brodsky, Phys. Rev. D **22**, 2157 (1980).

Few-Body Systems, Suppl. 7, 417—424 (1994)

Few-
Body
Systems
© Springer-Verlag 1994
Printed in Austria

# CHARMONIUM SPECTROSCOPY WITH ANTIPROTONS

Dimitri A. Dimitroyannis [1]
*Northwestern University, Evanston, IL 60208, USA*
on behalf of the E760 collaboration [2]

**Abstract**

We report on new measurements for the mass, width and branching ratios for the $J/\psi$, $\psi'$, $\chi_{c1}$ and $\chi_{c2}$. These charmonium states are formed in exclusive proton-antiproton annihilations at the Fermilab Antiproton Accumulator ring, where stochastically cooled antiprotons are brought into collision with the protons of an internal hydrogen gas jet target. The antiproton energy was precisely controlled and measured allowing for an accurate measurement of the resonance parameters.

# 1  Introduction

Quarkonia, mesonic bound states of a same flavour quark-antiquark, are to Quantum Chromodynamics (QCD) what positronium is to Quantum Electrodymanics. Such bound states provide the simplest known systems of QCD. Study of heavy quarkonia states is preferable to the light ones, since here non-perturbative effects are smaller and precision measurements can be used as means of determining accurately certain parameters of the theory or even test the validity of the theory itself.

Quarkonia of the charm and the bottom flavour have a rich spectrum with energy level spacing relatively small as compared to their masses, which implies the possibility to characterize these systems in a non-relativistic way. Indeed the spectra of these systems have been very well described by solving a Schrödinger equation with a potential incorporating the appropriate QCD asymptotic behaviour.

---

[1] Now at NIKHEF-K, Postbus 41992, NL-1009 DB Amsterdam, The Netherlands

[2] The E760 collaboration: T.A. Armstrong, D. Bettoni, V.K. Bharadwaj, C. Biino, D. Broemmelsiek, A. Buzzo, R. Calabrese, A. Ceccucci, R. Cester, M. Church, P. Dalpiaz, P.F. Dalpiaz, R. Dibenedetto, D. Dimitroyannis, M.G. Fabbri, J. Fast, A. Gianoli, C.M. Ginsburg, K. Gollwitzer, A. Hahn, M. Hasan, S. Hsueh, R. Lewis, E. Luppi, M. Macri, A.M. Majewska, M. Mandelkern, F. Marchetto, M. Marinelli, J. Marques, W. Marsh, M. Martini, M. Masuzawa, E. Menichetti, A. Migliori, R. Mussa, M. Pallavicini, S. Palestini, N. Pastrone, C. Patrignani, J. Peoples Jr., L. Pesando, M.G. Pia, S. Pordes, P.A. Rapidis, R. Ray, J. Reid, G. Rinaudo, B. Roccuzzo, J. Rosen, A. Santroni, M. Sarmiento, M. Savrie, A. Scalisi, J. Schultz, K.K. Seth, A. Smith, G.A. Smith, S. Trokenheim, M.F. Weber, S. Werkema, Y. Zhang, J. Zhao, G. Zioulas and M. Zito; **Fermilab, Ferrara, Genoa, Irvine, Northwestern, Pennsylvania, Turin**

418

The quarkonium states can be classified, exactly as in the case of positronium, by the usual spectroscopic notation $n^{2S+1}L_J$, where $n$ is the principal quantum number and $S$, $L$, $J$ are the spin, the orbital and the total angular momentum of the system. For quarkonia and any fermion-antifermion system for that matter, the parity, $P$ and the charge conjugation, $C$, quantum number are given as $P = (-1)^{L+1}$ and $C = (-1)^{L+S}$.

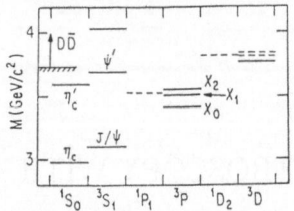

Figure 1: The charmonium states below the open-charm threshold. The $h_c(2^1P_1)$ state is indicated by a dash line.

# 2 Antiprotons: Why and How

Although charmonium was discovered almost 20 years ago, the study of its spectrum is not complete yet. The majority of the measurements on charmonium were obtained in electron-positron colliders [1]. With this technique one readily produces triplet states $n^3S_1$, that is $J/\psi$ and $\psi'$, which carry the quantum number of the photon ($J^{PC} = 1^{--}$). Other charmonium states were reached by radiative decays of these $\psi$ states; the precision study of these radiatively accessible states is limited by the instrumental resolution of the detector.

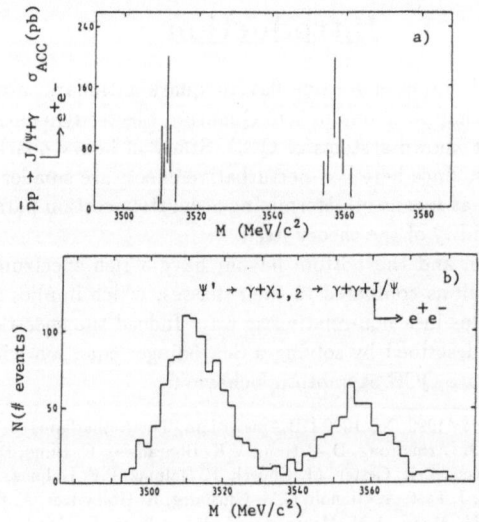

Figure 2: Study of the $\chi_{c1,c2}$ states: (a) via resonant $\bar{p}p$ production by R704[2] and (b) via radiative decay of the $\psi'$ by the Crystal Ball[1]

Even with this intrinsic difficulty a number of charmonium states have been seen in $e^+e^-$ colliders. To date the following charmonium states, below the open charm threshold, have been robustly identified : $\eta_c(1^1S_0)$, $J/\psi(1^3S_1)$, $\chi_{c0}(2^3P_0)$, $\chi_{c1}(2^3P_1)$, $\chi_{c2}(2^3P_2)$ and $\psi'(2^3S_1)$. The $\eta_c'(2^1S_0)$ has been seen only by one experiment and the $h_c(2^1P_1)$ has eluded detection until recently.

Charmonium states can also be formed in proton-antiproton annihilations, where they can be easily detected, despite the presence of a much larger hadronic background, by their characteristic decay into an electromagnetic final state, a high mass $e^+e^-$ or $\gamma\gamma$ for instance. A significant advantage of the study of charmonium in $\bar{p}p$ collisions is that the full spectrum of states can be resonantly produced, in contrast with the $e^+e^-$ collider case. This makes the direct study of the $\chi_c$ states and searches for previously unobserved states, such as the $h_c(2^1P_1)$ feasible.

Suitably intense antiproton beams with small energy spread became available during the last 10 years as "antiproton sources", feeders to high energy $\bar{p}p$ colliders. Antiproton sources are modest sized storage rings where antiprotons are collected and stochastically cooled. With the introduction of an internal hydrogen gas jet target in such a storage ring, one can create $\bar{p}p$ collisions at a center-of-mass energy equal to the mass of a charmonium state to be studied.

During such $\bar{p}p$ annihilations *all* the charmonium states can be produced directly and precision measurements of the mass and width of a state can be obtained by varying the energy of the circulating antiproton beam. In such a resonant formation experiment one measures the energy of the *initial* rather than the final state produced. This unique situation permits measurements of extremely good energy resolution since an antiproton source is an exquisite "spectrometer". In other words the precision of the measurement is no longer limited by the detector resolution.

The pioneering work of charmonium spectroscopy with antiprotons was carried at the Intersecting Storage Ring at CERN by the R704 collaboration[2]. The current effort by E760 is a second generation experiment at Fermilab[3].

In this experiment antiprotons circulating inside the Fermilab Antiproton Accumulator with momenta range of 3.5-6.0 GeV intercept an internal hydrogen gas jet target of thickness $5 \times 10^{13}$ atoms/cm². At the start of a run up to $4 \times 10^{11}$ antiprotons were circulating in the Accumulator, with a peak luminosity of $10^{31}$cm$^{-1}$s$^{-1}$. The small dimension of both the gas jet target and the circulating beam produced an effective interaction region of typical size of 0.5 cm.

The antiproton beam in the Accumulator was stochastically cooled during the experiment. Stochastic cooling counteracts the effects of the interaction of the beam with the gas jet target and with the residual gas inside the accelerator vacuum vessel. As a result the momentum spread of the beam is $\Delta p/p \simeq 2 \times 10^{-4}$(r.m.s.) which correspond to a resolution of 0.5 MeV (FWHM) at the center-of-mass energy.

## 3   Experimental Technique

A specific charmonium state was studied by adjusting the energy of the circulating antiprotons in small steps at the energy range of interest. The number of the characteristic electromagnetic decays of the state thus produced, high mass $e^+e^-$ or $\gamma\gamma$, as a function of the energy of the center-of-mass was used to extract the mass and the total width of the resonantly produced state.

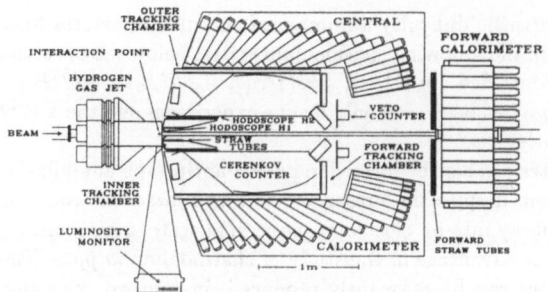

Figure 3: The E760 detector

The E760 detector is shown in Fig. 1. in a cross sectional projection perpendicular to the plane of the circulating antiproton beam[4].

The detector covers the polar angular range from 11-70 degrees and the full azimuth. It is a non-magnetic spectrometer optimized for efficient detection of electromagnetic final states and for discrimination against the hadronic background. Its major component is a cylindrical arrangement of 1280 lead glass counters, the central calorimeter (CCAL). A sandwich lead-scintillator calorimeter covers the forward polar range from 2-11 degrees. Two scintillator hodoscopes (H1, H2) identify charged particles and a threshold gas Čerenkov counter tags electrons. Cylindrical drift and wire chambers complete the inner part of the apparatus.

The luminosity was continuously monitored by measuring small angle $\bar{p}p$ elastic scattering with a set of solid state silicon detectors.

# 4   Analysis and Results

An $e^+e^-$ event is defined by the proper coincidence of the Čerenkov, the hodoscope and by the deposition of energy in the CCAL in two well defined clusters separated by more than 90 degrees in the azimuth. Similarly a $\gamma\gamma$ object is identified by imposing the appropriate veto conditions in the charge particle detectors while requiring two clusters in the CCAL.

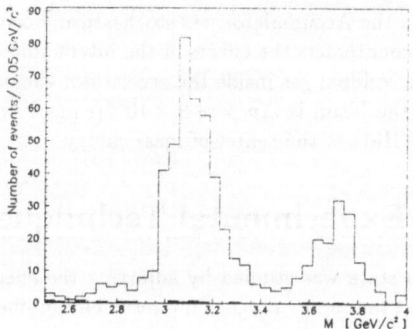

Figure 4: Invariant mass distribution for $e^+e^-$ events collected at the $\psi'$ for a typical "one-fill" run of E760 for a typical integrated luminosity $\simeq 1$ pb$^{-1}$

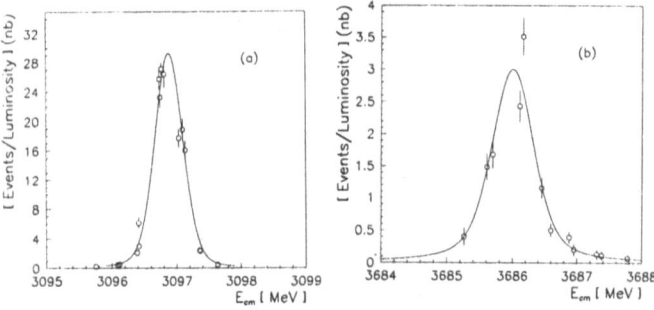

Figure 5: Excitation curves:(a) at the $J/\psi$ and (b) at the $\psi'$

The energy of the center of mass is calculated from the momentum of the circulating beam, which is turn is deduced from the revolution frequency of the stored antiprotons. We calibrated the length of the central orbit of the Accumulator by running at a resonance with precisely known mass. As reference we picked the $\psi'$, which mass has been measured to $\Delta M_{\psi'} = \pm 100$ keV[7]. With this technique we calibrated the 475 m long central orbit of the Accumulator to $\pm 0.67$ mm. This 0.67 mm uncertainty in the central orbit introduced in turn a systematic error of 33 keV at the measurement of the $J/\psi$ mass.

We kept the central orbit of the circulating beam constant to $\pm 1$ mm as we changed the momentum of the beam; this uncertainty introduced a 50 keV statistical error in the mass measurement of the $J/\psi$. The revolution frequency of the beam was measured to an accuracy of 1 part in $10^7$. The revolution frequency spectrum of the beam was used to determine the momentum spread of the circulating antiprotons[5]

The observed excitation curves is a convolution of the natural Breit-Wigner cross section of the resonant state with the energy distribution function of the beam[6]. Using the energy distribution of the beam, one can extract from the excitation curves the mass of the resonance, $M_R$, and its width, $\Gamma_R$. Furthermore, if the detector efficiency and acceptance is known, one can calculate the product $B_{\bar{p}p} \times B_{out}$, the branching ratio of the formation of the resonance times the branching ratio of the resonance decaying into the specific detection channel. We applied the techniques outlined above to study the charmonium states $J/\psi$, $\psi'$, $\chi_{c1}$ and $\chi_{c2}$[8][9]. We measured the excitation curves of these

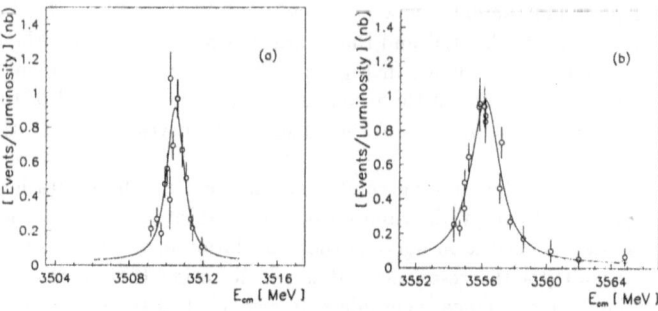

Figure 6: Excitation curves: (a) at the $\chi_{c1}$ and (b) at the $\chi_{c2}$

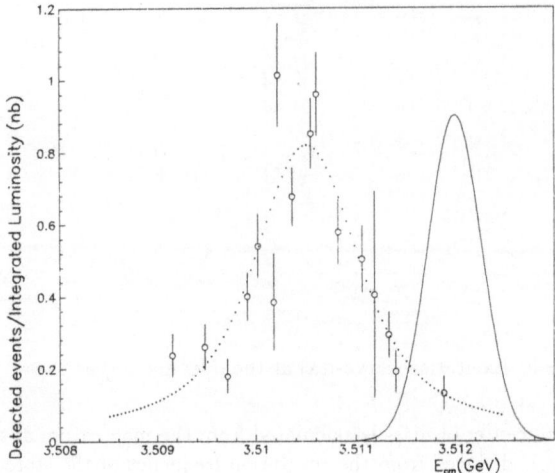

Figure 7: The $\chi_{c1}$ excitation curve. Open circles represent the data with their errors; the dotted line is the best fit. The solid curve on the right represents the typical energy distribution of the antiproton beam

resonances by the following decays:

$$J/\psi \to e^+e^-$$

$$\psi' \to e^+e^-$$

$$\psi' \to J/\psi X \to e^+e^-X$$

$$\chi_{c1,c2} \to J/\psi\gamma \to e^+e^-\gamma.$$

The decay channels take full advantage of the excellent discrimination power of our apparatus against hadronic background and have relatively large branching ratios.

The invariant mass distribution for $e^+e^-$ events collected at the region of the $\psi'$ is shown in Fig. 4. One sees a peak at 3.7 GeV due to the direct decay $\psi' \to e^+e^-$ and another peak at 3.1 GeV due to the inclusive decay channel $\psi' \to J/\psi X \to e^+e^-X$. The shaded area represents the expected background as determined from events collected at $\sqrt{s}$=3.667 GeV, far away from any resonances, normalized to the same integrated luminosity ($\simeq 1$ pb$^{-1}$) as the data show.

The excitation curves fo the $J/\psi$ and the $\psi'$ are shown in Fig. 5 while the excitation curves of the $\chi_{c1}$ and $\chi_{c2}$ are shown in Fig. 6. These curves are, as mentioned, the convolution of the natural width of the resonant with the energy spread of the probing beam. From a shape analysis of these of these excitation curves the mass, $M_R$ and the width, $\Gamma_R$ is extracted[6].

For illustration of the data taking conditions, in Fig. 7. the excitation curve of $\chi_{c1}$ along with a typical energy distribution of the antiproton beam, is shown. The experimentally obtained excitation curve is the convolution of the beam energy spread with the natural width of the resonance. It is worth noting that in the case of $J/\psi$ and $\psi'$, the width of the resonance is considerably *smaller* than the beam energy spread. This fact makes the excitation curves of these two resonances, in Fig. 5., to look very similar, widthwise. The measurements of the masses, widths and branching ratios for four charmonium resonances are summarized in Table 1. For comparison the previous to

Table 1: Charmonium resonance parameters as measured by E760. The quantities in square brackets refers to the previous state of the science measurements as reported by the Particle Data Group

| Resonance | $M_R$ (MeV) | $\Gamma_R$ (keV) | $B(R \to \bar{p}p) \times 10^{-4}$(keV) |
|-----------|-------------|------------------|------------------------------------------|
| $J/\psi$ | $3096.87 \pm 0.03 \pm 0.03$ | $99 \pm 12 \pm 6$ | $18.2^{+3.1}_{-2.6}$ |
|           | $[30096.93 \pm 0.09]$ | $85.5^{+6.1}_{-5.8}$ | $21.6 \pm 1.1$ |
| $\chi_{c1}$ | $351053 \pm 0.04 \pm 0.12$ | $880 \pm 110 \pm 80$ | $0.85 \pm 0.10 \pm 0.11$ |
|             | $[\ 3510.6 \pm 0.5\ ]$ | $[\leq 1300\ (CL\ 95\%)]$ | $[\geq 0.54\ (CL\ 95\%)]$ |
| $\chi_{c2}$ | $3556.15 \pm 0.07 \pm 0.12$ | $1908 \pm 170 \pm 70$ | $0.99 \pm 0.09 \pm 0.08$ |
|             | $[3556.3 \pm 0.4\ ]$ | $[2600^{+1200}_{-900}\ ]$ | $[0.97^{+0.52}_{-0.32}\ ]$ |
| $\psi'$ | input | $306 \pm 36 \pm 16$ | $2.61^{+0.39}_{-0.36}$ |
|         | $[3686.00 \pm 0.10]$ | $[278 \pm 32]$ | $[1.9 \pm 0.5]$ |

E760 state of the science is indicated by the figures in square brackets, as reported by the Particle Data Group[7]. For more details on the E760 measurement of these quantities see [8][9].

# 5  Conclusions

Heavy quarkonia are the simplest two body systems known in QCD. Advancements in accelerator technology made possible the use of intense antiproton beams for resonant production of charmonium states. With a carefully optimized but otherwise straighforward detection technique, the electromagnetic decays of these resonantly produced states can be clearly distinguished from the hadronic background. E760 has proven that the technique works well and produced a new set of very precise measurements. Apart from the results reported here with the same method and the same detector we have :

- Searched and reported evidence for the long sought $h_c(2^1P_1)$ charmonium state.

- Measured the partial width of the decay $\chi_{c2} \to \gamma\gamma$

- Measured the proton electromagnetic form factor in the time-like region by studying the reaction $\bar{p}p \to e^+e^-$

- Measured the exclusive two neutral meson production $\bar{p}p \to m_1m_2 \to$ multi-$\gamma$, where $m_{1,2} = \pi^0, \eta, \omega$.

- Searched for light neutral mesons $X$, produced via the reaction $\bar{p}p \to Xm_1 \to m_2m_3m_1 \to 6\gamma$

As it turns out, the mass region around charmonium is a fertile area for hadronic interactions but yet not well studied. During the study of $c\bar{c}$, the simple two body QCD system, there is hope that much more interesting objects can be found ( glue-glue, quark-quark-glue, mesonic molecules and other QCD exotica).

# 6  Acknowledgements

We thank the organizers for their invitation, hospitality and a very efficiently run Conference. E760 is the collaborative work of many talented and hard working people; they will understand why I choose to especially thank Professor Rosanna Cester for her tireless efforts. The research of E760 was supported in part by the U.S. Department of Energy, the U.S. National Science Foundationa and the Italian Istituto Nazionale di Fisica Nucleare.

# References

[1] For a comprehensive review of the state of the science on electron-positron physics circa 1980 see: "$e^+e^-$ Annihilations: New Quarks and Leptons", A volume in the Annual Reviews Special Collections Programme, Robert N. Cahn, ed., Benjamin-Cummings Co, Menlo Park, 1985.

[2] C. Baglin et al., Nucl. Phys. **B 286**, 592 (1987).

[3] J. Peoples, Jr., in Low Energy Antimatter: Proceedings of the Workshop on the Design of a Low Energy Antimatter Facility, D. Cline, ed., Madison,WI, 1985 ( World Scientific, Singapore, 1986), p 144.

[4] Details on the E760 detector can be found: L.Bartoszek et al., Nucl. Instrum Methods **A301**, 47 (1991); R. Ray et al., Nucl. Instrum. Methods **A307**, 254 (1991).

[5] T. Bagwell et al., "Antiproton Source Beam Position System", Fermilab Technical Memo 1254, 1984 ( unpublished).

[6] D. H. Wilkinson, Nucl. Instrum. Methods **95**, 259 (1971).

[7] K. Hikasa et al., " Review of Particle Properties" Phys. Rev. D, **45**, S1 (1992).

[8] T. A. Armstrong et al., (E760 Collaboration) Nucl. Phys. **B373**, 35, (1992)

[9] T. A. Armstrong et al., (E760 Collaboration) Phys. Rev D., **47**, 772, (1993).

Few-Body Systems, Suppl. 7, 425—432 (1994)

# QUARKS IN FEW HADRON SYSTEMS

P. González
Universitat de València and I.F.I.C.
E-46100 Burjassot (Valencia) Spain.

V. Vento [1]
Universitá degli Studi di Trento and I.N.F.N.
I-38050 Povo (Trento) Italy.

We make use of QCD-based quark models to analyze the observability of quark effects in few hadron systems. A hadron is described by a two phase picture. The dynamics of the interior (perturbative) phase consists of a mechanism to confine the elementary degrees of freedom and an asymptotic QCD type interaction. The dynamics of the exterior (non perturbative) phase is approximated by an effective mesonic lagrangian.

The aim of our presentation is to discuss effects that occur when the perturbative phases of two or more hadrons overlap. We describe the symmetry requirements in the two hadron wave function arising from the quark substructure. Due to the antisymmetrization at the quark level *Quark Exchange* phenomena become relevant giving rise to signatures whose observability we analyze. We look at their consequences in form factors and structure functions, paying special attention to exotic configurations where *Quark Pauli* blocking occurs and one might expect large effects. We study the contribution of *Quark Exchange* terms to the different pieces of the hadron-hadron interaction in a Constituent Quark Cluster Model, with specific application to $\Delta$ production mechanisms.

## 1. Introduction

Quantum Chromodynamics ($QCD$) is believed to be the theory of the strong interactions [1]. This theory is described in terms of quarks and gluons. However, these elementary degrees of freedom only become significant in the form of jets in the very high energy domain. At low and intermediate energy the theory is highly non perturbative and

---

[1] On leave of absence from Universitat de València and I.F.I.C.

effective theories in terms of non elementary degrees of freedom, i.e., constituent quarks, mesons, baryons, ..., have been developed to describe the experimental data.

Among the effective theories those based on the two phase scenario have met with a great deal of success. They assume that a hadron defines a hypertube which divides space-time in two regions. The interior region is described by a model for confined $QCD$, while the dynamics of the exterior region is attributed to an effective mesonic theory [2]. In this way, one aims at describing effects due to the short distance behavior of the theory, associated with the elementary degrees of freedom (color, quarks, gluons), and therefore difficult to describe by effective hadronic (flavor) theories.

Our dynamical description here will be based on the naive quark model [3] for the interior and the sigma model [4] for the exterior [5,6]. The coupling of quarks to pions will be included whenever the long range potential between hadrons becomes important. This complication will be avoided when the aim is the description of short distance (overlap) effects. The presentation will be centered on the quark degrees of freedom and how their elementary properties (spin, flavor and color) affect the structure and the dynamics of hadrons.

Knowing the present limitations of our understanding of $QCD$ in the non perturbative regime our goal, in the past years, has been the search for phenomena where the subnucleonic degrees of freedom become manifest and the more conventional effective theories break down. We have assumed as guiding principle, that effects associated with fundamental principles of the constituents, like the Pauli principle at the quark level, should be instrumental [7,8]. Therefore *Quark exchange* contributions in different realms of hadron physics will be the subject of our presentation.

## 2. The symmetrization principle

Microscopical physics is governed by the symmetrization principle which establishes that pure states of a system of identical particles have to be either totally symmetric or antisymmetric under the exchange of any two of them. A consequence of this principle and some very general assumptions is the spin-statistics connection theorem: pure states of a system of identical particles are totally symmetric (antisymmetric) if their spin is integer (half integer). The spin-statistics theorem is generally an accepted truth for all elementary particles, although not much experimental evidence exists for many of them [9].

Quarks are fermions and, as such, obey their own symmetrization principle. On the other hand hadrons are made up of quarks. Therefore within any scheme which describes the hadronic wave function in terms of quarks, their symmetrization principle is active. The wave function of any nuclear system becomes very complicated due the large number of particles involved (3A). The hadronic quark cluster decomposition (HQCD) is ideally suited for dealing with the large numbers of terms, ($O(3A!)$), that arise due to quark antisymmetrization. Moreover it has clarified the relation between hidden color components and the quark exchange terms [10].

At intermediate energies besides the nucleon, other baryonic components may be relevant, in particular the $\Delta$ isobar plays a relevant role in the description of nuclear processes [11] and the $\Lambda, \Sigma$ baryons in the study of hypernuclei [12]. Since these states are different from the nucleon no baryonic Pauli principle is active. On the other hand nucleons, deltas and the other baryonic states are made up of quarks and therefore their wave function has to obey the quark Pauli principle. We have analyzed the remnants of quark level antisymmetrization in systems composed of two non identical hadrons [13]. The basic result from our analysis is that the wave function of two non identical hadrons [14]

$$|\Psi^{ST}_{B_1 B_2}(R)> = \frac{A}{\sqrt{1+\delta_{B_1 B_2}}} \frac{1}{\sqrt{2}} \{ [\Phi_{B_1}(123, -\frac{R}{2})\Phi_{B_2}(456, +\frac{R}{2})]_{ST}$$

$$+ (-1)^{\mu}(-1)^{S_1+S_2-S+T_1+T_2-T}[\Phi_{B_2}(123, -\frac{R}{2})\Phi_{B_1}(456, +\frac{R}{2})]_{ST}\}, \tag{1}$$

vanishes unless $\mu + L = odd$. Here $A$ is the six quark antisymmetrizer and $B_1$ ($B_2$) denotes a baryon characterized by a definite spin $S_1$ ($S_2$) and isospin $T_1$ ($T_2$). Furthermore $\mu$ describes the symmetry in the spin-isospin degrees of freedom and $L$ is the total angular momentum. In the case of identical hadrons $\mu = S + T$ and we recover the familiar formulation $L + T + S = odd$. In the case of non identical baryons the remnant has been called Hidden Antisymmetrization Principle and has given rise to selection rules for decays [13] and a strong hard core like behavior of the phase shifts in hadron-hadron interactions [15]. The treatment can be immediatly extended to include strange baryons if one assumes SU(3) flavor symmetry and allows for the study of SU(3) breaking terms in a straightforward manner [13].

## 3. Unveiling the quark structure in nuclei

In order to study quark effects in nuclear systems a convenient mathematical tool is HQCD [10]. The natural implementation of the dynamics in this scheme is by considering that the center of the cluster is subject to the nuclear force, i.e., long range $QCD$ dynamics, and the motion of the quarks inside the cluster is governed by some potential whose origin is linked to short range $QCD$ dynamics and confinement.

Let us analyze the effects of quark antisymmetrization in a naive example [7,16]. We consider the elementary constituents to be quarks with spin and color but not flavor. Baryons are color singlets as in the real world. The dynamics governing the motion of the quarks is such that the baryon wave function turns out to have a symmetric spatial part. Finally, in order to simplify the algebra, we assume the world to have one spatial dimension. Despite these simplifications all the possible physical scenarios arise in this model.

The wave function for the lowest lying baryons are

$$P = R \otimes |\uparrow\uparrow\uparrow> \otimes C, \tag{2}$$

$$N = R \otimes \frac{1}{\sqrt{3}}\{|\uparrow\uparrow\downarrow> + |\uparrow\downarrow\uparrow> + |\uparrow\uparrow\downarrow>\} \otimes C, \tag{3}$$

$$M = R \otimes \frac{1}{\sqrt{3}}\{|\uparrow\downarrow\downarrow> + |\downarrow\uparrow\downarrow> |\downarrow\downarrow\uparrow>\} \otimes C, \tag{4}$$

$$Q = R \otimes |\downarrow\downarrow\downarrow> \otimes C, \tag{5}$$

where $C$ represents the singlet wave function in color space, $R$ the symmetric spatial wave function and we have written out explicitly the spin parts of the wave functions.

In terms of these baryonic states let us construct, for reasons which will be clear later on, the following nuclear systems of baryon number two,

$$D = |S = 2; S_z = +2> = \frac{1}{\sqrt{2}}(PN - NP), \tag{6}$$

$$E = |S = 0; S_z = 0 > \frac{1}{2}(PQ - QP + MN - NM). \tag{7}$$

Their wave functions in terms of quarks have a complicated structure with 6! terms each. The use of HQCD simplifies notably their description by taking benefit of the well known properties of the baryonic clusters. The calculation for different observables has been discussed in detail in refs.([7,16]). Let us study the expectation values of the following one body operator

$$\theta = \sum_i e_i \delta(x - x_i) \tag{8}$$

where $e_i = \frac{1}{6} + s_{zi}$. The result of the calculation is

$$<\theta>_D = 3\frac{\theta_1 - \frac{2}{9}\theta_{21} - \frac{7}{9}\theta_{22}}{N_1 - N_2} \tag{9}$$

$$<\theta>_E = \frac{\theta_1 + \frac{11}{9}\theta_{21} - \frac{7}{9}\theta_{22}}{N_1 + \frac{4}{9}N_2} \tag{10}$$

where $\theta_1, \theta_{21}, \theta_{22}, N_1$ and $N_2$ depend on the spatial structure of the wave function.

Eqs.(9) and (10) contain all the physics associated with one body observables. If we neglect the overlap terms (subscript 2) our results correspond to a conventional nuclear physics calculation where structure has been incorporated to the pointlike baryons by convoluting with the form factor determined by the model. The overlap terms ($\theta_{21}, \theta_{22}$ and $N_2$) arise due to the antisymmetrization in the quark degrees of freedom and therefore *cannot be introduced in a naive way into any conventional theoretical framework in terms of hadronic degrees of freedom.*

The $D$ and $E$ nuclei have been chosen because they lead to very different results. The $D$ nucleus consists of many quarks with the same spin quantum numbers: there are five quarks with spin projection $\uparrow$. This fact leads to a strong non-dynamical Pauli blocking phenomenon and, if the clusters overlap considerably, to a leakage of charge from the origin. On the contrary, in the $E$ nuclear system no such blocking occurs and, no matter how large the cluster overlap in the wave function, the observable effect is small. It is worth stressing that there is no dynamics associated with these divers behaviors, since in going from the state $D$ to the state $E$ only the spin changes.

This idea has been developed in a series of more realistic cases [17], e.g., form factors of light nuclei [7,8,16,18] and heavy nuclei [19], Coulomb energies (two body operators) [16] and exotic nuclei (delta states, hypernuclei) [20,21]. Unluckily we have not been able to find a reasonable experimental scenario where nuclear quark effects are undisputable. The search for a maximal Pauli blocked scenario, which cannot be described in terms of effective hadronic theories, still continues.

The formalism has been extended to describe deep inelastic lepton nucleus scattering. Quark exchange effects in the nuclear structure function have been analyzed [22,23,24]. They cannot be reproduced in the conventional nuclear convolution picture. Their magnitude is comparable to the other nuclear effects. Therefore they should not be neglected in the analysis. Moreover, in the region $x > 1$, they might play an important role [24].

## 4. Quark exchange in hadron collisions

The ideas just described have been applied to proton collisions with delta channels [15]. In this case however the actual calculation requires a dynamical input and model dependence is unavoidable. Our starting point is the constituent quark model [3] in an HQCD scenario. When baryons are considered as quark clusters, the HQCD, named in this dynamical context Constituent Quark Cluster Model (CQCM), can also be applied to the description of baryon-baryon interactions. In order to reproduce the nucleon-nucleon phenomenology the model needs the additional ingredient of the mesons in the external phase. A self-consistent way to incorporate the pion exchange between quarks can be provided, for example, by the instanton liquid model of the $QCD$ vacuum [25]. In this model the pion is associated with the spontaneous breakdown of chiral symmetry and a fundamental $\pi qq$ coupling arises. Using this long range interaction supplemented with a medium range attraction the model reproduces the properties of the nucleon and some hadron-hadron phenomenology [26].

We have derived in the framework of the CQCM and following the Born-Oppenheimer method an $NN - N\Delta$ transition potential which incorporates the underlying quark dynamics and quark antisymmetrization principle. In the $p(p, \Delta^{++})n$ case, the most relevant effect in the cross section arising from quark antisymmetry is the inclusion of projectile as well as target excitations. The same formalism allows a description of the $p(n, p)n$ reaction. In this case, an additional quantitative effect comes from quark exchanges that produce a sizeable modification of the central $OPE$ potential at short range, but mantain the tensor $OPE$ unchanged. This allows a simultaneous explanation of the $p(n, p)n$ and the $p(p, \Delta^{++})n$ cross sections, a feature which requires a rather artificial mechanism within the conventional mesonic approach [27].

## 5. Conclusions

During the past years we have come to understand why, despite the large size of baryons and mesons, the low and intermediate energy region is well described by effective hadronic theories. Color adds so many internal degrees of freedom to the quarks that situations, where Pauli blocking could signal the existence of substructure, are rare and difficult to explore experimentally. However, if we want to unveil the origin of nuclear structure and dynamics from the point of view of the more fundamental theory, the search for exotic behaviors has to continue.

Thus far, and for obvious reasons, the calculation of these processes has been rather primitive. Certainly to go beyond what has been done requires more support from the theory. The effective field theory formalism has been carried out with notable success for hadronic effective theories [28]. An effort of understanding $QCD$, without integrating out the color degrees of freedom, is required. This would imply a deep knowledge of its vacuum and confinement mechanism.

In the high energy regime, nuclear effects have opened up a whole new field of research. The observation of quark exchanges there looks very promising. In particular, in deep inelastic lepton scattering off nuclei, the region $x_{Bjorken} > 1$ seems ideal to study overlap effects. Moreover, a precise knowledge of the nuclear structure function around the two nucleon correlation region, may allow the separation of single particle mechanisms, i.e., *swelling*, from *in medium* effects on the nuclear interaction [24].

Although models of quark dynamics have proven quite succesful in describing both hadron structure and hadron-hadron collisions, one should not proceed indefinitely in this direction by adding *ad hoc* parameters and mechanisms. Solutions of $QCD$, obtained with well defined approximation schemes, should be used to generate the quark dynamics [25]. Recent developments in two dimensions have shown that this might be possible in the true theory [29].

## Acknowledgement

The presentation relies on work done in collaboration with F. Fernández, A. Ferrando, V. Sanjosé, M. Traini and A. Valcarce. I thank R. Bosch and G. Cognola for their assistance in preparing the manuscript. Research supported in part by Cicyt grant # AEN90-0040 and Dgicyt grant # PB91-0131-C02-01.

## References

1. Fritsch, H., Gell-Mann, M., Leutwyler, H. : Phys. Lett. 47B, 365 (1973); Muta, T. : Foundations of Quantum Chromodynamics. Singapore, World Scientific 1987.

2. Vento, V., Rho, M., Nyman, E.M., Jun, J.H., Brown, G.E. : Nulc. Phys. A470, 413 (1980).

3. De Rújula, A., Giorgi, H., Glashow, S.L. : Phys. Rev. D12, 147 (1975); Alvarez-Estrada, R. F., Fernández, F., Sánchez Gómez, J.L., Vento, V. : Lecture Notes in Physics 259, Heidelberg, Springer 1986.

4. Gell-Mann, M., Levy, M. : Nuov. Cim. 16, 705 (1958); de Alfaro, V., Fubini, S., Furlan, G., Rossetti, C. : Currents in Hadron Physics. Amsterdam, North-Holland 1973.

5. Navarro, J., Vento V. : Phys. Lett. 140B, 6 (1984); Phys. Lett. 141B, 28 (1984); Nucl. Phys. A440, 617 (1985).

6. Kvitsinskii, A.A., Kuperin Yu. A., Merkur'ev, S.P., Yarevskii, E.A. : Sov. Jour. Nucl. Phys. 51, 141 (1990).

7. González, P., Sanjosé, V., Vento, V. : Phys. Lett. B196, 1 (1987).

8. Toki, H., Suzuki, Y., Hecht, K.T. : Phys. Rev C26, 736 (1982); Maltman, K. : Nucl. Phys. A439, 648 (1985); Takeuchi, S., Shimizu, K., Yazaki, K. : Nucl. Phys. A449, 617 (1986); Hoodbhoy, P. : Nucl. Phys. A465, 637 (1987); de Forest, T., Mulders, P.J. : Phys. Rev. D35, 2849 (1987); Yazaki, K., Walcher, Th. (ed.) : Proc. Int. Conf. on a european hadron facility. Amsterdam: North Holland 1987.

9. Pauli, W. : Z. Phys. 31 765 (1925); A. Böhm : Quantum Mechanics. Berlin, Springer 1979; Streater R.F., Wightman, A.S. : PCT, spin, statistics and all that. New York, Benjamin 1964; Messiah, A.M.L., Greenberg. O.W. : Phys. Rev. B135, 1447 (1964); B138, 1115 (1968).

10. González, P., Vento, V. : Few Body Systems 2, 145 (1987); Nuovo Cim. 105A, 795 (1992).

11. Brown, G.E., Weise, W. : Phys. Reports 22, 280 (1975); Oset, E., Toki, H., Weise, W. : Phys. Reports 83, 281 (1982).

12. Oset, E., Fernández de Córdoba, P., Salcedo, L.L., Brockmann, R. : Physics Reports 180, 79 (1990).

13. González, P., Vento, V.: Phys. Rev. Lett. 60, 190 (1988). Fernández, F., Valcarce, A., González, P., Vento, V. : Phys. Rev. C47, 1807 (1993).

14. Suzuki, Y., Hecht K.T. : Nucl. Phys. A420, 525 (1984)

15. Fernández, F., Valcarce, A., González P., Vento, V. : Phys. Lett. B287, 35 (1992); Nucl. Phys. A (to appear).

16. González, P., Vento, V. : Nucl. Phys. A501, 710 (1989).

17. Mulders, P.J. : Phys. Rep. 185, 83 (1990).

18. Dieperink, A.E.L., Mulders, P.J. : Nucl. Phys. A483, 461 (1988); Nucl. Phys. A489, 627 (1988).

19. Spit, W.F.M., van Hees, A.G.M., Mulders, P.J. : Nucl. Phys. A510, 609 (1990).

20. González, P., Vento, V. : Nucl. Phys. A485, 413 (1987).

21. Yamazaki, T. : Nucl. Phys. A446, 467c (1987).

22. Hoodbhoy, P., Jaffe, R.L. : Phys. Rev D35, 113 (1987); Arizuzaman, Hoodbhoy P., Mahmood, S. : Nucl. Phys. A480, 469 (1988)

23. Spit, W.F.M., van Hees, A.G.M., Brussard, P.J., Mulders P.J. : FTUV/92-33; Meyer, H., Mulders, P.J., Spit, W.F.M. : nikhef 92.6.

24. González, P., Traini, M., Vento, V. : U.T.F./93 and work in preparation.

25. Shuryak, E. : Phys. Rep. 115, 151 (1984).

26. Faessler, A., Fernández, F., Lübeck, G., Shimizu, K. : Nucl. Phys. A402, 555 (1983); Bräuer, K., Faessler, A., Fernández, F., Shimizu, K. : Nucl. Phys. A507, 599 (1990); Buchmann, A., Hernández, E., Yazaki, K. : Phys. Lett. B269, 35 (1991).

27. Jain, B.K., Santra, A.B. : Nucl. Phys. A519, 697 (1990); Phys. Rev. C46, 1183 (1992).

28. Weinberg, S. : Physica (Amsterdam) 96A, 327 (1979); Phys. Lett. B251, 288 (1990); Gasser, J., Leutwyler, H. : Phys. Rep. 87, 77 (1982); Ann. Phys. (N.Y.) 158, 142 (1984); Nucl. Phys. B307, 763 (1988); Ecker, G., Gasser, J., Pich A., de Rafael, E. : Nucl. Phys. B312, 311 (1989); Meissner, U. : Phys. Rep. 161, 213 (1988); Espriu, D., de Rafael, E., Taron, J. : Nucl. Phys. B345, 22 (1990).

29. Ferrando, A. and Vento, V. : Z. Phys. C58, 133 (1993); FTUV92-3; Ellis, J., Frishman, Y., Hanany, A., Karliner, M. : Nucl. Phys. B382, 189 (1992).

Few-Body Systems, Suppl. 7, 433—440 (1994)

# PIONS AND NEUTRINOS AS PROBES OF THE NUCLEON AND NUCLEAR FEW-BODY SYSTEM

Cyrus M. Hoffman
Medium Energy Physics Division
Los Alamos National Laboratory
Los Alamos, NM 87545

A number of studies of few-body nucleon and nuclear systems are being pursued at LAMPF using pions and neutrinos as probes. These include: a high-statistics measurement of the cross section for the reaction $\pi^+ p \rightarrow \pi^+ \pi^0 p$ to determine the $I = 2$ $\pi\pi$ scattering length; precision measurements of the $\pi^- p$ charge exchange cross section in the region of the $\Delta$; measurements of the cross section for the $(\pi^{\pm}, \pi^{\pm} p)$ from D, $^3$He, and $^4$He as tests of charge symmetry and reaction mechanisms; measurements of the $\pi^+ p$ cross section below the $\Delta$; and studies of $\nu p$ elastic scattering at low $Q^2$ to determine the quark content of the proton spin. Some of these experiments acquired data last year, while others are presently running. LAMPF, the highest power proton accelerator in the world, is presently in the midst of its 1993 production run.

## The Reaction $\pi^+ p \rightarrow \pi^+ \pi^0 p$ Near Threshold and Chiral Symmetry Breaking

Quantum chromodynamics (QCD) is widely accepted as the theory of strong interactions mainly because of its success in describing processes at large momentum transfer (i.e., short interaction distances). At low energies, the coupling constant becomes large and QCD becomes nonperturbative so calculations become intractable. Chiral symmetry is an important feature of low-energy hadron dynamics and can form the framework for effective Lagrangian models that allow many low-energy properties of strongly-interacting systems to be calculated. The nature of chiral-symmetry breaking can be studied with measurements of the $\pi\pi$ scattering amplitudes at zero momentum, which vanish in the chiral limit. The $\pi\pi$ scattering amplitudes can be probed by studies of the reactions $\pi N \rightarrow \pi\pi N$ at threshold. Recently Olsson and Turner [1] showed that, with rather general assumptions, the nature of the chiral-symmetry breaking can be characterized by a single parameter, $\xi$. Several recent measurements of $\pi N \rightarrow \pi\pi N$ near threshold led Burkhardt and Lowe [2] to perform a comprehensive soft-pion analysis to extract a value for $\xi$. However, the available data for the $\pi^+ p \rightarrow \pi^+ \pi^0 p$ channel below 300 MeV/c were too poor in quality to constrain the global fit. Consequently, a group from Virginia, Stanford,

434

and Los Alamos [3] performed an experiment at the low-energy pion channel at LAMPF to study $\pi^+ p \to \pi^+ \pi^0 p$ for pion kinetic energies from 190 MeV to 260 MeV (the threshold is 164.8 MeV).

All of the final-state particles were detected. Photons from the neutral pion decays were detected in the LAMPF $\pi^0$ spectrometer [4], while the protons and charged pions were detected in a telescope array of plastic scintillator detectors. Each telescope consisted of thin (3 mm) and thick (25 mm) $\Delta E$ counters, and an absorption counter capable of stopping the pions from the desired reaction. In all, 56 distinct angular bins were covered simultaneously.

Events from $\pi^+ \pi^0$ and $\pi^0 p$ double coincidences and from $\pi^+ \pi^0 p$ triple coincidences were recorded. Detector acceptances, kinematics, and backgrounds are very different for these three types of events. The three data sets were analyzed separately, providing three nearly independent measurements and internal consistency checks. Comparisons with a Monte Carlo simulation show event distributions that are consistent with phase space at all energies, simplifying the evaluation of angle-integrated cross sections.

The total cross sections deduced from $\pi^+ \pi^0$, $\pi^0 p$, and $\pi^+ \pi^0 p$ coincidences are in good agreement with each other; a weighted average of these measurements is adopted as the final value. Figure 1(a) shows the measured total cross section as a function of beam energy. These data have been used in a new global fit to all five charge channels to obtain a new value of the chiral symmetry breaking parameter $\xi = -0.25 \pm 0.10$ [5]; the $\chi^2$ of the fit is 96 for 93 degrees of freedom. Figure 1(b) shows the constrained fit of the $\pi^+ \pi^0 p$ amplitudes: the four other channels show similar agreement with the data. Within the soft pion model, this fixes the $I = 0$

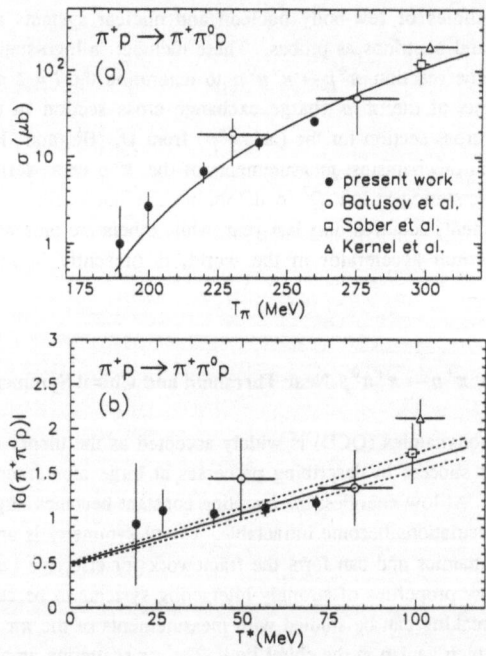

Fig. 1. (a) The total cross section for $\pi^+ p \to \pi^+ \pi^0 p$ measured in this work, and previously published results. The curve is the result of the new global fit of $\pi N \to \pi \pi N$ amplitudes. (b) The absolute value of the $\pi^+ \pi^0 p$ matrix element corresponding to the cross sections shown in (a), as a function of $T^*$, the total c.m. kinetic energy. The solid line is the global linear fit; the dashed lines show the uncertainties in the fit.

and 2 $\pi\pi$ s-wave scattering lengths to be $a_0^0(\pi\pi) = 0.177 \pm 0.006 \, \mathrm{m}_\pi^{-1}$, and $a_0^2(\pi\pi) = -0.041 \pm 0.003 \, \mathrm{m}_\pi^{-1}$, respectively.

Data analysis to obtain exclusive cross sections is proceeding. This should allow a model-independent $\pi\pi$ phase shift analysis.

### Precision Measurements of $\pi^- p$ Charge Exchange in the Region of the $\Delta$

The simple spin and isospin structure of the pion-nucleon system makes it an ideal source of information on fundamental hadron interactions. Assuming isospin invariance, the scattering amplitudes for the three observable reactions, ($\pi^+ p$ and $\pi^- p$ elastic scattering, and $\pi^- p$ charge exchange) are functions of only two independent isospin amplitudes, $F_{3/2}$ and $F_{1/2}$. The amplitudes can be expanded into partial waves. Each complex partial wave amplitude for isospin $I$, orbital angular momentum l, and total angular momentum $j = 1 \pm 1/2$, can be written in terms of two real variables, the phase shift and absorption parameter. $S$- and $P$-waves strongly dominate pion-nucleon scattering up to the $\Delta(1232)$. The phase shifts are strongly overdetermined in this regime, allowing a reliable treatment of electromagnetic corrections and a search for isospin-breaking effects.

The most extensive set of differential cross sections for $\pi N$ charge exchange in the region of the $\Delta$ were obtained by Jenefsky et al. [6], who made measurements from 128 to 246 MeV in the angular range $-0.98 \leq \cos\theta_{\pi^0} \leq 0.65$. This experiment detected neutrons only, and quoted total errors of 6–10%. Other data were obtained by several other experiments [7] over more restricted energy or angle regions. In addition, Bugg et al. [8] measured the total charge-exchange cross section from 90 to 290 MeV with quoted errors of 1%. The differential cross section data agree reasonably well with phase shift fits, but this is not true of the total cross section data. In addition, there are significant discrepancies in the elastic scattering data in this energy region. It would be particularly helpful to improve the accuracy of the charge exchange data to resolve the discrepancies [9].

Measurements of $\pi^- p$ charge exchange from 140 to 260 MeV are under way at LAMPF [10] using the new Neutral Meson Spectrometer (NMS) to detect neutral pions. The NMS is able to measure final-state $\pi^0$'s at all angles including 0°. The NMS is a large acceptance (2 msr), high resolution (<<1 MeV) device that consists of two arms used to detect the two photons from $\pi^0$ decay. The NMS allows a region of 40° in scattering angle to be covered at the same time. The measurement of $\pi^- p$ charge exchange does not require good $\pi^0$ energy resolution, so the photon converters are not used. The large acceptance and relatively large cross section (>1 mb/sr) imply that the statistical uncertainties will be small. Systematic uncertainties (such as the absolute beam normalization, detection efficiencies, etc.) should be no larger than a few percent.

Data were taken last year at 0° with pions of 138.8, 166, 190, 214.6, 237, and 263 MeV incident on $CH_2$ and C targets. Figure 2 shows very preliminary data from that run. Final corrections for the beam normalization and contamination, and the detection efficiency are still to be completed. Data taking is presently under way at 90° and 180°.

### Measurements of Exclusive $(\pi, \pi'x)$ Reactions on Light Nuclei

Studies of exclusive particle knockout reactions have been pursued to obtain information on particle clusters in nuclei and on reaction mechanisms. The strong isospin dependence of the

436

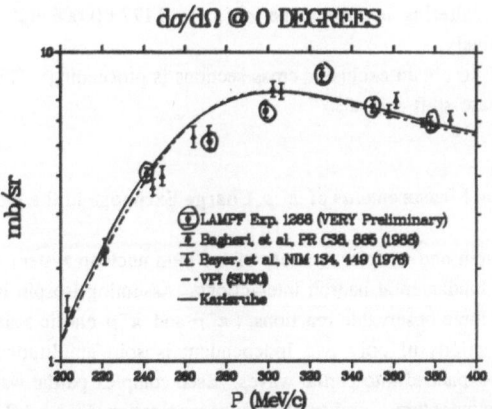

$d\sigma/d\Omega$ @ 0 DEGREES

P (MeV/c)

Fig. 2. Very preliminary data for the $\pi^- p$ charge exchange cross section in the forward direction as a function of pion energy from last year's run at LAMPF. The data points have been assigned systematic errors of 5%. Also shown are earlier data and predictions from two phase shift analyses.

elementary $\pi$-$N$ force in the region of the $\Delta$ allows measurements of the ratio of exclusive cross sections, $R_x = \sigma(\pi^+, \pi^{+'} x)/\sigma(\pi^-, \pi^{-'} x)$, to be used to uncover the reaction mechanism. If the $(\pi, \pi'p)$ reaction were to proceed exclusively by quasi-elastic $\pi p$ scattering, one would expect $R_p = 9$, reflecting the ratio of the elementary $\pi^+ p$ to $\pi^- p$ elastic scattering amplitudes. On the other hand, if the reaction were to proceed exclusively by quasi-elastic $\pi$-triton scattering, a ratio of nearly 0.5 would be expected; evidence for this latter mechanism in $\pi$-$^4$He scattering at 180 MeV was uncovered in an earlier experiment [11].

New measurements were made with incident $\pi^+$ and $\pi^-$ from 180 to 300 MeV on $^2$H and $^3$He, the simplest few-body nuclei. Some data on $^4$He were also taken to supplement earlier data. These targets minimize the effects due to nuclear structure while giving several different nuclei to compare reaction mechanisms. The experiment was performed by a group from Minnesota, Texas, LANL, Rutgers, New Mexico State, Colorado, and Tohuko, in the P$^3$ channel at LAMPF [12]. The experiment employed the Large Acceptance Spectrometer to detect the pions and a CsI-phoswich detector [13] for the recoil particles. The phoswich detector can identify recoil protons, deuterons, and tritons. Here I give some preliminary results for recoil protons.

Table 1 gives the ratio $R_p$ at the quasi-free knockout angle taken with a cooled-gas targets. $R_p$ is consistent with the quasi-elastic value of 9 for hydrogen but is significantly greater

Table 1. Preliminary results for $R_p = \sigma(\pi^+, \pi^{+'} p)/\sigma(\pi^-, \pi^{-'} p)$ with light targets.

| Target | $T_\pi$ (MeV) | $\theta_\pi^{lab}$ (deg) | $q$ (fm$^{-1}$) | $R_p$ |
|---|---|---|---|---|
| H | 240 | 32° | 0.95 | 10.2 ± 0.5 |
| H | 300 | 27° | 0.95 | 9.7 ± 0.7 |
| $^2$H | 180 | 40° | 0.95 | 13.4 ± 1.3 |
| $^2$H | 210 | 35° | 0.95 | 25.9 ± 3.1 |
| $^2$H | 240 | 32° | 0.95 | 23.7 ± 1.7 |
| $^2$H | 300 | 27° | 0.95 | 18.8 ± 1.8 |
| $^2$H | 300 | 44° | 1.50 | 8.4 ± 0.5 |
| $^4$He | 180 | 40° | 0.95 | 41.5 ± 16 |
| $^4$He | 240 | 32° | 0.95 | 31.5 ± 4.6 |

for both deuterium and helium between 180 MeV and 300 MeV for $q = 0.95$ fm$^{-1}$. Apparently, the residual core is not simply a spectator, even for the weakly-bound deuteron. $R_p \approx 9$ for the point with $q = 1.5$ fm$^{-1}$, indicating that $R_p$ is a function of momentum transfer, but it is apparently not a strong function of beam energy.

### Measurements of the Pion-Proton Integral Cross Section from 60 to 260 MeV

The first comprehensive measurements of $\pi^+ p$ scattering below 100 MeV [14] disagreed with predictions based upon extrapolations of established phase shifts. A number of subsequent measurements have failed to resolve this problem, as some measurements reported agreement with the predictions of the phase shift programs, and others reported differential cross sections substantially below these predictions. Recently, the group of Friedman *et al.* working at TRIUMF introduced a new technique of measuring "partial total" or integral cross sections to compare with the phase shift predictions [15]. These measurements involve the integrated scattering cross section for laboratory angles greater than some minimum angle. The measured integral cross section for low-energy $\pi^+ p$ scattering is quite insensitive to the accuracy with which the minimum scattering angle is known because the cross section is strongly peaked in the backward hemisphere. The integral cross section is an observable that can be directly compared with predictions of phase shift fits; there is no need to make an extrapolation through the Coulomb region to zero detector solid angle.

In an effort to resolve the experimental discrepancies, a group from Colorado, LANL, Minnesota, British Columbia, and Cal State Sacramento measured integral cross sections above 30° from 60 to 200 MeV at the P$^3$ channel at LAMPF [16]. Pions in the incident beam were cleanly identified by time-of-flight between beam counters and the accelerator RF signal. The experiment of Ref. [15] used a polyethylene target, necessitating a subtraction for events from carbon. The new experiment used a liquid hydrogen target, thus avoiding this subtraction. Scintillation counters identified events with pions scattered outside a 30° forward cone. Pions inside this cone were identified and separated from recoil protons by means of the pulse height and time-of-flight measured in a forward counter. Figure 3 compares the preliminary results for the integral cross section as a function of energy compared with the expectation from the KH80 [17] phase shift. Figure 4 shows the ratio of the preliminary data to the cross section calculated from the phase shifts KH80 [16]: the data of Friedman *et al.* are also shown. The preliminary results are consistently below the phase-shift expectations. This is in agreement with the early measurements and disagrees with the results of Ref. [15].

### Measuring the Quark Content of the Proton Spin with Neutrinos

It is not yet possible to calculate hadron structure functions in QCD. However, sum rules for the first moments of the $g_1$ structure functions have been derived from rather fundamental theoretical arguments. In particular, the derivation of the Ellis-Jaffe sum rule [18], $\int g_1^p(x)dx = 0.175 \pm 0.018$, relies on the assumptions that the strange quarks in the nucleon sea are unpolarized and that SU(3) symmetry holds in hyperon decays. The European Muon Collaboration (EMC) [19] and earlier SLAC experiments [20] measured $g_1^p$ for $0.01 < x < 0.7$. By extrapolating to the entire range of $x$, they find $\int g_1^p(x)dx = 0.126 \pm 0.01 \pm 0.015$, which differs significantly from the Ellis-Jaffe prediction. If this deviation is ascribed to the strange quarks, then the total quark spin projection along the direction of the nucleon spin is consistent

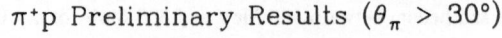

$\pi^+p$ Preliminary Results ($\theta_\pi > 30°$)

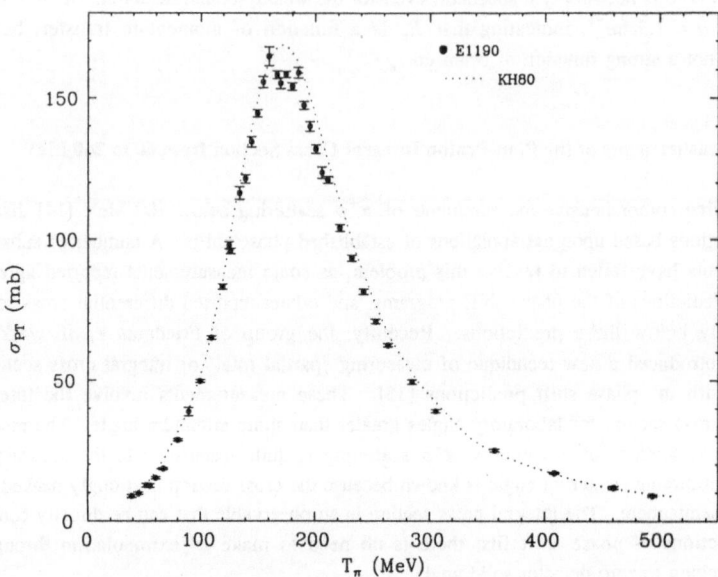

Fig. 3. The integral $\pi^+p$ cross section above 30° compared with the expectation from the KH80 phase shifts.

$\pi^+p$ Preliminary Results ($\theta_\pi > 30°$)

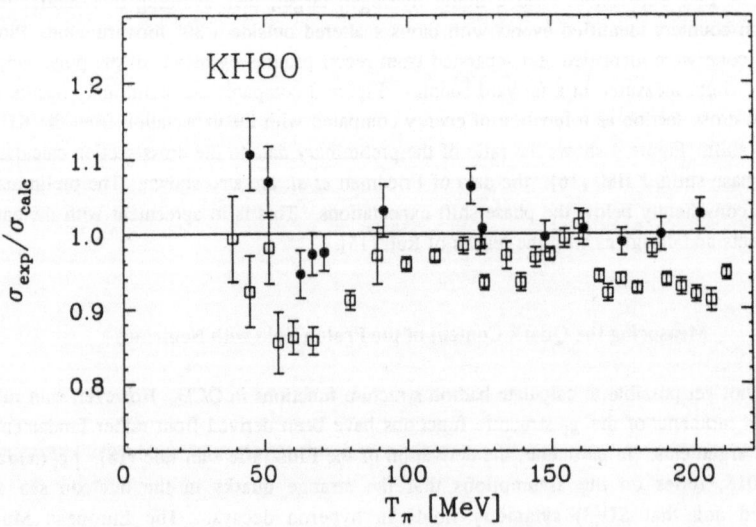

Fig. 4. The ratio between the preliminary results of the integral $\pi^+p$ cross section as a function of energy measured in this experiment above 30° to the results expected from the KH80 phase shift solution. The data are significantly below the expectations. The data of Friedman *et al.* are also shown.

with zero. This would be surprising but not inconsistent with any fundamental aspect of QCD. It may also be that the assumptions used to derive this conclusion (SU(3), the extrapolation to $x = 0$) are incorrect. In any event, it is clear that this deserves further study. $vp$ elastic scattering can be used to directly determine $\Delta s$, the strange quark contribution to the proton spin, without requiring SU(3) symmetry or an extrapolation in $x$.

The cross section for $vp$ elastic scattering at low $Q^2$ ($Q^2 << m_p^2$) is given by:

$$\frac{d\sigma}{dQ^2} \approx \frac{G_F^2}{2\pi}\left[G_1^2\left(1+\frac{Q^2}{4E_v^2}\right)+F_1^2\left(1-\frac{Q^2}{4E_v^2}\right)\right],$$

where $2G_1(0) = -g_A + G_1^s$, and $F_1(0) = -0.034 + F_1^s/2$. We know that $g_A = 1.26$ from neutron decay, and $F_1^s(0) = 0$ because $F_1^s$ is the strange charge radius. Thus, at low $Q^2$, the cross section determines $G_1^s = \Delta s$, the strange quark contribution to the proton spin.

An experiment has been built at LAMPF to extend the sensitivity of searches for neutrino oscillations. The experiment, the liquid scintillator neutrino detector [21] (LSND), will also provide an accurate measurement of the cross section for $vp$ elastic scattering at low $Q^2$. The detector consists of a cylindrical tank, approximately 6 m in diameter and 9 m long, containing 200 tons of dilute mineral-oil-based scintillator. The scintillator is viewed by 1224 8" photomultiplier tubes, which cover 22% of the surface of the tank and yields an energy resolution of 5% for 50-MeV protons. The different characteristics of Čerenkov light and scintillator light allow electrons, protons, and neutrons to be identified.

The detector is located inside a veto shield approximately 28 m downstream of the main proton beam stop at LAMPF. In the present beam stop, ~97.5% of the $\pi^+$ produced decay at rest and emit 30-MeV muon neutrinos; 2.5% of the pions decay in flight producing a muon-neutrino beam with an average energy of 150 MeV. A major upgrade to the LAMPF beam stop will greatly increase the decay-in-flight yield.

In $vp$ elastic scattering, the only observables are the energy and angle of the low-energy recoil proton. Only the proton energy can be measured in LSND. One can try to make an absolute measurement of the elastic scattering cross section, but this requires good knowledge of the neutrino flux. Another approach is to compare the relative yield of $(vp)$ and $(vn)$ events from carbon. This strategy gives a factor of 2 enhancement in the sensitivity to $\Delta s$ and is insensitive to the neutrino flux. However, one must understand possible nuclear effects and the detector response for neutrons. There has been an initial study of the nuclear effects [22]; Fig. 5 shows the sensitivity of the ratio of $(vp)$ to $(vn)$ events as a function of $\Delta s$ for a wide range of values of $F_2^s(0)$, the strange magnetism form factor; $F_2^s$ will be measured in the SAMPLE experiment at Bates [23]. In order to eliminate quasi-elastic events from free protons, only events with recoil nucleons above 60 MeV can be used. With the LAMPF beam stop in its present configuration, 15 $vp$ elastic scattering events per day with $T > 10$ MeV from free protons, 3.5 events with $T > 60$ MeV from bound protons, and 0.75 $vn$ with $T > 60$ MeV are expected. The principal background is cosmic-ray induced neutron events. After a minor upgrade to the beam stop, next year's run should result in approximately 750 (150) $vp$ $(vn)$ events yielding an uncertainty in $\Delta s$ of $\pm 0.1$. The major beam stop-upgrade, presently under consideration for 1995, would lower the expected uncertainty in $\Delta s$ to $\pm 0.04$ in 1995; if the major upgrade is not undertaken, $\Delta s$ will be determined to $\pm 0.06$.

This work is supported by the U.S. Department of Energy.

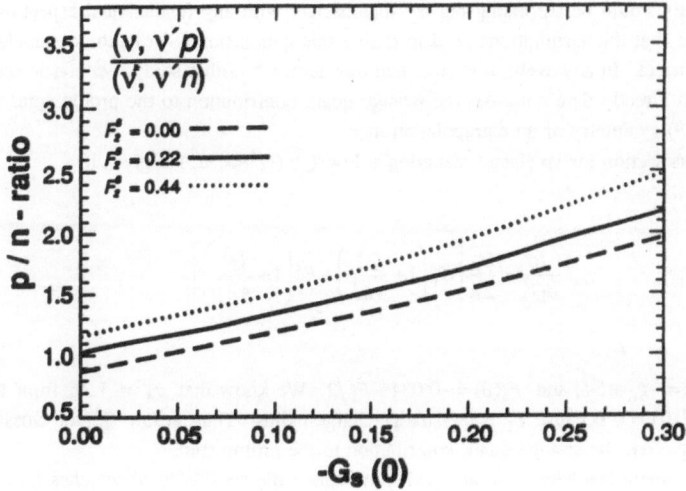

Fig. 5. The ratio of $(\nu p)$ to $(\nu n)$ elastic scattering events as a function of $G_1^s(0)$ for several values of $F_2^s$.

## References

1. Olsson, M.G., Turner, L.: *Phys. Rev. Lett.* **20**, 1127 (1968); *Phys. Rev.* **181**, 2141 (1969); and Turner, L.: Ph.D. Thesis, Univ. of Wisconsin, 1969 (unpublished).
2. Burkhardt, H., Lowe, J.: *Phys. Rev. Lett.* **67**, 2622 (1991).
3. LAMPF Experiment #1179, D. Počanić, spokesman.
4. Baer H., *et al.*: *Nucl. Instrum. Methods* **180**, 445 (1981).
5. Frlez, E.: Ph.D. Thesis, Univ. of Virginia, 1993 (unpublished).
6. Jenefsky, R.F., *et al.*: *Nucl. Phys.* **A290**, 407 (1977).
7. Comiso, J.C., *et al.*: *Phys. Rev.* **D12**, 738 (1975); Bayer, W., *et al.*: *Nucl. Instrum. Methods* **134**, 449 (1976).
8. Bugg, D.V., *et al.*: *Nucl. Phys.* **B26**, 588 (1971).
9. Hohler, G., Stahov, J.: $\pi N$ Newsletter No. 2, Hohler, G., Kluge, W., Nefkens, B.M.K. (eds.), **42**, May 1990.
10. LAMPF Experiment #1268, M. Sadler, spokesman.
11. Langenbrunner, J., *et al.*: *Phys. Rev. Lett.* **69**, 1508 (1992).
12. LAMPF Experiment #1216, J. Langenbrunner, spokesman.
13. An earlier CsI phoswich detector is described in Langenbrunner, J. L., *et al.*: *Nucl. Instrum. Methods* **A316**, 450 (1992). A faster detector was used in this experiment.
14. Frank, J.S., *et al.*: *Phys Rev.* **D28**, 1569 (1983).
15. Friedman, F., *et al.*: *Phys. Lett.* **B231**, 39 (1989); **B254**, 40 (1991), *Nucl. Phys.* **A514**, 601 (1990).
16. LAMPF Experiment #1190, C. L. Morris and R. A. Ristinen, spokesmen.
17. Koch, R., Pietarinen, E.: *Nucl. Phys.* **A336**, 331 (1980).
18. Ellis, J., Jaffe, R.: *Phys. Rev.* **D9**, 1444 (1974); **D10**, 1669 (1974).
19. Aschman, J., *et al.*: *Phys. Lett.* **B206**, 364 (1988); *Nucl. Phys.* **B328**, 1 (1989).
20. Alguard, M.J., *et al.*: *Phys. Rev. Lett.* **37**, 1261 (1976); *ibid.* **41**, 70 (1978); Baum, G., *et al.*: *Phys. Rev. Lett.* **51**, 1135 (1983).
21. LSND Proposal, Los Alamos National Laboratory Report LA-UR-3764 (1989) [unpublished].
22. Garvey, G.T., *et al.*: *Phys. Lett.* **B289**, 249 (1992).
23. McKeown, R.D., Beck, D.H., Spokesmen: Bates Proposal #89-06.

Few-Body Systems, Suppl. 7, 441–448 (1994)

© Springer-Verlag 1994
Printed in Austria

# SPIN–STRUCTURE FUNCTION OF THE NEUTRON ($^3$HE): SLAC RESULTS*

Z.-E. Meziani**
Department of Physics, Stanford University, Stanford, California 94305
Representing the E–142 Collaboration
Stanford Linear Accelerator Center, Stanford University, Stanford, California 94309

## Abstract

A first measurement of the longitudinal asymmetry of deep-inelastic scattering of polarized electrons from a polarized $^3$He target at energies ranging from 19 to 26 GeV has been performed at SLAC. The spin-structure function of the neutron $g_1^n$ has been extracted from the measured asymmetries allowing for a test of the Ellis-Jaffe and Bjorken sum rules. The Quark Parton Model (QPM) interpretation of the nucleon spin-structure function is examined in light of the new results.

## Introduction

In his pioneering work of 1966 and 1970, Bjorken [1] suggested that large asymmetries could be observed in deep-inelastic polarized-electron scattering off polarized-nucleon targets. Furthermore, he derived a fundamental relation known as the Bjorken sum rule. The test of the latter, described by Feynman [2] as one that would have a decisive influence on the future of high-energy physics, requires a measurement of both proton and neutron spin-structure functions. In the early seventies—given the perceived technical difficulties of polarized target developments—a measurement using a polarized-proton target was viewed as feasible, while that of a polarized-neutron target was, if not impossible, at least a very complicated task. Theoretical work initiated by Gilman [3], within the framework of SU(3) symmetry, focused on writing separate sum rules for the proton and the neutron. It was further developed by Ellis and Jaffe [4], who assumed that the strange sea in the nucleon was unpolarized, and derived what is known as the Ellis–Jaffe sum rule (E–J) for the proton and the neutron.

Two early experiments performed in 1976 (E–80) [7] and in 1983 (E–130) [8] by the Yale–SLAC collaboration at SLAC on a polarized proton target confirmed the suggestion of Bjorken giving grounds to the naive picture of the QPM. While a good

*Work supported in part by Department of Energy contracts DE–AC03–76SF00515 (SLAC) and DE–FG03–88ER40439 (Stanford University).
**Present address: Physics Department, Temple University, Philadelphia, PA 19122.

442

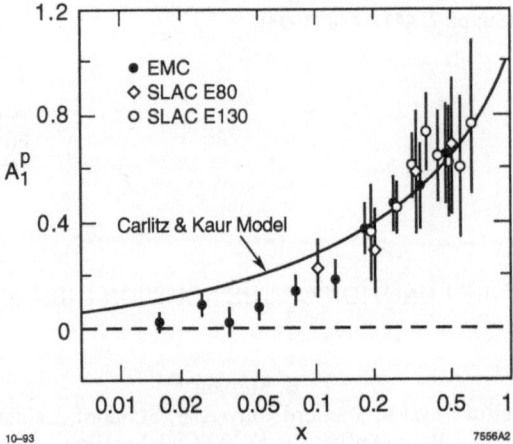

Fig. 1. World results for proton asymmetries $A_1^p$, and the QPM model [10]

agreement with the QPM prediction was observed in the $x$ region dominated by the valence quarks, no comparison was possible in the region of sea quarks due to a limited kinematic coverage. A first experimental test of the E–J sum rule found it to be fulfilled, but with a large uncertainty due the extrapolation uncertainty of $A_1^p$ in the unmeasured low $x$ region. The debate on the detailed spin structure of the proton was revived in 1988, when the European Muon Collaboration (EMC) [9] reported new results on polarized muon scattering off a polarized proton target extending the measurements of $A_1^p$ to low values of $x$. An evaluation of the E–J sum rule on the proton using the new proton data displayed a two standard and a half deviation from the predicted value. A QPM analysis of the spin structure of the proton in terms of its flavor components revealed a small net total spin contribution of the quarks, with a large negative strange-sea quarks component. It was clear that more experiments were needed to set limits on various speculations arising from these results, and to improve our understanding of the nucleon spin structure. The world proton asymmetry data are summarized in Fig. 1, with a QPM prediction [10] consistent with the E–J sum rule.

We first define the quantities of physics interest, following with a description of the $^3$He (neutron) spin structure function measurement carried out at SLAC by the E–142 collaboration. Finally, in light of the new results, we examine the spin structure of the nucleon, and present the crucial test of the Bjorken sum rule with a coherent set of assumptions.

### Asymmetries and Sum Rules

In deep-inelastic scattering, the measured longitudinal asymmetry $A^\parallel$ can be determined experimentally by measuring the difference over the sum in cross sections of polarized electrons on polarized nucleons between states where the spins are parallel and antiparallel [5,6],

$$A^\parallel = \frac{\sigma^{\uparrow\downarrow} - \sigma^{\uparrow\uparrow}}{\sigma^{\uparrow\downarrow} + \sigma^{\uparrow\uparrow}} = \frac{1-\epsilon}{(1-\epsilon R)\, W_1(Q^2,\nu)} \left[ M(E + E'\cos\theta)\, G_1(Q^2,\nu) - Q^2 G_2(Q^2,\nu) \right].$$

$$(1)$$

Here $\sigma^{\uparrow\uparrow}$ ($\sigma^{\downarrow\uparrow}$) is the inclusive $d^2\sigma^{\uparrow\uparrow}/d\Omega d\nu$ ($d^2\sigma^{\downarrow\uparrow}/d\Omega d\nu$) differential scattering cross section for longitudinal target spins parallel (antiparallel) to the incident electron spins. A corresponding relationship exists for scattering of longitudinally polarized electrons off a transversely polarized target where a transverse asymmetry is defined [6]:

$$A^{\perp} = \frac{\sigma^{\downarrow\leftarrow} - \sigma^{\uparrow\leftarrow}}{\sigma^{\downarrow\leftarrow} + \sigma^{\uparrow\leftarrow}} = \frac{(1-\epsilon)E'}{(1-\epsilon R)\,W_1(Q^2,\nu)}\,[(MG_1(Q^2,\nu) + 2EG_2(Q^2,\nu))\cos\theta]\,, \quad (2)$$

where

$$R = \frac{W_2}{W_1}\left(1 + \frac{\nu^2}{Q^2}\right) - 1\,; \qquad \epsilon = \left[1 + 2\left(1 + \frac{\nu^2}{Q^2}\right)\tan^2\frac{\theta}{2}\right]^{-1}\,. \quad (3)$$

Here $\sigma^{\downarrow\leftarrow}$ ($\sigma^{\uparrow\leftarrow}$) is the inclusive scattering cross section for beam-spin antiparallel (parallel) to the beam momentum, and for target-spin direction transverse to the beam momentum and towards the direction of the scattered electron. In all cases, $G_1$ and $G_2$ are the spin-dependent structure functions, whereas $W_1$ and $W_2$ are the spin-averaged structure functions; $R$ is the ratio of longitudinal-to-transverse virtual-photoabsorption cross sections; $\epsilon$ is the virtual photon polarization; $M$ is the mass of the nucleon; $Q^2$ is the square of the four-momentum of the virtual photon; $E$ is the incident electron energy; $E'$ is the scattered electron energy; $\nu = (E - E')$ is the electron energy loss; and $\theta$ is the electron scattering angle.

The system of Eqs. (1) and (2) allows for the separate determination of $G_1$ and $G_2$, knowing $W_2$ and $W_1$. In the scaling limit ($\nu$ and $Q^2$ large), these structure functions are predicted to depend only on the Bjorken variable $x = Q^2/2M\nu$, yielding

$$\begin{aligned} MW_1(\nu,Q^2) &\to F_1(x)\,, & \nu W_2(\nu,Q^2) &\to F_2(x)\,, \\ M^2\nu G_1(\nu,Q^2) &\to g_1(x)\,, & M\nu^2 G_2(\nu,Q^2) &\to g_2(x)\,. \end{aligned} \quad (4)$$

The experimental asymmetries $A^{\|}$ and $A^{\perp}$ are related to the virtual photon-nucleon longitudinal and transverse asymmetries, $A_1$ and $A_2$ respectively, via

$$\begin{aligned} A^{\|} &= D(A_1 + \eta A_2)\,, & A^{\perp} &= d(A_2 - \zeta A_1)\,, \\ D &= (1 - E'\epsilon/E)/(1 + \epsilon R)\,, & \eta &= \epsilon\sqrt{Q^2}/(E - E'\epsilon)\,, \\ d &= D\sqrt{2\epsilon/(1+\epsilon)}\,, & \zeta &= \eta(1+\epsilon)/2\epsilon\,. \end{aligned} \quad (5)$$

The proton (neutron) spin structure function is extracted in the finite $Q^2$ region following the relation

$$g_1^{p(n)} = \left[A_1^{p(n)}F_1^{p(n)} + A_2^{p(n)}F_1^{p(n)}\left(\frac{2Mx}{\nu}\right)^{1/2}\right]\Big/\left(1 + \frac{2Mx}{\nu}\right)\,, \quad (6)$$

where $F_1^{p(n)}$ is the spin averaged structure function of the proton ( neutron). Within the QPM interpretation, $F_1^{p(n)}(x)$ and $g_1^{p(n)}(x)$ are related to the momentum distribution of the constituents as

$$F_1(x) = \frac{1}{2} \sum_{i=1}^{f} z_i^2 \left[ q_i^\uparrow(x) + q_i^\downarrow(x) \right] , \qquad g_1(x) = \frac{1}{2} \sum_{i=1}^{f} z_i^2 \left[ q_f^\uparrow(x) - q_f^\downarrow(x) \right] , \tag{7}$$

where $i$ runs over the number of flavors, $z_i$ are the quark fractional charges, and $q_i^\uparrow$, $(q^\downarrow)_i$ are the quark plus antiquark momentum distributions for quark and antiquarks spins parallel (antiparallel) to the nucleon spin. Using the following set of assumptions—quark current algebra, isospin symmetry, SU(3) symmetry in the decay of the baryon octet, and zero net polarization for the strange-sea quarks—the Ellis–Jaffe sum rule on the proton (neutron) is expressed to first order correction in $\alpha_s$ as follows [11]:

$$I^{p(n)} = \int_0^1 g_1^{p(n)}(x) \, dx = \frac{1}{12} \left| \frac{g_A}{g_V} \right| \left\{ \left[ 1(-1) + \frac{5}{3} \left( \frac{3F-D}{F+D} \right) \right] - \frac{\alpha_s}{\pi} \left[ 1(-1) + \frac{7}{9} \left( \frac{3F-D}{F+D} \right) \right] \right\}, \tag{8}$$

where $\alpha_s$ is the QCD strong coupling constant, and $F$ and $D$ are the SU(3) invariant matrix elements of the axial vector current. From neutron $\beta$ decay, we obtain $(g_A/g_V) = F+D = 1.2573 \pm 0.0028$. Following [11], we use $F = 0.459 \pm 0.008$ and $D = 0.798 \mp 0.008$, giving $F/D = 0.575 \pm 0.016$. Within the QPM interpretation, we rewrite $I^n$ in terms of quark polarizations $\Delta q \equiv \int_0^1 dx [q^\uparrow(x) - q^\downarrow(x)]$ at finite $Q^2$ :

$$I^n = \frac{2}{9} \left( \Delta u - 2\Delta d + \Delta s \right) \left( 1 - \frac{\alpha_s}{\pi} \right) + \frac{1}{9} \left( \Delta u + \Delta d - 2\Delta s \right) \left( 1 - \frac{\alpha_s}{3\pi} \right) . \tag{9}$$

The primary motivation of the E–142 measurement of the neutron spin structure function is the test of the Bjorken sum rule. The latter is insensitive to the details of nucleon structure but depends solely on quark current algebra and isospin symmetry. It is expressed as the difference between the proton and the neutron spin structure function $g_1(x, Q^2)$ integrals. The Bjorken sum rule is expressed to first order in $\alpha_s$ as

$$I^p - I^n = \int_0^1 g_1^p(x, Q^2) - g_1^n(x, Q^2) \, dx = \frac{1}{12} \left| \frac{g_A}{g_V} \right| \left[ 1 - \frac{\alpha_s(Q^2)}{\pi} \right] . \tag{10}$$

Higher order PQCD [12], as well as higher twist [13] corrections, although not included in Eq. (10), are important in the analysis of the Bjorken sum rule and must be considered at low $Q^2$.

### E–142 Measurement

The experiment used the SLAC polarized electron beam at the three "magic" energies 19.4, 22.7, and 25.5 GeV, so that the electron spin is longitudinal as it enters End Station A. The electron beam helicity was reversed randomly on a pulse-to-pulse basis, allowing for the cancellation of many of the beam systematic errors. This was achieved by reversing the laser beam circular polarization used for photoemission from the AlGaAs photocathode in the electron source. The delivered beam polarization ($P_l$) was measured by a single arm Moller polarimeter and found to be stable at an average value of $(38.8 \pm 1.6)\%$, where the uncertainty is dominated by the measurement of the foil magnetization.

The target was a newly-built 30-cm-long, high-pressure double cell filled with a mixture of $^3$He, rubidium, and nitrogen [14]. With end windows approximately 0.012-cm thick, this target operated at number density of $2.3 \times 10^{20}$ atoms/cm$^2$ (8.6 atm at 0°C) . Polarization of $^3$He was achieved by optically pumping the rubidium vapor, which transfered its polarization to the $^3$He nuclei by spin exchange collisions. The small added quantity of nitrogen ($1.9 \times 10^{18}$ atoms/cm$^3$) increased the optical pumping efficiency. The $^3$He polarization ($P_t$) was measured with an NMR setup and observed to be variying slowly during the experiment, between 30% and 40% with a relative uncertainty $\Delta P_t/P_t$ of 7%. The polarization of the target was reversed frequently as a means to cancel systematic effects.

Data were collected using two single-arm spectrometers at scattering angles of 4.5° and 7° [15], covering a kinematical range of $0.03 < x < 0.6$ and $Q^2 > 1$ (GeV)$^2$. In each spectrometer arm, the electron detector package consisted of two threshold Čerenkov counters, six planes of hodoscopes, and a 24-radiation-length shower counter composed of 200 lead-glass blocks. The momentum resolution (rms) from hodoscope tracking was $\Delta E'/E' \sim 3\%$, and the shower energy resolution was typically $15\%/\sqrt{E'(\text{GeV})}$.

The experimental raw counting asymmetry $\Delta$ was converted to the experimental asymmetry $A^\parallel$, using the relation

$$\Delta = \frac{(N^{\uparrow\downarrow} - N^{\uparrow\uparrow})}{(N^{\uparrow\downarrow} + N^{\uparrow\uparrow})} \quad , \qquad A^\parallel = \frac{\Delta}{P_b P_t f} \quad , \tag{11}$$

where $N^{\uparrow\downarrow}$ ($N^{\uparrow\uparrow}$) represents the rate of scattered electrons for each bin of $x$ and $Q^2$ when the electron beam helicity is antiparallel (parallel) to the target spin, and $f$ is the dilution factor that corresponds to the fraction of events that originated from scattering off the neutron in $^3$He.

Small corrections for deadtime, pair-electron contamination, and misidentified pions were applied. These corrections are $x$ dependent, and dominate in the low $x$ region. The largest systematic uncertainty in the measurement of $A^\parallel$ comes from the determination of the dilution factor $f$. This factor was measured using glass cell runs, with variable pressures of $^3$He to separate the scattering contribution of $^3$He from that of glass, and was found to be $0.11 \pm 0.02$. False asymmetries were measured to be consistent with zero by comparing data with target spins in opposite directions.

External radiative corrections were evaluated using the Mo and Tsai method [17], and found to be small ($\sim 5\% A_1^n$) because of the relatively thin target ($\sim 0.3\%$ radiation length). Internal radiative corrections were more important, and were evaluated using the exact procedure of Kukhto and Shumeiko [16]. The total radiative corrections

446

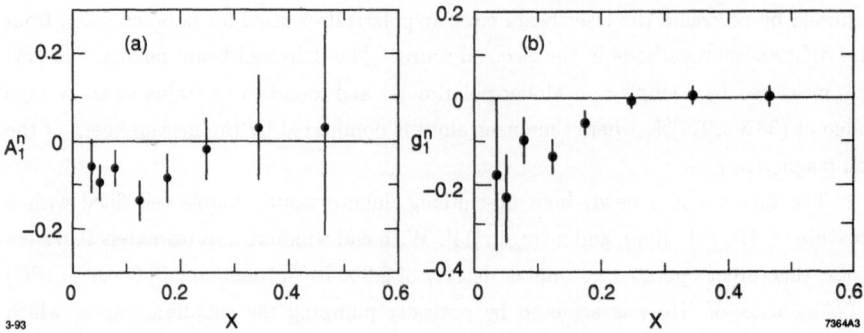

Fig. 2(a) Neutron asymmetries $A_1^n$ and (b) spin-structure function $g_1^n$ as a function of $x$.

amounted to a relative change of the asymmetry ranging from 30% at low $x$ to 15% at large $x$. Recent studies by several groups [18–21] have concluded that in deep-inelastic scattering a polarized $^3$He nucleus target can be regarded as a good model of a polarized neutron, provided a small correction for the $S'$ and $D$ states is applied. To extract the neutron asymmetry from the measured $^3$He asymmetry, we followed the method described in [20], allowing for a correction from the polarization of the two protons in $^3$He ($\sim -2.7\%$ per proton) and a correction for the polarization of the neutron in $^3$He ($\sim$87%).

Figure 2(a) shows the results of the physics asymmetry $A_1^n$ as a function of $x$. Statistical and systematic errors are presented, added in quadrature. Since no significant $Q^2$ dependence of the measurement was observed, data at a fixed $x$ bin were averaged over different $Q^2$. The extraction of $g_1^n$ used a measurement of the transverse asymmetry [Eq. (2)] which amounted to $A_2^n = 0.0 \pm 0.25$ over the full range in $x$. This measurement was performed by orienting the target spin transverse to the electron beam, in either direction. Figure 2(b) shows $g_1^n$ as a function of $x$, obtained using Eq. (6), where $F_1$ was derived from a global fit to the SLAC data for $R$ [22] and the recent NMC parametrization for $F_2$ [23]. Although small ($\sim 0.1$), there is a clear trend towards negative asymmetries $A_1^n$ in the region $0.03 < x < 0.2$.

## Sum Rules Tests and Nucleon Spin Structure

To test the sum rules and interpret the spin structure of the nucleon in terms of its constituents spin, $I^n$ is evaluated at a fixed average value $Q^2$. All $g_1^n$ data points are evolved to the average value of $Q^2$ assuming $A_1^n$ to be $Q^2$ independent. Integrating the measured range of $x$ we find

$$\int_{0.03}^{0.6} g_1^n \left[ \langle Q^2 \rangle = 2(\text{GeV})^2, x \right] \, dx = -0.019 \pm 0.007 \, (\text{stat}) \pm 0.006 \, (\text{syst}) \quad . \quad (12)$$

To evaluate the missing part of the integral, we consider the low- and high-$x$ regions separately. For $0 \leq x < 0.03$, we assume a plausible form of extrapolation of the spin-structure function $g_1^n(x) = g_1^n(x_0)(x/x_0)^\alpha$, as suggested by Regge theory [24], with $g_1(x_0 = 0.03) = -0.175$ and $0 \leq \alpha \leq 0.5$. For high-$x$ we extrapolate $A_1(x)$, using isospin arguments and the QPM. We assume that $A_1(x) \rightarrow +1$ as $x \rightarrow 1$. After adding the contribution from the unmeasured region, we find an experimental value $I^n = \int_0^1 g_1^n(x)dx = -0.022 \pm 0.011$ at an average $\langle Q^2 \rangle$ of $2(\text{GeV})^2$. Because of the low average value of the momemtum transfer, a serious consideration might be given to the contribution of higher twist effects and higher order in the PQCD corrections.

To have a consistent comparison with the EMC analysis of the proton, where $I^p$ was determined at a much larger average $Q^2$, we choose to evolve our data to the same $Q^2$. This was done by assuming once more that the physics asymmetry $A_1^n$ is $Q^2$ independent, which has to some extent been observed on the proton data [9]. Equivalently, this implies a common $Q^2$-dependence of both $g_1^n$ and $F_1^n$, such that $A_1^n$ is relatively constant as $Q^2$ varies. Although this choice is not unique, it is preferable, given the very poor low-$Q^2$ evaluation of higher twist effects at the present time. For example, in [25] it is argued that since the integral $\int_0^1 g_1^p(x)dx$ is very insensitive to $\langle Q^2 \rangle$, a better test of the Bjorken sum rule, as well as evaluation of the quark contributions to the nucleon spin, is performed by evolving the EMC proton results to low 4-momentum transfer. Uncertainties due to the lack of reliable calculation of higher twist effects makes this procedure not necessarily attractive.

In the QPM interpretation, we use Eq. (9) and the E–142 result at $Q^2 = 10.7$ $(\text{GeV})^2$; namely, $I^n = -0.031 \pm 0.007 \pm 0.009$, combined with the neutron $\beta$-decay relation $\Delta u - \Delta d = g_A/g_V = 1.257 \pm .0028$ and the SU(3) symmetry in the decay of the baryon octet $\Delta d - \Delta s = F - D = -0.34 \pm 0.17$ to find the net quark polarization $\Delta u + \Delta d + \Delta s \sim 0.5$, while $\Delta s \sim -0.03$. Notice that contrary to the proton results of EMC [9] and the Spin Muon Collaboration (SMC) [26], E–142 results agree with the Ellis-Jaffe sum rule, and predict a small strange-quark contribution to the net neutron polarization. This result is also consistent with the analysis of Ref. [27] where a bound on the strange-sea polarization $|\Delta s| \leq 0.021 \pm 0.001$ is argued.

We now turn to a fundamental test of the Bjorken sum rule, at a unified value for $Q^2$ of 10.7 $(\text{GeV})^2$, using results from the EMC and E–142 experiments:

$$
\begin{aligned}
\text{EMC} \quad & I^p(\langle Q^2 \rangle = 10.7) = 0.131 \pm 0.01 \pm 0.015 \\
\text{E} - 142 \quad & I^n(\langle Q^2 \rangle = 10.7) = -0.031 \pm 0.007 \pm 0.009
\end{aligned} \tag{13}
$$

with an "experimental" difference $I^p - I^n = 0.161 \pm 0.021$. This difference is now compared to the theoretical prediction of Bjorken, corrected for higher-order PQCD terms at the same value of $Q^2$ [12]:

$$
I^{p-n} = \frac{1}{6} \frac{g_A}{g_V} \left[ 1 - \frac{\alpha_s}{\pi} - 3.58 \left( \frac{\alpha_s}{\pi} \right)^2 - 20.4 \left( \frac{\alpha_s}{\pi} \right)^3 \cdots \right] = 0.185 \pm 0.004 .
$$

We observe that within approximately one standard deviation, the Bjorken sum rule is verified.

In conclusion, the Ellis-Jaffe sum rule is confirmed by the E-142 results to within one standard deviation. The QPM interpretation of E-142 results lead to a small (few percent at most) strange-sea quark contribution to the nucleon net polarization, but a large total quark contribution to the spin of the nucleon ($\sim 50\%$). Within the available uncertainty of the existing proton and the new neutron data, the Bjorken sum rule is verified to third-order PQCD corrections when the comparison is performed at $\rangle Q^2 \rangle = 10.7$ GeV . A more reliable and precise test at high-$Q^2$ is desirable. This should be achieved as we enter a new generation of proposed experiments that will be performed at CERN (SMC), HERA (Hermes), and SLAC (E-154, E-155) on the proton, deuteron, and $^3$He.

## References

1. Bjorken, J. D.: Phys. Rev. **148**, 1467 (1966); **D1**, 1376 (1970).
2. Feynman, R.P.: Photon–Hadrons Interactions, Frontiers in Physics Lecture Note Series, p. 159. Reading, Mass: Benjamen Inc.
3. Gilman, F.J. : Proc. Scottich Universities Summer School in Physics (1973).
4. Ellis, J., Jaffe, R.L.: Phys. Rev. **D9**, 1444 (1974). .
5. Carlson, C.E., Tung, W.-K. : Phys. Rev. **D5**, 721 (1972).
6. Hey, A.J.G. , Mandula, J.E.: Phys. Rev. **D5**, 2610 (1972).
7. Alguard, M.J., et al.: Phys. Rev. Lett. **37**, 1258 (1976); **37**, 1261 (1976).
8. Baum, G., et al.: Phys. Rev. Lett . **51**, 1135 (1983).
9. Ashman, J., et al.: Phys. Lett **B206**, 364 (1988); Nucl. Phys. **B328**, 1 (1989).
10. Carlitz, R. , Kaur, J. , Phys . Rev. Lett. **38**, 673 (1977); Erratum, ibid., p. 1102.
11. Close, F. E.: Rutherford preprint RAL–93–034 (1993).
12. Larin, S.A., Vermaseren, J.A.M.: Phys . Lett. **B259**, 345 (1991).
13. Balitsky, I.I., Braun, V.M., Kolesnichenko, A.V.: Phys. Lett . **B242**, 145 (1990).
14. Chupp, T.E., et al.: Phys. Rev. **C45**, 915 (1992); **36**, 2244 (1997).
15. Petratos, G.G., et al.: Stanford Linear Accelerator Center preprint SLAC–PUB–5678 (1991), unpublished.
16. Kukhto, T.V., Shumeiko, N.M.: Nucl. Phys. **219**, 412 (1983).
17. Mo, L.W., Tsai, Y.T.: Rev. Mod. Phys. **41**, 205 (1969).
18. Blankleider, B., Woloshyn, R.M.: Phys. Rev. **C29** 538 (1984).
19. Friar, J.L., Gibson, B.S., Payne, G.L., Bernstein, A.M., Chupp, T.E.: Phys. Rev. **C42** 2310 (1990).
20. Ciofi degli Atti, C., Scopetta, S., Pace, E., Salme, G. : University of Perugia preprint DFUPG–75/93 (to be published); see Salme, G., these proceedings.
21. Shulze, R.-W., Sauer, P.U.: Phys. Rev. **C48**, 38 (1993).
22. Whitlow, L.W., et al.: Phys. Lett. **B250**, 193 (1990); **282**, 475 (1992).
23. NMC Collaboration: Nucl. Phys. **B371**, 3 (1992).
24. Shafer, A.: Phys. Lett . **B208**, 175 (1988).
25. Ellis, J., Karliner, M.: Phys. Lett. **B313**, 131 (1993).
26. Adeva, B.; et al.: Phys. Lett. **B362**, 553 (1993).
27. Preparata, G., Ratcliffe, P.: Universita di Milano Preprint MITH–93/15, submitted to Phys. Lett. **B**.

Few-Body Systems, Suppl. 7, 449—457 (1994)

© Springer-Verlag 1994
Printed in Austria

# THE SPIN-DEPENDENT STRUCTURE FUNCTION OF THE DEUTERON

Stéphane Platchkov

*DAPNIA/Service de Physique Nucléaire*
*CEN Saclay, F-91191 Gif sur Yvette, France*
(On behalf of the Spin Muon Collaboration)

### Abstract

The SMC collaboration at CERN has recently measured the spin-dependent structure function $g_1^d(x)$ of the deuteron. The data taken cover the kinematical region 1 GeV$^2$ < $Q^2$ < 30 GeV$^2$ and 0.006 < $x$ < 0.6. The first moment $\Gamma_1^d$ of $g_1^d(x)$ is found to be two standard deviations below the prediction of the Ellis-Jaffe sum-rule. The inferred quark contribution to the proton spin is $\Delta\Sigma$ = 0.06 ± 0.20 ± 0.15. When combined with the result of EMC for the proton $\Gamma_1^p$, the difference $\Gamma_1^p - \Gamma_1^n = 0.20 \pm 0.05 \pm 0.04$ is deduced, in good agreement with the fundamental Bjorken sum rule. Comparison with other experiments shows that in the region of $x$ where the measurements overlap, the SMC deuteron data agree with the combination of EMC data for the proton and the recent E142 data for the neutron.

## 1   Introduction

Studies of unpolarised nucleon structure functions have shown[1] that the quarks have half-integral spin. The question how the quark spins account for the nucleon spin has been raised only recently, after the EMC collaboration at CERN has published a measurement[2] on the spin-dependent structure function of the proton, extending the previous measurements from SLAC [3,4] to lower x values. Interpretation of these data in the frame of the quark-parton model has shown that the quarks essentially do not contribute to the proton spin. This unexpected finding has stimulated a large amount of theoretical work providing several alternative explanations. On the experimental side, new data have been taken, measuring for the first time the deuteron[5] and the neutron[6] spin-dependent structure functions. Here I report the first results on the deuteron polarised structure function $g_1^d(x)$ obtained by the Spin Muon Collaboration (SMC) at CERN.

## 2   Polarised structure functions. Sum rules

For lepton scattering with both polarised beam and polarised target one has access to the two spin-dependent structure functions, $g_1(x)$ and $g_2(x)$. They are related to the virtual photon asymmetry $A_1(x)$ [8] by the relation $A_1 \simeq g_1/F_1$. In the quark-parton model $A_1(x)$ can also be expressed by [7] :

$$A_1 = \frac{\frac{1}{2}\sum_i e_i^2[q_i^\uparrow(x) - q_i^\downarrow(x)]}{\frac{1}{2}\sum_i e_i^2[q_i^\uparrow(x) + q_i^\downarrow(x)]} \tag{1}$$

In other words the $g_1(x)$ structure function can be interpreted as the difference between the probability to find partons of flavor $i$ having their spins parallel, and those having their spins opposite to the spin of the target nucleon respectively. For three quark flavors and after integrating over $x$, we obtain the first moment of the proton structure function $g_1^p(x)$:

$$\int_0^1 g_1^p(x)dx = \frac{1}{2}\left[\frac{4}{9}\Delta u + \frac{1}{9}\Delta d + \frac{1}{9}\Delta s\right] \tag{2}$$

where we have defined $\Delta q = \int_0^1 dx[q^\uparrow(x) - q^\downarrow(x)]$. Interchanging $u$ and $d$ quarks gives the corresponding expression for the neutron. Polarised lepton scattering thus measures the charge-squared weighted fraction of the nucleon spin carried by each quark flavor. Neglecting the deuteron D-state, the sum of the proton and neutron becomes:

$$\int_0^1 g_1^d(x)dx = \frac{1}{18}\left[4(\Delta u + \Delta d + \Delta s) + (\Delta u + \Delta d - 2\Delta s)\right] \tag{3}$$

The second term in the right-hand side $(\Delta u + \Delta d - 2\Delta s)$ is the flavor non-singlet component. Assuming SU(3) flavor symmetry, it is given by the combination $3F - D$, where F and D are the SU(3) couplings determined experimentally in low-energy hyperon beta decay. The singlet term $(\Delta u + \Delta d + \Delta s)$ is the fraction of the proton spin $\Delta\Sigma$ carried by the quarks. From the EMC data on the proton this number was found to be $\Delta\Sigma = 0.12 \pm 0.10 \pm 0.14$, which is consistent with the statement that the quarks do not contribute to the proton spin. This result has triggered an enormous interest, both theoretical and experimental. The theoretical situation has been reviewed in several recent publications[9,10].

The first moment of $g_1$ can be compared to sum rule predictions. Sum rules are integrals involving combinations of structure functions. They are particularly important since they are the most reliable theoretical calculations presently available. Some of the sum rules for weak and electromagnetic interactions have been derived using current algebra, before any experimental information was available. A sum rule for the the polarised structure functions has been obtained by Bjorken [11] using current algebra. This sum rule relates the difference of $g_1$ for the proton and the neutron to the axial vector coupling constant in nucleon beta decay:

$$\Gamma_{Bj} = \int_0^1 [g_1^p(x) - g_1^n(x)]dx = \frac{1}{6}g_A(1 - \frac{\alpha_s(Q^2)}{\pi}) \tag{4}$$

The term in parenthesis is the first order QCD correction. In order to test this sum rule one needs to know both $g_1^p(x)$ and $g_1^n(x)$. Separate sum rules for the proton and the neutron have been developed by Ellis and Jaffe [12]. In order to derive them they have further assumed exact SU(3) symmetry for the baryon-octet beta decay and zero polarisation of the strange quark sea of the nucleon. The sum rules are then written:

$$\Gamma^p = \int_0^1 g_1^p(x)dx = \frac{1}{12}\left[+1 + \frac{5}{3}\frac{3F/D - 1}{F/D + 1}\right](1 - \frac{\alpha_s(Q^2)}{\pi}) \tag{5}$$

$$\Gamma^n = \int_0^1 g_1^n(x)dx = \frac{1}{12}\left[-1 + \frac{5}{3}\frac{3F/D - 1}{F/D + 1}\right](1 - \frac{\alpha_s(Q^2)}{\pi}) \tag{6}$$

Here higher order QCD corrections have been neglected.

# 3 The SMC experiment

The SMC experiment uses a polarised muon beam, a polarised deuteron target, a spectrometer to measure the scattered muon, and a beam polarimeter. The target and the spectrometer are based on the apparatus built by the EMC collaboration [15], but have been upgraded to reduce systematic uncertainties. The polarimeter is a new equipment. During data taking, we have used a beam of positive muons with an average energy of 100 GeV[16], a spill length of 2 s and a period of 14 s. The beam intensity was $4 \times 10^7$ muons per spill. The CERN muon beam is naturally polarised due to the parity vio-

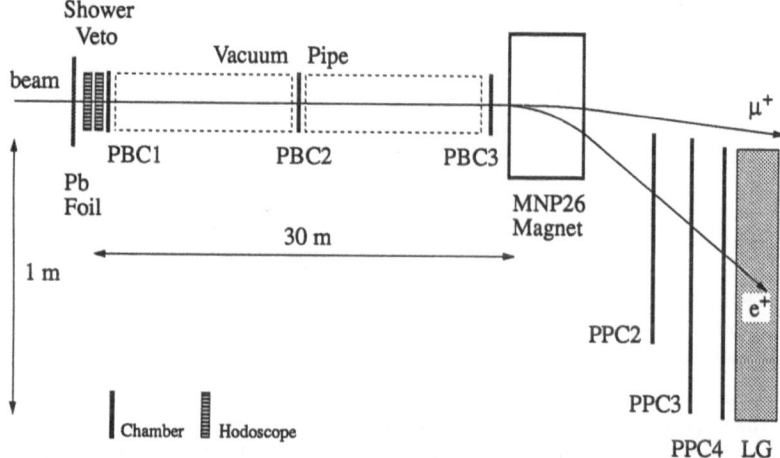

Figure 1: *The SMC decay polarimeter*

lating nature of the $\pi \rightarrow \mu\nu_\mu$ decay. Its polarisation can be determined from the shape of the energy spectrum of the positrons from the decay $\mu^+ \rightarrow e^+\nu_e\bar{\nu}_\mu$. Downstream of the muon spectrometer, we have built a muon decay polarimeter (Fig. 1). A 30 m long muon decay path is defined between a shower veto hodoscope (which identifies the $\mu^+$) and a dipole magnet. The parent and decay particles are tracked both before and after the analyzing magnet using multiwire proportional chambers. Identification of the decay positrons is done by measuring the energy deposited in a lead glass calorimeter. A measured spectrum, corrected for geometrical acceptance and radiative effects is shown in Fig. 2. The muon polarisation as determined by fitting the shape of this spectrum is found to be $P_\mu = -0.82 \pm 0.06$, in good agreement with Monte Carlo simulations of the beam transport[16]. The polarised target in the present experiment uses the same cryogenic components as the EMC target [2]. A superconducting solenoid provides a magnetic field of 2.5 T parallel to the beam direction. A dilution refrigerator cools the target to a temperature of about 500 mK during polarisation, and to 50 mK during frozen spin operation. The target is divided in two halves, each 40 cm long and 5 cm in diameter, separated by 20 cm. The longitudinal polarisations in the two halves are opposite in sign in order to record data with both polarisation directions simultaneously.

The target material is beads of deuterated butanol with an admixture of paramagnetic molecules. The target is polarised using the method of Dynamic Nuclear Polarisation (DNP), in which microwave power at a frequency close to the resonance of the paramagnetic electrons is applied. The typical deuteron polarisation was about 0.25 before it was discovered that a substantial increase in the polarisation can be obtained by modulating

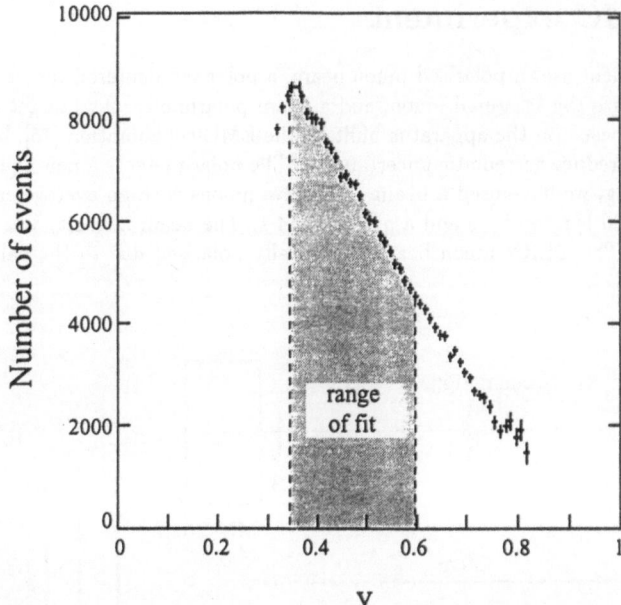

Figure 2: *Experimental muon decay spectrum as a function of the ratio*
*y of the decay positron and parent muon energiy. The spectrum was*
*folded with acceptance and radiative correction effects.*

rapidly the microwave frequency over a range of 30 MHz. In this way, deuteron polarisations larger than 0.40 have been routinely obtained. The polarisation is measured using 10 NMR coils embedded in the target material. The integrated NMR absorption signals were calibrated in thermal equilibrium at 1.1 K. Such absorption signals for two of the NMR coils are shown in Fig. 3. The final polarisation, averaged over the whole data taking period was $P_T = 0.35 \pm 0.02$.

In order to minimise systematic errors, we have reversed the polarisation directions in the two target halves at regular intervals. For most of our data the spins were reversed every 8 hours by rotating the field direction using the 0.2 T transverse field of a superconducting dipole coil wound on the microwave cavity. In addition, the relative orientation of the solenoid field and the target polarisation was changed once or twice per week with DNP.

The muon spectrometer (shown in Fig. 4) was originally designed and constructed by the EMC collaboration and was later upgraded by the NMC and SMC collaborations. Upstream of the target, veto counters and beam hodoscopes define the portion of the useful beam phase space. In the first stage of the spectrometer charged particles are momentum analysed with a conventional large aperture dipole magnet and several sets of proportional and drift chambers. Hadrons are absorbed in an iron wall. Downstream of this wall streamer tubes, drift tubes and scintillator hodoscopes are used for muon identification and triggering.

Off-line event reconstruction programs determine the kinematics of both the incident and scattered muon and the vertex position. The incident muon momentum is measured upstream of the SMC experimental hall in a detector consisting of a dipole magnet and four planes of fast scintillator hodoscopes. The scattered muon is first identified downstream of the absorber wall. Its trajectory is then reconstructed using the drift chambers located upstream of the absorber and extrapolated through the analysing

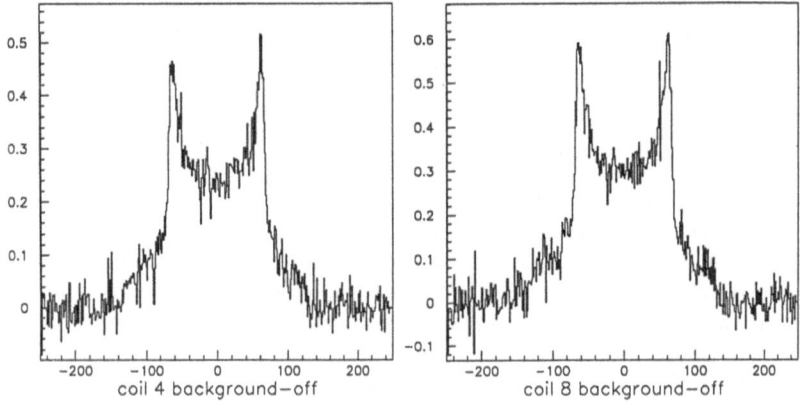

coil 4 background−off

coil 8 background−off

Figure 3: *NMR absorption signals used for the determination of the target polarisation.*

magnet up to the interaction point. The average vertex resolution achieved is 3 cm in the direction of the beam and 0.3 mm in the transverse plane. This permits a good separation between the events originating from the upstream and downstream target cells.

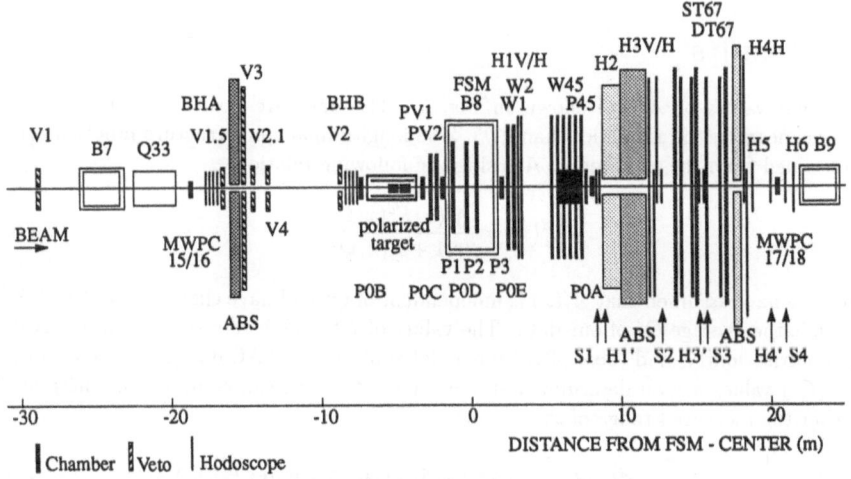

Figure 4: *The SMC muon spectrometer.*

The entire SMC experimental set-up was simulated using a Monte Carlo simulation program. This program takes into account the experimental resolution and efficiency of all detectors, as well as the distribution of incident muons recorded during the data taking. It has been used to test the event reconstruction, to determine resolution smearing and to estimate systematic errors due to efficiency variations of the various detectors.

Cuts were applied on kinematic variables in order to minimise smearing effects, to limit the size of radiative corrections and to reject muons originating from the decay of pions produced in the target. The final data cover the kinematic range $1 \text{ GeV}^2 < Q^2 < 30$ $\text{GeV}^2$ and $0.006 < x < 0.6$. The yield of each of the two target cells can be used to determine the muon-deuteron asymmetry $A_d$:

$$N_u = n_u \Phi a_u \sigma_0 (1 - f P_\mu P_u A^d) \tag{7}$$
$$N_d = n_d \Phi a_d \sigma_0 (1 - f P_\mu P_d A^d), \tag{8}$$

where the subscripts $u$ and $d$ refer to the upstream and downstream target cells, $n$ is the number of target nucleons, $\Phi$ the beam flux, $a$ the apparatus acceptance, $\sigma_0$ the unpolarised cross section, $f$ the fraction of the event yield from the deuterons in the target material (dilution factor), and $P_\mu$ and $P_{u,d}$ are the beam and target polarisations. From $A^d$ one computes the virtual photon asymmetry $A_1^d$ using the relation:

$$A_1^d = \frac{A^d}{D} - \eta A_2^d, \tag{9}$$

Here $D$ is a depolarisation factor that depends on the kinematics. The $\eta A_2^d$ term has been neglected and the resulting uncertainty included as a systematic error.

The combination of the two target cells for two data sets taken before and after a polarisation reversal provides four equations for the yields, from which $A_1^d$ was extracted under the assumption that the acceptance ratio $r = a_u/a_d$ remains constant. Particular care was taken to keep this ratio $r$ as constant as possible as a function of time by carefully monitoring the stability of the experimental set-up. The uncertainty on $r$ was computed by simulating the variation of the detector parameters with a Monte-Carlo code.

## 4 Results

The final values of $A_1^d$ are shown in Fig. 5. The data are positive for large $x$ and become negative for $x$ smaller than 0.05. The longitudinal spin structure function $g_1^d(x)$ is obtained from the asymmetry $A_1^d$ using the following relation:

$$g_1^d(x) = \frac{A_1^d(x) F_2^d(x, Q^2)}{2x[1 + R(x, Q^2)]}. \tag{10}$$

Here we have assumed that $A_1^d(x)$ is independent of $Q^2$ and have chosen $Q^2 = 4.6 \text{ GeV}^2$, which is the average $Q^2$ of our data. The values of $F_2^d(x, Q^2)$ were taken from the NMC parametrisation[17] and those of $R$ from a global fit of the SLAC data[18]. The resulting $g_1^d(x, Q^2)$ values are all determined at the same $Q^2$, so we can compute the integral of $g_1^d$ over the measured range of $x$:

$$\int_{0.006}^{0.6} g_1^d(x) dx = 0.024 \pm 0.020 \, (stat.) \pm 0.014 \, (syst.). \tag{11}$$

To estimate the integral in the unmeasured region at small $x$, the three lowest data points in $x$ were fitted assuming that the behaviour of $g_1^d(x)$ is proportional to $x^{-\alpha}$, with $-0.5 < \alpha < 0$ [19]. This contribution amounts to $-0.003 \pm 0.003$. For the extrapolation to $x > 0.6$, we use a phenomenological fit which is constrained to $g_1^d(x) = 0$ at $x = 1$. This contribution amounts to $0.002 \pm 0.004$. The final result for the first moment of $g_1^d(x)$ is thus (Fig. 6):

$$\Gamma_1^d = \int_0^1 g_1^d(x) dx = 0.023 \pm 0.020 \, (stat.) \pm 0.015 \, (syst.). \tag{12}$$

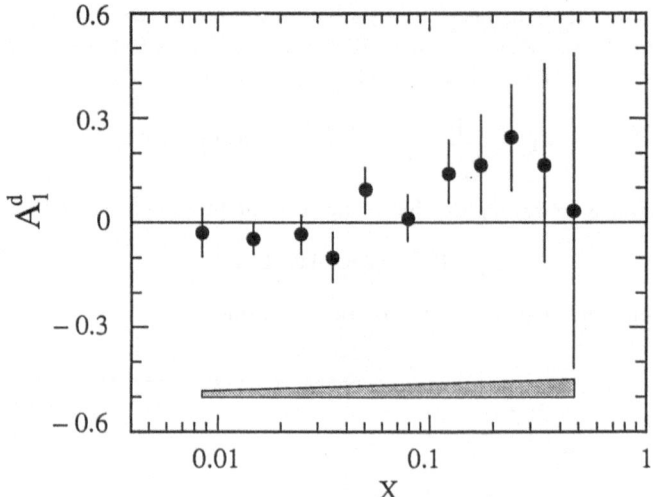

Figure 5: *Virtual photon-deuteron asymmetry as a function of the Bjorken variable x. The shaded area indicates the size of the systematic errors.*

The sum of the first moments of the spin structure functions of the proton and the neutron can be computed from $\Gamma_1^d$ using the relation $\Gamma_1^p + \Gamma_1^n \simeq 2\Gamma_1^d/(1 - 1.5\omega_D)$, where $\omega_D$ is the probability of the deuteron to be in a D-state. Using The Paris potential result $\omega_D = 0.058$ [20], we find $\Gamma_1^p + \Gamma_1^n = 0.049 \pm 0.044$ (*stat.*) $\pm 0.032$ (*syst.*). The Ellis-Jaffe sum rules [12] predict $\Gamma_1^p + \Gamma_1^n = 0.187 \pm 0.010$. This prediction is more than 2 standard deviations above the measured value.

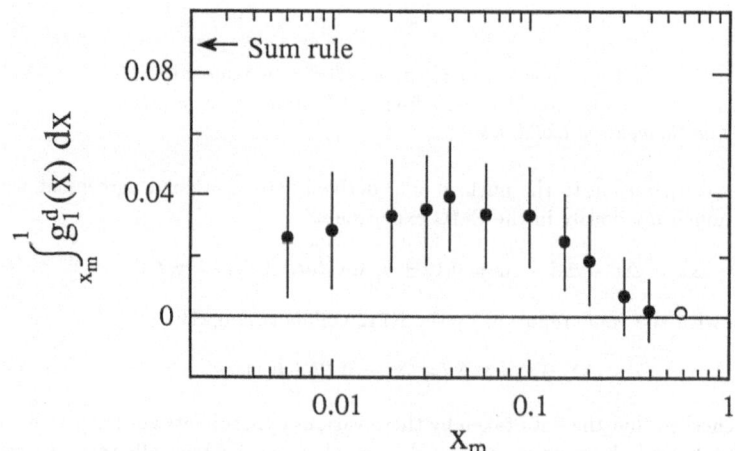

Figure 6: *Integral of $g_1^d(x)$ as a function of the lower integration limit $x_m$. The open circle represents the extrapolation to $x = 1$.*

The value of $\Gamma_1^d$ can also be expressed in terms of the quark spin polarisation distributions as shown in eq. (3). Using the value $1/\sqrt{3}(3F - D) = 0.397 \pm 0.020$ as derived from hyperon decay assuming SU(3) symmetry [13], we obtain:

$$\Delta\Sigma = \Delta u + \Delta d + \Delta s = 0.06 \pm 0.20 \pm 0.15. \tag{13}$$

This value indicates that the quark contribution to the proton spin is compatible with zero. The deduced strange quark contribution is $\Delta s = -0.21 \pm 0.09 \pm 0.07$. The result on $\Gamma_1^d$ allows us to test the Bjorken sum rule [11], which predicts at $Q^2 = 4.6$ GeV$^2$:

$$\Gamma_1^p - \Gamma_1^n = \frac{1}{6} \left| \frac{g_A}{g_V} \right| \left[ 1 - \frac{\alpha_s(Q^2)}{\pi} \right] = 0.191 \pm 0.002 \tag{14}$$

From our experiment and using the first moment of the proton [2], we find:

$$\Gamma_1^p - \Gamma_1^n = 0.20 \pm 0.05 \pm 0.04, \tag{15}$$

in agreement with the prediction of the Bjorken sum rule.

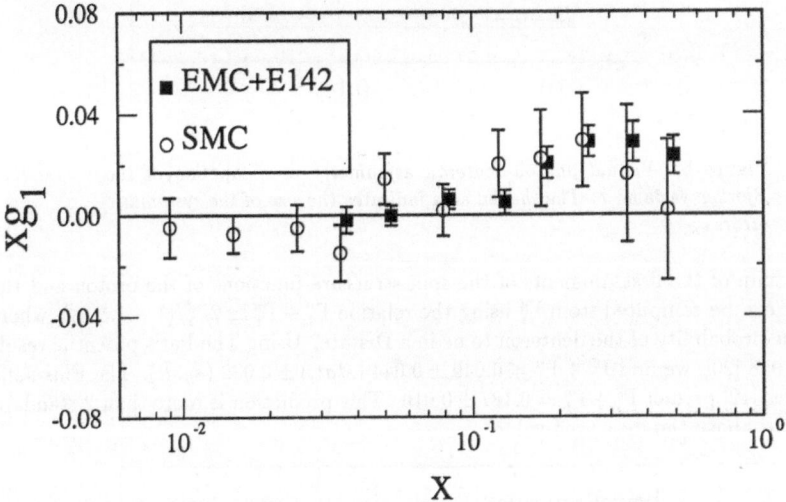

Figure 7: *The deuteron structure function $xg_1^d(x)$ as measured by SMC (open circles) and the combined data from EMC on the proton and from E142 on the neutron (solid boxes).*

The quark contribution to the nucleon spin derived here is in good agreement with the result obtained previously in the EMC experiment:

$$\Delta\Sigma = \Delta u + \Delta d + \Delta s = 0.12 \pm 0.10 \; (stat.) \pm 0.14 \; (syst.). \tag{16}$$

but disagrees with the value reported by the E142 collaboration[21]:

$$\Delta\Sigma = \Delta u + \Delta d + \Delta s = 0.57 \pm 0.11 \tag{17}$$

We have checked that the data taken by these various experiments are consistent. To do this we have used the relation $g_1^d = (g_1^p + g_1^n)(1 - \frac{3}{2}\omega_D)$ which allows us to construct "deuteron" data from the EMC experiment on the proton together with the E142 experiment on the neutron. We have further assumed that the measured virtual photon asymmetries scale with $Q^2$. Within this assumption we can use the $F_2(x, Q^2)$ parametrisation[17] to "rescale" the E142 and EMC data from 2 and 10.7 GeV$^2$ respectively, to 4.6 GeV$^2$, which is the average $Q^2$ of SMC. This procedure minimizes the uncertainty coming from the combination of data taken at different $Q^2$ values. The constructed deuteron $g_1^d(x)$ data thus obtained are shown in Fig. 7 together with the data from the present experiment. We conclude that in the region of overlap $(0.03 < x < 0.6)$ the data are in very good agreement.

Two possibilities remain to explain the difference between the SMC and E142 results. On the theoretical side, higher order perturbative QCD contributions [23] and higher twist corrections [24] have been shown to significantly modify the Bjorken sum rule. These effects are important at 2 GeV$^2$ but become negligible at 10 GeV$^2$, so they mainly affect the E142 data. On the experimental side the extrapolations to $x = 0$ are done in all experiments assuming Regge behaviour for the shape of the $g_1(x)$ structure function at low $x$. The exact value of $x$, up to which the Regge form is a good approximation to $g_1(x)$, is not known. The contribution of the extrapolation to the first moment of $g_1(x)$ thus strongly depends on the value of the lowest-$x$ data point. Further data on the proton and the deuteron are expected to measure the low $x$ region with better accuracy.

In conclusion we have measured for the first time the polarised structure function $g_1^d$ of the deuteron. We have found that the first moment of $g_1^d$ is smaller than the prediction of Ellis-Jaffe sum rule. The contribution of the quark spins to the nucleon spin is compatible with zero. The deduced difference between the first moment of the proton and the neutron is in good agreement with the fundamental Bjorken sum rule.

# References

[1] T. Sloan, G. Smadga and R. Voss, Physics Reports, **162** (1988) 45.

[2] J. Ashman et al., Phys. Lett. **B206** (1988) 1167;
    J. Ashman et al., Nucl. Phys. **B328** (1989) 1

[3] M.J. Alguard et al., Phys. Rev. Lett. **37**, (1976) 1261; ibid. **41** (1978) 70;

[4] G. Baum et al., Phys. Rev. Lett. **51** (1983) 1135

[5] B. Adeva et al., Phys. Lett. **B302** (1993) 533.

[6] P.L.Anthony et al., Phys. Rev. Lett. **71** (1993) 959.

[7] V. W. Hughes and J. Kuti, Ann. Rev. Nucl. Part. Sci. **33** (1983) 611.

[8] E. Leader and E. Predazzi, Gauge theories and the new physics, Cambridge Univ. press, 1985.

[9] S. D. Bass and A. W. Thomas, J.Phys. **G19** (1993) 925.

[10] E. Reya, University of Dortmund preprint DO-TH 93/09 (1993).

[11] J. D. Bjorken, Phys. Rev. **148** (1966) 1467; Phys. Rev. **D1** (1970) 1376

[12] J. Ellis and R. L. Jaffe, Phys. Rev. **D9** (1974) 1444; **D10** (1974) 1669.

[13] M. Bourquin et al., Z. Phys. **C21** (1983) 27.

[14] Z. Dziembowski and J. Franklin, J. Phys. G: Nucl. Part. Phys. **17** (1991) 213.

[15] O.C. Allkofer et al., Nucl. Instr. Meth. **179** (1981) 445.

[16] L. Gatignon et al., The muon beam at the SPS, to be published in Nucl. Instr. Meth.

[17] P. Amaudruz et al., Nucl. Phys. **B371** (1992) 3.

[18] L.W. Whitlow et al., Phys. Lett. **B250** (1990) 193.

[19] J. Ellis and M. Karliner, Phys. Lett. **B213** (1988) 73.

[20] M. Lacombe et al., Phys. Lett. **B101** (1981) 139.

[21] Z. E. Meziani, invited talk at this conference.

[22] J. Ellis and M. Karliner, CERN preprint CERN/TH-2052-93.

[23] S. A. Larin and J. A. M. Vermaseren, Phys. Lett. **B259** (1991) 345.

[24] I. I. Balitsky, V. M. Braun and A. V. Kolesnichenko, Phys. Lett. **B242** (1990) 245.

Few-Body Systems, Suppl. 7, 458—465 (1994)

Few-
Body
Systems
© Springer-Verlag 1994
Printed in Austria

# INCLUSIVE QUASIELASTIC AND DEEP INELASTIC SCATTERING OF POLARIZED ELECTRONS BY POLARIZED $^3$He

C. Ciofi degli Atti and S. Scopetta

*Department of Physics, University of Perugia, and INFN, Sezione di Perugia,*
*Via A. Pascoli, I-06100 Perugia, Italy*

E. Pace

*Dipartimento di Fisica, Università di Roma "Tor Vergata", and INFN,*
*Sezione Tor Vergata, Via E. Carnevale, I-00173 Roma, Italy*

G. Salmè

*INFN, Sezione Sanità, Viale Regina Elena 299, I-00161 Roma, Italy*

## Abstract

A comprehensive treatment of the theoretical approach for describing nuclear effects in inclusive scattering of polarized electrons by polarized $^3$He is presented.

## 1. Introduction

The advent of new experimental facilities, allowing systematic measurements with polarized $^3$He and polarized electron beams, are substantially increasing the amount of information on the electromagnetic properties of the neutron, in a wide range of the kinematical variables. As is well known, polarized $^3$He represents a good candidate as an effective neutron target [1]. The effects of nuclear structure, however, have to be carefully investigated in order to reliably extract information on the neutron from the data both in the quasielastic (qe) [2,3] and in the deep inelastic [4] region. In what follows, we will illustrate how the proton contribution affects the extraction of the neutron elastic form factors [5-7] and spin structure functions [8].

## 2. The polarized inclusive cross section

The inclusive cross section describing the scattering of a longitudinally polarized lepton of helicity $h = \pm 1$ by a polarized hadron of spin $J = 1/2$, is given in one photon exchange approximation by [1a]

$$\frac{d^2\sigma(h)}{d\Omega_2 d\nu} \equiv \sigma_2\left(\nu, Q^2, \vec{S}_A, h\right) = \frac{4\alpha^2}{Q^4}\frac{\epsilon_2}{\epsilon_1} m^2 L^{\mu\nu}W_{\mu\nu} = \frac{4\alpha^2}{Q^4}\frac{\epsilon_2}{\epsilon_1} m^2 \left[L_s^{\mu\nu}W_{\mu\nu}^s + L_a^{\mu\nu}W_{\mu\nu}^a\right] \quad (1)$$

Presented by G. Salmè

where $L^{\mu\nu}_{s(a)}$ and $W^{s(a)}_{\mu\nu}$ are the symmetric $(s)$ (antisymmetric $(a)$) leptonic and hadronic tensors, respectively. The antisymmetric hadronic tensor is given by

$$W^a_{\mu\nu} = i\epsilon_{\mu\nu\rho\sigma}q^\rho V^\sigma \tag{2}$$

where $V^\sigma$ is a pseudovector that can be expressed as follows

$$V^\sigma \equiv S^\sigma_A \frac{G^A_1}{M_A} + (P_A \cdot q \, S^\sigma_A - S_A \cdot q \, P^\sigma_A) \frac{G^A_2}{M^3_A} \tag{3}$$

In the above equations, the index A denotes the number of nucleons composing the target; $G^A_1$ and $G^A_2$ are the polarized structure functions; $k^\mu_{1(2)} \equiv (\epsilon_{1(2)}, \vec{k}_{1(2)})$ and $P^\mu_A \equiv (M_A, 0)$ are electron and target four-momenta; $q^\mu \equiv (\nu, \vec{q})$ is the four-momentum transfer, $Q^2 = -q^2$; $g_{\mu\nu}$ is the symmetric metric tensor, $\epsilon_{\mu\nu\rho\sigma}$ the fully antisymmetric tensor and $S^\mu_A$ the polarization four-vector (in the rest frame $S^\mu_A \equiv (0, \vec{S}_A)$).

The antisymmetric hadronic tensor, $W^a_{\mu\nu}$, is constrained to the very general expression (2) by invariance principles (Lorentz, gauge, parity and time reversal invariance); moreover the antisymmetric tensor $\epsilon_{\mu\nu\alpha\beta}$ in Eq.(2) cancels out any contribution to $W^a_{\mu\nu}$ arising from possible terms proportional to $q^\mu$ ($\epsilon_{\mu\nu\alpha\beta} q^\alpha q^\beta = 0$) in $V^\sigma$. This fact has to be carefully taken into account once a model is adopted for obtaining the polarized structure functions [6].

The polarized structure functions $G^A_1$ and $G^A_2$ have to be obtained by expressing them in terms of the components of $W^a_{\mu\nu}$. Thus, assuming in the rest frame of the target the z-axis along the momentum transfer ($\hat{q} \equiv \hat{u}_z$), and using Eq.(3), one has (cf. Refs.[6,7])

$$\frac{G^A_1}{M_A} = -i\left(\frac{Q^2}{|\vec{q}|^3}\frac{W^a_{02}}{S_{Ax}} + \frac{\nu}{|\vec{q}|^2}\frac{W^a_{12}}{S_{Az}}\right) \qquad \frac{G^A_2}{M^2_A} = -i\frac{1}{|\vec{q}|^2}\left(\frac{\nu}{|\vec{q}|}\frac{W^a_{02}}{S_{Ax}} - \frac{W^a_{12}}{S_{Az}}\right) \tag{4}$$

It should be pointed out that in Ref.[1a] $G^A_{1(2)}$ have been obtained by another procedure, namely using the components of the pseudovector $V^\sigma$, Eq.(3); in this case one has

$$\frac{G^A_1}{M_A} = -\frac{(V \cdot q)}{|\vec{q}|S_{Az}} \qquad \frac{G^A_2}{M^2_A} = \frac{V_0}{|\vec{q}|S_{Az}} \tag{5}$$

Given the form (3) for $V^\sigma$, Eq.(4) are totally equivalent to Eq.(5). However, such an equivalence will *not hold* if a term proportional to $q^\mu$ is explicitely added to the r.h.s. of Eq.(3), since Eq.(4) will be unaffected by the added term, whereas Eq. (5) will be; therefore $G^A_1$ and $G^A_2$ obtained from Eq.(5) will not be correct in this case [6].

After contracting the two tensors in Eq.(1) one has

$$\frac{d^2\sigma(h)}{d\Omega_2 d\nu} = \Sigma + h\,\Delta \tag{6}$$

where $\Sigma$ and $\Delta$ describe the unpolarized and the polarized scattering, respectively. The polarized term is

$$\Delta = \sigma_{Mott} \, 2\tan^2\frac{\theta_e}{2}\left[\frac{G^A_1(Q^2, \nu)}{M_A}(\vec{k}_1 + \vec{k}_2) + 2\frac{G^A_2(Q^2, \nu)}{M^2_A}(\epsilon_1 \vec{k}_2 - \epsilon_2 \vec{k}_1)\right] \cdot \vec{S}_A \tag{7}$$

Experimentally one measures the following asymmetry

$$A = \frac{\sigma_2\left(\nu, Q^2, \vec{S}_A, +1\right) - \sigma_2\left(\nu, Q^2, \vec{S}_A, -1\right)}{\sigma_2\left(\nu, Q^2, \vec{S}_A, +1\right) + \sigma_2\left(\nu, Q^2, \vec{S}_A, -1\right)} = \frac{\Delta}{\Sigma} \tag{8}$$

In order to obtain the theoretical asymmetry, one has to introduce some approximations; in particular, till now, the Plane Wave Impulse Approximation (PWIA) has been adopted. Within such a framework the polarized structure functions $G_1^A$ and $G_2^A$ are given by (cf. Refs.[6-7] for the qe case and [8-9] for the deep inelastic one)

$$
\frac{G_1^A(Q^2,\nu)}{M_A} = \sum_{N=p,n} \int dz \int dE \int d\vec{p} \, \frac{1}{E_p \, M} \left\{ \hat{G}_1^N\left(z,\nu,Q^2\right) \left[ M \, P_\parallel^N(|\vec{p}|,E,\alpha) + \right. \right.
$$
$$
\left. \left. -|\vec{p}| \left( \frac{\nu}{|\vec{q}|} - \frac{|\vec{p}| \cos\alpha}{M+E_p} \right) \mathcal{P}^N(|\vec{p}|,E,\alpha) \right] - \frac{Q^2}{|\vec{q}|^2} \, \mathcal{L}^N \right\} \delta\left( z + \frac{M^2 - p \cdot p}{2M\nu} - \frac{q \cdot p}{M\nu} \right) \quad (9)
$$

$$
\frac{G_2^A(Q^2,\nu)}{M_A^2} = \sum_{N=p,n} \int dz \int dE \int d\vec{p} \, \frac{1}{E_p \, M} \left\{ \left[ \hat{G}_1^N\left(z,\nu,Q^2\right) \frac{|\vec{p}|}{|\vec{q}|} \, \mathcal{P}^N(|\vec{p}|,E,\alpha) + \right. \right.
$$
$$
\left. + \frac{\hat{G}_2^N(z,\nu,Q^2)}{M} \left( E_p \, P_\parallel^N(|\vec{p}|,E,\alpha) - \frac{|\vec{p}|^2 \cos\alpha}{M+E_p} \, \mathcal{P}^N(|\vec{p}|,E,\alpha) \right) \right] - \frac{\nu}{|\vec{q}|^2} \, \mathcal{L}^N \right\}
$$
$$
\delta\left( z + \frac{M^2 - p \cdot p}{2M\nu} - \frac{q \cdot p}{M\nu} \right) \quad (10)
$$

where E is the removal energy, $p \equiv (M_A - \sqrt{(E + M_A - M)^2 + |\vec{p}|^2}, \vec{p})$, $E_p = \sqrt{M^2 + |\vec{p}|^2}$ and $\hat{G}_{1(2)}^N(z,\nu,Q^2)$ are the nucleon structure functions to be used in the energy transfer region considered (cf. Sects. 4 and 5, below). The function $\mathcal{L}^N$ is

$$
\mathcal{L}^N = \left[ \hat{G}_1^N\left(z,\nu,Q^2\right) \mathcal{H}_1^N + |\vec{q}| \frac{\hat{G}_2^N(z,\nu,Q^2)}{M} \, \mathcal{H}_2^N \right] \quad (11)
$$

where $\mathcal{H}_1^N$ and $\mathcal{H}_2^N$ are proper combinations of $P_\parallel^N(p,E,\alpha)$ and $P_\perp^N(p,E,\alpha)$ [6], and $\mathcal{P}^N(p,E,\alpha) = \cos\alpha \, P_\parallel^N(p,E,\alpha) + \sin\alpha \, P_\perp^N(p,E,\alpha)$. The quantities $P_\parallel^N(p,E,\alpha)$ and $P_\perp^N(p,E,\alpha)$, already used in a previous paper of ours [5], are related to the elements of the 2x2 matrix, representing the spin-dependent spectral function of a nucleon inside a nucleus with polarization $\vec{S}_A$. The elements of this matrix are

$$
P_{\sigma,\sigma',\mathcal{M}}^N(\vec{p},E) = \sum_{f_{A-1}} {}_N\langle \vec{p},\sigma; \psi_{A-1}^f | \psi_{JM} \rangle \langle \psi_{JM} | \psi_{A-1}^f; \vec{p},\sigma' \rangle_N \, \delta(E - E_{A-1}^f + E_A) \quad (12)
$$

where $|\psi_{JM}\rangle$ is the ground state of the target nucleus polarized along $\vec{S}_A$, $|\psi_{A-1}^f\rangle$ is an eigenstate of the (A-1) nucleon system, $|\vec{p},\sigma\rangle_N$ is the plane wave for the nucleon $N \equiv p(n)$.

### 4. The asymmetry in the quasielastic region

As is well known, the experiments in the qe region [2,3] are aimed at investigating the neutron elastic form factor. In this kinematical region, the nucleon structure functions $\hat{G}_{1(2)}^N(z,\nu,Q^2)$ are related to the Sachs electromagnetic form factors $G_E^N$ and $G_M^N$ as follows

$$
\hat{G}_1^N\left(z,\nu,Q^2\right) = -\frac{1}{\nu}\delta\left(z - \frac{Q^2}{2M\nu}\right) \frac{G_M^N}{2} \frac{(G_E^N + \tau \, G_M^N)}{(1+\tau)} \quad (13)
$$

$$
\hat{G}_2^N\left(z,\nu,Q^2\right) = \frac{1}{\nu}\delta\left(z - \frac{Q^2}{2M\nu}\right) \frac{G_M^N}{4} \frac{(G_M^N - G_E^N)}{(1+\tau)} \quad (14)
$$

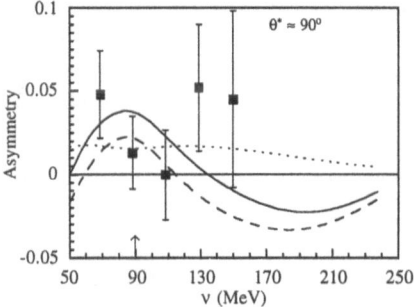

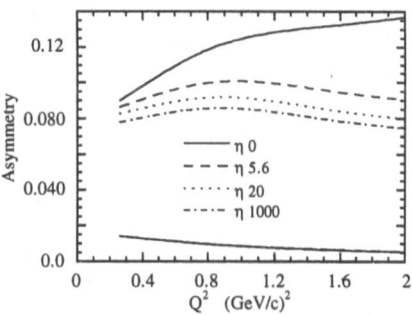

Fig. 1. The asymmetry corresponding to $\epsilon_1 = 574\ MeV$ and $\theta_e = 44°$, vs. the energy transfer $\nu$ calculated by Eqs. (10) and (11) (solid line) and using the spin-dependent spectral function of Ref.[5]; the dotted (dashed) line represents the neutron (proton) contribution. The nucleon form factors of Ref. [10] have been used and the experimental data are from Ref.[2]. The arrow indicates the position of the qe peak. (After Ref.[6])

Fig. 2. The total asymmetry at the top of the qe peak, vs. $Q^2$, for $\theta_e = 75°$ and $\beta = 95°$, using Eqs. (10) and (11). The Galster form factors [11] have been used. The curves in the lower part of the figure represent the corresponding proton contributions. (After Ref.[6])

with $\tau = Q^2/(4M^2)$.

By substituting Eqs.(13) and (14) in Eqs.(9) and (10) one obtains the expressions of $G_{1(2)}^A$ given in Ref.[6]. Present experimental results aim at measuring the quantities $R_{T'}^A$ and $R_{TL'}^A$, given by

$$R_{T'}^A(Q^2,\nu) = -2\left(\frac{G_1^A(Q^2,\nu)}{M_A}\nu - Q^2\frac{G_2^A(Q^2,\nu)}{M_A^2}\right) = i\,2\,\frac{W_{12}^a}{S_{Az}} \qquad (15)$$

$$R_{TL'}^A(Q^2,\nu) = 2\sqrt{2}\,|\vec{q}|\left(\frac{G_1^A(Q^2,\nu)}{M_A} + \nu\frac{G_2^A(Q^2,\nu)}{M_A^2}\right) = -i\,2\sqrt{2}\,\frac{W_{02}^a}{S_{Ax}} \qquad (16)$$

The interest in these quantities is due to the fact that $R_{T'}^A$ and $R_{TL'}^A$, at the top of the qe peak, are proportional to $(G_M^n)^2$ and $G_E^n\,G_M^n$, respectively, *provided the proton contribution can be disregarded* [2,3].

In Fig. 1 the asymmetry, measured by the MIT-Caltech collaboration [2], corresponding to $\epsilon_1 = 574\ MeV$ and $\theta_e = 44°$ and averaged over three different values of the polarization angles around $\theta^* \approx 90°$, $(\cos\theta^* = \vec{S}_A \cdot \hat{q})$ is shown together with the neutron (dotted line) and proton (dashed line) contributions. It is worth noting that, in these kinematical conditions, the measured asymmetry reduces to $R_{TL'}$ only at the top of the qe peak $(A_{qe}^{exp} \propto R_{TL'}^{exp})$. Therefore, as previously explained, one could have access to $G_E^n\,G_M^n$, provided the proton contribution can be disregarded; but unfortunately, it is shown that this is not the case, for the proton contribution is relevant at the top of the qe peak. A comparison with the experimental value obtained after averaging over the polarization angle and over a 100 $MeV$ interval for the energy transfer yields

$$A_{qe}^{exp} = 2.41 \mp 1.29 \mp 0.51\ \%\quad MIT-Caltech^2$$
$$A^{th} = 1.65\ \%\quad (3.74\ \%)$$

The theoretical value in the brackets is obtained without averaging over the energy, i.e. at $\nu = \nu_{peak}$.

In correspondence to a different choice of kinematical variables, $\epsilon_1 = 574\ MeV$ and $\theta_e = 51.1^o$, only one experimental point has been obtained, just for the asymmetry averaged over both three polarization angles around $\theta^* \approx 0^o$ and a 50 $MeV$ interval for the energy transfer around the top of the qe peak, where $A_{qe}^{exp} \propto R_{T'}^{exp}$ [2]. The comparison reads as follows

$$A_{qe}^{exp} = -3.79 \mp 1.37 \mp 0.67\ \%\ \ MIT - Caltech^2$$
$$A^{th} = -4.30\ \%\ \ \ \ (-3.43)\ \%$$

It should be pointed out that our results are only slightly different from the ones obtained in Ref.[7], where a spin-dependent Faddeev spectral function for $^3He$ and the nucleon form factors of Ref.[11] have been used.

The results presented in Fig. 1 show that the proton contribution to the measured asymmetry is sizeable. However, as shown in Refs. [5,6], one can minimize or even make vanishing the proton contribution. As a matter of fact, it turns out that it is possible to find a polarization angle $\beta = \beta_c$ ($cos\ \beta = \vec{S}_A \cdot \hat{k}_1$) in correspondence of which the proton contribution at the top of the qe peak vanishes within a wide range of values of the incident electron energy, and even for different models of the nucleon form factors. Moreover, in order to investigate the sensitivity of the asymmetry upon $G_E^n$, such a quantity has been calculated [6] in the range $0.3 \leq Q^2 \leq 2\ (GeV/c)^2$, at fixed values of $\beta_c = 95^o$ and $\theta_e = 75^o$, using the Galster form factors [11], since within such a model $G_E^n$ can be changed independently of $G_M^n$. In fact one has

$$G_M^n = \mu_n\ G_E^p \qquad\qquad G_E^n = \frac{-\tau\ \mu_n}{(1 + \eta\ \tau)}\ G_E^p \qquad\qquad (17)$$

where $G_E^p = 1/(1 + Q^2/B)^2$, $B = 0.71(GeV/c)^2$ and $\eta$ is a parameter. The resulting asymmetry and the proton contribution are shown in Fig. 2 for different values of $\eta$. Therefore Fig.2 illustrates how the total asymmetry can depend upon $G_E^n$, having a vanishing proton contribution.

It should be stressed that the proposed kinematics, which minimizes the proton contribution, corresponds to the qe peak, where the final state interaction is expected to play only a minor role.

## 5. The asymmetry in the deep inelastic region

Deep inelastic scattering (DIS) of longitudinally polarized electrons off polarized $^3He$ is aimed at measuring the spin structure functions (SSF) of the neutron, $g_1^n$ and $g_2^n$. Knowledge of this SSF provides information on the spin distribution among the nucleon partons and can allow a very important test of QCD, such as the check of the Bjorken Sum Rule [12]. As is well known, only recently $g_1^n$ has become experimentally available from two different experiments [13-14] on polarized deuterons and $^3He$ nuclei.

The longitudinal ($\beta = 0$) asymmetry can be recast in the following form, suitable for the analysis of DIS (see, e.g., Refs.[9,14])

$$A_{||} = 2x[1 + R(x, Q^2)]\frac{g_1^A(x, Q^2) - \frac{Q^2}{\nu(\epsilon_1 + \epsilon_2\ cos\ \theta_e)}g_2^A(x, Q^2)}{F_2^A(x, Q^2)} \qquad\qquad (18)$$

where $x = Q^2/2M\nu$ is the Bjorken variable, $g_1^A$ and $g_2^A$ are the nuclear SSF, $F_2^A$ is the spin-independent structure function of the target $A$, $R(x, Q^2) = \sigma_L(x, Q^2)/\sigma_T(x, Q^2)$.

The expressions for $g_1^A$ and $g_2^A$ [9] can easily be obtained from Eqs.(9) and (10) by using the following replacements

$$g_1^A = M\nu \frac{G_1^A}{M_A} \qquad\qquad g_2^A = M\nu^2 \frac{G_2^A}{M_A^2} \qquad (19)$$

$$g_1^N(x, z, Q^2) = \frac{p \cdot q}{M} \hat{G}_1^N\left(z, \nu, Q^2\right) \qquad g_2^N(x, z, Q^2) = \frac{(p \cdot q)^2}{M^3} \hat{G}_2^N\left(z, \nu, Q^2\right) \qquad (20)$$

In the Bjorken limit ($\nu/|\vec{q}| \to 1$, $Q^2/|\vec{q}|^2 \to 0$), the asymmetry reduces to

$$A_{\parallel} = 2x \frac{g_1^A(x)}{F_2^A(x)} \qquad (21)$$

and $g_1^A$ becomes a function of $g_1^N$ only; namely it reads as follows

$$g_1^A(x) = \sum_N \int_x^A dz \frac{1}{z} g_1^N\left(\frac{x}{z}\right) G^N(z) \;, \qquad (22)$$

with the spin-dependent light cone momentum distribution for the nucleon given by

$$G^N(z) = \int dE \int d\vec{p} \left\{ P_{\parallel}^N(p, E, \alpha) - \left[1 - \frac{p_{\parallel}}{E_p + M}\right] \frac{|\vec{p}|}{M} \mathcal{P}^N(p, E, \alpha) \right\} \delta\left(z - \frac{p^+}{M}\right) \qquad (23)$$

where $p^+ = p^0 - p_{\parallel}$ is the light cone momentum component.

The $^3\vec{H}e$ asymmetry (Eq.(18)) and the SSF $g_1^3$, Eq.(22), calculated in the Bjorken limit using the SSF $g_1^N$ of Ref.[15], are shown in Figs. 3a and 3b, respectively. We would like to stress the following point: the non-vanishing proton contribution to the asymmetry shown in Fig. 3a hinders in principle the extraction of the neutron structure function from the $^3\vec{H}e$ asymmetry. It can be seen from Fig. 3b that for $0.01 \le x \le 0.3$ the neutron contribution, $g_1^{3,n}$, differs from the neutron structure function $g_1^n$ by a factor of about 10%; since this factor is generated by nuclear effects, one might be tempted to consider it as the theoretical error on the determination of $g_1^n$; however, it should be recalled that the difference between $g_1^n$ and $g_1^{3,n}$ is in principle model dependent through the way nuclear effects are introduced and the specific form of $g_1^n$ used in the convolution formula. In order to investigate in detail such a question, Eq.(22) has been extensively analysed in Refs.[8,9], where it has been shown that a factorized formula for $g_1^3$ (see Ref.[16] for the qe region) represents a reliable approximation of the Eq.(22) at least for $x \le 0.9$. The factorized formula can be heuristically obtained by expanding $\frac{1}{z} g_1^N\left(\frac{x}{z}\right)$ in Eq.(22) around $z = 1$ and by disregarding the term proportional to $\mathcal{P}^N$ in Eq.(23), which gives anyway a very small contribution, being of the order $|\vec{p}|/M$. Thus one has

$$g_1^3(x) \approx 2p_p g_1^p(x) + p_n g_1^n(x) \qquad (24)$$

where $p_p$ and $p_n$ are the effective nucleon polarizations, produced by the $S'$ and $D$ waves in the ground state of $^3$He, and given by

$$p_N = \int dE \int d\vec{p} P_{\parallel}^N(p, E, \alpha) \qquad (25)$$

Our calculations yield $p_p = -.030$ and $p_n = 0.88$ in agreement with world values $p_p = -0.028 \pm 0.004$ and $p_n = 0.86 \pm 0.02$ reported in Ref.[16]. In Fig. 4, the relevant nuclear effects, due to the effective nucleon polarizations induced by $S'$ and $D$ waves,

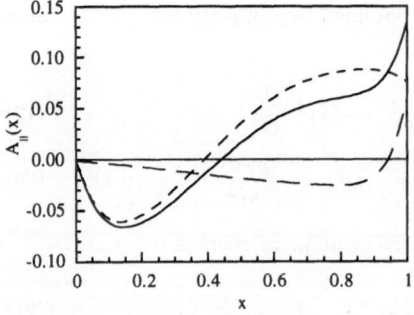

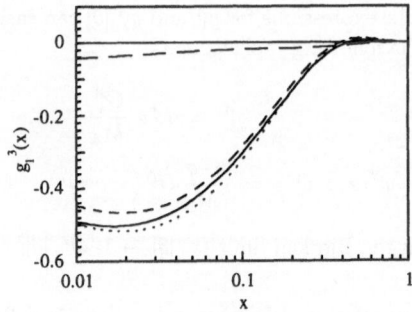

Fig. 3a. The $^3$He asymmetry [Eq. (21)] calculated within the convolution approach [Eq. (22)] (full line). Also shown are the neutron (short-dashed line) and proton (long-dashed line) contributions. (After Ref.[8])

Fig. 3b. The SSF $g_1^3$ of $^3$He (full line); also shown are the neutron (short-dashed line) and proton (long-dashed line) contributions. The dotted curve represents the free neutron structure function $g_1^n$. The difference between the dotted and short-dashed lines is due to nuclear structure effects. (After Ref.[8])

Fig. 4. The free neutron structure function $g_1^n$ (dotted line) compared with the neutron structure function given by Eq. (26)(dashed line). The difference between the two curves is due to Fermi motion and binding effects. The SSF $g_1^N$ of Ref.[15] has been used. (After Ref.[8])

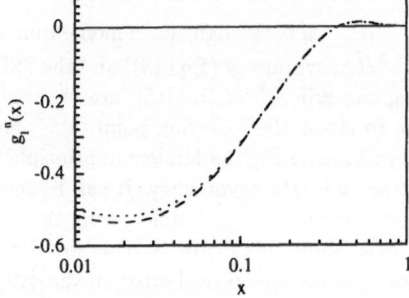

are illustrated through the comparison between the free neutron structure function and the quantity

$$\tilde{g}_1^n(x) = \frac{1}{p_n}\left[g_1^3(x) - 2p_p g_1^p(x)\right] \qquad (26)$$

calculated using the convolution formula for $g_1^3(x)$. It can be seen that the two quantities are very close to each other, differing, because of binding and Fermi motion effects, by at most 4%. Such a small difference is rather independent of the form of any well behaved $g_1^N$ [8], and therefore Eq.(24) can be considered a workable formula for extracting $g_1^n(x)$ from the experimental $g_1^3(x)$.

## 6. Summary and conclusion

The analysis of the asymmetry, based on the correct expression of $G_1^A$ and $G_2^A$ given by Eqs.(9) and (10), respectively, has put in evidence : i) the relevance of the proton both in the qe [6,7] and the DIS regions [8,9], ii) the possibility of selecting a polarization angle, which leads at qe peak to an almost vanishing proton contribution for a wide range of the kinematical variables [6], and therefore making feasible the analysis of the sensitivity of the asymmetry to the electric neutron form factor; iii) the reliability of the factorized formula, represented by Eq.(24), for extracting $g_1^n(x)$ from the experimental $g_1^3(x)$.

Calculations of the final state effects are in progress.

## 6. References

1. a) B. Blankleider and R.M. Woloshyn, *Phys. Rev.* **C 29**, 538 (1984); b) R.M. Woloshin, *Nucl. Phys.* **A495**, 749 (1985).

2. a) C. E. Jones-Woodward et al., *Phys. Rev.* **C 47**, 110 (1993) and references quoted therein; b) R.D. McKeown, this Conference.

3. A. K. Thompson et al., *Phys. Rev. Lett.* **68**, 2901 (1992); A. M. Bernstein, *Few-Body Systems Suppl.* **6**, 485 (1992).

5. C. Ciofi degli Atti, E. Pace and G. Salmè, *Phys. Rev.* **C 46**, R1591 (1992).

6. C. Ciofi degli Atti, E. Pace and G. Salmè, INFN-ISS 93-3, Proceedings of the VI Workshop on "Perspectives in Nuclear Physics at Intermediate Energies", ICTP, Trieste, May 3-7, 1993 (World Scientific, Singapore) to be published.

7. R.W. Schultze and P.U. Sauer, *Phys. Rev.* **C 48**, 38 (1993).

8. C. Ciofi degli Atti, E. Pace, S. Scopetta and G. Salmè, *Phys. Rev.* **C 48**, R968 (1993).

9. C. Ciofi degli Atti, S. Scopetta, E. Pace, and G. Salmè, Proceedings of the VI Workshop on "Perspectives in Nuclear Physics at Intermediate Energies", ICTP, Trieste, May 3-7, 1993 (World Scientific, Singapore) to be published.

10. M. Gari and W. Krumpelmann, *Z. Phys.* **A322**, 689 (1985); *Phys. Lett.* **B 173**, 10 (1986).

11. S. Galster et al., *Nucl. Phys.* **B32**, 221 (1971).

12. J. D. Bjorken, *Phys. Rev.* **D 1**, 1376 (1971).

13. a) SMC Collaboration, B. Adeva et al.,*Phys. Lett.* **B 302**, 553 (1993); b) S. Platchkov, this Conference.

14. a) E142, P. Anthony et al.,*Phys. Rev. Lett.* **71** 95 (1993); b) Z.E. Meziani, this Conference.

15. A. Schäfer, *Phys. Lett.* **B 208**, 175 (1988).

16. J. L. Friar, B. F. Gibson, G. L. Payne, A. M. Bernstein, and T. E. Chupp, *Phys. Rev.* **C 42**, 2310 (1990).

**D.I. Abramov**
St-Petersburg State Univ
St-Petersburg 198904
Russia
*present address*
Dept. of Scientific Visualization
Postfach 1316
Schloss Birlinghoven
D-5205 Sankt Augustin 1
Germany
Phone:    +49-02241-142357
Fax:       +49-02241-142040
E-mail:   gusev@viswiz.gmd.de


**F.D.S.S. dos Aidos**
Dept. de Fisica
Universidade de Coimbra
P-3000 Coimbra
Portugal
Phone:    +351-39-23675
Fax:       +351-39-29158
E-mail:   fcfiaidos@ciuc2.uc.pt
Telex:     52601 defiuc p


**K. Allaart**
Dept. of Physics and Astronomy
Free University
De Boelelaan 1081
1081 HV Amsterdam
The Netherlands
Phone:    +31-20-548 4754
Fax:       +31-20-646 1459
E-mail:   allaart@nat.vu.nl


**E.O. Alt**
Institut für Physik
Universität Mainz,
Staudingerweg 7
D-55099 Mainz
Germany
Phone:    +49-6131-392874
Fax:       +49-6131-394611
E-mail:   alt@vipmzf.physik.uni-mainz.de
Telex:     4187155 phmz d


**C.H.M. van Antwerpen**
Flinders University of South Australia
School of Physical Sciences
Bedford Park, 5042 South Australia
Australia
Phone:    +61-8-201 3811
Fax:       +61-8-201 3035
E-mail:   phchv@pippin.cc.flinders.edu.au


**M.G. Bachman**
Physics Department
University of Texas
Austin, Texas 78712
USA
Phone:    +1-512-471 6980
E-mail:   bachman@utaphy.ph.utexas.edu


**B.L.G. Bakker**
Dept. of Physics and Astronomy
Free University
De Boelelaan 1081
1081 HV Amsterdam
The Netherlands
Phone:    +31-20-548 4732
Fax:       +31-20-646 1459
E-mail:   blgbkkr@nat.vu.nl


**V.B. Belyaev**
Department of Theoretical Physics
Joint Institute for Nuclear Research
(JINR)
Dubna
141980 Moscow district
Russia
Phone:    +7-096-21 65900
Fax:       +7-096-21 65084
E-mail:   belyaev@thsun1.jinr.dubna.su
          belyaev@theor.jinrc.dubna.su
Telex:     911 621 dubna su


**M. Beyer**
Institut für Theor. Kernphysik
Universität Bonn
Nussallee 14-16
D-5300 Bonn
Germany
Phone:    +49-228-752374
Fax:       +49-228-753728
E-mail:   beyer@itkp.uni-bonn.de


**R. Bijker**
Utrecht University
R.J. van de Graaff Laboratorium
P.O. Box 80000
3508 TA Utrecht
The Netherlands

Phone: +31-30-532210
Fax: +31-30-518689
E-mail: bijker@fys.ruu.nl

**F. Blaazer**
Dept. of Physics and Astronomy
Free University
De Boelelaan 1081
1081 HV Amsterdam
The Netherlands
Phone: +31-20-548 4125
Fax: +31-20-646 1459
E-mail: ferry@nat.vu.nl

**B. Blankleider**
School of Physical Sciences
Flinders University
Bedford Park, 5042 S.A.
Australia
Phone: +61-8-201 2802
Fax: +61-8-201 3035
E-mail: phbb@cc.flinders.edu.au

**H.P. Blok**
Department of Physics and Astronomy
Free University
De Boelelaan 1081
1081 HV Amsterdam
The Netherlands
Phone: +31-20-548 6224
Fax: +31-20-646 1459
E-mail: henkb@nat.vu.nl

**L.D. Blokhintsev**
Nucl. Phys. Inst.
Moscow State University
Moscow
Russia
Fax: +7-095-939 0896
E-mail: ldblokh@compnet.msu.su

**K. Bodek**
Federal Institute of Technology,
Institute of Intermediate Energy Physics
CH-8093 Zuerich
Switzerland
Phone: +41-1-377 2044
Fax: +41-1-371 2665
E-mail: bodek@cageir5a.bitnet

**H.J. Boersma**
Department of Physics and Astronomy
Free University
De Boelelaan 1081
1081 HV Amsterdam
The Netherlands
Phone: +31-20-548 4735
Fax: +31-20-646 1459
E-mail: harrie@nat.vu.nl

**L.N. Bogdanova**
Institute for Theoretical and
Experimental Physics (ITEP)
B. Cheremushkinskaya 25
Moscow 117 259
Russia
Phone: +7-095-123 0292
Fax: +7-095-123 6584
E-mail: bogdanova@vxitep.itep.msk.su
Telex: 411059 cerii su

**K.Bräuer**
Institut für Theoretische Physik
Universität Tübingen
Auf der Morgenstelle 14
D-72076 Tubingen
Baden/Württemberg
Germany
Phone: +49-7071-29-6377
Fax: +49-7071-29-6400   E-mail:
ptibr01@mailserv.zdv.uni-tuebingen.de

**J.F.J. van den Brand**
Department of Physics
University of Wisconsin-Madison
1150 University Avenue
1508 Sterling Hall
Madison, 53706 Wisconsin
USA
*present address*
NIKHEF
P.O. Box 41882
NL-1009 DB Amsterdam
The Netherlands
Phone: +31-20-592 2142
Fax: +31-20-592 2165
E-mail: jo@nikhefk.nikhef.nl
Telex: 11538 iko nl

**J.S. Briggs**
Fakultät für Physik
Albert Ludwigs Universität Freiburg
Hermann Herderstr. 3
D-7800 Freiburg
Germany
E-mail:  sod@ibm.ruf.uni-freiburg.de

**E.E.W. Bruins**
Utrecht University
P.O. Box 80000
NL-3508 TA Utrecht
The Netherlands
Phone:   +31-30-533822
Fax:     +31-30-518689
E-mail:  bruins@fys.ruu.nl

**P.J. Brussaard**
Faculty of Physics and Astronomy
Utrecht University
P.O. Box 80.000
NL-3508 TA Utrecht
The Netherlands
Phone:   +31-30-532516
Fax:     +31-30-518689
E-mail:  brussaar@fys.ruu.nl

**J. Carbonell**
Institut des Sciences Nucléaires
53, Avenue des Martyrs
F-38026 GRENOBLE
France
Phone:   +33-76 28 40 34
Fax:     +33-76 28 40 04
E-mail:  carbonel@frcpn11.in2p3.fr
Telex:   320 301 isngren f

**J.A. Carlson**
Los Alamos National Lab
T-5 MS B283, LANL
Los Alamos, 87545 NM
USA
Phone:   +1-505-667 6245
Fax:     +1-505-667 1931
E-mail:  carlson@qmc.lanl.gov

**S. Cherubini**
INFN-LNS (Catania)
*and*
University of Catania
V.le A. Doria Angolo via S. Sofia
I-95125 Catania
Italia

**Phone:**   +39-95-542286
**Fax:**     +39-95-7141815
**E-mail:**  cherubini@lns.infn.it
**Telex:**   971432 lnsi

**H.A. Clement**
Physikalisches Institut
Universität Tübingen
Auf der Morgenstelle 14
D - 72076 Tübingen
Germany
Phone:   +49-7071-29 6352/6297
Fax:     +49-7071-29 6296
E-mail:  clement@pit.physik.
                        uni-tuebingen.de
Telex:   7262867 utna d

**J.G. Congleton**
Institute for Theoretical Physics
Utrecht University
P.O. Box 80.006
3508 TA Utrecht
The Netherlands
Phone:   +31-30-531869
Fax:     +31-030-531601
E-mail:  jimcon@fys.ruu.nl
Telex:   40048 fylut nl

**A. Csoto**
Institute of Nuclear Research
Hungarian Academy of Sciences
P. O. Box 51
H-4001 Debrecen
Hungary
Phone:   +36-52-17266
Fax:     +36-52-16181
E-mail:  h988cso@huella.bitnet
Telex:   hungary-72210

**R. van Dantzig**
NIKHEF
P.O. Box 41882
NL-1009 DB Amsterdam
The Netherlands
Phone:   +31-20-592 2129
Fax:     +31-20-592 2165
E-mail:  rvd@nikhef.nl
Telex:   11538 iko nl

470

**J.B. Delos**
Physics Department
College of William and Mary
Williamsburg, 23185 Virginia
USA
Phone:   +1-804-221 3500
Fax:      +1-804-221 3549
E-mail:  delos@atoms.physics.wm.edu

**K.S. Dhuga**
Dept. of Physics
George Washington University
Washington, DC 20052
USA
E-mail:  dhuga@gwuvm.gwu.edu

**J. Pierre Didelez**
Institut de Physique Nucléaire
IPN-Orsay
Bat 100
F-91406 Orsay CEDEX
France
E-mail:  didelez@fripn51.bitnet

**A.E.L. Dieperink**
KVI
Zernikelaan 25
NL-9747 AA Groningen
The Netherlands
E-mail:  dieperink@kvi.nl

**D. Dimitroyannis**
NIKHEF
P.O. Box 41882
NL-1009 DB Amsterdam
The Netherlands
Phone:   +31-20-592 2090
Fax:      +31-20-592 2165
E-mail:  ddimitri@nikhefk.nikhef.nl
Telex:    11538 iko nl

**M. Dineykhan**
Laboratory of Theoretical Physics
Joint Institute for Nuclear Research
Dubna
Head Post Office Box 79
MOSCOW 101000
Russia

Phone:   +7-221-63334
Fax:      +7-095-9752381
E-mail:  dineykh@theor.jinrc.dubna.su
Telex:    911621 dubna su

**D. Drechsel**
Institut für Kernphysik
Universität Mainz
J.J. Becherweg 33
D-55099 Mainz
Germany
E-mail:  klotter@vkpmzd.kph.
            uni-mainz.de

**J.E. Ducret**
DAPNIA/SPhN, C.E. Saclay
*and*
Institut für Kernphysik
Mainz University
D-55099 Mainz
Germany

**M. Fabre de la Ripelle**
Université de Paris-Sud
Institut de physique Nucléaire
Division de Physique Théorique
F-91406 Orsay CEDEX
France
Phone:   +33-1-6941 7921
Fax:      +33-1-6928 5897
E-mail:  fabre@frcpn11.in2p3.fr

**H. Fiedeldey**
Department of Physics
University of South Africa
P.O. Box 392
Pretoria 0001
South Africa
Phone:   +27-12-429 8027
Fax:      +27-12-429 3434
E-mail:  parrips@risc3.unisa.ac.za
Telex:    350068

**A.C. Fonseca**
Centro Fisica Nuclear
University of Lisboa
Av. Gama Pinto 2
Lisboa 1699
Portugal
E-mail:  fonseca@ptifm.bitnet

**B.F. Gibson**
Los Alamos National Laboratory
Group T-5, MS-B283
Los Alamos, 87545 New Mexico
USA
Phone:   +1-505-667 5059
Fax:     +1-505-667 1931
E-mail:  gibson@lampf.lanl.gov

**L.Ya. Glozman**
Alma-Ata Power Engineering Inst.
*and*
Institut für Theoretische Physik
Universität Tübingen
Auf der Morgenstelle 14
D-7400 Tübingen
Germany
Phone:   +49-7071-296785
Fax:     +49-707 -296400
E-mail:  glozman@mailserv.zdv.
                uni-tuebingen.de

**J. Golak**
Institute of Physics
Jagellonian University
PL 30-059 Cracow
Poland
E-mail:  ufgolak@cyf-kr.edu.pl

**P.F.A. Goudsmit**
Federal Institute of Technology (ETHZ)
Institute of Intermediate Energy Physics
CH-5232 Villigen PSI
Switzerland
Phone:   +41-56-99 32 77
Fax:     +41-56-99 32 94
E-mail:  goudsmit@cvax.psi.ch
                matsinos@vxcern.cern.ch
Telex:   82 74 19

**F. Gross**
CEBAF and William and Mary MS-12H,
Physics Division CEBAF 12000 Jefferson
Ave. Newport News, 23606 VA USA
Phone:   +1-804-249 7537
Fax:     +1-804-249 7002
E-mail:  gross@cebaf.gov
                gross@cebth2.cebaf.gov

**S. Gurvitz**
Weizmann Institute
Rehovot
Israel
E-mail:
     fngurvtz@weizmann.weizmann.ac.il

**H. Haberzettl**
George Washington University
Physics Department
Washington, DC 20052
USA
Phone:   +1-202-994 0886
Fax:     +1-202-994 3001
E-mail:  helmut@gwuvm.bitnet

**J. Haidenbauer**
IKP KFA Juelich
Postfach 1913
D-5170 Juelich
Germany
*and*
Institute for Theoretical Physics
University of Graz
Universitätsplatz 5
A-8010 Graz
Austria
E-mail:  jhaidenb@bkfusc.kfunigraz.ac.at

**W. Heil**
Institut für Physik, EXAKT
Johannes-Gutenberg-Universität Mainz
Staudingerweg 7
D-55099 Mainz
Germany
Phone:   +49-6131-39 2279
Telex:   4 187 155 phmz d

**J.P.T. Hersbach**
Institute for Theoretical Physics
Utrecht University.
Princetonplein 5
P.O. Box 80006
3508 TA Utrecht
The Netherlands
Phone:   +31-30-533059
Fax:     +31-30-531601
E-mail:  hersbach@ruunta.fys.ruu.nl

**C.M. Hoffman**
Medium Energy Division
Mailstop H 844
Los Alamos National Laboratory
Los Alamos, NM 87545

USA
E-mail:  hoffman@lampf.lanl.gov

**L.L. Howell**
Physics Department
University of South Africa
P.O. Box 392
Pretoria 0001
Transvaal
South Africa
Phone:    +27-12-4298027
Fax:      +27-12-4293434
E-mail:  howelll@risc7.unisa.ac.za

**D. Hüber**
Institut für Theoretische Physik 2,
Ruhr-Universität Bochum
Universitätstr.. 150
D-44780 Bochum
Germany
Phone:    +49-234-700 3716
Fax:      +49-234-7094 248
E-mail:  hbo002@zam001.zam.
                  kfa-juelich.de
Telex:    17-234356

**P. Hummel**
Institute for Theoretical Physics
University of Graz
Universitätsplatz 5
A-8010 Graz
Austria
E-mail:  hummel@bkfug.kfunigraz.ac.at

**D.G. Ireland**
NIKHEF
P.O. Box 41882
NL-1009 DB Amsterdam
The Netherlands
Phone:    +31-20-592 2065
Fax:      +31-20-592 2165
E-mail:  dave@nikhefk.nikhef.nl
Telex:    11538 iko nl

**C.W. de Jager**
NIKHEF
P.O. Box 41882
Amsterdam
NL-1009 DB
The Netherlands

Phone:    +31-20-592 2065
Fax:      +31-20-592 2165
E-mail:  kees@nikhefk.nikhef.nl
Telex:    11538 iko nl

**E. Jans**
NIKHEF
P.O. Box 41882
NL-1009 DB Amsterdam
The Netherlands
Phone:    +31-20-592 2085
Fax:      +31-20-592 2165
E-mail:  eddy@nikhefk.nikhef.nl
Telex:    11538 iko nl

**C.E. Jones**
Physics Division
Argonne National Lab
9700 S. Cass Ave.
Argonne, 60439 Illinois
USA
Phone    +1-708-252 7267
Fax:      +1-708-252 3903
E-mail:  cjones@anlphy.phy.anl.gov
              cjones@anlmep.phy.anl.gov

**Yu.S. Kalashnikova**
Institute for Theoretical and
Experimental Physics (ITEP)
B. Cheremushkinskaya 25
Moscow 117 259
Russia
Phone:    +7-095-1230292
Fax:      +7-095-1236584
E-mail:  yulia@vxitep.itep.msk.su
Telex:    411059 cerii su

**H. Kamada**
Institut für theoretische Physik II
Ruhr-Universität Bochum,
NB6/125, Universitätstr. 155
D-44780 Bochum,
Germany
Phone:    +49-234-700 3715
E-mail:  hbo003@djukfa11.bitnet

**W.J. Kasdorp**
NIKHEF
P.O. Box 41882
NL-1009 DB Amsterdam
The Netherlands

Phone: +31-20-5902074
Fax: +31-20-592 2165
E-mail: wkasdo@nikhefk.nikhef.nl
Telex: 11538 iko nl

**R.Ya. Kezerashvili**
Institute of Physics
Academy of Sciences of the
Republic of Georgia
Ul. Tamarashvili 6
SU-380077 Tbilisi
Republic of Georgia
Phone: +7-8832-743 822
Fax: +7-3832-352163
E-mail: gurvich@physics.aod.ge
        rukh@compmath.aod.georgia.su
Telex: 133116 atom su

**A. Kievsky**
Dipartimento di Fisica
Istituto Nazionale di Fisica Nucleare
(INFN)
Piazza Torricelli 2
I-56100 Pisa
Italy
Phone: +39-50-45221
Fax: +39-50-48277
E-mail: kievsky@mvxpi1.difi.unipi.it
Telex: 500319 psafis

**K.K. Kilian**
KFA Juelich
Postfach 1913
D-5170 Juelich
Germany

**L.G. Kiseleva**
Institute of Astronomy
Madingley Road
Cambridge CB3 0HA
UK
Phone: +44-0223-337548
Fax: +44-0223-337523
E-mail: lgk@mail.ast.cam.ac.uk
Telex: 17297 astron g

**F. Kleefeld**
Institut für Theoretische Physik III
Universität Erlangen-Nürnberg
Staudtstr. 7
D-91051 Erlangen
Germany

**R.A.M.M. Klomp**
Institute for Theoretical Physics
Nijmegen University
Toernooiveld 1
NL- 525 ED Nijmegen
The Netherlands
Phone: +31-80-652800
Fax: +31-80-553450
E-mail: renek@wn3.sci.kun.nl
Telex: 48228 wina nl

**W.G. Koepf**
Department of Physics, FM-15
University of Washington
Seattle, 98195 Washington
USA
Phone: +1-206-685 3971
Fax: +1-206-685 0635
E-mail: koepf@alpher.npl.washington.edu

**J.L. de Kok**
Institute for Theoretical Physics
University of Nijmegen
Toernooiveld 1
Nijmegen
6525 ED
The Netherlands
Phone: +31-80-652801
E-mail: jlk@kaon.sci.kun.nl

**L.P. Kok**
Institute for Theoretical Physics
Groningen University
Nijenborgh 4
9747 AG Groningen
The Netherlands
Phone: +31 50 634955
Fax: +31 50 634947
E-mail: lpkok@rugth3.th.rug.nl

**J. Konijn**
NIKHEF
P.O. Box 41882
NL-1009 DB Amsterdam
The Netherlands

474

Phone: +31-20-590 2121
Fax: +31-20-592 2165
E-mail: joop@nikhefk.nikhef.nl
Telex: 11538 iko nl

**T.I. Kopaleishvili**
High Energy Physics Institute
Tbilisi State University
University St. 9
SU-380086 Tbilisi
Republic of Georgia
Fax: +7-8832-352163
E-mail: kopal@hepitu.kheta.georgia.su

**G.Ya. Korenman**
Institute of Nuclear Physics
Moscow State University
Moscow 119899
Russia
Phone: +7-095-939 2513
Fax: +7-095-939 0896
E-mail: korenman@compnet.msu.su
Telex: 411483 mgu su

**H. Peter Kotz**
Institut für Theoretische Physik
Universität Graz
Universitätsplatz 5
8010 Graz
Austria
E-mail: hpk@majestix.kfunigraz.ac.at
hpk@magnix.kfunigraz.ac.at

**G.A. Kozlov**
Bogoliubov Theoretical Laboratory
Dubna 141980
Russia
Fax: +7-096-216 5084
E-mail: kozlov@theor.jinrc.dubna.su
Telex: 911621 dubna su

**R. Krivec**
Institute J. Stefan
Jamova 39
SLO-61111 Ljubljana
Slovenia
Phone: +38-61-159-199
Fax: +38-61-161-029
E-mail: rajmund.krivec@ijs.si
Telex: 31-296 jostin si

**R.A. Kunne**
CNRS/IN2P3
Laboratoire National Saturne
CE-Saclay
F-91191 Gif-sur-Yvette Cedex
France
Phone: +33-1-6908 3358
Fax: +33-1-6908 2970
E-mail: kunne@frcpn11.in2p3.fr

**A. Lande**
Institute for Theoretical Physics
Groningen University
Nijenborgh 4
9747 AG Groningen
The Netherlands
E-mail: lande@th.rug.nl

**L. Lapikas**
NIKHEF
P.O. Box 41882
NL-1009 DB Amsterdam
Netherlands
Phone: +31-20-5922067
Fax: +31-20-5922165
E-mail: louk@nikhefk.nikhef.nl
Telex: 11538 iko nl

**K.J.J. van Leeuwe**
NIKHEF
P.O. box 41882
NL-1009 DB Amsterdam
The Netherlands
Phone: +31-20-5922096
Fax: +31-20-5922165
E-mail: koos@nikhefk.nikhef.nl
Telex: 11538 iko nl

**J.S. Levinger**
Dept. of Physics
Rensselaer Polytech. Inst.
Troy, 12180 N.Y.
USA
Phone: +1-518-276 8410
Fax: +1-518-276 6680
E-mail: usera0uq@mts.rpi.edu

**R. Van Lieshout**
Messchaertlaan 4
NL-1272 NZ Huizen
Phone: +31-2159-45759

**E.L. Lomon**
MIT
Room 6-304 MIT
Cambridge, 02139 MA
USA
Phone:   +1-617-253 4877
Fax:     +1-617-253 8674
E-mail:  lomon@pierre.mit.edu

**R.A.R.L. Malfliet**
KVI
Zernikelaan 25
9747 AA Groningen
The Netherlands
E-mail:  malfliet@kvi.nl

**V.B. Mandelzweig**
Racah Institute of Physics
Hebrew University
Jerusalem 91904
Israel
Fax:     9722-584437
E-mail:  victor@vms.huji.ac.il

**C. Marchand**
CEN Saclay
DAPNIA/SPhN, Batiment 703
F-91191 GIF-SUR-YVETTE
France
Phone:   +33-1-6908 8659
Fax:     +33-1-6908 7584
E-mail:  marchand@frsac12.bitnet

**L.C. Maximon**
Physics Department
The George Washington Univ.
Washington, DC 20052
USA
Phone:   +1-202-994 6264
Fax:     +1-202-994 3001
E-mail:  max@gwuvm.gwu.edu

**R.D. McKeown**
Caltech
106-38 Kellogg
Pasadena, CA 91125
USA
Phone:   +1-818-356 4316
Fax:     +1-818-564 8708
E-mail:  bmck@erin.caltech.edu

**R.L.J. van der Meer**
NIKHEF
P.O. Box 41882
NL-1009 DB Amsterdam
The Netherlands
Phone:   +31-20-592 2146
Fax:     +31-20-592 2165
E-mail:  robvdm@nikhefk.nikhef.nl
Telex:   11538 iko nl

**Yu.P.Mel'nik**
Kharkov Institute of Physics and
Technology
Akademicheskaya St. 1
Kharkov 310108
Ukraine
Phone:   +7-57-235 1993
Fax:     +7-57-235 1758
Telex:   311052 dekan su

**V.S. Melezhik**
Joint Inst. for Nuclear Research (JINR)
Dubna, Moscow Region 141980
Russia
Phone:   +7-095-926 2220
Fax:     +7-09621-66 666
E-mail:  melezh@lcta7.jinr.dubna.su
Telex:   911621 su

**Z.-E. Meziani**
Physics Department
Stanford University
Stanford, CA 94305-4060
USA
E-Mail:
    zwang@forsythe.stanford.edu
*present address*
Physics Department
Temple University
Philadelphia, PA 19122
USA

**G. van Middelkoop**
Department of Physics and Astronomy
Free University
De Boelelaan 1081
1081 HV Amsterdam
The Netherlands
Phone:   +31-20-548 2468
Fax:     +31-20-646 1459
E-mail:  gervanm@nat.vu.nl

476

**W.H. Miller**
Dept. of Chem.
Univ. of California
Berkeley, CA 94720
USA
E-mail:   miller@neon.cchem.berkeley.edu


**B. Mosconi**
Dipartimento di Fisica
Università di Firenze
Largo E. Fermi 2
I-50125 Firenze
Italy
Phone:    +39-55-2298141
Fax:      +39-55-229330
E-mail:   mosconi@fi.infn.it
Telex:    572570

**A. Mukhamedzhanov**
Institute for Nuclear Physics
Uzbek Academy of Sciences
and
Institut für Physik
Universität Mainz
Postfach 39 80
D-55099 Mainz
Germany
Phone:    +49-6131-393383
Fax:      +49-6131-394611
E-mail:   fewbody@vipmzf.physik.
                uni-mainz.de

**D.V.Y. Van Neck**
KVI
Zernikelaan 25
NL-9747 AA Groningen
The Netherlands
Phone:    +31-50-633568
E-mail:   vneck@kvi.nl

**T.E. Nieuwenhuis**
Institute for Theoretical Physics
Buys Ballotlaboratorium
Princetonplein 5, P.O. Box 80.006
NL-3508 TA Utrecht
Netherlands
Phone:    +31-30-533059
Fax:      +31-030-531601
E-mail:   nieuwenh@ruunte.fys.ruu.nl

**H. Pierre Noyes**
Stanford Linear Accelerator Center
SLAC, M.S. No. 81
P.O. Box 4349
Stanford, CA 94309
USA
Phone:    +1-415-926 2665
Fax:      +1-415-926 4500
E-mail:   noyes@slacvm.slac.stanford.edu
Telex:    3722871 stanuniv

**W.T.H. van Oers**
University of Manitoba
and
TRIUMF
4004 Wesbrook Mall
Vancouver, V6T 2A3 B.C.
Canada
Phone:    +1-604-222 1047
Fax:      +1-604-222 1074
E-mail:   vanoers@reg.triumf.ca


**S. Oryu**
Science University of Tokyo
Department of Physics
Faculty of Science and Technology
2641 Yamazaki
Neda 278
Chiba
Japan
E-mail:   ph5290%jpnsut30.bitnet
                @pucc.princeton.edu


**E. Pace**
Dipartimento di Fisica
Università di Roma "Tor Vergata"
Viale Della Ricerca Scientifica
I-00133 Rome
Italy
Phone:    +39-6-72594563
Fax:      +39-6-2023507
E-mail:   pace@roma2.infn.it
                fisteor@irmiss.bitnet


**Z. Papp**
Institute of Nuclear Research
Bem ter 18/c P.O. Box 51
4001 Debrecen
Hungary

Telex:    hungary-72210 atom h
Phone:   +36-52-317266
Fax:     +36-52-316181
E-mail:   h3182pap@huella.bitnet

**M.T. Peña**
Centro de Fisica Nuclear
Lisboa
Portugal
*and*
CEBAF
12000 Jefferson Avenue
Newport News, VA 23606
USA
Phone:   +1-804-249 7412
Fax:     +1-804-249 7002
E-mail:   teresa@cebth0.cebaf.gov

**S. Platchkov**
CEN Saclay
DAPNIA, CEN Saclay
F-91191 Gif-sur-Yvette
France
Phone:   +33-1-6908 7209
Fax:     +33-1-6908 7584
E-mail:   spl@na47sun05.cern.ch

**W. Plessas**
Institute for Theoretical Physics
University of Graz
Universitätsplatz 5
A-8010 Graz
Austria
Phone:   +43-316-380 5231
Fax:     +43-316-384091
E-mail:   b6241dac@awiuni11.edvz.
                 univie.ac.at
                 plessas@edvz.kfunigraz.ac.at
Telex:    311662 ubgraz a

**H. Postma**
Delft University of Technology
Dept. of Applied Physics
Lorentzweg 1
NL-2628 CJ Delft
The Netherlands
E-mail: postma@duttncb.tn.tudelft.nl

**K. Rabitsch**
Inst. f. Kernphysik
TU Wien
Wiedner Hauptstr. 8-10/142

A-1040 Vienna
Austria
Phone:   +43-222-58801/5575
Fax:     +43-222-564203
E-mail:   rabitsch@ekpds1.tuwien.ac.at
Telex:    (61)3222467

**M.C.M. Rentmeester**
Inst. for Theoretical Physics
Nijmegen University
Toernooiveld 1
NL-6525 ED Nijmegen
The Netherlands
Phone:   +31-80-652800
Fax:     +31-80-553450
E-mail:   martr@sci.kun.nl

**G.A. Retzlaff**
Saskatchewan Accelerator Laboratory
University of Saskatchewan
Saskatoon S7N 0W0 Saskatchewan
Canada
Phone:   +1-306-966 6319
Fax:     +1-306-966 6058
E-mail:   greg@skatter.usask.ca

**M.M. Rigney**
Physikalisches Institut
Università Bonn
Nussallee 12
D-53115 Bonn
Germany
Phone:   +49-228-753229
Fax:     +49-228-757869
E-mail:   rigney@pib1.physik.uni-bonn.de

**T.A. Rijken**
Institute for Theoretical Physics
Nijmegen University
Toernooiveld 1
NL-6525 ED Nijmegen
The Netherlands
Phone:   +31-30-533062/532284
            +31-80-652982
Fax:     +31-80-553450
E-mail:   u634999@hnykun11.bitnet
            u634002@hnykun11.bitnet
Telex:    48228 wina nl

**L. Rosier**
IPN Orsay
F-91406 Orsay CEDEX
France

478

Phone:    +33-1-6941 5111
Fax:      +33-1-6941 6470
E-mail:   rosier@fripn51.earn
          rosier@frcpn11.bitnet

**F.H. Roudot**
Institut de Physique Nucléaire
Université de Paris XI
F-91406 ORSAY CEDEX
France
E-mail:   roudot@ipncls.in2p3.fr

**F.J. Rubio-Melon**
Institute for Theoretical Physics
Nijmegen University
Toernooiveld 1
NL-6525 ED Nijmegen
The Netherlands
Phone:    +31-80-652803
Fax:      +31-80-553450
E-mail:   jesus@sci.kun.nl
Telex:    48228 wina nl

**Th.W. Ruijgrok**
Instituut voor Theoretische Fysica
Universiteit Utrecht
Princetonplein 5, P.O. Box 80.006
NL-3508 TA Utrecht
The Netherlands
Phone:    +31-30-532288
Fax:      +31-30-531601
E-mail:   ruijgrok@ruunts.fys.ruu.nl

**G. Rupp**
Centro de Fisica da Materia Condensada
Technical University of Lisbon
Av. Gama Pinto, 2
P-1699 Lisboa CODEX
Portugal
Phone:    +351-1-7950790
Fax:      +351-1-7954288
E-mail:   george@ptifm2.fc.ul.pt
Telex:    62593 iifm p

**G.Salmè**
INFN-Sezione Sanità
Viale Regina Elena 299
I-00161 Rome
Italy

Phone:    +39-6-4941086
Fax:      +39-6-4462872
E-mail:   fisteor@irmiss.bitnet
          gslm@sanita.infn.it
Telex:    610071

**M. Sambataro**
INFN - Sezione di Catania
Corso Italia 57
I-95129 Catania
Italy
Phone:    +39-95-7195435
Fax:      +39-95-371600
E-mail:   samba@catania.infn.it
Telex:    971554 infnct

**A. Sandacz**
Institute for Nuclear Studies
ul. Hoza 69
PL 00-681 Warsaw
Poland
Phone:    +48-22-212804
Fax:      +48-22-213829
E-mail:   sandacz@fuw.edu.pl

**W. Sandhas**
Physikalisches Institut
Universität Bonn
Endenicher Allee 11-13
D-53115 Bonn
Germany
Phone:    +49-228-753750
Fax:      +49-228-757869
E-mail:   unp064@ibm.rhrz.uni-bonn.de
Telex:    8869693 phyb d

**A.M. Sandorfi**
Brookhaven National Laboratory
Physics Department, bldg. 510A
Upton, 11975 N.Y.
USA
Phone:    +1-516-282 7951
Fax:      +1-516-282 5568
E-mail:   sandorfi@bnlcl1.bitnet
Telex:    6852516 bnl doe

**P. Saracco**
INFN - sez. Genova
Via Dodecaneso, 33
I-16146 Genova
Italy

Phone:   +39-10-3536210
Fax:     +39-10-313358
E-mail:  saracco@genova.infn.it

**P.U. Sauer**
Technical University Hannover
Physics Department
Appelstrasse 2
D-30167 Hannover
Germany
E-mail:   sauer@cdc2.itp.
                uni-hannover.dbp.de

**W. Schadow**
Physikalisches Institut
Universität Bonn
Endenicher Allee 11-13
D-53115 Bonn
Germany
Phone:   +49-228-732344
Fax:     +49-228-737869
E-mail:  schadow@avzw.physik.
                uni-bonn.de

**N.W. Schellingerhout**
Inst. for Theor. Phys.
Univ. of Groningen
Nijenborgh 4
NL-9747 AG Groningen
The Netherlands
Phone:   +31-50-63 4950
Fax:     +31-50-63 4947
E-mail:  schellin@rugth3.th.rug.nl

**S. Scherer**                  •
TRIUMF
4004 Wesbrook Mall
Vancouver, V6T 2A3 British Columbia
Canada
Phone:   +1-604-222 1047 ext. 453
Fax:     +1-604-222 1074
E-mail:  scherer@triumfcl.bitnet
Telex:   (0)-4508503

**F. Schlumpf**
Stanford Linear Accelerator Center
Stanford University
Theory Group, MS 81
P.O. Box 4349
Stanford, 94309 California
USA

Phone:   +1-415-926 4429
Fax:     +1-415-926 4500
E-mail:  schlumpf@slac.stanford.edu

**E.W. Schmid**
Institute for Theoretical Physics
University of Tübingen
Auf der Morgenstelle 14
D-72076 Tübingen
Germany
Phone:   +49-7071-292996
Fax:     +49-7071-295400
E-mail:
    eschmid@mailserv.zdv.uni-tuebingen.de
    ptisc02@mailserv.zdv.uni-tuebingen.de
Telex:   7 262 867 utna d

**B. Schoch**
Universitat Bonn
Physikalisches Institut
Nussallee 12
D-5300 Bonn
Germany
Phone:   +49-228-732344
Fax:     +49-228-737869
E-mail:  schoch@pib1.physik.uni-bonn.de

**R. Wolfgang Schulze**
Institut für Theoretische Physik
University Hannover
Appelstr.2
D-30161 Hannover
Germany
Phone:   +49-511-762 4834
Fax:     +49-511-762 3023
E-mail:
    schulze@kastor.itp.uni-hannover.d400.de

**J.J. Schut**
Institute for Theoretical Physics
Nijenborgh 4
NL-9700 AV Groningen
The Netherlands
Phone:   +31-50-634965
Fax:     +31-50-634947
E-mail:  schut@rugth3.th.rug.nl

**S. Scopetta**
Dipartimento di Fisica
Università di Perugia

via A. Pascoli
I-06100 Perugia
Italy
Phone:    +39 75 5853060
Fax:       +39 75 44666
E-mail:   scopetta@perugia.infn.it


**C.N.G. Semay**
Université de Mons-Hainaut
19 Avenue Maistriau
B-7000 Mons
Belgium
Phone:    +32-65-37 34 67
Fax:       +32-65-37 30 54
E-mail:   ssemay@bmsuem11.bitnet
Telex:     57764 uemons b


**P.M.C. Serra**
Departamento de Fisica Teorica
Universidade de Coimbra
P-3000 Coimbra
Portugal
Phone:    +351-39-23675
Fax:       +351-39-29158
E-mail:   fcfiaidos@ciuc2.uc.pt
Telex:     52601 defiuc p


**I. Sick**
Universität Basel
Klingelbergstr. 82
CH-4056 Basel
Switzerland

**D.M. Skopik**
Saskatchewan Accelerator Laboratory
University of Saskatchewan
Saskatoon S7N 0W0 Saskatchewan
Canada
Phone:    +1-306-966 6054
Fax:       +1-306-966 6058
E-mail:   skopik@skatter.usask.ca


**S.A. Sofianos**
Department of Physics, P.O. Box 392,
University of South Africa
Pretoria 0001
South Africa

Phone:    +27-12-429 8027
Fax:       +27-12-429 3434
E-mail:   sofiasa@risc3.unisa.ac.za
            parrips@risc3.unisa.ac.za
Telex:     350068


**J. Sowinski**
Indiana University Cyclotron Facility
2401 Milo B. Sampson Lane
Bloomington, 47405 Indiana
USA
Phone:    +1-812-855 9365
Fax:       +1-812-855 6645
E-mail:   sowinski@venus.iucf.indiana.edu


**A. Stadler**
Department of Physics
College of William and Mary
Williamsburg, VA 23185
USA
Phone:    +1-804-221 3531
Fax:       +1-804-221 3540
E-mail:   stadler@cebaf

**M. Staszel**
Institute of Experimental Physics
Warsaw University
Hoza 69
00681 Warsaw
Poland
Phone:    +48-2-6283031
E-mail:   staszel@fuw.edu.pl

**J. Stepaniak**
Institute for Nuclear Studies
Hoza 69
00681 Warsaw
Poland
Phone:    +48-22-219404
            +48-22-121012 (home)
Fax:       (4822) 213829
E-mail:   joste@fuw.edu.pl


**H.T.C. Stoof**
Utrecht University (after July 31, 1993)
Eindhoven University of Technology
P.O. Box 513
Nl-5600 MB Eindhoven
The Netherlands

Phone:   +31-40-474019
Fax:     +31-40-475253
E-mail:  henk@qmrs.ni.phys.tue.nl
*present address*
Instituut voor Theoretische Fysica
Universiteit Utrecht
Princetonplein 5, P.O. Box 80.006
NL-3508 TA Utrecht
The Netherlands

**Y. Surya**
Physics Dept.
College of William & Mary
Williamsburg, VA 23185
USA
Phone:   +1-804-565 0322
Fax:     +1-804-249 7002
E-mail:  surya@cebaf.gov

**J.J. de Swart**
Institute for Theoretical Physics
Nijmegen University
Toernooiveld 1
NL-6525 ED Nijmegen
The Netherlands
Phone:   +31-80-653426
Fax:     +31-80-553450
E-mail:  swart@sci.kun.nl
Telex:   48228 wina nl

**F. Tabakin**
Dept. of Physics and Astronomy
University of Pittsburgh
Pittsburgh, 15260 PA
USA
Phone:   +1-412-624 9025
Fax:     +1-412-624 9163
E-mail:  tabakin@pittvms
         frankt@tabakin.phyast.pitt.edu

**S. Taddei**
Dipartimento di Fisica
Università di Firenze
Largo Enrico Fermi, 2 - (Arcetri)
I-50125 Firenze
Italy
Phone:   +39-586-211141
Fax:     +39-55-229330
E-mail:  taddei@fi.infn.it
Telex:   572570

**N.Zh. Takibayev**
Institute of Nuclear Physics
Alma-Ata-82, 480082
Republic of Kazakhstan
E-mail:  ntakib@inp.alma-ata.su

**J.A. Templon**
NIKHEF
P.O. Box 41882
NL-1009 DB Amsterdam
The Netherlands
Phone:   +31-20-592 2075
E-mail:  templon@nikhefk.nikhef.nl
Telex:   11538 iko nl

**J.A.P. Theunissen**
NIKHEF
P.O. Box 41882
NL-1009 DB Amsterdam
The Netherlands
E-mail:  jean@nikhefk.nikhef.nl
Telex:   11538 iko nl

**P.C. Tiemeijer**
Institute for Theoretical Physics
Utrecht University
Pox 80.006
NL-3508 TA Utrecht
The Netherlands
Phone:   +31-30-531740
Fax:     +31-30-531601
E-mail:  tmeijer@fys.ruu.nl

**R. Timmermans**
Los Alamos National Laboratory
T-5 Mailstop B283
Los Alamos, NM 87545
USA
Phone:   +1-505-667 4916
Fax:     +1-505-667 1931
E-mail:  timmer@t5zia.lanl.gov

**J.A. Tjon**
Instituut voor Theoretische Fysica
Universiteit Utrecht
Princetonplein 5, P.O. Box 80.006
NL-3508 TA Utrecht
The Netherlands
Phone:   +31-30-533062
Fax:     +31-30 - 531601
E-mail:  tjon@ruuntm.fys.ruu.nl

**M. Tokarev**
Joint Institute for Nuclear Research
Head Post Office P.O. Box 79
Moscow 101000
Russia
Fax:    +7-095-975 2381
E-mail:  tokarev@main1.jinr.dubna.su
Telex:   dubna 911621

**V. Vento**
Dept. Fisica Teorica
Universitat de Valencia
c/ Moliner, 50
Burjassot
E-46100 Valencia
Spain
Phone:   +34-6-3864349
            +39-461-881522
Fax:     +34-6-3642345
            +39-461-881696
E-mail:  vento@vm.ci.uv.es
           vento@itnvax.cineca.it
*present address*
Università degli Studi di Trento
*and*
I.N.F.N.
I-38050 Povo (Trento)
Italy

**S.I. Vinitsky**
Laboratory of Theoretical Physics,
Joint Institute for Nuclear Research
Dubna, near Moscow
Postal/Zip code: 141980
Russia
Fax:     +7-095-975 2381
           +7-09621-65084
E-mail:  smirnov@lcta6.jinr.dubna.su
        pogosyan@theor.jinrc.dubna.su
Telex:   91-16-21 dubna su

**J.L. Visschers**
NIKHEF
P.O. Box 41882
NL-1009 DB Amsterdam
The Netherlands
Phone:   +31-20-5922152
Fax:     +31-20-5922165
E-mail:  janv@nikhef.nl
Telex:   11538 iko nl

**M. Viviani**
Physics Department
Pisa University
Piazza Torricelli, 2
I-56100 Pisa
Italy
Phone:   +39-50-45221
Fax:     +39-50-48277
E-mail:  viviani@pisa.infn.it
Telex:   500319 psafis

**H. de Vries**
NIKHEF
P.O. Box 41882
NL-1009 DB Amsterdam
The Netherlands
Phone:   +31-20-592 2148
Fax:     +31-20-592 2165
E-mail:  hansv@nikhefk.nikhef.nl
Telex:   11538 iko nl

**R. de Vries**
NIKHEF
Kruislaan 409
NL-1098 SJ Amsterdam
The Netherlands
Phone:   +31-20-592 2039
Fax:     +31-20-592 2165
E-mail:  robertdv@nikhefk.nikhef.nl
Telex:   11538 iko nl

**R. van Wageningen**
Cannenburgh 7
Amsterdam
The Netherlands
Phone:   +31-20-6440293

**Th. Walcher**
Institute for Nuclear Physics
Mainz University
Johann-Joachim-Becher-Weg 45
D-55099 Mainz
Germany
Phone:   +49-6131-395196
Fax:     +49-6131-392964
E-mail:
    Walcher@vkpmzq.kph.uni-mainz.de

**N.R. Walet**
Dept. of Physics
David Rittenhouse Lab
University of Pennsylvania
Philadelphia, PA 19104-6396
USA
Phone:    +1-215-898 8148
Fax:      +1-215-898 2010
E-mail:   walet@walet.physics.upenn.edu

**S.J. Wallace**
Department of Physics
University of Maryland
College Park, MD 20742 USA
USA
Phone:    +1-301-405 6121
Fax:      +1-301-405 6114
E-mail:   wallace@quark.umd.edu

**A.H. Wapstra**
NIKHEF
P.O. Box 41882
NL-1009 DB Amsterdam
The Netherlands
Phone:    +31-20-592 2079
Fax:      +31-20-592 2165
Telex:    11538 iko nl

**S.Y. Werf van der**
KVI
Zernikelaan 25
NL-9747 AA Groningen
The Netherlands
E-mail:   vdwerf@kvi.nl

**H.J. Weyer**
Basel Univ and PSI
PSI
PD/3
CH-5232 Villigen PSI
Switzerland
Phone:    +41-56-99 3494 (PSI)
          +41-61-267 3700 (Basel)
Fax:      +41-56-99 3294 (PSI)
          +41-61-267 3784 (Basel)
E-mail:   weyer@cageir5a.bitnet
          weyer@psiclu.cern.ch

**P. Wilhelm**
Inst. f. Kernphysik
Univ. Mainz
Joh.-Joachim-Becher-Weg 45

D-55099 Mainz
Germany
Phone:    +49-6131-395850
Fax:      +49-6131-392964
E-mail:
  wilhelmp@vkpmzp.kph.uni-mainz.de
Telex:    04 187718

**P.K.A. de Witt Huberts**
NIKHEF
P.O. Box 41882
NL-1009 DB Amsterdam
The Netherlands
Phone:    +31-20-592 2162
Fax:      +31-20-592 2165
Telex:    11538 iko nl
E-mail:   marijke@nikhefk.nikhef.nl

**R.L. Workman**
Dept. of Physics
Virginia Tech
Blacksburg, Virginia 24061
USA
Phone:    +1-703-231 5533
Fax:      +1-703-231 7511
E-mail:   workman@vtvm1.cc.vt.edu

**D.M. Yeomans**
Dept. of Physics and Astronomy
Utrecht University
P.O. Box 80000
NL-3508 TA Utrecht
The Netherlands
Phone:    +31-30-534599
Fax:      +31-30-518689
E-mail:   yeomans@fys.ruu.nl

**L.S. Zolin**
Joint Institute for Nuclear Research JINR
Dubna
Head Post Office P.O. Box 79
Moscow 101000
Russia
Phone:    +8-221-62031
Fax:      +7-095-9752381
E-mail:   zolin@lshe8.jinr.dubna.su
Telex:    911621 dubna su

# Index

C. Ciofi degli Atti, E. Pace, G. Salmè, S. Simula (eds.)

# Few-Body Problems in Physics

Proceedings of the XIIIth European Conference on
Few-Body Physics, Marciana Marina, Isola d'Elba, Italy,
September 9-14, 1991

(Few-Body Systems / Supplementum 6)

1992. 260 figures and 1 frontispiece. XVI, 635 pages.
Cloth DM 220,–, öS 1540,–
Reduced price for subscribers to "Few-Body Systems":
Cloth DM 198,–, öS 1386,–
ISBN 3-211-82343-3

*Prices are subject to change without notice.*

This book collects all of the invited papers and contributions to the Discussion
Sessions, presented at the 13th European Conference on Few-Body Problems
in Physics, and is addressed to senior and young researchers and students
interested in the field of few-body problems in elementary particle and nuclear
physics, as well as in atomic and molecular physics. The volume contains a
survey of recent, and not yet published results  on theoretical and experimental investigations of the structure of hadrons and hadronic systems, novel
theoretical methods suitable for an accurate treatment of the few-body
problems in different fields, present status and future developments in muon
catalysed  fusion. A detailed illustration of the few-body physics programs of
running (MIT-Bates, CEBAF, CERN, HERA, Mainz, HIKHEF, SATURNE,
Saskatchewan, SLAC, TRIUMF) and proposed (European Electron Facility
Project, Indiana cooler beam) experimental facilities represents a valuable
feature of the book.

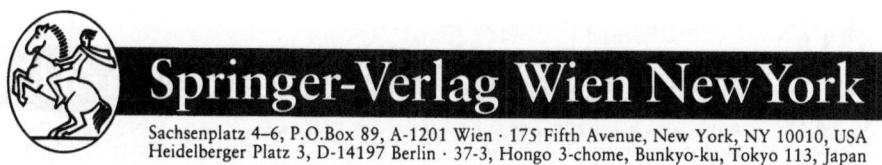

## Springer-Verlag Wien New York

Sachsenplatz 4–6, P.O.Box 89, A-1201 Wien · 175 Fifth Avenue, New York, NY 10010, USA
Heidelberger Platz 3, D-14197 Berlin · 37-3, Hongo 3-chome, Bunkyo-ku, Tokyo 113, Japan

E. Truhlik, R. Mach (eds.)

# Mesons and Light Nuclei

Proceedings of the 5th International Symposium,
Prague, September 1-6, 1991

(Few-Body Systems / Supplementum 5)

1992. 239 figures and 1 frontispiece. XVIII, 525 pages.
Cloth DM 200,–, öS 1400,–
Reduced price for subscribers to "Few-Body Systems":
Cloth DM 180,–, öS 1260,–
ISBN 3-211-82342-5

*Prices are subject to change without notice.*

Giving emphasis on electroweak nuclear interactions the book collects more
than 60 papers presented at the 5th International Symposium on Mesons and
Light Nuclei, Prague, September 1-6, 1991.
Further topics covered are: nuclear physics with pions and antiprotons,
nuclear physics with strange particles, relativistic nuclear physics, and quark
degrees of freedom. They are viewed in their theoretical as well es experimental aspects.

Sachsenplatz 4–6, P.O.Box 89, A-1201 Wien · 175 Fifth Avenue, New York, NY 10010, USA
Heidelberger Platz 3, D-14197 Berlin · 37-3, Hongo 3-chome, Bunkyo-ku, Tokyo 113, Japan